Communications
in Computer and Information Science   1064

Commenced Publication in 2007
Founding and Former Series Editors:
Phoebe Chen, Alfredo Cuzzocrea, Xiaoyong Du, Orhun Kara, Ting Liu,
Krishna M. Sivalingam, Dominik Ślęzak, Takashi Washio, and Xiaokang Yang

More information about this series at http://www.springer.com/series/7899

Tatjana Welzer · Johann Eder ·
Vili Podgorelec · Robert Wrembel ·
Mirjana Ivanović · Johann Gamper ·
Mikołaj Morzy · Theodoros Tzouramanis ·
Jérôme Darmont · Aida Kamišalić Latifić (Eds.)

New Trends in Databases and Information Systems

ADBIS 2019 Short Papers, Workshops
BBIGAP, QAUCA, SemBDM, SIMPDA, M2P, MADEISD
and Doctoral Consortium
Bled, Slovenia, September 8–11, 2019
Proceedings

 Springer

Editors
Tatjana Welzer ⓘ
University of Maribor
Maribor, Slovenia

Vili Podgorelec ⓘ
University of Maribor
Maribor, Slovenia

Mirjana Ivanović ⓘ
University of Novi Sad
Novi Sad, Serbia

Mikołaj Morzy ⓘ
Poznań University of Technology
Poznan, Poland

Jérôme Darmont ⓘ
Université Lumière Lyon 2
Lyon, France

Johann Eder ⓘ
Alpen Adria University Klagenfurt
Klagenfurt am Wörthersee, Austria

Robert Wrembel ⓘ
Poznan University of Technology
Poznan, Poland

Johann Gamper ⓘ
Free University of Bozen-Bolzano
Bolzano, Italy

Theodoros Tzouramanis ⓘ
University of Thessaly
Lamia, Greece

Aida Kamišalić Latifić ⓘ
University of Maribor
Maribor, Slovenia

ISSN 1865-0929   ISSN 1865-0937 (electronic)
Communications in Computer and Information Science
ISBN 978-3-030-30277-1   ISBN 978-3-030-30278-8 (eBook)
https://doi.org/10.1007/978-3-030-30278-8

This Springer imprint is published by the registered company Springer Nature Switzerland AG
The registered company address is: Gewerbestrasse 11, 6330 Cham, Switzerland

Preface

The European Conference on Advances in Databases and Information Systems (ADBIS) celebrated its 23rd anniversary. Previous ADBIS conferences were held in St. Petersburg (1997), Poznan (1998), Maribor (1999), Prague (2000), Vilnius (2001), Bratislava (2002), Dresden (2003), Budapest (2004), Tallinn (2005), Thessaloniki (2006), Varna (2007), Pori (2008), Riga (2009), Novi Sad (2010), Vienna (2011), Poznan (2012), Genoa (2013), Ohrid (2014), Poitiers (2015), Prague (2016), Nicosia (2017), and Budapest (2018). After 20 years the conference returned to Slovenia, and was organized in Bled.

ADBIS can be considered one of the most established and recognized conferences in Europe, in the broad field of databases and information systems. The conference aims at: (1) providing an international forum for presenting research achievements on database theory and practice, development of advanced DBMS technologies, and their applications; (2) promoting the interaction and collaboration between the database and information systems research communities from European countries and the rest of the world; (3) offering a forum for a less formal exchange of research ideas by means of affiliated workshops; and (4) activating young researchers from all over the world by means of a doctoral consortium.

ADBIS workshops have been associated with the ADBIS conference since 2005 and doctoral consortia have been associated since 2008. This long tradition continued this year. Therefore, the program of ADBIS 2019 included keynotes, research papers, thematic workshops, and a doctoral consortium (DC). The main conference, workshops, and DC had their own international Program Committees.

This volume contains 19 short research papers from the main conference, 31 workshops papers, and 5 DC papers, which were all presented at ADBIS 2019, held during September 8–11, 2019, in Bled, Slovenia.

The selected short papers span a wide spectrum of topics related to the ADBIS conference. Most of them are related to database and information systems technologies for advanced applications. Typical applications are text databases, streaming data, and graph processing. In addition, there are also papers covering the theory of databases.

The main conference received a total of 103 submissions. After a rigorous reviewing process 27 papers were accepted as full papers and 19 were selected as short papers for presentation and publication in this volume, giving an acceptance rate for short papers of 45%.

The following six workshops were run at ADBIS 2019:

- International Workshop on BI & Big Data Applications (BBIGAP), chaired by: Fadila Bentayeb (Université Lyon 2, France) and Omar Boussaid (Université Lyon 2, France)

- International Workshop on Qualitative Aspects of User-Centered Analytics (QAUCA), chaired by: Nicolas Labroche (Université de Tours, France), Patrick Marcel (Université de Tours, France), and Veronika Peralta (Université de Tours, France)
- A joint workshop Semantics in Big Data Management (SemBDM) and Data-Driven Process Discovery and Analysis (SIMPDA), chaired by: Paolo Ceravolo (Universita degli Studi di Milano, Italy), Florence Sedes (Toulouse Institute of Computer Science Research, France), Maria Teresa Gomez Lopez (University of Seville, Spain), and Maurice van Keulen (University of Twente, The Netherlands)
- Modelling is going to become Programming (M2P), chaired by: Ajantha Dahanayake (Lappeenranta University of Technology, Finland) and Bernhard Thalheim (Kiel University, Germany)
- Modern Approaches in Data Engineering and Information System Design (MADEISD), chaired by: Ivan Luković (University of Novi Sad, Serbia) and Slavica Kordić (University of Novi Sad, Serbia)

In total, 67 papers were submitted to these workshops, out of which 31 were selected for presentation and publication in this volume, giving an acceptance rate of 46%.

The ADBIS 2019 DC was a forum where PhD students had a chance to present their research ideas to the database research community, to receive inspiration from their peers and feedback from senior researchers, and to tie cooperation bounds. DC papers aim at describing the current status of the thesis research. The DC Committee accepted five papers that were presented at the DC. Two main topics emerged this year: big data integration and big data analytics.

ADBIS chairs would like to express their sincere gratitude to everyone who contributed to make ADBIS 2019 successful:

- All the organizers of the previous ADBIS workshops and conferences. They made ADBIS a valuable trademark and we are proud to continue their work.
- The authors, who submitted papers of high quality to the conference.
- The members of the international Program Committee for dedicating their time and expertise to assure a high-quality program.
- The members of ADBIS Steering Committee for proven trust and conferred organization of the conference.
- Springer for publishing these proceedings.
- Last but not least, to all the helping hands from the webmaster, programmers to technicians, and administration, without whom the organization of such a conference would not have been possible.

- Finally, we would like to express our special thanks to the local chair Lili Nemec Zlatolas for her continuous coordinating activities that ensured the success of ADBIS 2019.

July 2019

Tatjana Welzer
Johann Eder
Vili Podgorelec
Robert Wrembel
Mirjana Ivanović
Johann Gamper
Mikołaj Morzy
Theodoros Tzouramanis
Jérôme Darmont
Aida Kamišalić Latifić

Organization

Steering Committee Chair

Yannis Manolopoulos   Open University of Cyprus, Cyprus

Steering Committee

Ladjel Bellatreche	Laboratory of Computer Science and Automatic Control for Systems, France
Andras Benczur	Eötvös Loránd University, Hungary
Maria Bielikova	Slovak University of Technology, Slovakia
Barbara Catania	University of Genoa, Italy
Johann Eder	Alpen-Adria-Universität Klagenfurt, Austria
Theo Haerder	University of Kaiserslautern, Germany
Mirjana Ivanović	University of Novi Sad, Serbia
Hannu Jaakkola	Tampere University, Finland
Marite Kirikova	Riga Technical University, Latvia
Yannis Manolopoulos	Open University of Cyprus, Cyprus
Rainer Manthey	University of Bonn, Germany
Manuk Manukyan	Yerevan State University, Armenia
Tadeusz Morzy	Poznan University of Technology, Poland
Pavol Navrat	Slovak University of Technology, Slovakia
Boris Novikov	Saint Petersburg State University, Russia
George Angelos Papadopoulos	University of Cyprus, Cyprus
Jaroslav Pokorny	Charles University Prague, Czech Republic
Boris Rachev	Technical University of Varna, Bulgaria
Bernhard Thalheim	Kiel University, Germany
Goce Trajcevski	Iowa State University of Science and Technology, USA
Tatjana Welzer	University of Maribor, Slovenia
Robert Wrembel	Poznan University of Technology, Poland
Ester Zumpano	Università Della Calabria, Italy

Program Committee Chairs

Johann Eder	University of Klagenfurt, Austria
Vili Podgorelec	University of Maribor, Slovenia

Program Committee

Syed Sibte Raza Abidi	Dalhousie University, Canada
Bernd Amann	Sorbonne University, France

Costin Badica	University of Craiova, Romania
Marko Bajec	University of Ljubljana, Slovenia
Rodrigo Coelho Barros	Pontifical Catholic University of Rio Grande do Sul, Brazil
Andreas Behrend	University of Bonn, Germany
Ladjel Bellatreche	LIAS, ENSMA, France
András Benczür	Eötvös Loránd University, Hungary
Maria Bielikova	Slovak University of Technology in Bratislava, Slovakia
Nikos Bikakis	University of Ioannina, Greece
Zoran Bosnić	University of Ljubljana, Slovenia
Drazěn Brdjanin	University of Banja Luka, Bosnia and Herzegovina
Albertas Caplinskas	Vilnius University, Lithuania
Christos Doulkeridis	University of Piraeus, Greece
Johann Eder	University of Klagenfurt, Austria
Markus Endres	University of Passau, Germany
Werner Esswein	TU Dresden, Germany
Flavio Ferrarotti	Software Competence Centre Hagenberg, Austria
Flavius Frasincar	Erasmus University Rotterdam, The Netherlands
Jänis Grabis	Riga Technical University, Latvia
Francesco Guerra	University of Modena and Reggio Emilia, Italy
Giancarlo Guizzardi	Federal University of Espirito Santo, Brazil
Hele-Mai Haav	Tallinn University of Technology, Estonia
Theo Härder	University of Kaiserslautern, Germany
Tomáš Horváth	Eötvös Loránd University, Hungary
Marko Hölbl	University of Maribor, Slovenia
Andres Iglesias	University of Cantabria, Spain
Mirjana Ivanović	University of Novi Sad, Serbia
Hannu Jaakkola	Tampere University, Finland
Lili Jiang	Umea University, Sweden
Aida Kamišalić Latifić	University of Maribor, Slovenia
Mehmed Kantardzic	University of Louisville, USA
Dimitris Karagiannis	University of Vienna, Austria
Sašo Karakatič	University of Maribor, Slovenia
Zoubida Kedad	University of Versailles, France
Marite Kirikova	Riga Technical University, Latvia
Attila Kiss	Eötvös Loránd University, Hungary
Margita Kon-Popovska	Ss. Cyril and Methodius University in Skopje, North Macedonia
Harald Kosch	University of Passau, Germany
Michal Kratky	VSB-Technical University of Ostrava, Czech Republic
Ralf-Detlef Kutsche	TU Berlin, Germany
Julius Köpke	University of Klagenfurt, Austria
Dejan Lavbič	University of Ljubljana, Slovenia
Sebastian Link	The University of Auckland, New Zealand
Audrone Lupeikiene	Vilnius University, Lithuania

Federica Mandreoli	University of Modena, Italy
Yannis Manolopoulos	Open University of Cyprus, Cyprus
Manuk Manukyan	Yerevan State University, Armenia
Karol Matiasko	University of Žilina, Slovakia
Goran Mausă	University of Rijeka, Croatia
Bálint Molnár	Eötvös University of Budapest, Hungary
Angelo Montanari	University of Udine, Italy
Tadeusz Morzy	Poznan University of Technology, Poland
Boris Novikov	St. Petersburg University, Russia
Kjetil Nørvåg	Norwegian University of Science and Technology, Norway
Andreas Oberweis	Karlsruhe Institute of Technology, Germany
Andreas L. Opdahl	University of Bergen, Norway
Eneko Osaba	Tecnalia Research & Innovation, Spain
Odysseas Papapetrou	Eindhoven University of Technology, The Netherlands
András Pataricza	Budapest University of Technology and Economics, Czech Republic
Tomas Pitner	Masaryk University, Czech Republic
Vili Podgorelec	University of Maribor, Slovenia
Jaroslav Pokorný	Charles University Prague, Czech Republic
Giuseppe Polese	University of Salerno, Italy
Boris Rachev	Technical University of Varna, Bulgaria
Miloš Radovanović	University of Novi Sad, Serbia
Heri Ramampiaro	Norwegian University of Science and Technology, Norway
Stefano Rizzi	University of Bologna, Italy
Peter Ruppel	Technical University of Berlin, Germany
Gunter Saake	University of Magdeburg, Germany
Petr Saloun	VSB - Technical University of Ostrava, Czech Republic
José Luis Sánchez de la Rosa	University of La Laguna, Spain
Shiori Sasaki	Keio University, Japan
Kai-Uwe Sattler	TU Ilmenau, Germany
Miloš Savić	University of Novi Sad, Serbia
Timos Sellis	Swinburne University of Technology, Australia
Bela Stantic	Griffith University, Australia
Kostas Stefanidis	Tampere University, Finland
Claudia Steinberger	University of Klagenfurt, Austria
Sergey Stupnikov	Russian Academy of Sciences, Russia
Bernhard Thalheim	Kiel University, Germany
Raquel Trillo-Lado	Universidad de Zaragoza, Spain
Muhamed Turkanović	University of Maribor, Slovenia
Olegas Vasilecas	Vilnius Gediminas Technical University, Lithuania
Goran Velinov	Ss. Cyril and Methodius University, North Macedonia
Peter Vojtas	Charles University Prague, Czech Republic

Isabelle Wattiau ESSEC and CNAM, France
Tatjana Welzer University of Maribor, Slovenia
Robert Wrembel Poznan University of Technology, Poland
Jaroslav Zendulka Brno University of Technology, Czech Republic

Additional Reviewers

Nabila Berkani
Dominik Bork
Andrea Brunello
Loredana Caruccio
Stefano Cirillo
Victoria Döller
Peter Gašpar
Sandi Gec
Yong-Bin Kang
Selma Khouri

Haridimos Kondylakis
Ilya Makarov
Riccardo Martoglia
Matteo Paganelli
Marek Rychlý
Victor Sepulveda
Paolo Sottovia
Nicola Vitacolonna
Farhad Zafari

General Chair

Tatjana Welzer University of Maribor, Slovenia

Honorary Chair

Ivan Rozman University of Maribor, Slovenia

Proceedings Chair

Aida Kamišalić Latifić University of Maribor, Slovenia

Workshops Chairs

Robert Wrembel Poznan University of Technology, Poland
Mirjana Ivanović University of Novi Sad, Serbia
Johann Gamper Free University of Bozen-Bolzano, Italy

Doctoral Consortium Chairs

Jerome Darmont Université Lumière Lyon 2, France
Mikolay Morzy Poznan University of Technology, Poland
Theodoros Tzouramanis University of the Aegean, Greece

Local Chair

Lili Nemec Zlatolas   University of Maribor, Slovenia

Organizing Committee

Marko Hölbl   University of Maribor, Slovenia
Luka Hrgarek   University of Maribor, Slovenia
Aida Kamišalić Latifić   University of Maribor, Slovenia
Marko Kompara   University of Maribor, Slovenia
Lili Nemec Zlatolas   University of Maribor, Slovenia
Tatjana Welzer   University of Maribor, Slovenia
Borut Zlatolas   University of Maribor, Slovenia

Workshops

Workshops Chairs

Robert Wrembel	Poznan University of Technology, Poland
Mirjana Ivanović	University of Novi Sad, Serbia
Johann Gamper	Free University of Bozen-Bolzano, Italy

Modelling Is Going to Become Programming – M2P

Chairs

Ajantha Dahanayake	Lappeenranta University of Technology, Finland
Bernhard Thalheim	Kiel University, Germany

Program Committee Members

Witold Abramowicz	Poznan University of Economics, Poland
Igor Fiodorov	Plekhanov Russian University of Economics, Russia
Holger Giese	University of Potsdam, Germany
Hannu Jaakkola	University of Tampere, Finland
Heinrich C. Mayr	University of Klagenfurt, Austria
John Mylopoulos	University of Toronto, Canada
Bernhard Rumpe	RWTH Aachen University, Germany
Klaus-Dieter Schewe	Zhejiang University, China
Veda Storey	Georgia State University, USA
Marina Tropmann-Frick	Hamburg University of Applied Sciences, Germany

Additional Reviewers

Manuela Dalibor	RWTH Aachen University, Germany
Nico Jansen	RWTH Aachen University, Germany
Jörg Christian Kirchhof	RWTH Aachen University, Germany
Judith Michael	RWTH Aachen University, Germany
Lukas Netz	RWTH Aachen University, Germany

Modern Approaches in Data Engineering and Information System Design – MADEISD

Chairs

Ivan Luković	University of Novi Sad, Serbia
Slavica Kordić	University of Novi Sad, Serbia

Program Committee Members

Paulo Alves	Polytechnic Institute of Bragança, Portugal
Marko Bajec	University of Ljubljana, Slovenia
Zoran Bosnić	University of Ljubljana, Slovenia
Moharram Challenger	University of Antwerp, Belgium
Boris Delibašić	University of Belgrade, Serbia
Joaõ Miguel Lobo Fernandes	University of Minho, Portugal
Krešimir Fertalj	University of Zagreb, Croatia
Krzysztof Goczyla	Gdánsk University of Technology, Poland
Ralf-Christian Härting	Aalen University, Germany
Dušan Jakovetić	University of Novi Sad, Serbia
Miklós Krész	InnoRenew CoE and University of Primorska, Slovenia
Dragan Maćoš	Beuth University of Applied Sciences Berlin, Germany
Zoran Marjanović	University of Belgrade, Serbia
Sanda Martinčić-Ipšić	University of Rijeka, Croatia
Cristian Mihaescu	University of Craiova, Romania
Nikola Obrenović	Ecole Polytechnique Fédérale de Lausanne, Switzerland
Maxim Panov	Skolkovo Institute of Science and Technology, Moscow, Russia
Rui Humberto Pereira	Polytechnic Institute of Porto, Portugal
Aleksandar Popović	University of Montenegro, Montenegro
Patrizia Poščić	University of Rijeka, Croatia
Adam Przybylek	Gdánsk University of Technology, Poland
Sonja Ristić	University of Novi Sad, Serbia
Sergey Rykovanov	Skolkovo Institute of Science and Technology, Russia
Igor Rožanc	University of Ljubljana, Slovenia
Nikolay Skvortsov	Russian Academy of Sciences, Russia

A Joint Workshop on Semantics in Big Data Management – SemBDM and Data-Driven Process Discovery and Analysis – SIMPDA

Chairs

Paolo Ceravolo	Universita degli Studi di Milano, Italy
Florence Sedes	Toulouse Institute of Computer Science Research, France
Maria Teresa Gomez Lopez	University of Seville, Spain
Maurice van Keulen	University of Twente, The Netherlands

Program Committee Members

Pnina Soffer	University of Haifa, Israel
Kristof Böhmer	University of Vienna, Austria
Luisa Parody	Loyola University Andalusia, Spain
Gabriel Tavares	Londrina State University, Brazil
Roland Rieke	Fraunhofer Institute for Secure Information Technology, Germany
Angel Jesus Varela Vaca	University of Seville, Spain
Massimiliano de Leoni	University of Padova, Italy
Faiza Allah Bukhsh	University of Twente, The Netherlands
Robert Singer	FH JOANNEUM, Austria
Christophe Debruyne	Trinity College Dublin, Ireland
Antonia Azzini	Consorzio per il Trasferimento Tecnologico (C2T), Italy
Mirjana Pejic-Bach	University of Zagreb, Croatia
Marco Viviani	University of Milano-Bicocca, Italy
Carlos Fernandez-Llatas	Technical University of Valencia, Spain
Richard Chbeir	University of Pau and Pays de l'Adour, France
Manfred Reichert	University of Ulm, Germany
Valentina Emilia Balas	Aurel Vlaicu University of Arad, Romania
Mariangela Lazoi	University of Salento, Italy
Maria Leitner	Austrian Institute of Technology, Austria
Karima Boudaoud	University of Nice Sophia Antipolis, France
Chiara Di Francescomarino	Fondazione Bruno Kessler, Italy
Haralambos Mouratidis	University of Brighton, UK
Helen Balinsky	Hewlett Packard Laboratories, USA
Mark Strembeck	Vienna University of Economics and Business, Austria
Tamara Quaranta	40Labs, Italy
Yingqian Zhang	Eindhoven University of Technology, The Netherlands
Edgar Weippl	SBA Research, Austria

International Workshop on BI & Big Data Applications – BBIGAP

Chairs

Fadila Bentayeb	University of Lyon 2, France
Omar Boussaid	University of Lyon 2, France

Program Committee Members

Thierry Badard	Laval university of Quebec, Canada
Hassan Badir	University of Tanger, Morocco
Ladjel Bellatreche	National Engineering School for Mechanics and Aerotechnics, France
Nadjia Benblidia	University of Blida, Algeria

Sandro Bimonte   National Research Institute of Science and Technology
                              for Environment and Agriculture, France
Azedine Boulmalkoul   University of Mohammadia, Morocco
Laurent d'Orazio   University of Rennes 1, France
Abdessamad Imine   University of Lorraine, France
Daniel Lemire   University of Quebec in Montreal, Canada
Gérald Gavin   University of Lyon, France
Rokia Missaoui   University of Quebec in Gatineau, Canada
Rim Moussa   University of Carthage, Tunisia
Abdelmounaam Rezgui   New Mexico Tech, USA
Olivier Teste   University of Toulouse, France
Gilles Zurfluh   University of Toulouse, France

International Workshop on Qualitative Aspects of User-Centered Analytics – QAUCA

Chairs

Nicolas Labroche   University of Tours, France
Patrick Marcel   University of Tours, France
Veronika Peralta   University of Tours, France

Program Committee Members

Julien Aligon   University of Toulouse, France
Ladjel Bellatreche   National Engineering School for Mechanics
                              and Aerotechnics, France
Laure Berti-Equille   Aix-Marseille University, France
Ismael Caballero   University of Castilla-La Mancha, Spain
Silvia Chiusano   Polytechnic University of Turin, Italy
Jérôme Darmont   University of Lyon, France
Matteo Golfarelli   University of Bologna, Italy
Magdalini Eirinaki   San Jose State University, USA
Lorena Etcheverry   University of Uruguay, Uruguay
Zoubida Kedad   Versailles Saint-Quentin University, France
Natalija Kozmina   University of Latvia, Latvia
Nicolas Labroche   University of Tours, France
Marie Jeanne Lesot   Sorbonne University, France
Patrick Marcel   University of Tours, France
Adriana Marotta   University of Uruguay, Uruguay
Christophe Marsala   Sorbonne University, France
Elsa Negre   Paris Dauphine University, France
Veronika Peralta   University of Tours, France
Franck Ravat   University of Toulouse, France
Samira Si-said   National Conservatory of Arts and Craftsč, France
Alejandro Vaisman   Buenos Aires Institute of Technology, Argentina

| Panos Vassiliadis | University of Ioannina, Greece |
| Gilles Venturini | University of Tours, France |

Additional Reviewers

Ghislain Atemezing	MONDECA, France
Faten Atigui	Conservatoire National des Arts et Métiers (CNAM), France
Soufiane Mir	Orange Bank, France
Anton Dignös	Free University of Bozen-Bolzano, Italy
Marko Tkalčič	Free University of Bozen-Bolzano, Italy

Doctoral Consortium

Doctoral Consortium Chairs

Jerome Darmont	Université Lumière Lyon 2, France
Mikolay Morzy	Poznan University of Technology, Poland
Theodoros Tzouramanis	University of the Aegean, Greece

Doctoral Consortium Program Committee

Kamel Boukhalfa	USTHB, Algeria
Mahfoud Djedaini	University of Tours, France
Imen Megdiche	ISIS Albi, France
Jiefu Song	University of Toulouse, France

Abstract of Invited Talk

Databases Meet Blocks, Ledgers and Contracts

Muhamed Turkanović

Faculty of Electrical Engineering and Computer Science,
University of Maribor, Koroška cesta 46, 2000 Maribor, Slovenia
muhamed.turkanovic@um.si

Introduction and Motivation

Considering the impact on society blockchain technology has, it is falsely to predict the technology is adequately understood. Even in professional circles, there exists the notion, that this technology is only connected with crypto currencies and finances, whereby neglecting the fact that in its core, it is a novel type of a database technology, having its specifics, which make it suitable for special use cases [8]. As it was with the NoSQL databases in the early 21st century, the public tends to perceive it as a solution for everything, while slowly realising the contrary [5]. Even though the technology is still in it early stages, it should not be neglected, as indicated also by Gartner in its top 10 strategic technology trends for 2019 report, where Blockchain is within the mesh section [3].

Furthermore, Blockchain is also the cornerstone of smart contracts as well as decentralized application. Where the latter is a novel Web technology, based on the Web3 notion, which emphasizes the decentralized and distributed nature of the new Internet [2], the former is the core of it, whereby pushing the business logic onto the data layer [4, 6].

As aforementioned, blockchain technology can be categorized as kind of a database, however, it has several specific characteristics, which consequently effects the possible use cases, making it hard to apply without thinking out of the box. Furthermore, due to its distributed nature and its focus on data integrity, it is interesting to see how it falls into the CAP theorem [1].

Breaking Down Blockchain

lockchain technology is part of the bigger family of distributed ledger technologies. Its main characteristic is storing transaction-based data into a chronological structure in multiple virtual locations simultaneously, while relying on the decentralized environment and not on a central authority to manage the process [7]. Another important characteristic is the fact that it ensures the data integrity on the highest level, even though it is operating in a distributed environment. The blockchain protocols store data into predefined size of data blocks, which stores only a limited size of transactions-based data that is cryptographically signed. Each block is referenced with the previous one, forming an unbreakable chain of blocks, distributed throughout the network in a fully replicated manner, i.e. a digital ledger between network nodes. The data in the ledger cannot be queried as per se, whereas depending on the blockchain platform, it has to be structured or can be unstructured in some degree. The latter is the case with smart contracts, which are a collection of programmatic instruction, stored in blocks of data, which can later on reference newly added data. Smart contracts are programmable code, which is execute in a dedicated Database Management Systems, which includes a specific and dedicated virtual environment (e.g. Ethereum Virtual Machine) [4]. As such smart contracts can host the business logic, which is stored on the data layer, consequently in a distributed and decentralized environment, accessible by decentralized applications. Such feature puts blockchain on the top of the database world, where it requires much more attention from the professional community.

The focus of this work will thus be on the features, possibilities, as well as on the challenges of blockchain technology, breaking it down in terms of data structures, topology, as well as current open research topics.

References

1. Brewer, E.: CAP twelve years later: how the "rules" have changed. Computer **45**(2), 23–29 (2012). https://ieeexplore.ieee.org/document/6133253/
2. Cai, W., Wang, Z., Ernst, J.B., Hong, Z., Feng, C., Leung, V.C.M.: Decentralized applications: the blockchain-empowered software system. IEEE Access **6**, 53019–53033 (2018). https://ieeexplore.ieee.org/document/8466786/
3. Cearley, D., Burke, B.: Top 10 Strategic Technology Trends for 2019: A Gartner Trend Insight Report. Technical report, https://emtemp.gcom.cloud/ngw/globalassets/en/doc/documents/383829-top-10-strategic-technology-trends-for-2019-a-gartner-trend-insight-report.pdf
4. Chinchilla, C.: A Next-Generation Smart Contract and Decentralized Application Platform. https://github.com/ethereum/wiki/wiki/White-Paper
5. Leavitt, N.: Will NoSQL databases live up to their promise? Computer **43**(2), 12–14 (2010). http://ieeexplore.ieee.org/document/5410700/
6. Mohanta, B.K., Panda, S.S., Jena, D.: An overview of smart contract and use cases in blockchain technology. In: 2018 9th International Conference on Computing, Communication and Networking Technologies (ICCCNT), pp. 1–4. IEEE, July 2018. https://ieeexplore.ieee.org/document/8494045/
7. Suciu, G., Nadrag, C., Istrate, C., Vulpe, A., Ditu, M.C., Subea, O.: Comparative analysis of distributed ledger technologies. In: 2018 Global Wireless Summit (GWS), pp. 370–373. IEEE, November 2018. https://ieeexplore.ieee.org/document/8686563/
8. Turkanovic, M., Holbl, M., Kosic, K., Hericko, M., Kamisalic, A.: EduCTX: a blockchain-based higher education credit platform. IEEE Access **6**, 5112–5127 (2018). http://ieeexplore.ieee.org/document/8247166/

Abstract of the Invited Talk on International Workshop on Qualitative Aspects of User-Centered Analytics – QAUCA

Analytical Metadata Modeling

Oscar Romero

Universitat Politècnica de Catalunya, BarcelonaTech

Abstract. Decision-making is based on thorough analyses of the available data. However, due to the data growth and diversification, data exploration becomes a challenging and complex task. Next generation systems aim at assisting the user and facilitating this process. Complex metadata constructs sit at the core of these systems and they are the key component to automate and support data analysts tasks. In this talk, we will discuss how metadata is managed and exploited to provide strategic advantage to organizations by lowering the high-entry barriers to complex data analysis.

Keywords: Big data · Data variety · Metadata · User-centric Analysis

Introduction

Traditional Decision Support Systems (DSS) already underlined the relevance of metadata for decision making. Some approaches, such as [4], described several aspects of the data lifecycle (e.g., schemata, provenance, etc.) in terms of metadata. However, such metadata was not used to support the user in their day-by-day tasks (rather, it was used as internal information to manage the BI system). Also, the advent of Big Data emphasized the need to extract information from data sources out of our control, which typically expose unstructured or, at most, semi-structured data. As such, the Data Variety challenge [1] was coined, which claims for crossing and integrating as much relevant data, of any type, as possible. This includes text, images, video and audio. which require a pre-processing stage to extract relevant features from them prior to integrating such data with other sources. The confluence of both trends reinforced the need to annotate the processes and data generated with metadata in order to automate the data lifecycle and support the users in their analytical tasks regardless of the potentially complex data management processes in the background.

The so-called next generation BI systems (e.g., [2, 3], among others), acknowledge the wealth of data available in diverse formats and therefore, in addition, claim for expanding the use of metadata to other fields such as source discovery, data integration/crossing, recommendation, personalisation and, in general, to support the user exploiting the available data. Indeed, there is a strong common backbone in this shift. Next generation systems should automate repetitive and burdensome tasks (e.g., creating a catalog of variables available) and support and facilitate those tasks where a full automation is harder to achieve (e.g., entity resolution).

Analytical Metadata for Next Generation Systems

In this talk we will claim for a strict methodology when creating, annotating and managing metadata. We will present SM4AM [5], *a Semantic Model for Analytical Metadata* based on RDF, and jointly designed by the DTIM[1] and DAISY[2] research groups. SM4AM is grounded on ontological metamodeling, which we claim to be the proper solution to model and store metadata in these scenarios. This way, we avoid imposing a fixed universal model, which is unfeasible due to the (meta)data models heterogeneity in the current wealth of data. Further, RDF supports sharing and flexible metadata representation, which opens the door for advanced interoperability between semantic-aware systems. We will also discuss how to properly instantiate and exploit SM4AM. In this sense, metadata annotation should be a manual task but fully automatic to avoid errors and facilitate a common understanding of the RDF metamodel.

We will wrap-up this talk by presenting novel ideas about how to exploit a metadata repository compliant with SM4AM. First, we will introduce a novel way to cross data through what we call *metamodel-driven (meta)data exploration*, where SM4AM is asserted on top of existing RDF models to facilitate the automatic extraction and processing of already existing data repositories. We will compare our approach to other prominent solutions such as KBPedia[3]. Next, we will present a next generation Big Data system that uses SM4AM at its core. This system uses metadata to automatically bootstrap sources (for now, CSV, JSON, XML and relational sources), and extract and align their schemata in an automatic manner. From such information, the system is able to generate a global view of the sources in a bottom-up approach. Finally, the system provides high-level user-friendly querying mechanisms over the global view generated for non-IT people. All in all, we will show that there are nowadays solid foundations as to pave the road to build semantic-aware systems able to efficiently manage the data variety challenge while providing advanced support for data analysts.

References

1. Bean, R.: Variety, not volume, is driving big data initiatives, March 2016
2. Deng, D., et al.: The data civilizer system. In: 8th Biennial Conference on Innovative Data Systems Research CIDR 2017 (2017)
3. Nadal, S., Romero, O., Abelló, A., Vassiliadis, P., Vansummeren, S.: An integration-oriented ontology to govern evolution in big data ecosystems. Inf. Syst. (2018). https://doi.org/10.1016/j.is.2018.01.006
4. Object Management Group.: Common Warehouse Metamodel Specification 1.1. http://www.omg.org/spec/CWM/1.1/PDF/. Accessed Sep 2016
5. Varga, J., Romero, O., Pedersen, T.B., Thomsen, C.: Analytical metadata modeling for next generation BI systems. J. Syst. Softw. **144**, 240–254 (2018)

---

[1] http://www.essi.upc.edu/dtim/.

[2] https://www.daisy.aau.dk/.

[3] http://kbpedia.org/.

Contents

**ADBIS 2019 Workshop: Modern Approaches in Data Engineering
and Information System Design – MADEISD**

ADBIS 2019 Workshop: Semantics in Big Data Management – SemBDM and Data-Driven Process Discovery and Analysis – SIMPDA

ADBIS 2019 Workshop: International Workshop on BI and Big Data Applications – BBIGAP

ADBIS 2019 Workshop: International Workshop on Qualitative Aspects of User-Centered Analytics – QAUCA

ADBIS 2019 Doctoral Consortium

ADBIS 2019 Short Papers

Distributed Computation of Top-k Degrees in Hidden Bipartite Graphs

Panagiotis Kostoglou[1(✉)], Apostolos N. Papadopoulos[1], and Yannis Manolopoulos[2]

[1] Aristotle University of Thessaloniki, Thessaloniki, Greece
{panakost,papadopo}@csd.auth.gr
[2] Open University of Cyprus, Latsia, Cyprus
yannis.manolopoulos@ouc.ac.cy

Abstract. Hidden graphs are flexible abstractions that are composed of a set of known vertices (nodes), whereas the set of edges are not known in advance. To uncover the set of edges, multiple edge probing queries must be executed by evaluating a function $f(u, v)$ that returns either true or false, if nodes u and v are connected or not respectively. Evidently, the graph can be revealed completely if all possible $n(n - 1)/2$ probes are executed for a graph containing n nodes. However, the function $f()$ is usually computationally intensive and therefore executing all possible probing queries result in high execution costs. The target is to provide answers to useful queries by executing as few probing queries as possible. In this work, we study the problem of discovering the top-k nodes of a hidden bipartite graph with the highest degrees, by using distributed algorithms. In particular, we use Apache Spark and provide experimental results showing that significant performance improvements are achieved in comparison to existing centralized approaches.

Keywords: Graph mining · Hidden networks · Top-k degrees

1 Introduction

Nowadays, graphs are used to model real-life problems in many diverse application domains, such as social network analysis, searching and mining the Web, pattern mining in bioinformatics and neuroscience. Graph mining is an active and growing research area, aiming at knowledge discovery from graph data [1]. In this work, we focus on *simple bipartite graphs*, where the set of nodes is composed of two subsets B (the *black* nodes) and W (the *white* nodes), such as $V = B \cup W$ and $B \cap W = \emptyset$.

Bipartite graphs have many interesting applications in diverse fields. For example, a bipartite graph may be used to represent product purchases by customers. In this case, an edge exists between a product p and a customer c when p was purchased by c. As another example, in an Information Retrieval or Text Mining application a bipartite graph may be used to associate different types

© Springer Nature Switzerland AG 2019
T. Welzer et al. (Eds.): ADBIS 2019, CCIS 1064, pp. 3–10, 2019.
https://doi.org/10.1007/978-3-030-30278-8_1

of tokens that exist in a document. Thus, an edge between a document d and a token t represents the fact that the token t appears in document d.

In conventional bipartite graphs, the sets of vertices B and W as well as the set of edges E are known in advance. Nodes and edges are organized in a way to enable efficient execution of fundamental graph-oriented computational tasks, such as finding high-degree nodes, computing shortest paths, detecting communities. Usually, the adjacency list representation is being used, which is a good compromise between space requirements and computational efficiency. However, a concept that has started to gain significant interest recently is that of *hidden graphs*. In contrast to a conventional graph, a hidden bipartite graph is defined as $G(B, W, f())$, where B and W are the subsets of nodes and $f()$ is a function $B \times W \rightarrow \{0, 1\}$ which takes as an input two vertex identifiers and returns true or false if the edge exists or not respectively.

Formally, for n nodes, there is an exponential number of different graphs that may be generated (the exact number is $2^{\binom{n}{2}}$). To materialize all these different graphs demands significant space requirements. Moreover, it is unlikely that all these graphs will be used eventually. In contrast, the use of a hidden graph enables the representation of any possible graph by the specification of the appropriate function $f()$. It is noted that the function $f()$ may require the execution of a complex algorithm in order to decide if two nodes are connected by an edge. Therefore, interesting what-if scenarios may be tested with respect to the satisfaction of meaningful properties. It is evident, that the complete graph structure may be revealed if all possible $n_b \times n_w$ edge probing queries are executed, where $n_b = |B|$ and $n_w = |W|$. However, this solution involves the execution of a quadratic number of probes, which is expected to be highly inefficient, taking into account that the function $f()$ is computationally intensive and that real-world graphs are usually very large. Therefore, the target is to provide a solution to a graph-oriented problem by using as few edge probing queries as possible. The problem we are targeting is the discovery of the k nodes with the highest degrees. The degree of a node v is defined as the number of neighbors of v. This problem has been addressed previously by [6] and [8]. In this work, we are interested in solving the problem by using distributed algorithms over Big Data architectures. This enables the analysis of massive hidden graphs and provides some baseline for executing more complex graph mining tasks, overcoming the limitations provided by centralized approaches. In particular, we study distributed algorithms for the discovery of the top-k degree nodes in Apache Spark [9] over YARN and HDFS [7] and we offer experimental results based on a cluster of 32 physical machines. Experiments demonstrate that the proposed techniques are scalable and can be used for the analysis of hidden networks.

2   Related Research

A hidden graph is able to represent arbitrary relationship types among graph nodes. A research direction that is strongly related to hidden graphs is *graph property testing* [4]. In this case, one is interested in detecting if a graph satisfies a property or not by using fast approximate algorithms. Evidently, detecting graph properties efficiently is extremely important. However, an ever more useful and more challenging task is to detect specific subgraphs or nodes that satisfy specific properties, by using as few edge probing queries as possible.

Another research direction related to hidden graphs focuses on *learning* a graph or a subgraph by using edge probing queries using pairs or sets of nodes (group testing) [2]. A similar topic is the *reconstruction* of subgraphs that satisfy certain structural properties [3]. An important difference between graph property testing and analyzing hidden graphs is that in the first case the graph is known whereas in the second case the set of edges is unknown and must be revealed gradually. Moreover, property testing focuses on checking if a specific property holds or not, whereas in hidden graph analysis we are interested in answering specific queries exactly. The problem we attack is the discovery of the top-k nodes with the highest degrees. This problem was solved by [6] and [8] assuming a centralized environment.

The basic algorithm used in [6] was extended in [5] for unipartite graphs, towards detecting if a hidden graph contains a k-core or not. Again, the algorithm proposed in [5] is centralized and its main objective is to minimize the number of edge probing queries performed.

In this paper, we take one step forward towards the design and performance evaluation of distributed algorithms for detecting the top-k nodes with the highest degrees in hidden bipartite graphs. We note that this is the first work in the literature to attack the specific problem in a distributed setting.

3   Proposed Approach

The detection of the top-k nodes with the highest degrees in a bipartite graph has been studied in [6]. In that paper, the authors present the *Switch-On-Empty* (SOE) algorithm which provides an optimal solution with respect to the required number of probing queries. SOE receives as input a hidden bipartite graph, with bipartitions B and W. The output is composed of the k vertices from B or W with the highest degrees. Without loss of generality, we are focusing on vertices of B. For a vertex $b_1 \in B$, selects a vertex $w_1 \in W$ and executes $f(b_1, w_1)$. If the edge (b_1, w_1) is solid, it continues to perform probes between b_1 and another vertex $w_2 \in W$, but if it's empty the algorithm proceed with $b_2 \in B$. A round is complete when all vertices of B have been considered. After each round, some vertices can be safely included in the result set R and they are removed from B. SOE continues until the upper bound of vertex degrees in B is less than the current k-th highest degree determined so far.

In this work, we proceed one step further in order to detect high-degree nodes in large graphs using multiple resources. Our algorithms are implemented in the Apache Spark engine [9] using the Scala programming language. In the sequel, we focus on the algorithms developed to solve the distributed detection of top-k nodes in a hidden bipartite graph. The first algorithm, Distributed Switch-On-Empty (DSOE), is a distributed version of the SOE algorithm and the second algorithm, DSOE*, is an improved and more flexible version of DSOE.

3.1   The DSOE Algorithm

The main mechanism, for DSOE, has the following rationale. For all vertices $b \in B$ we are executing, in parallel, edge-probing queries until we get a certain amount of negative results f. The value of f is relevant to the size of the graph and is being reduced exponentially in every iteration, because the majority of the degree distribution in real life graphs follows a Power law distribution, we do not expect to find many vertices with a high degree. The closer we get to the point of exhausting all $w \in W$ the smaller we set f, because we want to avoid as many unnecessary queries as possible. When a vertex $b \in B$ completes all possible probes is added in the answer-set R. If $|R| \geq k$, then we need one last batch of routines to finalize R. Algorithm 1 contains the pseudocode of DSOE.

Algorithm 1: DSOE

Result: Return a set R of vertices that have the highest degree

1   **Routine** $v : vertex, limit : Int$ **is**
2   **while** $(Falseprobes) \leq limit$ **do**
3   Execute the next query ;
4   **if** $edge\text{-}probing\ query{==}False$ **then**
5 | $Falseprobes + +$;
6   **end**
7   **end**
8   f=2;
9   **while** $|R| \leq k$ **do**
10   map(routine($\forall b \in B, |W|/f$)) ;
11     $\forall b \in B$ we check if b has probe all the vertices $\forall w \in W$, each vertices that did it is being removed from B and is being added to R;
12     $f = f * 2$
13   **end**
14   map(routine($\forall b \in B, f$));
15   $\forall b \in B$ we check if b has probed all the vertices $\forall w \in W$, each vertices that did it, is being removed from B and is being added to R;
16   return R;

In this part we will prove that our conditions regarding R are correct. For a vertex $b_1 \in B$, we assume that b_1 completes all possible probes in the routine r_1. So we will have $d(b_1) = |W| - r_1$ or $d(b_1) = |W| - r_1 - 1$ in case the last probe was negative. For a routine r_2 with $r_2 > r_1$ we will have $d(b_1) = |W| - r_2$ or $d(b_1) = |W| - r_2 - 1$. We can conclude that $d(b_1) \geq d(b_2)$ and the equality is possible only if $r_2 = r_1 + 1$. For this reason we run a last routine when $R \geq k$.

3.2   The DSOE* Algorithm

DSOE is the natural extension of SOE. However, we advance one step further in order to improve the runtime as much as possible. Inevitably we will have a big number of probes for big graphs so by adding a small amount of probes in favor of execution time may be beneficial.

DSOE* requires an initial prediction about the degree of the vertices with sample. Then we calculate all possible edge-probing queries for the vertices we predict to have a large degree. This calculation provides a threshold for the remaining vertices, in order not to waste time on low degree vertices.

More specifically, DSOE* initially performs some random edge-probing queries $\forall b \in B$. The number for these queries is set to $\log(\log(|W|))$. Then we execute repeatedly a batch of routines $\forall b \in B$, with the difference that, this time a single routine for a vertex $b \in B$ is complete after N negative edge-probing queries, where N is an outcome of the prediction performed by the prediction process.

When a vertex exhausts all possible queries, then it is added to a temporary set M. The contents of M provide a threshold T which is set to the minimum degree among the vertices in M. Probing queries are executed $\forall b \in B$, until $maxPossibleD(b) < T$. The vertices that have completed all possible probes are added to M. Finally, the k best vertices from M with respect to the degree are returned as the final answer. Algorithm 2 contains the pseudocode of DSOE*.

Algorithm 2: DSOE*

Result: Return a set R of vertices that have the highest degree

1 **Prediction** *(v : vertex, probes : int) : int* **is**
2   **while** *counter < probes* **do**
3   **if** *edge-probing query==True* **then**
4   prediction++;
5   return prediction;

6 **Routine** *(v: vertex) : void* **is**
7   **while** *negatives < v.prediction* **do**
8   **if** *edge-probing query==False* **then**
9   negatives++;

10 **Exhaust** *(v: vertex, T: int) : void* **is**
11   **while** $T \leq v.maxPossibleDegree$ **do**
12   execute the next edge-probing query;

13 map(prediction($\forall b \in B$,log(log(| W |))));
14 **while** $| M |\leq k$ **do**
15   map(routine($\forall b \in B$));
16    $\forall b \in B$ we check if b has probe all the vertices $\forall w \in W$,each vertices that did it is being removed from B and is being added to M;
17 threshold=min(M);
18 map(exhaust($\forall b \in B$,threshold));
19 $\forall b \in B$ we check if b has probe all the vertices $\forall w \in W$,each vertices that did it is being removed from B and is being added to M;
20 R=top-k from M;
21 return R;

4 Performance Evaluation

In this section, we present performance evaluation results depicting the efficiency of the proposed distributed algorithms. All experiments have been conducted in a cluster of 32 physical nodes (machines) running Hadoop 2.7 and Spark 2.1.0. One node is used as the *master* whereas the rest 31 nodes are used as *workers*. The data resides in HDFS and YARN is being used as the resource manager.

All datasets used in performance evaluation correspond to real-world networks. The networks used have different number of black and white nodes as well as different number of edges. More specifically, we have used three networks: DBLP, YOUTUBE, and WIKI. All networks are publicly available at the Koblenz Network Collection (http://konect.uni-koblenz.de/).

DBLP. The DBLP network is the authorship network from the DBLP computer science bibliography. With $|B| = 4,000,150$, $|W| = 1,425,813$, the highest degree for set B is 114, the highest degree for set W is 951 and the average degree for B is 6.0660 and for W is 2.1622.

YOUTUBE1. This is the bipartite network of YouTube users and their group memberships. With $|B| = 94,238$ and $|W| = 30,087$, the highest degree of set B is 1,035, the highest degree of set W is 7,591, the average degree for B is 3.1130 and the average degree of W is 9.7504.

YOUTUBE2. This network represents friendship relationships between YouTube users. Nodes are users and an undirected edge between two nodes indicates a friendship. We transform this graph to bipartite by duplicate it's vertices and the set B is one clone of the nodes whereas W is the other. Two nodes between B and W are connected if their corresponding nodes are connected in the initial graph. Evidently, for this graph the statistics for the two sets B and W are exactly the same: $|B| = |W| = 1,134,890$, the maximum degree is 28,754 and the average degree is 5.2650.

WIKIPEDIA. This is the bipartite network of English Wikipedia articles and the corresponding categories they are contained in. For this graph we have $|B| = 182,947$, $|W| = 1,853,493$, the highest degree for B is 11,593 and for W is 54, the average degree for B is 2.048 while for set W is 20.748.

In the sequel, we present some representative experimental results demonstrating the performance of the proposed techniques. First, we perform a comparison of DSOE and DSOE* with respect to runtime (i.e., time duration to complete the computation in the cluster) by modifying the number of Spark executors running. For this experiment, we start with 8 Spark executors and gradually we keep on increasing their number keeping $k = 10$. The DBLP graph has been used in this case. Figure 1(a) shows the runtime of both algorithms by increasing the number of executors. It is evident that both algorithms are scalable, since there is a significant speedup as we double the number of executors. Moreover, DSOE* shows better performance than DSOE.

Another important factor is the number of probes vs k. We are running our algorithms in a significantly smaller dataset (YOUTUBE1) for different values of

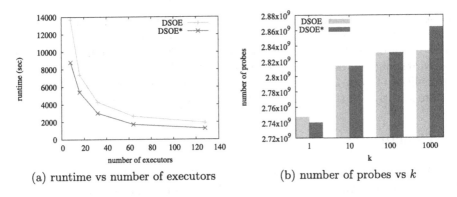

(a) runtime vs number of executors   (b) number of probes vs k

Fig. 1. Comparison of algorithms with respect to runtime and number of probes.

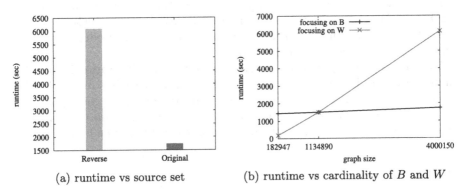

(a) runtime vs source set   (b) runtime vs cardinality of B and W

Fig. 2. Performance of DSOE* with respect to the size of node sets B and W.

k (i.e., 1, 10, 100 and 1000). The corresponding results are given in Fig. 1(b). The first observation is that both algorithms perform a significant amount of probes. However, this was expected, taking into account that if the average node degree is very small or very large, then many probes are required before the algorithms can provide the answer, as it has been shown in [6]. Also, in general DSOE* requires less probes in comparison to DSOE for small values of k. One more very interesting aspect is the performance with respect to the graph size. For this reason we apply DSOE* on the DBLP graph twice: for the first execution we have $|B| = 4,000,150$ and $|W| = 1,425,813$ and for the second one we reverse the direction of the queries from W to B. This way the graphs that we compare have the exact same number of edges and vertices but the differ significantly in the cardinality of B and W. Figure 2(a) presents the corresponding results. It is observed that in the reverse case the runtime is significantly higher, since for every node in W there are more options to perform probes on B. In general, the cost of the algorithm drops if the cardinality if B is larger than that of W.

The goal of the last experiment is to test scalability. First we focus on B and then on W. For this experiment DBLP, YOUTUBE2 and WIKIPEDIA datasets have been used. These graphs have almost the same cardinality in one set and they differ on the cardinality of the other. For all tests we have used 64 Spark executors and the implementation of the DSOE* algorithm with $k = 10$. The corresponding results are given in Fig. 2(b). It is evident that although the cardinalities of both B and W have an impact on performance, execution time is more sensitive on the cardinality of W when used as the source set.

5 Conclusions

In this work, we study for the first time the distributed detection of the top-k nodes with the highest degrees in a hidden graph. Since the set of edges is not available a-priori, edge probing queries must be applied in order to be able to explore the graph. We have designed two algorithms to attack the problem (DSOE and DSOE*) and evaluate their performance based on real-world networks. In general the algorithms are scalable, showing good speedup factors by increasing the number of Spark executors.

By studying the experimental results, one important observation is that the number of probes is in general large. Therefore, more research is required to be able to significantly reduce the number of probes preserving the good quality of the answer. Also, it is interesting to evaluate the performance of the algorithms in even larger networks containing significantly larger number of nodes and edges.

References

1. Aggarwal, C.C., Wang, H.: Managing and Mining Graph Data. Springer, New York (2010). https://doi.org/10.1007/978-1-4419-6045-0
2. Alon, N., Asodi, V.: Learning a hidden subgraph. SIAM J. Discrete Math. **18**(4), 697–712 (2005)
3. Bouvel, M., Grebinski, V., Kucherov, G.: Combinatorial search on graphs motivated by bioinformatics applications: a brief survey. In: Kratsch, D. (ed.) WG 2005. LNCS, vol. 3787, pp. 16–27. Springer, Heidelberg (2005). https://doi.org/10.1007/11604686_2
4. Goldreich, O., Goldwasser, S., Ron, D.: Property testing and its connection to learning and approximation. J. ACM **45**(4), 653–750 (1998)
5. Strouthopoulos, P., Papadopoulos, A.N.: Core discovery in hidden graphs. CoRR (to appear in Data and Knowledge Engineering) abs/1712.02827 (2017)
6. Tao, Y., Sheng, C., Li, J.: Finding maximum degrees in hidden bipartite graphs. In: Proceedings ACM International Conference on Management of Data (SIGMOD), Indianapolis, IN, pp. 891–902 (2010)
7. White, T.: Hadoop: The Definitive Guide, 4th edn. O'Reilly Media Inc., Sebastopol (2015)
8. Yiu, M.L., Lo, E., Wang, J.: Identifying the most connected vertices in hidden bipartite graphs using group testing. IEEE Trans. Knowl. Data Eng. **25**(10), 2245–2256 (2013)
9. Zaharia, M., et al.: Apache spark: a unified engine for big data processing. Commun. ACM **59**(11), 56–65 (2016)

Applying Differential Evolution with Threshold Mechanism for Feature Selection on a Phishing Websites Classification

Lucija Brezočnik$^{(\boxtimes)}$ ⓘ, Iztok Fister Jr. ⓘ, and Grega Vrbančič ⓘ

Institute of Informatics, Faculty of Electrical Engineering and Computer Science,
University of Maribor, Koroška cesta 46, 2000 Maribor, Slovenia
{lucija.brezocnik,iztok.fister1,grega.vrbancic}@um.si

Abstract. The rapid growth of data and the need for its proper analysis still presents a big problem for intelligent data analysis and machine learning algorithms. In order to gain a better insight into the problem being analyzed, researchers today are trying to find solutions for reducing the dimensionality of the data, by adopting algorithms that could reveal the most informative features out of the data. For this purpose, in this paper we propose a novel feature selection method based on differential evolution with a threshold mechanism. The proposed method was tested on a phishing website classification problem and evaluated with two experiments. The experimental results revealed that the proposed method performed the best in all of the test cases.

Keywords: Classification · Differential evolution · Feature selection · Intelligent data analysis

1 Introduction

Big data, Blockchain, Internet-of-Things (IoT), and Data Science are just a few buzzwords that depict the modern technological world. All of them are linked with an amount of data that is rapidly increasing every day. At present, some influential people in the world say that data is the new oil. Interestingly, data is not useful unless it is explored with methods that are tailored to wards knowledge discovery.

There are currently numerous methods for knowledge discovery of data, e.g., classification, association rule mining, and clustering. Classification problems are commonly solved by machine-learning algorithms where the first step is data pre-processing. Data pre-processing is considered to be the hardest and most complex step in the whole machine learning ecosystem. Here, we are confronted with data that will play a role in the following steps of a pipeline. Sometimes data can be very ugly, such as in cases of missing data, non-standardized data, etc. whilst, sometimes even raw data that is well collected can be problematic.

© Springer Nature Switzerland AG 2019
T. Welzer et al. (Eds.): ADBIS 2019, CCIS 1064, pp. 11–18, 2019.
https://doi.org/10.1007/978-3-030-30278-8_2

But what happens if the data that we want to analyze consists of hundreds of instances where each of those instances has thousands, or tens of thousands, of features? Such high-dimensional data constitutes a serious problem for modern machine-learning algorithms because of the so-called curse of dimensionality. To overcome such a problem, it is necessary to find a way to reduce the number of features. Generally, two techniques are often used: feature selection and feature extraction. The latter creates new variables as a combination of others to reduce the dimensionality, while feature selection works by removing features that are not relevant or are redundant [1].

Feature selection (FS) can also be modeled as an optimization problem. However, the biggest problem is the significant time complexity. On the other hand, in order to tackle this problem, researchers have recently utilized some stochastic population-based nature-inspired algorithms that can find pseudo-optimal solutions in real-time [2]. There exist various feature selection methods that are based on stochastic population-based nature-inspired algorithms [3,9,12].

Fister et al. have recently proposed a new self-adaptive differential evolution with a threshold mechanism for feature selection. The authors introduced a new threshold mechanism which extends the basic self-adaptive differential evolution by adding another feature threshold and thus mechanically control the presence/absence of a particular feature in the solution [2]. In this way, the optimal threshold as well as the optimal features are searched for during the optimization process. Inspired by previous studies [2], here we apply a threshold mechanism in canonical differential evolution [8] as well as apply proposed methods on a phishing website classification.

Altogether, the main contributions of this paper can be summarized as follows:

- a novel differential evolution for feature selection where a threshold mechanism (DEFSTH) is proposed,
- the proposed method is evaluated on a phishing dataset, and
- the performance comparison study is conducted on the most commonly used conventional classifiers.

The structure of this paper is as follows. Section 2 outlines the fundamentals of the proposed DEFSTH method, which is later tested by the experiment presented in Sect. 3. Section 4 depicts the results, while Sect. 5 concludes the paper and outlines directions for future work.

2 The Differential Evolution for Feature Selection with a Threshold Mechanism

A proposed differential evolution with a threshold mechanism for feature selection (DEFSTH) extends the Differential Evolution (DE) algorithm with threshold mechanism and utilizes it for feature selection. The method is explained in detail in the following subsections. Subsection 2.1 comprises steps made in the initialization phase, while the evaluation process of the method is covered in Subsects. 2.2 and 2.3.

2.1 Initialization

In the initialization phase of the method, the following parameters are set: the lower (*Lower*) and upper (*Upper*) bounds of the search space, the threshold (*TH*), the population size (*NP*), the number of function evaluations (*nFES*), the scaling factor (*F*), the crossover rate (*CR*), and the number of folds (*k*). Those parameters together with the initialization process of individuals are presented in detail in Subsect. 2.2.

2.2 Differential Evolution with Feature Selection

The core of the method DEFSTH is explained in detail in the following two subsections. Subsection 2.2 presents the basics of the DE algorithm while Subsect. 2.2 defines the FS problem.

Fundamentals of Differential Evolution. Differential Evolution is an evolutionary algorithm used widely in solving many combinatorial, continuous, as well as real-world problems. DE was proposed by Storn and Price in 1997 [8]. A population in DE consists of individuals that are represented as real-value-coded vectors representing the candidate solutions:

$$\mathbf{x}_i^{(t)} = (x_{i,1}^{(t)}, \dots, x_{i,n}^{(t)}), \quad \text{for } i = 1, \dots, NP, \tag{1}$$

where each element of the solution is in the interval $x_{i,1}^{(t)} \in [x_i^{(L)}, x_i^{(U)}]$, and $x_i^{(L)}$ and $x_i^{(U)}$ denote the lower and upper bounds of the i-th variable. The DE is composed of three variation operators, i.e., mutation, crossover, and selection.

Mutation in DE is expressed as follows:

$$\mathbf{u}_i^{(t)} = \mathbf{x}_{r1}^{(t)} + F \cdot (\mathbf{x}_{r2}^{(t)} - \mathbf{x}_{r3}^{(t)}), \quad \text{for } i = 1, \dots, NP, \tag{2}$$

where $r1$, $r2$, $r3$ are randomly selected values in the interval $[1 \dots NP]$.

Crossover in DE is expressed as follows:

$$w_{i,j}^{(t+1)} = \begin{cases} u_{i,j}^{(t)} & \text{rand}_j(0,1) \leq CR \vee j = j_{rand}, \\ x_{i,j}^{(t)} & \text{otherwise}, \end{cases} \tag{3}$$

Selection in DE is expressed as:

$$\mathbf{x}_i^{(t+1)} = \begin{cases} \mathbf{w}_i^{(t)} & \text{if } f(\mathbf{w}_i^{(t)}) \leq f(\mathbf{x}_i^{(t)}), \\ \mathbf{x}_i^{(t)} & \text{otherwise}. \end{cases} \tag{4}$$

Feature Selection Mechanism. In order to apply the FS mechanism into DE, some modifications of the latter were necessarily introduced.

Individuals in the DEFSTH method are represented as a vector containing real values:

$$\mathbf{x}_i^{(t)} = (x_{i,0}^{(t)}, \dots, x_{i,M}^{(t)}, TH_i^{(t)}), \tag{5}$$

for $i = 0, \ldots, NP$, where each feature $x_{i,0}^{(t)}$ for $i = 0, \ldots, n$ is drawn from the interval $[0, 1]$. $TH^{(t)}$ determines if the corresponding feature is present or absent in the solution. This mapping is expressed as follows:

$$a_{i,j}^{(t)} = \begin{cases} 0, \text{ if } x_{i,j}^{(t)} \leq TH^{(t)} \\ 1, \quad \text{otherwise,} \end{cases} \tag{6}$$

Vector $\mathbf{a}_i$ presents a matrix, determining the presence or absence of the observed j-th feature in the i-th solution. Let us mention that value 1 means that the feature is present, while the value 0 means that the feature is absent in the solution.

There is a theoretical chance that vector a_i would have only zero values, meaning that no feature is selected. If such a marginal case occurs, the proposed method returns the maximum value of the fitness function of that individual, i.e., 1.

2.3 Fitness Function Evaluation

To evaluate the fitness value of each solution produced by DEFSTH, we utilized the Logistic Regression (LR) classifier applying the Limited-memory Broyden-Fletcher-Goldfarb-Shanno (L-BFGS or LM-BFGS) [5] optimization algorithm with the following settings: Maximum iterations (max_iter) was set to 100, tolerance for stopping criteria to $tol = 10^{-4}$, and inverse of regularization strength $C = 1.0$.

The evaluation of the fitness function was conducted using the Logistic Regression classifier calculating the accuracy against the test subset of the initial training set and can be formally expressed as presented in Eq. (7), where $test\_acc$ stands for the previously mentioned calculated accuracy. Given the nature of the DE algorithm, which is basically designed to search for the global minimum, we are converting the problem of searching the best-evaluated individual using fitness function, to the problem of searching for the global minimum via the subtraction of the accuracy from a value of 1.

$$f(test\_acc) = 1 - test\_acc \tag{7}$$

The fitness function evaluation is performed for each produced individual until the stopping criteria, in our case, the $nFES$, is reached.

3 Experiment

The experimental approach was utilized to show the performance of the proposed DEFSTH method on a dataset presented in detail in Subsect. 3.2. How the experiment was carried out along with the evaluation method and metrics are shown in Subsect. 3.3. In the following Subsection, the used setup is explicitly listed.

The method was implemented in the Python programming language where two external libraries were used, i.e. NiaPy [10], a micro-framework for building nature-inspired algorithms, and Scikit-learn [7], a Python module integrating a vast number of machine-learning algorithms.

3.1 Setup

The experiment was conducted using a quad-core Intel Core i7-6700 CPU with a base clock speed at 3.4 GHz and 16 GB of DDR4 memory, running the Linux Mint 19 operating system.

To initialize the DE algorithm, the following optimal parameter settings were manually determined after an extensive tuning of parameters. The population size NP was set to 40, the number of function evaluations $nFES$ was set to 1000, the scaling factor F and crossover rate CR were set to 0.5 and 0.9, respectively.

3.2 Test Data

For this experiment, we composed a dataset on our own. Using the Phishtank website [6], we collected a list of 30,647 community-confirmed URLs of phishing websites and 58,000 legitimate website URLs. The legitimate URLs are gathered from a list of community-labeled and organized URLs containing the objectively reported news and top Alexa ranking websites and are thus legitimate. Using a total of 88,647 URLs, we extracted mostly address-bar based and domain-based features, extracting a total of 111 features, without the target class (phishing or legitimate).

In order to extract the previously mentioned address-bar based features, we performed a count of special characters or symbols of different parts of a URL such as the whole URL, domain part of the URL, the parameters part of the URL, etc. More information about the dataset is available at the URL [11].

3.3 Evaluation Method and Metrics

To exhaustively evaluate the performance of the proposed DEFSTH method, we conducted an evaluation using three predictive performance measures: the accuracy (ACC), $F1$ score and area under the ROC curve (AUC). The accuracy measures the ratio of correctly classified website instances regardless of their class, while the $F1$ score presents a harmonic mean of the precision and recall, averaged over all decision classes, whereas precision refers to the positive predictive value and recall refers to the true positive rate. And the aggregate measure of performance across all possible classification thresholds is presented by an AUC metric.

To objectively evaluate the performance of the proposed DEFSTH method and compare the performance against the conventional classification methods, we conducted a gold standard 10-fold cross validation [4] procedure. The 10-fold cross validation devises a given dataset into train and test sets at a ratio of 90:10.

In the same manner, the procedure is repeated for a total of 10 times, each time using a different test set for validation.

In the first experiment, we measured the performance of Logistic Regression, Naive Bayes (NB), k-Nearest-Neighbors (KNN), Decision Tree (DT) and Multilayer Perceptron (MLP) classifiers without any feature selection utilization or any kind of pre-processing. Also, while conducting the second experiment, measuring the performance of the proposed DEFSTH method, we used the same classifiers as in the first experiment to adequately compare the obtained results. All performance classifiers were initialized with the Scikit-learn [7] default settings.

All the presented results in the following section are averaged ACC, $F1$ score and AUC values, obtained on the test websites instances over all runs for each of 10 folds, if not specified otherwise.

4 Results

A 10-fold comparison of average metrics (ACC, AUC, and $F1$) with a subset of features and without a whole set of features, via the DEFSTH method, is presented in Table 1 and in Fig. 1. The results show that the proposed method obtained better results in all of the cases. After the utilization of the DEFSTH, the highest accuracy was achieved by the NB classification method (96.82%), closely followed by the KNN (96.07%). The accuracies of the remaining classification methods DT and MLP were 93.65% and 92.93%, respectively. A similar description could also be used for the results of AUC and $F1$. After a utilization of the DEFSTH algorithm, NB again performed the best in both cases, by obtaining results of 96.38% and 95.38% for AUC and $F1$, respectively.

Table 1. Comparison of average accuracies, areas under the curves and $F1$ scores, conducted 10-fold, with and without the utilization of DEFSTH.

Metrics		ACC [%]	AUC[%]	$F1$[%]
Logistic Regression	Without FS	92.06 ± 0.42	92.03 ± 0.42	89.07 ± 0.64
	DEFSTH	**92.3 ± 1.91**	**92.1 ± 2.33**	**89.12 ± 2.79**
Naive Bayes	Without FS	92.17 ± 0.36	91.80 ± 0.35	88.89 ± 0.48
	DEFSTH	**96.82 ± 0.21**	**96.38 ± 0.25**	**95.38 ± 0.31**
k-Nearest Neighbors	Without FS	84.29 ± 0.30	79.22 ± 0.42	73.43 ± 0.62
	DEFSTH	**96.07 ± 0.51**	**95.54 ± 0.62**	**94.29 ± 0.75**
Decision Tree	Without FS	87.0 ± 0.60	89.45 ± 0.43	83.82 ± 0.62
	DEFSTH	**93.65 ± 0.32**	**92.96 ± 0.23**	**90.81 ± 0.32**
Multilayer Perceptron	Without FS	88.06 ± 0.22	86.56 ± 0.28	82.55 ± 0.34
	DEFSTH	**92.93 ± 0.42**	**92.45 ± 0.44**	**89.89 ± 0.59**

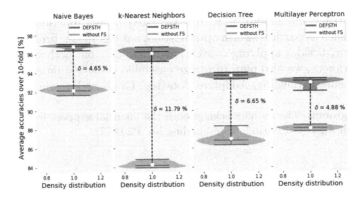

Fig. 1. Comparison of ACC over 10-fold, with and without utilization of DEFSTH.

As explained in the aforementioned results, LR was exempted since the performance of it was used to evaluate each individual (subset of features) in the process of the optimization. Thus, the results on the test data, when using DEF-STH or not, are practically the same, as expected (see row Logistic Regression in Table 1).

Since the experiment used 10-fold cross-validation, we wanted to see which features were selected in all of the folds, which in nine of the ten folds, etc. Feature 100 was chosen in all 10-folds. This feature represents the time (in days) of domain activation. Six features (3, 18, 19, 20, 40, and 83) were chosen nine times or in 5.41% of all cases. Those features represent the number of slashes (/) in the URL, URL length, number of dots (.) in the domain, number of hyphens (-) in the domain, number of dots (.) in the directory, and number of "and" signs (&) in the parameters. In eight out of ten folds, again six features were selected (23, 34, 41, 42, 47, and 106). Those features represent the number of question marks (?) in the domain, number of dollar signs ($) in the domain, number of hyphens (-) in the directory, number of underlines (_) in the directory, number of "and" signs (&) in the directory, and is a site that has a valid TLS/SSL Certificate.

Features that were never selected are features 11 and 101, representing Autonomous system (AS) number and the time (in days) of the domain expiration, respectively.

5 Conclusion

In this work, we proposed a novel method based on DE with a threshold mechanism for FS. Two experiments were conducted using a phishing dataset, to assess its performance. The findings show that the proposed method performed best in all of the test cases and extracted highly informative features.

Based on these encouraging results, we plan to extend our research, apply the DEFSTH method to various problems including the datasets with multiple classes. Since the proposed method is implemented modularly, we plan to extend it further and modify it with a weighted fitness function or different algorithms, such as particle swarm optimization, bat algorithm, firefly algorithm, etc. Moreover, besides applying conventional classification methods, as we did in our experiment, we also plan to utilize ensemble classification methods such as Random Forest, Bagging, Adaptive Boosting, Gradient Boosting, etc.

Acknowledgment. The authors acknowledge the financial support from the Slovenian Research Agency (Research Core Funding No. P2-0057).

References

1. Brezočnik, L., Fister, I., Podgorelec, V.: Swarm intelligence algorithms for feature selection: a review. Appl. Sci. **8**(9) (2018). https://doi.org/10.3390/app8091521
2. Fister, D., Fister, I., Jagric, T., Fister Jr., I., Brest, J.: A novel self-adaptive differential evolution for feature selection using threshold mechanism. In: IEEE SSCI2018 Symposium Series on Computational Intelligence, pp. 17–24 (2018)
3. Khushaba, R.N., Al-Ani, A., AlSukker, A., Al-Jumaily, A.: A combined ant colony and differential evolution feature selection algorithm. In: Dorigo, M., Birattari, M., Blum, C., Clerc, M., Stützle, T., Winfield, A.F.T. (eds.) ANTS 2008. LNCS, vol. 5217, pp. 1–12. Springer, Heidelberg (2008). https://doi.org/10.1007/978-3-540-87527-7_1
4. Kohavi, R.: A study of cross-validation and bootstrap for accuracy estimation and model selection. In: Proceedings of the 14th International Joint Conference on Artificial Intelligence, IJCAI 1995, vol. 2, pp. 1137–1143. Morgan Kaufmann Publishers Inc., San Francisco (1995)
5. Liu, D.C., Nocedal, J.: On the limited memory BFGS method for large scale optimization. Math. Program. **45**(1–3), 503–528 (1989)
6. OpenDNS: PhishTank data archives. https://www.phishtank.com/. Accessed 21 Feb 2019
7. Pedregosa, F., et al.: Scikit-learn: machine learning in python. J. Mach. Learn. Res. **12**, 2825–2830 (2011)
8. Storn, R., Price, K.: Differential evolution-a simple and efficient heuristic for global optimization over continuous spaces. J. Global Optim. **11**(4), 341–359 (1997)
9. Unler, A., Murat, A.: A discrete particle swarm optimization method for feature selection in binary classification problems. Eur. J. Oper. Res. **206**(3), 528–539 (2010)
10. Vrbančič, G., Brezočnik, L., Mlakar, U., Fister, D., Fister Jr., I.: NiaPy: python microframework for building nature-inspired algorithms. J. Open Source Softw. **3** (2018). https://doi.org/10.21105/joss.00613
11. Vrbančič, G.: Phishing dataset (2019). https://github.com/GregaVrbancic/Phishing-Dataset. Accessed 23 May 2019
12. Zorarpacı, E., Özel, S.A.: A hybrid approach of differential evolution and artificial bee colony for feature selection. Expert Syst. Appl. **62**, 91–103 (2016)

Document Data Modeling: A Conceptual Perspective

David Chaves[(⊠)] and Elzbieta Malinowski

Department of Computer Science and Informatics, University of Costa Rica,
San José, Costa Rica
{david.chavescampos, elzbieta.malinowski}@ucr.ac.cr

Abstract. The growing availability of data and the increased popularity of NoSQL databases, that support the idea of managing unstructured or semi-structured data, motivate implementers to skip the phase of a conceptual view of data. However, document data stores belonging to the NoSQL group show a clear tendency of looking for some common feature among documents creating collections. This aspect motivates us to propose a model for the conceptual representation of a document data store based on UML class diagrams and mapping rules for its implementation. We also include a case study using Twitter data and show implementation using three data stores: MongoDB, CouchDB, and ArangoDB.

Keywords: Document data stores · NoSQL databases ·
Conceptual data modeling

1 Introduction

References to big data mention the common characteristics of being unstructured. This opened the possibility of skipping a conceptual modeling phase. On the other hand, the benefits of using database conceptual models have been acknowledged for decades; however, the conceptual design domain for NoSQL repositories is still at a research stage leading to poorly-designed systems. The modeling does not aim to enforce structure over data, but it helps to understand how data is organized for analysis [1].

Document data stores are the second most popular data model [2] and are similar to a key-value model with a difference of having self-describing, hierarchical, and examinable value. The usual practice in document datastores is to skip the conceptual design taking directly implementation aspects into account. Even though this practice can give positive results for small systems, it becomes more difficult for more complex ones.

In this paper, we propose an extension of UML class diagrams for representing document stores and mapping rules, showing the implementation in the three document stores [2]. In our approach, we look for simplicity and for bridging the gap between academics and practitioners.

This paper is organized as follows: Sect. 2 refers to related works, Sect. 3 introduces our proposal for conceptual modeling. Sect. 4 shows mapping rules for the implementation and Sect. 5 includes a case study using the proposed conceptual

© Springer Nature Switzerland AG 2019
T. Welzer et al. (Eds.): ADBIS 2019, CCIS 1064, pp. 19–27, 2019.
https://doi.org/10.1007/978-3-030-30278-8_3

model and its implementation in document stores; lastly, Sect. 6 gives conclusions and future work.

2 Related Work

Currently, there are few systematic studies on data modeling for NoSQL databases, e.g., [1, 3]. Some works propose particular solutions that can be used for conceptual modeling of NoSQL databases [4, 5]; however, since the complexity of these approaches is high, they can be difficult to infiltrate into real-world applications. On the other hand, several studies refer explicitly to modeling documents in MongoDB using UML notation [6]; other approaches refer to the JSON format to represent the documents for different NoSQL databases [7]. Others develop automatic tools to map from JSON [8] and applying reverse engineering to already deployed systems [9].

In this work, we do not consider performance evaluation; however, this aspect is important and we plan to extend this research since different reports are contradicting. For example, [10] shows a better performance for indexed referenced documents compared to embedded ones, but the results of [11] demonstrate the opposite conclusions.

3 Conceptual Representation of Document Data Stores

Using a conceptual model in a document data store provides the advantage of representing data in a way that helps to understand, access, and analyze it from the beginning of the implementation process. Lacking the model forces the implementers to retain the details of data "structure" considering an implementation level that can be complex in the presence of different document collections.

Conceptual modeling is a product-independent design allowing its creators to focus on user requirements and implement the system, if adequate logical/physical mapping rules are established [12]. The proposed conceptual model uses the UML class diagram in a similar way as the conceptual modeling is done in the relational databases.

3.1 Document with Fields

A document is the main element and presents a set of data in an organized form, even though its structure can differ from other documents. Each document contains fields; one of them is reserved for document id. Since documents can have different fields, we propose to choose as a representative document the one that includes all fields indicating some fields as optional. Figure 1b shows an example where two fields are included in all documents, e.g., *movieId*, and *movieTitle*, and one is optional, e.g., *language*, indicating this by the symbol of "~" before the name. Other fields group elements in an array, e.g., *genres*; this data type with its cardinalities is indicated in square parenthesis.

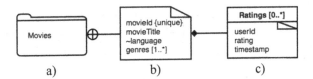

Fig. 1. Document representation: (a) collection, (b) document itself, and (c) an embedded one.

3.2   Document Collection

The collection represents a grouping of similar documents. Compared to relational databases, a document could correspond to a row and a collection of documents to a table. Figure 1a shows a graphical representation for a collection using the UML package. We use the symbol of contention relationship ($\oplus$) to indicate that documents form part of a collection, i.e., its membership [13].

3.3   Embedded Documents

The field in the document can refer to another document forming nested documents. We propose two different UML notations to represent this: a composition relationship and an aggregation relationship. Figure 1c shows a general form for representing the composition relationship with an example of movies and their ratings. This embedded document includes the name, its multiplicity (0..* in Fig. 1c), and the specification of its fields. This kind of relationship is required when a nested document existence depends on its container document, e.g., movie rating is part of a specific movie. Our approach is different from [6] since the last one, additionally, requires class inheritance that increases unnecessarily the complexity of the model.

On the other hand, when a nested document depends on its container but, if necessary for the further extension, they can be converted to standalone documents, we propose to use the aggregation relationship ($\Diamond$), e.g., the movie storyline is closely related but not strictly dependent to the movie.

3.4   Referenced Documents

When a collection is related to two or more other collections, it is necessary to define a relationship between them to avoid data repetition. To represent this relationship, we propose to use a bi-directional association as it can be seen in Fig. 2 (*Directed by* relationship). This representation includes multiplicity values in the (min, max) form to indicate the number of documents that should participate (min value) in the association and number of documents from one collection that can be associated with documents from another collection (max value).

Figure 2 shows two document collections representing movies and directors. Since a movie can be directed by one or more directors and the director can lead some other movies, the cardinality is many-to-many (indicated by the * symbol). Further, not all movies have a specified director (min value is 0), but all directors have associated at least one movie (min value is 1).

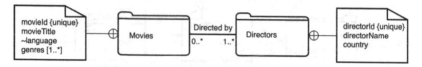

Fig. 2. An example of a referenced collection.

4 Mapping Rules

After outlining the conceptual proposal, we define the following mapping rules using the JSON markup language. It emulates an intermediate stage for the design of the data store, similar to the logical representation in relational databases that allows one to specify relations before their deployment in the particular DBMS. The translation from JSON to the physical level according to the specific system is a straightforward task and may consider other aspects, such as indexing, sharding, replication (if available), among other features. Furthermore, it is possible to automate the mapping process from UML to JSON based on already existing tools, such as crowd [8].

4.1 Document with Fields and Document Collection

To represent a document, we use the JSON specification as shown in Fig. 3 for a document conceptual representation in Fig. 1. Each field is represented in a key-value fashion with the key (field name) between quotes, followed by the associated value. In addition, to represent an array (*genres* in the figure), values separated by commas are included in square parentheses. Additionally, JSON file can include many documents arranged in an array forming a collection.

```
{"id": "321","movieId": "123","movieTitle": "Science", "language": "English",
"genres": ["Comedy", "Fiction"],
"Ratings": [{"userId": "453", "rating": "4","timestamp": "2017-11-09 T 11:20 UTC"},
            {"userId": "784","rating": "5","timestamp": "2016-10-08 T 13:40 UTC"}]}
```

Fig. 3. JSON file representing a document from Fig. 1.

4.2 Embedded Documents

The mapping of embedded documents is based on the commonly-known principles used for object-relational databases [14]. Even though document data stores do not belong to this group, general practice demonstrates the use of this mapping [13]. The following rules are applied considering the multiplicity shown on the conceptual level:

- (0..1) or (1..1): indicates the existence of none or only one embedded document; this document can be represented as such or its fields can be merged with the fields of the main document.
- (0..*) or (1..*): indicates the existence of none, one, or many embedded documents; these nested documents are organized as an array stored in one field of the main document. Each related document is an element of the array.

Notice that we do not consider a many-to-many relationship between main and embedded documents since it would indicate that some "external" documents are referencing an embedded document. We consider that if the embedded document must be accessed by other "external" documents, it should be modeled as a collection of documents with the corresponding association relationship.

4.3 Referenced Documents

Mapping of referenced documents, similar to the previous case, is based on known principles from object-relational databases [14] according to the following rules:

- One-to-one cardinality: the document key is included as a field in another document.
- One-to-many cardinality is mapped in two ways: each document on the n-side cardinality stores the key of the document from the one-side cardinality or each document on the one-side cardinality stores an array of keys from the n-side cardinality.
- Many-to-many cardinality is also be mapped in two ways: documents in one or both collections include an array of identifiers of corresponding documents from another collection.

Figure 4 illustrates the case of two documents with a many-to-many cardinality between collection of documents: Fig. 4a represents a movie with an array of corresponding director IDs, while Fig. 4b includes arrays with associated movie IDs.

```
{"id": "564", "movieId": "417",          {"id": "650", "directorId": "853",
"movieTitle": "Night",                    "directorName": "James",
"Directors": [{"nameId":"853"},           "country": "United Kingdom",
            {"nameId":"384"}]}            "Movies": [{"movieId":"417"},
                                                    {"movieId":"932"}]}
```

a) Movie document with several   b) Director document with several
        directors.   movies.

Fig. 4. Implementation in JSON for referenced many-to-many documents.

5 Case Study

After presenting the conceptual model and its mapping rules, we propose a case study to demonstrate the use of the notation and its deployment according to [15] guidelines.

5.1 Case Delimitation

Objective. We evaluate the application of the proposed notation in different open source products for document storing. The main hypothesis is that the implementation is common among systems without incurring to particular additional requirements that could change the proposed conceptual schema.

Related Cases. Many academic works include examples of using document stores as relational implementations (e.g. [3, 13]) or modeling, particularly, in MongoDB [1].

Methodology. Using a qualitative approach, we design a conceptual model for a document data store and implement it in selected data stores. We overview the raw data [16] to define requirements for data store design. Afterward, we develop the schema (Sect. 3) and map it (Sect. 4), showing the implementation differences.

Limitations. This model does not include some possible optimizations for each system, such as indexes and buckets that could be included after the model is deployed.

5.2   Conceptual Representation

Considering Twitter messages, it is possible to identify data referring to users and messages, leading to conceptual schema showed partially in Fig. 5.

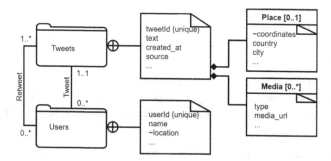

Fig. 5. An extract of a conceptual representation of Twitter data.

As can be seen in Fig. 5, the *Tweets* and *Users* collections include their own fields, e.g., *tweetId*, *text* or *userId*, *name*, among others. Furthermore, the tweet document refers to two optional embedded documents: *Place* that may appear at most one time and *Media* that may consist of several documents representing different media types. In addition, the *Tweets* collection is related to the *Users* collection through two relationships: *Tweet* and *Retweet*. According to shown multiplicity, no all users publish tweets, but every published tweet must have an associated (and only one) user. The *Retweet* association shows the possibility that a message can be republished by many users and these users can republish many messages.

5.3   Applied Mapping for Implementation

MongoDB Implementation. Figure 6 shows an example of a tweet in MongoDB mapped according to rules in Sect. 4. Fields of embedded document *Place* are included in the main document (lines 5–8) with an optional field (coordinates, line 5). The composition *Media* object is embedded as an array (lines 9–10) considering its multiplicity (only two elements are shown). In addition, *userId* (line 3) represents a one-to-many association relationship *Tweet* between *Tweets* and *Users* collections.

```
1      { "_id":ObjectId("5b21705709429256cd1bc75c"),
2        "tweetId":NumberLong("934592617924276225"),
3        "userId":NumberLong("839337948323713028"),
4        ...,
5        "Place_coordinates":{"coordinates":[40.417416, -79.993930],"type":"Point"}
6        "Place_id":"5db3a841345615",
7        "Place_country":"United States",
8        "Place_city":"Pittsburgh, Pennsylvania",
9        "Media":[ { "id":NumberLong("934421567307747328"),"type":"photo", ...},
10               { "id":NumberLong("908781363528110080"),"type":"photo", ...}]}
```

Fig. 6. Example in MongoDB of embedded (*Place* and *Media*) and referenced (*Users*) documents from Fig. 5.

CouchDB Implementation. This store does not include the concept of collection; therefore, it is necessary to insert the documents in the same database with a field identifying its type, e.g., *Tweets* or *Users*. This adaptation is minimal and does not affect the conceptual model. Figure 7 shows an example of a tweet document in CouchDB with the *Retweet* relationship. This many-to-many cardinality is shown as an array of users (the field *retweetUserId*, line 8). Also, we include the field *type* (line 3) to define its collection.

```
1      {"_id": "339f175188d7c1d587ab41af873b9fcb",
2        "_rev": "1-2a34bf1359537ef283dcc2633a4395fc",
3        "type": "Tweets",
4        "tweetId": "934592567584411649",
5        "userId": 157772888,
6        "text": "RT @mnrothbard: https://t.co/aoJqK0eJz7",
7        "created_at": "Sun Nov 26 01:20:06 +0000 2017",
8        "retweetUserId": [{157772888}, {2367997502}],
9        ...}
```

Fig. 7. Example in CouchDB of documents with many-to-many cardinality.

ArangoDB Implementation. ArangoDB supports a user-defined unique *_key* to identify each document. In addition, it includes the *_id* field, which is the combination of the collection name and the document key (Fig. 8, lines 1 and 2). Particularly, both CouchDB and ArangoDB include a revision value (Fig. 7, line 2 and Fig. 8, line 3) in order to support concurrency control, which does not affect the conceptual level design.

```
1      { "_key": "2816",
2        "_id": "tweets/2816",
3        "_rev": "_X-RUUJG-_x",
4        "tweetId":"934592680847327232",
5        "userId": 159104114,
6        ...}
```

Fig. 8. Example in ArangoDB with keys combination and revision value support.

6 Conclusions and Future Work

The growing use of NoSQL databases and the increasing amount of data in these repositories make the understanding of their "structure" increasingly difficult. The affirmation that NoSQL databases, in particular, document data stores, manage semi-structured data opens the possibility of skipping the conceptual phase; this phase is important since it helps in understanding the nature of data and relationships existing between different elements. As a consequence, it facilitates the expression of queries to analyze data. Although it is well-known that documents can have different fields, there is a clear tendency to create a document collection in order to group "similar" documents.

In this paper, we propose the use of UML class diagrams to represent document stores on a conceptual level. We also include mapping rules that facilitate the document data stores implementation, showing examples of three data stores. Even though the proposed model and mapping rules can be extended, we expect that the simplicity of this conceptual proposal may be appealing to a wide forum of document data stores implementers.

References

1. Vera, H., Boaventura, W., Holanda, M., Guimarães, V., Hondo, F.: Data modeling for NoSQL document-oriented databases. In: 2nd Annual International Symposium on Information Management and Big Data, Cusco (2015)
2. DB-Engines: DB-Engines Ranking category, May 2019. https://db-engines.com/en/ranking_categories. Accessed 21 May 2019
3. Imam, A., Basri, S., Ahmad, R., Aziz, N., González-Aparicio, M.: New cardinality notations and styles for modeling NoSQL document-store databases. In: IEEE Region 10 Conference (TENCON), Malaysia (2017)
4. Abdelhedi, F., Brahim, A., Atigui, F., Zurfluh, G.: MDA-based approach for NoSQL databases modelling. In: International Conference on Big Data Analytics and Knowledge Discovery, Lyon (2017)
5. Bugiotti, F., Cabibbo, L., Atzeni, P., Torlone, R.: Database design for NoSQL systems. In: Yu, E., Dobbie, G., Jarke, M., Purao, S. (eds.) ER 2014. LNCS, vol. 8824, pp. 223–231. Springer, Cham (2014). https://doi.org/10.1007/978-3-319-12206-9_18
6. Poveda, J.: Propuesta de Notación Gráfica para el Modelo Orientado a Documentos de MongoDB. Universidad Distrital Francisco José de Caldas, Bogotá (2013)
7. Zola, W.: 6 Rules of Thumb for MongoDB Schema Design, 29 May 2014. https://bit.ly/2FUb3cp. Accessed 21 May 2019
8. Braun, G., Gimenez, C., Fillottrani, P., Cecchi, L.: Towards Conceptual Modelling Interoperability in a Web Tool for Ontology Engineering in Simposio Argentino de Ontologías y sus Aplicaciones, Córdoba (2017)
9. Hernández, A., Feliciano, S., Sevilla, D., García-Molina, J.: Exploring the visualization of schemas for aggregate-oriented NoSQL databases. In: Proceedings of the ER Forum 2017 and the ER 2017 Demo track, Valencia (2017)
10. Reis, D.G., Gasparoni, F.S., Holanda, M., Victorino, M., Ladeira, M., Ribeiro, E.O.: An evaluation of data model for NoSQL document-based databases. In: Rocha, Á., Adeli, H., Reis, L.P., Costanzo, S. (eds.) WorldCIST'18 2018. AISC, vol. 745, pp. 616–625. Springer, Cham (2018). https://doi.org/10.1007/978-3-319-77703-0_61

11. Calvo, K., Durán, J., Quirós, E., Malinowski, E.: MongoDB: alternativas de implementar y consultar documentos. In: IX Congreso Internacional de Computación y Telecomunicaciones, COMTEL, Lima (2017)
12. Gulden, J., Reijers, H.: Toward advanced visualization techniques for conceptual modeling. In: Proceedings of the CAiSE 2015 Forum at the 27th International Conference on Advanced Information Systems Engineering, Stockholm, pp. 33–40 (2015)
13. Lima, C., Santos, R.: A workload-driven logical design approach for NoSQL document databases. In: 17th International Conference on Information Integration and Web-based Applications & Services, Brussels (2015)
14. Dietrich, S., Urban, S.: An Advanced Course in Database Systems: Beyond Relational Databases. Prentice Hall, New Jersey (2004)
15. Runeson, P., Höst, M.: Guidelines for conducting and reporting case study research in software engineering. Empir. Softw. Eng. **14**, 131–164 (2009)
16. Scott, J.: Archive Team: The Twitter Stream Grab, 6 December 2012. https://archive.org/details/twitterstream. Accessed 21 May 2019

Towards Automated Visualisation of Scientific Literature

Evelina Di Corso[1]([✉])[iD], Stefano Proto[1][iD], Tania Cerquitelli[1][iD],
and Silvia Chiusano[2][iD]

[1] Dipartimento di Automatica e Informatica, Politecnico di Torino, Turin, Italy
{evelina.dicorso,stefano.proto,tania.cerquitelli}@polito.it
[2] Dipartimento Interateneo di Scienze, Progetto e Politiche del Territorio,
Politecnico di Torino, Turin, Italy
silvia.chiusano@polito.it

Abstract. Nowadays, an exponential growth in biological data has been recorded, including both structured and unstructured data. One of the main computational and scientific challenges in the modern age is to extract useful information from unstructured textual corpora to effectively support the decision making process. Since the emergence of topic modelling, new and interesting approaches to compactly represent the content of a document collection have been proposed. However, the effective exploitation of the proposed strategies requires a lot of expertise.

This paper presents a new scalable and exploratory data visualisation engine, named ACE-HEALTH (AutomatiC Exploration of textual collections for HEALTH-care), whose target is to easily analyse medical document collections through the Latent Dirichlet Allocation. To streamline the analytics process and enhance the effectiveness of data and knowledge exploration, a variety of data visualisation techniques have been integrated in the engine to provide navigable informative dashboards without requiring any a-priori knowledge on the analytics techniques.

Preliminary results obtained on a real PubMed collection show the effectiveness of ACE-HEALTH in correctly capturing the high-level overview of textual medical collections through innovative visualisation techniques.

Keywords: Textual data visualisation · Topic modelling · Informative dashboards

1 Introduction

In the recent years, we have been witnessing the exponential growth of biological data, including both structured and unstructured data. One of the main computational and scientific challenges in the modern age is to extract useful information from unstructured texts with minimal user intervention [4]. Their

© Springer Nature Switzerland AG 2019
T. Welzer et al. (Eds.): ADBIS 2019, CCIS 1064, pp. 28–36, 2019.
https://doi.org/10.1007/978-3-030-30278-8_4

value is severely undermined by the inability to translate them into knowledge and, ultimately, actions [2].

This paper presents a new scalable and exploratory data visualisation engine, named ACE-HEALTH (AutomatiC Exploration of textual collections for HEALTH-care), whose target is to analyse medical document collections. The framework brings many different analytic techniques to help non-expert users to make sense of large medical data collections. The framework exploits the Latent Dirichlet Allocation [1], a generative probabilistic model, to divide a given corpus into correlated groups of documents with a similar topic. ACE-HEALTH includes several kinds of visualisations to show knowledge at different granularity levels to easily capture the high-level overview of textual medical collections, and drill-down the knowledge to the single document.

We have experimentally evaluated the engine on a real textual medical collection. The performed experiments highlighted ACE-HEALTH's ability to autonomously identify homogeneous groups of medical documents and efficiently represent a manageable set of human readable results.

The next Sections are organised as follows. Section 2 presents ACE-HEALTH architecture and its main building components, Sect. 3 shows the tests run to assess ACE-HEALTH's performances and discusses the experimental results. Lastly, Sect. 4 draws the research conclusions and presents future works.

2 The ACE-HEALTH Framework

ACE-HEALTH (AutomatiC Exploration of textual collections for HEALTH-care) has been tailored to analyse any medical textual collection. This new framework is able to automatically extract and graphically represent multiple knowledge items from textual collections, minimising the user intervention. ACE-HEALTH includes three main components: (i) *Data processing and characterisation*, (ii) *Self-tuning topic modelling*, and (iii) *Knowledge visualisation*.

2.1 Data Processing and Characterisation

In the data processing and characterisation, ACE-HEALTH computes five steps which are carried out conclusively. (1) *Document splitting*, in which documents can be split into paragraphs or analysed in their total content. (2) *Tokenisation* represents the process of segmenting a text into words. (3) *Case normalisation*, in which each word is converted completely to lower-case characters. (4) *Stemming* maps each word into its own root form, and (5) *Stopword removal*, which discards the words which do not bring any additional information (e.g. articles, prepositions). These steps lead to the Bag-Of-Words (BOW) representation.

Statistics Definition and Computation. To describe the data distribution [3] of each corpus, ACE-HEALTH computes a set of statistical features able to characterise the lexical richness of the corpus:

- *Avg frequency*: the average frequency of word occurrence in the corpus;
- *Max frequency*: the maximum frequency of word occurrence in the corpus;
- *Min frequency*: the minimum frequency of word occurrence in the corpus;
- *# documents*: the number of textual documents in the corpus;
- *# terms*: number of terms in the corpus, with repetitions;
- *Avg document length*: the average length of documents in the corpus;
- *Dictionary*: the number of different terms in the corpus, without repetition;
- *TTR*: the ratio between the Dictionary and # terms;
- *Hapax %*: the ratio between the number of terms with one occurrence in the whole corpus and the cardinality of the Dictionary;
- *Guiraud Index*: the ratio between the cardinality of the Dictionary and the square root of the number of tokens (*# terms*).

Weighting Schemas. To highlight the relevance of specific words in the collections, ACE-HEALTH includes two weighting strategies. A weight is a positive real number associated with each term of the collection that quantifies its degree of importance. In literature, several weighting schemas have been proposed [8]. Each corpus is represented as a matrix, named *document-term* matrix, in which each row corresponds to a document of the collection and each column corresponds to a distinct term. Each weight is computed as the product of a local weight (which measures the relevance of each word in each document) and a global one (which measures the relevance of each word in the entire collection). In ACE-HEALTH, two local weights, *Term-Frequency* (TF) and *Logarithmic Term-Frequency* (LogTF), and the global weight *Inverse Document Frequency* (IDF) are integrated.

2.2   Self-tuning Topic Modelling

To cluster each collection into well-separated topics, ACE-HEALTH includes the Latent Dirichlet Allocation (LDA), an unsupervised generative probabilistic model for corpora [1]. Each document is described by a distribution of topics and each topic by a distribution of words. To generate the clusters, ACE-HEALTH selects the topic with the highest probability for each document. To automatically identify a suitable value for the number of topics, ACE-HEALTH includes the ToPIC-SIMILARITY strategy proposed in [9].

Given a lower and an upper bound for the number of clusters, a new LDA model is generated for each K value. For each partition, ToPIC-SIMILARITY computes three steps:

1. *topic characterisation*, to describe each topic with its most representative words;
2. *similarity computation*, to assess how the topics in the same partition are similar;
3. *K identification*, to find good clustering configurations to be proposed.

Steps (1) and (2) are repeated for each topic in every probabilistic model. For each weighting strategy, ACE-HEALTH reports to the analyst the top-3 values that represent good quality partitions. ACE-HEALTH includes three quality metrics: (i) *Perplexity*, (ii) *Entropy*, and (iii) *Silhouette*. The perplexity [1] describes how well the probabilistic model depicts a sample. The lower the perplexity value, the better the model. The entropy [1] is defined as the amount of information in a transmitted message. The larger the entropy, the more uncertainty is contained in the message. Lastly, the Silhouette [10] measures the similarity of a document with respect to its own cluster (cohesion) compared to other clusters (separation). It assumes values in $[-1, 1]$, where higher values indicate that the document is well matched to its own topic.

2.3   Knowledge Visualisation

A dynamic and visual exploration of both data and information hidden in the scientific literature may significantly improve the knowledge understanding and its real exploitation. To this aim, ACE-HEALTH enriches the cluster set to provide two forms of human-readable knowledge: (i) *document-topic distribution* and (ii) *topic-term distribution*.

Document-topic distribution characterises the document distribution over the topics. It includes the exploitation of (i) *topic cohesion/separation* in terms of document distribution and (ii) *coarse-grained versus fine-grained* groups through the analysis of the impact of the different weighting schemas. (i) focuses on the characterisation of the documents distribution through *pie charts* and *t-Distributed Stochastic Neighbour Embedding* (t-SNE) [7]. t-SNE allows representing high-dimensional data into lower dimensional map. The colouring of the points reflects the assignment to a specific topic after the LDA model. (ii) carries out the analysis of the weights impact in terms of coarse versus fine grained groups, through the analysis of the correlation matrix. Documents belonging to the same macro category tend to be more similar to each other than those belonging to different ones.

Topic-term distribution describes the distribution over words for each latent topic. Specifically, ACE-HEALTH includes the characterisation of (i) *topic-term distribution* through the analysis of the top-k relevant words in terms of probabilities, and (ii) the *topic cohesion/separation* in terms of relevant words. For task (i), the most probable top-k terms are extracted from each topic and plotted using word-clouds [5]. The comparison of the word-clouds obtained is left to the human analyst judgement. For task (ii), we propose the graph representation. We introduce two types of nodes: *topic nodes*, which are green nodes one for each topic, and *term nodes*, which are pink nodes one for each distinct term. Then, for each topic we add an edge for each term linked with that topic. ACE-HEALTH extracts only the top-k most relevant words for each topic (the default value is 40). If a word appears in more than one topic, we colour that node in red. The framework computes the connectivity of the graph, since if a topic is characterised by words that are not used in other topics, that topic will be disconnected by the others.

3   Experimental Results

Experimental validation has been designed to address two main issues: (i) the effectiveness of ACE-HEALTH in discovering good document partitions and (ii) the ability of visualising and making the information and the extracted knowledge easy to be interpreted at different detail levels by various non-expert analysts.

ACE-HEALTH has been developed to be distributed and implemented in Python. All experiments have been performed on the BigData@PoliTO cluster[1] running Apache Spark 2.3.0. The virtual nodes, the driver and the executors, have a 7 GB main memory and a quad-core processor each.

Table 1. Statistics for the PubMed collection

Features	WH	WoH
# documents	1,000	
Max frequency	775	
Min frequency	1.0	2.0
Avg frequency	15	18
Avg document length	3600	3469
# terms	3,600,153	3,469,305
Dictionary $|V|$	227,210	96,362
TTR	0.06	0.05
Hapax %	57.02	0
Guiraud index	119.75	51.73

Dataset. We experimentally assessed ACE-HEALTH on a real collection of medical articles extracted from PubMed[2], which is the largest bio-medical literature database in the world. From this collection, we extract 1000 papers which statistics are reported in Table 1. The dataset contains documents characterised by a great lexical richness (defined by the *Guiraud index*). Moreover, removing Hapax (i.e. column WoH) does not change the main corpus features and statistics, but allows better LDA modelling. The number of expected categories is not a-priori known.

3.1   Performance

Table 2 includes a row for each K obtained using our proposed clustering methodology and the three quality metrics. For each weighting strategy, the best solution found by ACE-HEALTH is reported in bold. The number of clusters found by

---

[1] https://bigdata.polito.it/content/bigdata-cluster.
[2] https://www.ncbi.nlm.nih.gov/pubmed/.

Table 2. Experimental results

Weight	K	Perplexity	Silhouette	Entropy
TF-IDF	3	9.715	0.285	0.208
	6	9.511	0.28	0.314
	8	**9.432**	**0.276**	**0.352**
LogTF-IDF	3	10.318	0.258	0.251
	4	10.229	0.283	0.293
	6	**10.164**	**0.301**	**0.301**

the TF-IDF schema is greater than the one obtained with LogTF-IDF. This means that the two weighting schemas characterise the same dataset in a different way.

ACE-HEALTH integrates the Adjusted Rand Index (ARI) [6] to compare the solutions obtained using the two weighting schemas. A larger ARI means a higher agreement between the two partitions. For the dataset, ARI is 0.487, which means that the partitions are different leading to different partitions.

To analyse the cardinalities of each cluster set, ACE-HEALTH integrates the pie chart. Figure 1 shows the document frequency for each weighting strategy. The TF-IDF finds a large number of clusters with respect to the LogTF-IDF.

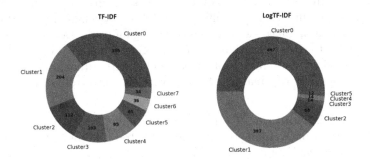

Fig. 1. Cardinality of cluster set though pie chart representation, weighting via TF-IDF weighting schema (Left) and LogTF-IDF weighting schema (Right)

3.2   Knowledge Visualisation

ACE-HEALTH characterises the discovered topics through two types of interesting knowledge: *document distribution* and *topic-term distribution*. For the document distribution, two different visualisations are reported: (i) t-SNE and (ii) correlation matrix. For the topic-term distribution, ACE-HEALTH integrates (i) the word-clouds and (ii) the graph visualisation.

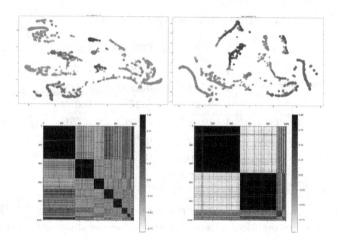

Fig. 2. t-SNE representation (Top) and Correlation matrix (Bottom), TF-IDF (Left) $K = 8$ and LogTF-IDF (Right) $K = 6$ weighting schemas.

Document-Topic Distribution. To analyse the topic cohesion/separation in terms of documents distribution, ACE-HEALTH includes the t-SNE representation. Figure 2 (Top) reports the t-SNE representation, weighting via TF-IDF (Left) and LogTF-IDF (Right). Each colour point is based on the assignment to a specific topic. In both case the colours are well separated and not overlapped.

To analyse the impact of the different weighting schemas, Fig. 2 (Bottom) shows the correlation matrices. Documents belonging to the same topic tend to be more similar to each other than those belonging to different ones. The dark rectangles highlight that the topics are well-separated, although some correlations can be found by the analysis of topics with a lower cardinality.

Topic-Term Distribution. To interpret the content of each cluster ACE-HEALTH derives the word-clouds to analyse the top-k words in terms of probabilities and the graph representation to evaluate the topic cohesion/separation in terms of words.

As shown in Fig. 3, Cluster0 includes documents concerning the analysis of problems related to shock or a feeling of anxiety. The most frequent words (i.e., those reported with a larger size), are related to terms such as *PTSD*, *paediatric*, *geriatric*, *adolescent* and *ASQ*. On the other side, Cluster7 (see Fig. 3) includes documents addressing the cancer analysis. The fact that all the clusters are well separated is also confirmed by the graph representation shown in Fig. 3 (Right). Specifically, it includes the graph representations using the top-40 words. The clusters are well-separated, and the graph is not connected.

Fig. 3. TF-IDF weighting schemas with K = 8. Word-clouds of Cluster0 and Cluster7 (Left) and graph visualisation using the top-40 frequent words (Right)

4   Conclusion and Future Works

This paper presented ACE-HEALTH (AutomatiC Exploration of textual collections for HEALTH-care) a new scalable and exploratory data visualisation engine, whose target is to analyse textual medical collections through the LDA topic modelling and graphically represents the results using innovative visualisation techniques. As future directions, we are currently extending ACE-HEALTH with (i) a querying engine able to assign topics to a new document using the generated LDA models.

Acknowledgements. The research leading to these results was partially funded by Project CANP, POR-FESR 2014-2020 - Technology Platform "Health and Wellness" - Piedmont Region, Italy.

References

1. Blei, D.M., Ng, A.Y., Jordan, M.I.: Latent dirichlet allocation. J. Mach. Learn. Res. **3**(Jan), 993–1022 (2003)
2. Cerquitelli, T., Baralis, E., Morra, L., Chiusano, S.: Data mining for better healthcare: a path towards automated data analysis? In: 2016 IEEE 32nd International Conference on Data Engineering Workshops (ICDEW), pp. 60–63. IEEE (2016)
3. Cerquitelli, T., Di Corso, E., Ventura, F., Chiusano, S.: Data miners' little helper: data transformation activity cues for cluster analysis on document collections. In: Proceedings of the 7th International Conference on Web Intelligence, Mining and Semantics, p. 27. ACM (2017)
4. Di Corso, E., Cerquitelli, T., Ventura, F.: Self-tuning techniques for large scale cluster analysis on textual data collections. In: Proceedings of the Symposium on Applied Computing, pp. 771–776. ACM (2017)
5. Heimerl, F., Lohmann, S., Lange, S., Ertl, T.: Word cloud explorer: text analytics based on word clouds. In: 2014 47th Hawaii International Conference on System Sciences (HICSS), pp. 1833–1842. IEEE (2014)
6. Hubert, L., Arabie, P.: Comparing partitions. J. Classif. **2**(1), 193–218 (1985)
7. van der Maaten, L., Hinton, G.: Visualizing data using t-SNE. J. Mach. Learn. Res. **9**, 2579–2605 (2008)
8. Nakov, P., Popova, A., Mateev, P.: Weight functions impact on LSA performance. In: EuroConference RANLP 2001 Recent Advances in NLP, pp. 187–193 (2001)

9. Proto, S., Di Corso, E., Ventura, F., Cerquitelli, T.: Useful ToPIC: self-tuning strategies to enhance latent dirichlet allocation. In: 2018 IEEE International Congress on Big Data (BigData Congress), pp. 33–40. IEEE (2018)
10. Rousseeuw, P.J.: Silhouettes: a graphical aid to the interpretation and validation of cluster analysis. J. Comput. Appl. Math. **20**, 53–65 (1987)

Metadata Management for Data Lakes

Franck Ravat[1] and Yan Zhao[1,2](✉) ⓘ

[1] Institut de Recherche en Informatique de Toulouse, IRIT-CNRS (UMR 5505),
Université Toulouse 1 Capitole, Toulouse, France
{Franck.Ravat,Yan.Zhao}@irit.fr
[2] Centre Hospitalier Universitaire (CHU) de Toulouse, Toulouse, France

Abstract. To prevent data lakes from being invisible and inaccessible to users, an efficient metadata management system is necessary. In this paper, we propose a such system based on a generic and extensible classification of metadata. A metadata conceptual schema which considers different types (structured, semi-structured and unstructured) of raw or processed data is presented. This schema is implemented in two DBMSs (relational and graph) to validate our proposal.

Keywords: Data lake · Metadata management · Metadata classification

1 Introduction

The concept of Data Lake (DL) was created by Dixon [4] and extended by various authors [5,8,20]. DL allows to ingest raw data from various sources, store data in their native format, process data upon usage, ensure the availability of data and provide accesses to data scientists, analysts and BI professionals, govern data to insure the data quality, security and data life cycle.

DLs facilitate different types of analysis such as machine learning algorithms, statistics, data visualisation... (unlike Data Warehouses (DW) [16]). The main characteristic of DL is 'schema-on-read' [5], data are only processed upon usage. Compared to DWs, which are structured data repositories dedicated to predetermined analyses, DLs have great flexibility and can avoid losing information.

However, a data lake that contains a great amount of structured, semi-structured and unstructured data without explicit schema or description can easily turn into a data swamp which is invisible, inaccessible and unreliable to users [18]. To prevent data lakes from turning into data swamps, metadata management is essential [1,8,20]. Metadata can help users find data that correspond to their needs, accelerate data accesses, verify data origin and processing history to gain confidence and find relevant data to enrich their analyses [1,14].

Nevertheless, many papers are focused on a single zone (especially ingestion zone) or a single data type of data lakes. Therefore, the goal of this paper is to propose a metadata management system dedicated to data lakes and applied to the whole life-cycle (multiple zones) of data. The set of the paper is as follows:

© Springer Nature Switzerland AG 2019
T. Welzer et al. (Eds.): ADBIS 2019, CCIS 1064, pp. 37–44, 2019.
https://doi.org/10.1007/978-3-030-30278-8_5

the second section introduces related work on metadata. In the third section, we propose our metadata conceptual schema with a classification. The fourth section describes the implementation of a metadata management system.

2   Related Work

DL metadata, inspired by the DW classifications [6,7], are classified into two ways. A first classification includes three categories [12,14]: *Technical* metadata concern data type, format and structure (schema). *Operational* metadata concern data processing information and *Business* metadata concern business objects and descriptions. A second classification includes not only the information of each dataset (intra-metadata) but also the relationships between datasets (inter-metadata). Intra-metadata are classified into data characteristics, definitional, navigational, activity, lineage, rating and assessment [2,6,19]. Inter-metadata describe relationships between datasets, they are classified into dataset containment, provenance, logical cluster and content similarity [9].

Compared to the first classification, the second one is more specific. Nevertheless, the second classification can be improved. Some sub-categories are not adapted to data lakes. For instance, the *rating* subcategory that concerns user preferences [19] needs to be removed. Because in data lakes, datasets can be processed and analysed by different users [5], a dataset that makes no sense to BI professionals can be of great value to data scientists. What's more, this classification can be extended with more sub-categories. For instance, data sensitivity and accessibility also need to be controlled in data lakes.

Concerning metadata management, various solutions for data lakes are presented with different emphases [1,9,15,17]. Regarding all the solutions of metadata management, authors mainly focused on a few points. Firstly, the detection of relationships between different datasets is always presented [1,9,15]. Relationships between datasets can help users find as many relevant datasets as possible to enrich data analysis. While we want to find a metadata model that shows not only the relationships between datasets but also the information of each single dataset. Secondly, authors often focused on unstructured data (mostly textual data) [15,17] for the difficulty of extracting information. However, in a data lake, there are various types of data (images, pdf files...). Thirdly, data ingestion is the most considered phase to extract metadata [1,9,17]. Nevertheless, the information that is produced during process and access phases has value too [6,17].

Until now, there isn't a generic metadata management system that works on both structured and unstructured data for the whole data life-cycle in data lakes. The objective of this paper is to define a metadata management system that addresses these weaknesses.

3   Metadata Model

Considering the diversity of data structural type and different processes that applied on datasets, our solution is based on intra- and inter-metadata.

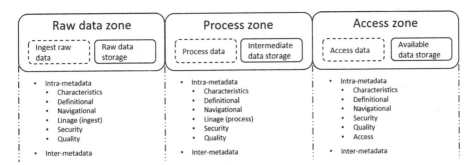

Fig. 1. Meta data classification

3.1 Metadata Classification

Our metadata classification has the advantage of integrating both intra-metadata and inter-metadata for all datasets. Intra-metadata allow users to understand datasets with their characteristics, meaning, quality and security level [2,19]. Inter-metadata help users find relevant datasets that can answer their requirements to make their data discovery easier [9,17].

- *Inter-metadata.* We complete the classification of [9] and obtain 5 subcategories. *Dataset containment* signifies that a dataset is contained in other datasets. *Partial overlap* signifies that some attributes with corresponding data in some datasets overlap. For instance, in a hospital, health care and billing databases contain the same attributes and data about patients, prescriptions and stays. But these two databases also contain their own specific data. *Provenance* signifies that one dataset is the source of another dataset. *Logical clusters* signifies that some datasets are in the same domain. For example, different versions, duplication of the same logical dataset. *Content similarity* signifies that different datasets share the same attributes.
- *Intra-metadata.* We extend the classification of [2,19] to include access, quality and security.
  - *Data characteristics* consist of information such as identification, name, size, structural type and creation date of datasets. This information helps users to have a general idea of a dataset.
  - *Definitional metadata* specifies datasets' meanings. In the original taxonomy, there are vocabulary and schema subcategories. We classify definitional metadata into *semantic* and *schematic* metadata. Structured and unstructured datasets can be semantically described by a text or by some keywords (vocabularies). Schematically, a structured dataset can be presented by a database schema.
  - *Navigational metadata* concerns the location of datasets, for instance, file paths and database connection URLs.
  - *Lineage* presents data life-cycle. It consists of the original source of datasets and the processing history. Information on datasets sources and process history makes datasets more reliable.

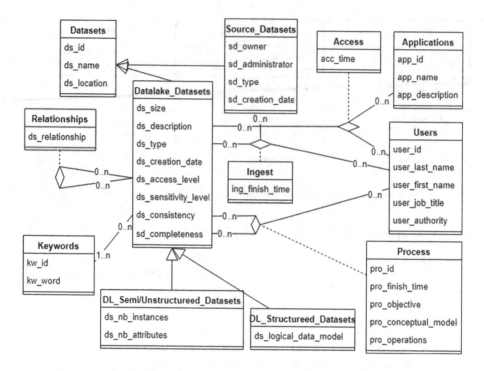

Fig. 2. Class diagram of metadata conceptual model

- *Access metadata* present access information, for example, name of the users who accessed datasets and the access tools. This information helps users to find relevant datasets by accessed users and to trust data by other users' access histories.
- *Quality metadata* consist of data consistency and completeness [10] to ensure datasets' reliability.
- *Security metadata* consist of data sensitivity and access level. Data lakes store datasets from various sources. Some datasets may contain sensitive information that can only be accessed by certain users. Security metadata can support the verification of access. This information ensures the safety of sensitive data.

3.2   Metadata Conceptual Schema

From the functional architecture point of view [5, 11, 13, 20], a data lake contains four essential zones. A *raw data zone* allows to ingest data without processing and stores raw data in their native format. A *process zone* allows to process raw data upon usage and provides intermediate storage areas. The *access zone* stores refined data and ensures data availability. And a *governance zone* is in charge of insuring data quality, security and data life-cycle.

Dataset Name	Description
ORBIS	Raw dataset of all the health care information.
SURGERY	Intermediate dataset originating from ORBIS, concerns information on surgery.
OPTIMISME_TC	Available dataset generated from SURGERY aims to optimise head trauma procedures.
COMEDIMS	Available issued by SURGERY concerns prescriptions of expensive drugs.
COMPTE_RENDU	Raw dataset consists of scans of medical reports (pdf).

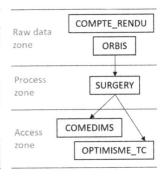

Fig. 3. List of datasets

Our metadata classification is applied on the multi-zones functional architecture of data lakes (see Fig. 1). Based on this classification, we propose a metadata conceptual schema (see Fig. 2). A dataset, structured or unstructured, is ingested from one or plural sources by one or more users. Datasets can be processed by users to transform to new datasets. Users can access datasets with some tools for their analyses. Datasets that are stored in a data lake can have relationships.

This metadata conceptual schema has several advantages: (i) Data sources (external datasets) are recorded. (ii) Both structured and unstructured datasets are considered. (iii) All the work (ingest, process, access) that has been done in a data lake is registered. (iv) Information of each single dataset and relationships between different datasets are stored. And (v) datasets' quality, sensibility and access level are controlled.

4   Metadata Implementation

The University Hospital Center (UHC) of Toulouse is the largest hospital center in the south of France. All medical, financial and administrative data are stored in the information system of this center. The UHC of Toulouse plans to launch a data lake through an iterative process. The aim of this project is to ingest (i) all the internal relational databases and medical e-documents (including scans of hand written medical reports), (ii) external data coming from other French UHCs and (iii) some public medical data. The objective of this data lake is to combine these different data sources in order to allow data analysts and BI professionals to analyse available data to improve medical treatments. The first step of this project concerns 5 datasets (4 structured and 1 unstructured datasets). These 5 datasets are in different functional zones of a data lake (see Fig. 3).

We have implemented two proofs of concept in the UHC of Toulouse in order to validate our proposal. Regarding metadata management systems, there are metadata stored in key-value [9], XML documents [15,17], relational databases [17] or by ontology [1]. We have chosen to implement a relational database and a graph database for the fallowing reasons: relational databases have a standard

SELECT
   ddsource.*
FROM
   datalakedatasets dds,
   datalakedatasets ddp,
   process p
WHERE
   ddp.dsid = p.dsid AND
   p.dsiddatalakedatasets = dds.dsid AND
   lower(ddp.dsname) = 'comedims';

(a)

SELECT ddprocessed.*
FROM datalake_datasets ddraw,
   datalake_datasets ddprocessed,
   process p, have h, keywords kw
WHERE h.kw_id = kw.kw_id AND
   ddraw.ds_id = h.ds_id AND
   ddraw.ds_id = p.ds_id AND
   p.ds_id_datalake_datasets = ddraw.ds_id AND
   p.ds_id = ddprocessed.ds_id AND
   lower(kw.kw) = 'medicine';

(b)

Fig. 4. Logical data model

query language (SQL) and a high security level insured by many RDBMSs (Relational Database Management System); graph databases ensure scalability and flexibility. Moreover, these systems are currently used in the UHC of Toulouse.

4.1 Relational Database

We firstly implemented the conceptual schema of metadata on a relational DBMS. After the implantation, we collected the needs of data scientists from a metadata point of view. The first questions were about data trust and data lineage analysis. To validate our proposal, we have written several queries to compare the feasibility and usability of different environments. In the following paragraphs, you will find two examples.

(i) When a user works on a dataset, he may wants to know where does the data come from to have more confidence on the dataset. There is an example to find the original dataset of 'COMEDIMS' (see Fig. 4 (a)). (ii) Besides finding the origin of one dataset, users may also want to find relevant datasets that come from the same origin of the dataset. For example, users want to find out all the datasets that used the data of the original dataset of COMEDIMS (see Fig. 4 (b)).

4.2 Graph Database

The second solution of implementation is graph database. We firstly introduce a Neo4j model for the 5 datasets of UHC of Toulouse. In addition, 2 queries that answer the same questions in the last subsection will be executed.

We extended the mapping from UML class diagram to property graphs that proposed by [3] to Neo4j Cypher query language. Based on this mapping, we implemented a graph database with neo4j (Fig. 5). To test the implementation, we also answered the same questions than the relational database.

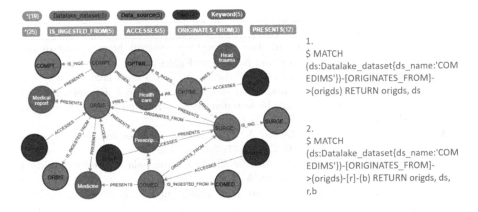

1.
$ MATCH
(ds:Datalake_dataset{ds_name:'COM
EDIMS'})-[ORIGINATES_FROM]-
>(origds) RETURN origds, ds

2.
$ MATCH
(ds:Datalake_dataset{ds_name:'COM
EDIMS'})-[ORIGINATES_FROM]-
>(origds)-[r]-(b) RETURN origds, ds,
r,b

Fig. 5. Neo4j data model

5   Conclusion and Future Work

To prevent a data lake from turning into a data swamp, metadata management
is recommended. In this paper, we firstly proposed a generic and extensible classification of metadata based on a multi-zones of data lake functional architecture. The classification considers not only the metadata on each dataset (intra-
metadata) but also the relationships between datasets (inter-metadata). Based
on the classification, we presented a metadata conceptual schema for data lakes.
What's more, for validating the conceptual schema, we implemented a graph
DBMS and a relational DBMS for metadata management system in UHC of
Toulouse.

Our next plan concerns the automatic extraction of metadata. For this automatic extraction, we plan to adapt to the context of existing works such as automatic detection of relationships between datasets [1] and automatic extraction of
data structure, metadata proprieties and semantic data [1]. Nevertheless, there
isn't a system which can extract automatically inter-metadata, intra-metadata
from different types (structured, semi-structured, unstructured) of datasets.

Our long term goal is to accomplish a metadata management system which
integrates automatic extraction of data, effective researches of metadata, automatic generation of dashboards or other analyses.

References

1. Alserafi, A., Abelló, A., Romero, O., Calders, T.: Towards information profiling:
   data lake content metadata management. In: 2016 IEEE 16th International Conference on Data Mining Workshops (ICDMW), pp. 178–185. IEEE (2016)
2. Bilalli, B., Abelló, A., Aluja-Banet, T., Wrembel, R.: Towards intelligent data
   analysis: the metadata challenge. In: Proceedings of the International Conference
   on Internet of Things and Big Data, Rome, Italy, pp. 331–338 (2016)

3. Delfosse, V., Billen, R., Leclercq, P.: Uml as a schema candidate for graph databases. NoSql Matters 2012 (2012)
4. Dixon, J.: Pentaho, Hadoop, and Data Lakes, October 2010. https://jamesdixon. wordpress.com/2010/10/14/pentaho-hadoop-and-data-lakes/
5. Fang, H.: Managing data lakes in big data era: what's a data lake and why has it became popular in data management ecosystem. In: 2015 IEEE International Conference on Cyber Technology in Automation, Control, and Intelligent Systems (CYBER), pp. 820–824. IEEE (2015)
6. Foshay, N., Mukherjee, A., Taylor, A.: Does data warehouse end-user metadata add value? Commun. ACM **50**(11), 70–77 (2007)
7. Gabriel, R., Hoppe, T., Pastwa, A.: Classification of metadata categories in data warehousing-a generic approach. In: AMCIS, p. 133 (2010)
8. Hai, R., Geisler, S., Quix, C.: Constance: an intelligent data lake system. In: Proceedings of the 2016 International Conference on Management of Data, pp. 2097–2100. ACM (2016)
9. Halevy, A.Y., et al.: Managing Google's data lake: an overview of the Goods system. IEEE Data Eng. Bull. **39**(3), 5–14 (2016)
10. Kwon, O., Lee, N., Shin, B.: Data quality management, data usage experience and acquisition intention of big data analytics. Int. J. Inf. Manag. **34**(3), 387–394 (2014)
11. LaPlante, A., Sharma, B.: Architecting Data Lakes, March 2016
12. Lopez Pino, J.L.: Metadata in Business Intelligence, January 2014. https://www. slideshare.net/jlpino/metadata-in-business-intelligence
13. Menon, P.: Demystifying Data Lake Architecture, July 2017. https://medium.com/ @rpradeepmenon/demystifying-data-lake-architecture-30cf4ac8aa07
14. Oram, A.: Managing the Data Lake. OReilly Media, Inc., Sebastopol (2015)
15. Quix, C., Hai, R., Vatov, I.: Metadata extraction and management in data lakes With GEMMS. Complex Syst. Inf. Model. Q. **9**, 67–83 (2016)
16. Ravat, F., Song, J.: A unified approach to multisource data analyses. Fundam. Inf. **162**(4), 311–359 (2018)
17. Sawadogo, P., Kibata, T., Darmont, J.: Metadata management for textual documents in data lakes. In: 21st International Conference on Enterprise Information Systems (ICEIS 2019) (2019)
18. Thor, O.: 3 keys to keeping your data lake from becoming a data swamp, June 2017. https://www.cio.com/article/3199994/3-keys-to-keep-your-data-lake-from-becoming-a-data-swamp.html
19. Varga, J., Romero, O., Pedersen, T.B., Thomsen, C.: Towards next generation BI systems: the analytical metadata challenge. In: Bellatreche, L., Mohania, M.K. (eds.) DaWaK 2014. LNCS, vol. 8646, pp. 89–101. Springer, Cham (2014). https:// doi.org/10.1007/978-3-319-10160-6_9
20. Walker, C., Alrehamy, H.: Personal data lake with data gravity pull. In: 2015 IEEE Fifth International Conference on Big Data and Cloud Computing, pp. 160–167. IEEE (2015)

Correlation Between Students' Background and the Knowledge on Conceptual Database Modelling

Lili Nemec Zlatolas[(⊠)] [ID], Aida Kamišalić [ID],
and Muhamed Turkanović [ID]

Faculty of Electrical Engineering and Computer Science, University of Maribor,
Koroška cesta 46, 2000 Maribor, Slovenia
{lili.nemeczlatolas, aida.kamisalic,
muhamed.turkanovic}@um.si

Abstract. Students taking a Databases Course have different pre-knowledge of Conceptual Database Modelling. At the beginning of the course, they were asked to self-evaluate their knowledge of database design concepts. We have compared the results of self-evaluation of students with Technical and General High School education, as well as the results they achieved at the final exam. All together 132 students have collaborated in both parts of the survey. The results show that the students with a technical background have a better pre-knowledge of database design constructs, but, on the other hand, received lower results in the exam where the knowledge on Database Modelling was tested.

Keywords: Database Modelling · Comparison · Database design learning

1 Introduction

Conceptual Database Modelling is an essential step in creating a good database. Teaching the students how to create a good conceptual model is also a challenge for all teachers, since there is no straight line for the perfect solution of any specific scenario [1, 2]. Studies show that differences reside between basic and expert database (model) designers, whereby many conceptual mistakes are produced by the former, while modelling a database [3–5]. Becoming a good database designer is a cumbersome process, requiring a lot of studying and practicing.

Students with different secondary education level backgrounds are joining Universities to study Computer Science or similar ICT related studies. The Database course usually serves as a fundamental course in such studies, due to the importance of understanding database fundamentals, including modelling, as soon as possible. However, many students come to the University with some pre-knowledge of databases and tend to overestimate their actual knowledge. The studies have shown that pre-knowledge can have an effect on students, so that they believe already having enough knowledge on the topic and not needing to learn any more [6, 7]. This might lead to the effect that such students do not achieve a level, which is required or desired by University standards.

© Springer Nature Switzerland AG 2019
T. Welzer et al. (Eds.): ADBIS 2019, CCIS 1064, pp. 45–51, 2019.
https://doi.org/10.1007/978-3-030-30278-8_6

In our previous studies, we examined the effectiveness of learning database fundamentals, depending on the notation used for conceptual design [8]. A multi-level experimental study was set up. Considering the influencing factors, students' achievements were examined throughout the learning process. Additionally, the influence of notation used for conceptual design on student knowledge perception [7] as well as the influence of students' educational background on their learning outcomes [9] were examined.

In this work, we present another aspect of the aforementioned study, where we analysed if students with different secondary education level backgrounds, thus different pre-knowledge of Database Modelling and their self-perception of this knowledge, influences their results. The analysis was performed with active students, after they had taken the course lessons and completed the practical laboratory work.

The structure of the paper is as follows. Subsection 1.1 presents the research questions set up for this study. In Sect. 2, information on research methods is provided. The main contribution of the paper is presented in Sect. 3 where results and discussion are detailed. Finally, the conclusions are presented in Sect. 4.

1.1 Research Questions

We have set up the following two research questions:

RQ1: How does previous education affect the knowledge on conceptual design concepts', between students with technical or general secondary education backgrounds?

RQ2: How does the pre-knowledge of conceptual design concepts affect the students' final grade gained for Database Modelling?

2 Research Methods

For the purposes of the study we used a paper format questionnaire that was designed to answer the research questions. Survey questions were designed based on existing literature and discussion with fellow University members.

2.1 Data Collection and Participants

At the beginning of the semester, students attending the Databases Course were asked to fill in a self-evaluation questionnaire on the topic of their knowledge on conceptual database design concepts. After the course was finished and they had attended lectures on Database Modelling within the course, we tested their knowledge on a conceptual database design. The sample of demographics is presented in Table 1. All together 132 students collaborated in both parts of the survey – the self-evaluation questionnaire and the test after the course was finished. Since students who are taking the Databases Course are signed into the Study Programmes at the Faculty of Electrical Engineering and Computer Science, there are more male than female students taking this course. Most of the students are in their first or second year of studies, so the average age concurs with this. The previous secondary education (High School) of students was either general or technical.

Table 1. Sample of demographics (n = 132).

Variable	Sample results
Gender	Male 84.10% Female 15.90%
Age mean	M 19.86 SD 1.23
Study programme	Computer Science and Information Technologies 62.1% Informatics and Technologies of Communication 37.9%
Previous education	General 55.30% Technical 44.70%

2.2 Measures

Measurement items were tested with a 5-point Likert-scale ranging from 1 to 5. The measurement items are presented in Table 2. There were altogether 3 demographical questions, 1 question on previous education and 5 questions regarding the Database Modelling concepts. The grade for Database Modelling was awarded by a teacher in the course after the students had already taken the lectures. To connect the pre-knowledge of each student and the grade for Database Modelling, we used a unique ID number that a student had to enter when filling in the self-evaluation questionnaire on his/her knowledge and on the exam at the end of the course.

Table 2. Measurement of variables.

Questions	Answers
Previous education	General education/Technical education
Perception of knowledge: • Entity (E) • Relationship (R) • Attribute (A) • Primary Key (PK) • Foreign Key (FK)	Possible answers: • I am not familiar with the term (1) • I am familiar with the term, but not with the meaning (2) • Undefined (3) • I am familiar with the meaning but I do not know how to use it (4) • I am familiar with the meaning and I know how to use it (5)
Grade for Database Modelling	0–100%

3 Data Analysis and Results

A data analysis was performed with the use of IBM SPSS Statistics 24.0. First, we did a comparison of means for the pre-knowledge between general and technical students, which is presented in Table 3. To test the significance of the results, we used the Mann-Whitney U test presented in Table 4 [10]. The students who finished a general secondary education self-evaluated their knowledge of database design concepts lower than the students with a technical secondary education level.

Table 3. Comparing means between groups with different high school education on Database Modelling knowledge

	Previous education	N	Mean rank	Sum of ranks
Entity	General	73	61.94	4521.50
	Technical	59	72.14	4256.50
Relationship	General	73	54.06	3946.50
	Technical	59	81.89	4831.50
Attribute	General	73	54.82	4002.00
	Technical	59	80.95	4776.00
Primary key	General	73	60.38	4407.50
	Technical	59	74.08	4370.50
Foreign key	General	73	51.36	3749.00
	Technical	59	85.24	5029.00
Mean for all database properties	General	73	53.84	3930.00
	Technical	59	82.17	4848.00

The Mann-Whitney U test in Table 4 shows that there are significant differences between the two groups with different higher education for almost all the database design concepts, except for the Entity.

Table 4. Mann-Whitney U test using grouping variable of previous education

	E	R	A	PK	FK	Mean of database properties
Mann-Whitney U	1820.5	1245.5	1301.0	1706.5	1048.0	1229.0
Asymp. Sig. (2-tailed)	.11	.00	.00	.03	.00	.00

Next, we divided the respondents into groups based on their self-evaluated pre-knowledge on Database Modelling. We calculated the mean for each participant for perception of knowledge of Entity, Relationship, Attribute, Foreign and Primary Key. The mean was between 1 and 5. Then, we divided the participants into 4 similarly sized groups. The first, with no pre-knowledge, rated their pre-knowledge with 1–2.4. The second group, with some pre-knowledge, rated it with 2.6–3.4. The third group, with good pre-knowledge, rated it with 3.6–4.8. The last group rated their knowledge high with a mean of 5. As can be observed in Fig. 1, the students with general pre-education received higher results in the exam at the end of the lectures of the Databases Course. The lowest grade that the students with high pre-knowledge of Database Modelling received was 28%, which is actually the lowest grade among all students. One possible explanation of this outcome could be that some students might think that they already know a lot and do not pay much attention to the Course itself.

In Table 5, the mean results of the grade for Database Modelling is presented for two groups of students with different secondary education background and different self-evaluation categories on their knowledge. In the majority of cases, students with general secondary education background wrote the exam for Database Modelling better than the ones with a technical background. Some of the students with general background and very high pre-knowledge on database design concepts achieved a bit lover results at the exam than the students with a technical background. If we combine all students together, then the results show that the more pre-knowledge the students had, the higher the grade they received for Database Modelling, but, further analysis shows that the students with general background and high pre-knowledge were probably too confident with their knowledge, since they received worse results than the ones with good pre-knowledge. On the other hand, the students with a technical background and some or good pre-knowledge, were also a bit too confident in their knowledge of database design, and received worse exam results for Database Modelling than the rest of the students with no or high pre-knowledge.

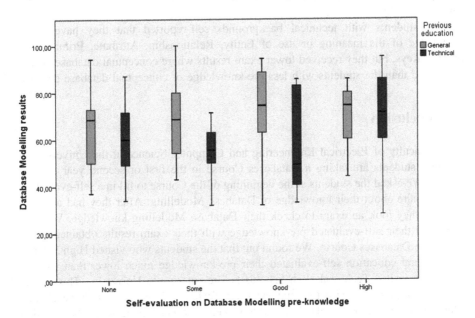

Fig. 1. Comparison of test results of Database Modelling for groups of students with technical or general high school education and different self-evaluation categories.

Table 5. The results of the grade for the Database Modelling results for different groups.

	All students			General background			Technical background		
	Mean	N	SD	Mean	N	SD	Mean	N	SD
No pre-knowledge	63,51	30	15,54	64,47	23	14,84	60,36	7	18,55
Some pre-knowledge	63,81	34	17,74	67,80	24	18,13	54,23	10	13,05
Good pre-knowledge	66,36	33	20,17	75,00	17	16,02	57,18	16	20,48
High pre-knowledge	71,15	35	16,20	69,01	9	14,63	71,88	26	16,92
Total	66,32	132	17,60	68,58	73	16,38	63,54	59	18,77

3.1   Answering the Research Questions

The students with technical backgrounds reported higher pre-knowledge of conceptual database design concepts in comparison to students with a general high education background. The students with technical background reported to have higher knowledge of what Relationship, Attribute, Primary Key and Foreign Key mean in conceptual database design. There are no statistically significant differences for Entity as a concept of conceptual database design among the students with technical or general education backgrounds.

The students with technical backgrounds self-reported that they have better knowledge of the meaning or use of Entity, Relationship, Attribute, Primary and Foreign Keys, but they received lower exam results where conceptual database design was tested than the students with less pre-knowledge of conceptual database design.

4   Conclusion

At the Faculty of Electrical Engineering and Computer Science at the University of Maribor, students are taking a Databases Course in the first or second year of their studies. We asked the students at the beginning of the Course to fill in a self-evaluation questionnaire about their knowledge of Database Modelling. After they had attended lectures, they took an exam to check their Database Modelling knowledge. We then compared their self-evaluated pre-knowledge with their exam results obtained at the end of the Databases Course. We found out that the students who visited High Schools with general education self-evaluated their pre-knowledge much lower than the students who visited Technical High Schools before entering to University. This should not be a surprise, because the students from Technical High Schools have already attended some lectures on databases within their secondary education. On the other hand, we would also expect that the students with technical backgrounds would get higher grades for Database Modelling after attending the lessons of the Databases Course at the University. However, we found out that the students with general backgrounds actually achieved higher grades in the Database Modelling part of the Course than the ones with technical backgrounds. There could be a number of reasons for this. Our limitation was the small sample, which cannot be generalized to all the population. Another point is that the students with technical backgrounds could have had a different pre-education on conceptual model language, and might mix some basic concepts of Conceptual Modelling.

Acknowledgements. The authors acknowledge the financial support from the Slovenian Research Agency (Research Core funding No. P2-0057) and from the European Union's Horizon 2020 Research and Innovation program under the Cybersecurity CONCORDIA project (GA No. 830927). We would also like to thank the participants of this research project.

References

1. Embley, D.W., Thalheim, B.: Handbook of Conceptual Modeling: Theory, Practice, and Research Challenges. Springer, Heidelberg (2012). https://doi.org/10.1007/978-3-642-15865-0
2. Mylopoulos, J.: Conceptual modelling and telos. In: Conceptual Modelling, Databases, and CASE: An Integrated View of Information System Development, pp. 49–68 (1992)
3. Batra, D., Davis, J.G.: Conceptual data modelling in database design: similarities and differences between expert and novice designers. Int. J. Man Mach. Stud. **37**, 83–101 (1992)
4. Batra, D., Antony, S.R.: Novice errors in conceptual database design. Eur. J. Inf. Syst. **3**, 57–69 (1994)
5. Engels, G., et al.: Conceptual modelling of database applications using an extended ER model. Data Knowl. Eng. **9**, 157–204 (1992)
6. Sewell, A.: Constructivism and student misconceptions: why every teacher needs to know about them. Aust. Sci. Teach. J. **48**, 24 (2002)
7. Kamišalić, A., Turkanović, M., Heričko, M., Welzer, T.: Knowledge perception influenced by notation used for conceptual database design. In: Heričko, M. (ed.) 21st International Multiconference Information Society - IS 2018, pp. 19–22. Institut "Jožef Stefan", Ljubljana, Slovenia (2018)
8. Kamišalić, A., Heričko, M., Welzer, T., Turkanović, M.: Experimental study on the effectiveness of a teaching approach using barker or bachman notation for conceptual database design. Comput. Sci. Inf. Syst. **15**, 421–448 (2018)
9. Kamišalić, A., Turkanović, M., Welzer, T., Heričko, M.: Influence of educational background on the effectiveness of learning database fundamentals. In: Heričko, M. (ed.) The 29th International Conference on Information Modelling and Knowledge Bases, EJC 2019, Lappeenranta, Finland (2019)
10. Ruxton, G.D.: The unequal variance t-test is an underused alternative to Student's t-test and the Mann-Whitney U test. Behav. Ecol. **17**, 688–690 (2006)

Smart Caching for Efficient Functional Dependency Discovery

Anastasia Birillo[1,2] and Nikita Bobrov[1,2]

[1] JetBrains Research, Saint-Petersburg, Russia
anastasia.i.birillo@gmail.com, nikita.v.bobrov@gmail.com
[2] Saint Petersburg State University, Saint-Petersburg, Russia

Abstract. Functional dependency (FD) is a concept for describing regularities in data, that traditionally was used for schema normalization. Recently, a different problem has emerged: for a given dataset, find FDs that are contained in it. Efficient FD discovery is of great importance due to FD usage in many tasks such as data cleaning, schema recovery, and query optimization.

In this paper we consider a particular class of FD discovery algorithms that rely on partition intersection. We present a simple approach for caching intermediate results that are generated during run times of these algorithms. Our approach is essentially a heuristic that selects which partitions to cache based on measures of disorder in data. For this purpose, we adopt such well-known measures as Entropy and Gini Impurity, and propose a novel one—Inverted Entropy. Our approach has a negligible computational overhead, and can be used with a number of FD discovery algorithms, both exact and approximate. Our experiments demonstrate that our heuristic allows to decrease algorithm run times by up to 40% while simultaneously requiring up to an order of magnitude less space compared to the state-of-the-art caching approaches.

Keywords: Functional dependency · Partition intersection ·
Partition caching · Caching · Entropy · Gini impurity · PYRO · TANE

1 Introduction

Functional dependency (FD) states that some attributes of a table functionally determine the value of some other attribute, or a set of other attributes. Since their introduction by Codd [2] in 1969, they have been widely applied, mainly for schema normalization, i.e. in this case they were known in advance.

However, in the 90's the focus of research community shifted to different scenarios, where one has to discover FDs for a given dataset. This problem

This is a student work, authored solely by students. Advisors: George Chernishev (chernishev@gmail.com) and Kirill Smirnov (kirill.k.smirnov@gmail.com). This work is partially supported by RFBR grant 16-57-48001.

T. Welzer et al. (Eds.): ADBIS 2019, CCIS 1064, pp. 52–59, 2019.
https://doi.org/10.1007/978-3-030-30278-8_7

formulation gained even more popularity with the advent of big data era, because the task of data exploration became much more common.

The problem of FD discovery is very computationally demanding, since the number of potential irreducible FDs grows as $O(N * 2^N)$, where N is the number of involved attributes. Moreover, algorithm run times depend not only on the number of attributes, but on the number of rows as well. In the 90's, the state-of-the-art algorithms on the average desktop PC could handle [4] datasets of up to 20 attributes and tens of thousands of rows within reasonable time (up to several hours). Nowadays [7,8], contemporary algorithms run on a server-class many-core multiprocessor computer discover dependencies for datasets consisting of hundreds of attributes (up to 200) and tens of thousands of rows (up to 200K) within roughly the same time limits.

No doubt, this is substantial progress. Nevertheless, it is insufficient: these volumes are unacceptable for the majority of real-life applications. Therefore, efficient FD mining remains an acute problem that requires designing new and speeding up existing algorithms.

Many algorithms for FD discovery (including some of the recent ones, e.g. [5]) rely on the notion of partition. Informally, a partition for a given attribute X is a collection of lists where each list contains row numbers that have the same value for X. Next, this notion can be straightforwardly generalized for an arbitrary number of attributes. The core of algorithms that employ this concept is the operation of partition intersection. It is rather costly and more importantly, it is performed frequently. However, it can greatly benefit from caching since the algorithm may intersect the same partitions multiple times. Therefore, several existing algorithms use partition caching.

In this paper we propose a novel caching mechanism that employs a heuristic to guide cache population. Our idea is the following: during partition intersection, check the heuristic for the result and if it passes, add it to the cache. This heuristic is based on a measure of "uniqueness" of values of an attribute set. For this purpose, we have adopted several existing measures such as Entropy, Gini Impurity, and developed a new one, which we named Inverted Entropy. Our technique has a negligible computational overhead and can be used with a number of FD discovery algorithms, both exact [4,9] and approximate [5]. We have also proposed a novel cache eviction policy, that captures usefullness of partitions and tries to retain the most useful ones.

Evaluation performed on a number of real datasets has demonstrated that our caching approach allows to improve run times up to 40%, while simultaneously decreasing the required memory by up to an order of magnitude.

2   Background: Partitions and Caching

Many existing FD discovery algorithms employ the partition intersection approach [7]. Formally, a *stripped partition* or *position list index* (PLI) is defined as follows [5] (for the ease of comprehension, we will follow the Kruse and Naumann notation throughout our paper):

Definition 1. *Let r be a relation with schema R, and let $X \subseteq R$ be a set of attributes. A cluster is a set of all tuple indices in r that have the same value for X, i.e., $c(t) = \{i | t_i[X] = t[X]\}$. The PLI of X is the set of all such clusters except for singleton clusters:*

$$\overline{\pi}(X) := \{c(t) | t \in r \wedge |c(t)| > 1\}$$

Next, the *size* of a PLI is defined as $\|\overline{\pi}(X)\| = \sum_{c \in \overline{\pi}(X)} |c|$. Essentially, it is the number of included tuple indices. Finally, *arity* of a partition is defined as the number of involved attributes and we *atomic partition* a partition that has arity of 1.

PLIs are extensively used for FD validation, i.e. for checking whether a given FD holds. A frequently used criterion [4] is the following: $X \to A\ A, X \subseteq R$ holds $\Leftrightarrow \|\overline{\pi}(X)\| = \|\overline{\pi}(X \cup A)\|$. To compute $\overline{\pi}(X \cup A)$, a procedure called PLI intersection (also known as partition product) is performed. Its idea is to avoid the costly rescan of the necessary attributes for computing $\overline{\pi}(X \cup A)$ via reusing the already known $\overline{\pi}(X)$ and $\overline{\pi}(A)$. This procedure significantly contributes to algorithm run time, since it is invoked many times during the search space traversal. However, it is often called for overlapping attribute sets. Therefore, PLIs are promising for caching.

Several recent FD discovery algorithms employ caching [1,5]. In study [5] authors use a set-trie data structure, which is essentially a prefix tree where each node stores the corresponding PLIs. In this paper, the authors have proposed a sophisticated algorithm for cache lookup, but they relied on coin flip for cache population (see algorithm 2). In our study, we argue that such an important part should not be left to chance.

3   Our Approach

3.1   Preliminaries

Consider the following. Let $D = X \to Y$ be an arbitrary FD, and for simplicity, let X and Y be single attributes. The following observations are true:

1. If X contains unique values, then D holds regardless of Y.
2. If Y contains unique values, then D holds only if X contains only unique values too.
3. If X contains equal values, then D holds only if contains Y equal values too.
4. If X contains equal values, but Y has at least one non-equal value, then D is violated.

Therefore, the more "unique" values are in X, the more probable that D will hold. For Y it is vice versa: the more similar values are in Y, the more probability that the dependency will hold.

Next, note that partition-based FD discovery algorithms share a common pattern: when a dependency is violated, they try to fix it via introducing additional attributes to the left part (in some papers, this procedure is called specialization [4]).

Our idea is the following: PLIs obtained during the course of the algorithm are not equally worthy of caching. There is no need to cache PLIs for dependencies that are almost satisfied, i.e. that have a high probability of being satisfied during the next specialization. Contrary to this, PLIs that would require many rounds of specialization should be cached.

As shown above, this probability is directly linked to the degree of "uniqueness" of values of dependency's left hand side. Note that this probability is a monotonic, non-decreasing function with regard to specialization steps. In other words, after a specialization step, the probability that this new (specialized) dependency holds is at least the same as before.

There are several ways to calculate the degree of "uniqueness" of a set of values:

1. Shannon Entropy is a classic measure of disorder in data. It is given as follows:

$$H = -\sum_i P_i \log P_i,$$

   where P_i is the probability mass function. The entropy is 0 if there is a single value for all entries for a given attribute. If all records contain different values for this attribute (e.g. the attribute is the primary key) then entropy is $\log n$, where n is the number of records.

2. Gini impurity is a measure used for construction of decision trees. It is computed as follows:

$$I_G = 1 - \sum_{i=1}^{J} P_i^2,$$

   where J is the number of classes and P_i is the probability of correctly classifying i-th item.

3. We have also proposed our metric, which is essentially a minor modification of Shannon Entropy:

$$H^{-1} = -\sum_i (1 - P_i) \log(1 - P_i).$$

Such a metric inherits the behaviour of the previous ones (it also grows as a set of values becomes more unique), and its sensitivity to "uniqueness" can be compared to the other metrics as follows: $I_G \leq H^{-1} \leq H$.

3.2 Caching Algorithm and Cache Population

In order to evaluate our ideas, we have decided to modify the PYRO algorithm [5] by introducing our heuristic. In their paper, cache lookup and cache population were contained in a single algorithm as described below. Suppose that we have to compute the result of partition intersection $\overline{\pi}(X)$, where X is an attribute set.

1. If the required $\overline{\pi}(X)$ is not present in cache, try to compute it by efficiently reusing cached results for subsets of X. The idea is to minimize the number of intersections and to keep the size of intermediate results small.
2. While incrementally constructing $\overline{\pi}(X)$, for each intermediate result $\overline{\pi}(C)$ decide whether to cache it or not. The decision is made randomly, depending on the outcome of the coin toss.

We propose to use the following heuristic to guide caching:

$$if(PLIUniqueness(C) \leq medianUniqueness * (1 + mod_1(C) + mod_2)) \ then \ cache.$$

Here, $PLIUniqueness(C)$ and $medianUniqueness$ are "uniqueness" degrees of newly computed PLI and the median of "uniqueness" degrees for individual attributes. In our study we experimentally evaluate all three "uniqueness" measures that were described above.

The modifiers are as follows:

$$mod_1(C) = \frac{\max_{c \in \overline{\pi}(C)} |c|}{N}, \ mod_2 = \frac{maxHeapSize - availableMemory}{maxHeapSize}.$$

Here, $|c|$ is cluster size and N is relation size, both in records. The mod_1 modifier indicates how close our newly computed PLI is to being primary key. It is $1/N$ when it is primary key (all values are unique) and 1 when its values are all the same. This way, we stimulate caching of PLIs that are far from being primary key, because dependencies with such left hand sides are likely to fail.

In the second modifier, $availableMemory$ and $maxHeapSize$ represent currently available and maximum available memory respectively (in megabytes). This modifier allows to keep track of cache occupancy and allows to favour population while there is plenty of free space.

It is necessary to note that the proposed heuristic is computationally lightweight: everything that is required can be computed during partition intersection.

3.3   Cache Eviction

During the course of the algorithm there is always a chance to run out of available memory, making it impossible to add more PLIs into the cache. The authors of PYRO suggest to perform cache eviction (shrink procedure) each time when the cache size is more than $modifer * maxHeapSize$ ($modifier$ set to 0.85 by default setting). The original procedure is performed as follows: (1) construct a priority queue based on arity of all cached PLIs, starting from PLIs with the arity of 2; (2) remove PLIs from the cache until target size requirement of the cache is met (half of the original size by default). Obviously, such an approach does not take into consideration the real usefulness of PLIs.

We propose a PLI eviction strategy that is based on the number of PLI accesses. This statistic in combination with the cache population strategy described above can help to collect and retain useful PLIs. Therefore, we modified the original eviction algorithm as follows: (1) every time a new PLI is constructed, increment the number of accesses for each cached PLI that was used for its construction; (2) when the shrinking is run, calculate the median of a PLI's number of accesses; (3) add all PLIs that have a less than median number of accesses into a priority queue; (4) clean cache by polling the queue and disposing elements on a per PLI basis until the target size of the cache is met; (5) set the number of accesses of remaining PLIs to zero.

4   Experiments

We have implemented our approach in the Metanome [6] system and used the PYRO algorithm (with agree-set sample size parameter of 10,000) for comparison. In our experiments, we have used the following hardware and software configuration: Intel(R) Core(TM) i5-4670K CPU 3.40GHz (4 CPUs) RAM 16GB, Ubuntu Linux 18.04 (64 bit). The characteristics of the used tables are presented in the left part of Table 1.

We have conducted two experiments. In Fig. 1 we present the results of the run time experiment with our cache population approach and PYRO's eviction policy. We can see that heuristic cache population always improves run times compared to the default approach (Coin). Depending on the dataset, improvement ranges from 10% to 40%. However, there is no clearly superior data disorder measure. In this experiment we have also measured the volume of data which was put into cache (presented in Table 1, right part) during the whole run time, including evictions. We can see that our heuristic allows to drastically reduce the amount of cached data, sometimes up to an order of magnitude.

The second experiment assessed our cache population approach and our eviction policy. To force frequent invocations, we have set a tight memory limit of 512 MBs. Due to that, run times increased by almost two times. We do not present a bar chart or table with figures due to space constraints, but in this experiment heuristic population approach was also superior to the coin-based one in terms of run times. This experiment also did not identify the best measure, i.e. some measures excel on different tables, but all of them are superior to baseline.

5   Related Work

There are several dozen of algorithms for FD discovery. We will not survey all of them due to space constraints, instead referring an interested reader to the study [7].

Most of these algorithms employ partition intersection and some of those can benefit from partition caching. However, to the best of our knowledge, there are only two algorithms that implement such caching techniques.

The first one is PYRO [5], where caching is performed randomly (via coin toss) for every partition that was computed during intersection.

Table 1. Used tables, their details and memory consumption (MB) of different caching approaches.

Table	Records	Attrs	#FD	Coin	Entropy	Gini	Inverted entropy
ditag	3,960,124	13	171	1363	665	681	649
iowa	27,373,453	24	1202	15625	9281	5430	7656
structSheet	664,128	32	1058	837	180	1,267	13
pdbx	29,787	35	108001	71,503	59,103	21,302	35,340
apogeeStar[a]	277,370	49	71066	170,126	165,578	73,351	41,553
plista	1,000	63	316866	10,312	1435	4512	570
flight	1,000	109	144185	1,220	409	125	943

[a]Sloan Digital Sky Survey. https://www.sdss.org/dr14/

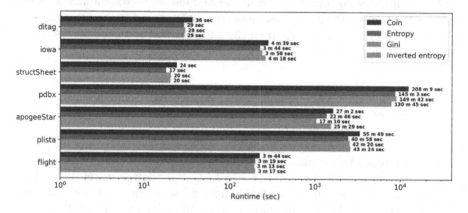

Fig. 1. Runtime comparison of different caching approaches.

The other one is DFD [1] that has adopted caching from the DUCC [3] algorithm. Similarly to PYRO, its cache population strategy is also straightforward: the authors propose to cache all obtained partitions. However, their cache eviction strategy is more sophisticated. First, during the course of the algorithm, calculate the number of uses for each partition. Second, whenever the number of stored partitions exceeds a threshold, run cache clean-up. During this process, remove all non-atomic partitions whose use counts are below the median value of the currently stored partitions.

6   Conclusion

In this paper, we have addressed the problem of intermediate result caching for partition-based FD discovery algorithms. We have proposed to use three measures of disorder in data to guide the cache population algorithm. Next, we have designed a simple cache eviction algorithm to conserve memory. To experimentally validate our approach, we have implemented it inside the Metanome system and modified its PYRO. Experimental results show that our approach decreases run times up to 40% on a number of datasets, while simultaneously decreasing the volume of used memory by up to an order of magnitude. The proposed approach has a negligible computational overhead and, more importantly, can be used with a different partition-based FD discovery algorithms such as TANE.

References

1. Abedjan, Z., Schulze, P., Naumann, F.: DFD: efficient functional dependency discovery. In: Proceedings of the 23rd ACM International Conference on Conference on Information and Knowledge Management, CIKM 2014, pp. 949–958. ACM, New York (2014)
2. Codd, E.F.: Further normalization of the data base relational model. IBM Research Report, San Jose, California, RJ909 (1971)
3. Heise, A., Quiané-Ruiz, J.-A., Abedjan, Z., Jentzsch, A., Naumann, F.: Scalable discovery of unique column combinations. Proc. VLDB Endow. 7(4), 301–312 (2013)
4. Huhtala, Y., Kärkkäinen, J., Porkka, P., Toivonen, H.: TANE: an efficient algorithm for discovering functional and approximate dependencies. Comput. J. 42, 100–111 (1992)
5. Kruse, S., Naumann, F.: Efficient discovery of approximate dependencies. Proc. VLDB Endow. 11(7), 759–772 (2018)
6. Papenbrock, T., Bergmann, T., Finke, M., Zwiener, J., Naumann, F.: Data profiling with metanome. Proc. VLDB Endow. 8(12), 1860–1863 (2015)
7. Papenbrock, T., et al.: Functional dependency discovery: an experimental evaluation of seven algorithms. Proc. VLDB Endow. 8(10), 1082–1093 (2015)
8. Papenbrock, T., Naumann, F.: A hybrid approach to functional dependency discovery. In: Proceedings of the 2016 International Conference on Management of Data, SIGMOD 2016, pp. 821–833. ACM, New York (2016)
9. Wang, S.-L., Shen, J.-W., Hong, T.-P.: Incremental discovery of functional dependencies using partitions. In: Proceedings Joint 9th IFSA World Congress and 20th NAFIPS International Conference (Cat. No. 01TH8569), vol. 3, pp. 1322–1326, July 2001

The Agents' Selection Methods for a Consensus-Based Investment Strategy in a Multi-agent Financial Decisions Support System

Marcin Hernes[1] , Adrianna Kozierkiewicz[2]([⊠]) ,
and Marcin Pietranik[2]

[1] Wrocław University of Economics, ul. Komandorska 118/120,
53-345 Wrocław, Poland
marcin.hernes@ue.wroc.pl
[2] Faculty of Computer Science and Management,
Wroclaw University of Science and Technology, Wybrzeże Wyspiańskiego 27,
50-370 Wrocław, Poland
{adrianna.kozierkiewicz,marcin.pietranik}@pwr.edu.pl

Abstract. Modern financial decision support systems are often based on a multi-agent approach to make advice for investors. However, having a large set of different decisions, collected from agents participating in the process, may entail problems related to data integration and its computational complexity. In this paper, we present some algorithms for selecting agents from a set of all available participants to be included in the eventual decision-making process. All algorithms have been experimentally verified using the a-Trader - a prototype of a multi-agent financial decision support systems on a Forex market.

Keywords: Consensus · Multi-agent system · Forex · Investment strategies · Agents' performance

1 Introduction

Contemporary financial decisions are often made in accordance with distributed information systems, including multi-agent systems. More importantly, these decisions must be taken in near real time, especially in case of High-Frequency Trading. The agents running in such a supporting system generate buy/sell decisions based on different methods and factors, such as technical, fundamental or behavioral analysis. It may highly increase their number within systems, frequently reaching hundreds of participating agents. Using their decisions, investment strategies are built. For example, in the Forex market, these strategies provide signals for open/close short/long positions.

Collecting decisions from distributed and independent agents entails two problems. The first concerns issues related to integrating partial outcomes into a unified, final decision that can be further utilized. This problem is widely known in the literature, and a plethora of different approaches, with their advantages and disadvantages, exist.

© Springer Nature Switzerland AG 2019
T. Welzer et al. (Eds.): ADBIS 2019, CCIS 1064, pp. 60–67, 2019.
https://doi.org/10.1007/978-3-030-30278-8_8

The second problem is directly related to the efficiency of both collecting participating agents' decisions and their further processing. There are many approaches to this issue, spanning from, architectural requirements, through parallelization of computations. The question appears: how much and which intelligent agents should participate in making the decision process to give the most profitable investment strategy?

In our earlier research, we showed that it is not required to have a large number of agents to assert a high quality of their decision integration. The still open problem is, the selection of agents to be included in the decision process.

In this paper, we want to investigate how reducing this number of participating agents, based on different selection methods, impacts the quality of strategies created by agents in a financial application (Fintech). Therefore, the main aim of this paper is to implement and evaluate different methods for agents' selection, to build a consensus-based strategy in a-Trader- a prototype of a multi-agent financial decision support systems on a Forex market.

The paper is structured as follows. In the next section, an overview of publications covering similar topics is given. Section 3 contains some basic information about the a-Trader system. The core of the article can be found in Sect. 4, where we present a set of algorithms for selecting agents in a multi-agent decision support system from a pool of available ones. The experimental verification of these algorithms is given in Sect. 5. Section 6 concludes that paper and sheds some light on our upcoming research plans.

2   Related Works

Incorporating agent-based solutions within contemporary financial applications is not an uncommon approach. In [8] authors give some characteristics and requirements, that such tools must fulfill. A starting point is a detailed description of the downfalls of a centralized approach. This included (among others) an enormous amount of knowledge that needed to be processed, dealing with bottlenecks and complex tasks related to data filtering. Distributing these tasks among a set of independent agents may be a good solution for given problems.

A practical example of a system fulfilling overviewed expectations is described in [9], where an application of a multi-agent based model in the stock exchange has been described. This system provides an environment capable of both performing efficient distributed computations and, by using a fuzzy expert system, intelligent decision making.

In [1] authors propose to use agents in a Forex market. The paper contains a description of an implementation of a set of independent currency trading agents, based on classification and regression models. Six trading agents were developed, each responsible for trading different currency pairs in the Forex market.

Authors of [4] describe the design of TACtic, which is a multi-agent based system built upon multi-behavioral techniques. It was developed to fulfill several goals: (i) provide bidding decisions in uncertain environments, (ii) to perform predictions about the outcomes of different auctions, (iii) to modify agents' bidding strategies in response to trending market conditions. The architecture of the other client and server-side agents are described in [3]. Conducted experiments show the correctness of proposed ideas.

An important issue related to all approaches is their efficiency. In modern applications (especially, while processing high traffic financial data) computational complexity of knowledge processing cannot be a bottleneck. Therefore, there are many attempts to overcome related problems. One of them is using a multi-level approach to partial results integration [7, 8], which has been proved useful in terms of quality of obtained outcomes and its time-efficiency. More issues related to this problem can also be found in [2].

3   A-Trader System

The a-Trader system has been broadly described in [1]. Due to the page limit, in this paper, we provide only general information about this system. The a-Trader consists of about 1500 agents divided into the following groups: Notification Agent, Market Communication Agents, Basic Agents, Intelligent Agents, and Supervisor Agents. The strategies are built based on agents generating buy or sell a financial instrument. Thus, in our system a single agent's decision is defined as follows [6]:

Definition 1. *Decision D about finite set of financial instruments* $E = \{e_1, e_2, \ldots, e_N\}$ *is defined as a set:*

$$D = EW^+, EW^\pm, EW^-, Z, SP, DT \tag{1}$$

where: $EW^+ = \{\langle e_o, pe_o \rangle, \langle e_q, pe_q \rangle, \ldots, \langle e_p, pe_p \rangle\}$. A couple $\langle e_x, pe_x \rangle$, where: $e_x \in E$ and $pe_x \in [0, 1]$ denote a financial instrument and this instrument's participation in the set EW^+, $x \in \{o, q, \ldots, p\}$. The financial instrument $e_x \in EW^+$ is denoted by e_x^+ $EW^\pm = \{\langle e_r, pe_r \rangle, \langle e_s, pe_s \rangle, \ldots, \langle e_t, pe_t \rangle\}$.   A   couple $\langle e_x, pe_x \rangle$,   where: $e_x \in Epe_x \in [0, 1]$ and denote a financial instrument and this instrument's participation in $EW^\pm$ the set, $x \in \{r, s, \ldots, t\}$. The financial instrument $e_x \in EW^\pm$ will be denoted by $e_x^\pm$, $EW^- = \{\langle e_u, pe_u \rangle, \langle e_v, pe_v \rangle, \ldots, \langle e_w, pe_w \rangle\}$.   The   couple $\langle e_x, pe_x \rangle$,   where: $e_x \in Epe_x \in [0, 1]$ and, denote a financial instrument and this instrument's participation in a set EW^-, $\in \{u, v, \ldots, w\}$. The financial instrument $e_x \in EW^-$ will be denoted by e_x^-. $Z \in [0, 1]$ - predicted rate of return, $SP \in [0, 1]$ - degree of certainty of rate Z, DT- date of a decision.

A-Trader system is a multi-agent system which gives some suggestion about buying or selling some financial instruments. Different agents in the system determine decisions based on their knowledge. The set of decisions provided by each Base and Intelligent Agents are understood as the knowledge profile. The Supervisor Agent provide the final advice to the trader. In our previous work [5] we have proposed an algorithm for generating a final decision of a multi-agent system. For this purpose, a Consensus Theory has been applied.

4   Approaches to Agents' Selection

The effectiveness of the A-Trader system has been proved in our previous works [5, 6]. It has been pointed out that the strategies implemented in Intelligent Agents have the biggest influence on the profitability of the final decision. Thus, to compare the agents' performance, the following evaluation function was used [5]:

$$y = (a_1x_1 + a_2x_2 + a_3x_3 + a_4x_4 + a_5x_5 + a_6x_6 + a_7x_7 + a_8x_8 + a_9x_9) \tag{2}$$

where: x_1 - rate of return, x_2 - gross profit, x_3 - 1-gross loss, x_4 - the number of profitable transactions, x_5 - the number of profitable transactions in a row, x_6 - 1-the number of unprofitable transactions in a row, x_7 - Sharpe ratio, x_8 - 1-the average coefficient of volatility, x_9 - the average rate of return per transaction, counted as the quotient of the rate of return and the number of transactions.

All values of x_i (where i $\in$ (1, 2, ..., 9)) are normalized. It was adopted in the test that coefficients a_1 to a_9= 1/9. The final investment strategy is established based on partial decisions of Intelligent Agents. However, the question which appears: how much and which Intelligent Agents should participate in making the decision process to give the most profitable investment strategy? In this paper, we propose some heuristic algorithms for choosing agents, which can assert the best final decision. The result of the proposed algorithms is a generated set of chosen agents BA, based on input data: the performance of M agents: $y^{(1)}, y^{(2)}, ..., y^{(M)}$, M is the number of agents, $k \leq M$- the number of classes, $x_1^{(1)}, ..., x_9^{(1)}, x_1^{(2)}, ..., x_9^{(2)}, ... x_1^{(M)}, ..., x_9^{(M)}$.

Algorithm 1. Simple agents' choosing	Algorithm 2. The ranking based agents' choosing
Begin 1: Order of the performance values $y^{(1)}$, $y^{(2)}$, ..., $y^{(M)}$ in descending order. 2: Put into the set BA them agents with the highest value of performance. **End** **Return BA**	**Begin** 1: Choose the biggest value $y^{(max)}$ from $y^{(1)}, y^{(2)}$, ..., $y^{(M)}$. 2: Calculate the mean value of the performance: $\mu = \frac{\sum_{i=1}^{M} y^{(i)}}{M}$. 3: Calculate the standard deviation of the performance: $\sigma = \sqrt{\frac{\sum_{i=1}^{M}(y^{(i)}-\mu)^2}{M}}$. 4: For each agent i$\in$ {1, ..., M} if $y^{(i)} \geq y^{(max)} - \sigma$ put i-th agent into set BA. **End Return BA**

Algorithm 3. The k-means based agents' choosing	Algorithm 4. Modification of k-means based agents' choosing
Begin 1: $S^{(1)}=\emptyset$, $S^{(2)}=\emptyset$, ..., $S^{(k)}=\emptyset$ 2: Randomly choose k initial centroid: $c^{(1)},c^{(2)},...c^{(k)}$ of class $S^{(1)}$, $S^{(2)}$,..., $S^{(k)}$ 3: Assign each agent x to the nearest cluster $S^{(i)}$ based on the following assumptions: arg $\min_{c^{(i)}} \sum_{j=1}^{9} \lvert c_j^{(i)} - x_j \rvert$ 4: Recompute the centroids by taking the mean of all agents assigned to that centroid cluster: $c_j^{(i)} = \frac{1}{\lvert S^{(i)}\rvert}\sum_{x \in S^{(i)}} x_j$ 5: Repeat steps 3 and 4 until the centroids no longer move. 6: Calculate the mean performance for each class: $y_s = \frac{1}{\lvert S^{(i)}\rvert}\sum_{x \in S^{(i)}} \sum_{j=1}^{9} \alpha_j * x_j$ 7: Put into the set BA the agents from class S with the highest value of the performance y_s **End Return BA**	**Begin** 1: Run the algorithm 4. 2: Calculate $y_{max}= \frac{1}{\lvert BA\rvert}\sum_{x \in BA} \sum_{j=1}^{9} \alpha_j * x_j$ 3: For each agent z from the set BA if $\frac{1}{\lvert BA\setminus\{z\}\rvert}\sum_{x \in BA\setminus\{z\}} \sum_{j=1}^{9} \alpha_j * x_j \geq y_{max}$ then BA=BA\{z} **End** **Return BA**

The simplest is Algorithm 1, which serves as the reference algorithm to compare with others. Algorithm 2 depends on choosing agents with the best performance value. However, in comparison with Algorithm 1, this algorithm does not need to get the number of agents in an arbitrarily way. Algorithm 3 depends on a well-known from the literature *k-means* algorithm. This algorithm is very simple; however, it requires to get the number of classes. Algorithm 4 is a simple modification of Algorithm 3. It depends on removing agents that spoil the performance of the final decision.

5 Research Experiment

For the purpose of performing the research experiment, the following assumptions were made:

1. GBP/PLN M1 quotations were randomly selected for four periods: 16-04-2018, 00:00 to 17-04-2018, 23:59, 19-04-2018, 00:00 to 20-04-2018, 23:59, 07-05-2018, 00:00 to 08-05-2018, 23:59, 21-05-2018, 00:00 to 22-05-2018, 23:59.
2. Simple look-back window is used for testing algorithms, i.e. the first period is used for selection agents, which determine decisions in the second period.
3. Signals for open long/close short position equals 1, close long/open short position equals −1) were generated by the Supervisor agents, which implements certain algorithms. The Buy and Hold (B&H) strategy were used as a benchmark (trader buys on the beginning of a given period and sell at the end of this period).
4. The unit of the agents' performance analysis was pips.
5. The transaction costs were directly proportional to the number of transactions.
6. The investor engages 100% of the capital held in each transaction. 7.
7. The measures (ratios) and the evaluation function presented in Sect. 4 were used in performance analysis, which can be found in Table 1.

Taking into consideration the values of the evaluation function, in the second period, Algorithm 1 was the best one. In the third period, Algorithm 4 was the best. In the fourth period the Algorithm 2 was the best. The Buy and Hold benchmark were ranked lowest in all periods. The values of obtained ratios are characterized by a very large distribution. For example, the number of transactions in the period 1 in case Standard Algorithm of consensus determining (without a selection of agents) equals 65 and in case Algorithm 3 it equals 103. Therefore, a transactions' cost in case of using Algorithm 3 is twice as high as the transactions' cost in case of using the Standard Algorithm.

The very important ratio is the number of unprofitable consecutive transactions. If it is higher than 2 than most often the strategy must perform a market exit. Algorithms characterized by the highest rate of the return not always were evaluated as the best, because also other ratios, such as risk ratios or an average rate of the return per transaction, were taken into consideration.

Reducing the risk level is a very important task of investment strategies. All Algorithms from 1 to 4 are characterized by a higher number of transactions, in comparison with the standard consensus determination algorithm. It may be a result of a fact, that the convergence of agents' decisions is not taken into a consideration in proposed algorithms. After the selection, the set of agents may consist of a large number of agents characterized by a very high level of decision's convergence. Agents with the highest level of this convergence could be removed during the selection process. Therefore, the evaluation function value and ratio values cannot be the only selection criterion. However, for most of the investors, the plausible method of evaluating a stock exchange strategy is a final rate of return. Utilizing any of the proposed algorithms (1–4) gives better results than a strategy without agents' selection. A total rate of return is the highest for Algorithm 4 (and equal 6212) for periods 2–4. The second contestant (with a slightly lower total rate of return equal to 6149) is Algorithm 1. The result obtained by the Buy and Hold (B&H) strategy is only 5892.

Table 1. The trading performance

Ratio	Standard Algorithm				Algorithm 1			Algorithm 2			Algorithm 3			Algorithm 4			B & H			
	Period 1	Period 2	Period 3	Period 4	Period 2	Period 3	Period 4	Period 2	Period 3	Period 4	Period 2	Period 3	Period 4	Period 2	Period 3	Period 4	Period 1	Period 2	Period 3	Period 4
Rate of return [Pips]	153	2020	3563	-136	2576	3676	-103	2283	3620	-98	2638	3489	-315	2736	3592	-116	-1336	2697	3454	-259
Number of transactions	56	65	105	44	83	126	56	78	123	47	103	138	67	98	112	48	1	1	1	1
Gross profit [Pips]	78	158	163	144	186	174	168	164	189	148	179	185	192	193	239	151	0	2697	3454	0
Gross loss [Pips]	-93	-110	-154	-236	-129	-103	-207	-132	-147	-117	-127	-154	-147	-147	-175	-138	-1336	0	0	-259
Number of profitable transactions	38	47	28	23	59	64	39	52	68	28	68	75	37	69	74	23	0	1	1	0
Number of profitable consecutive transactions	11	6	5	5	9	13	8	7	10	6	14	13	4	16	17	6	0	1	1	0
Number of unprofitable consecutive transactions	2	2	1	4	2	3	3	2	3	2	3	4	3	2	3	4	1	0	0	1
Sharpe ratio	0.68	0.52	0.46	0.39	0.36	0.32	0.34	0.34	0.29	0.37	0.33	0.21	0.34	0.48	0.40	0.27	0	0	0	0
Average coefficient of volatility	0.27	0.34	0.27	0.15	0.42	0.35	0.27	0.38	0.42	0.29	0.29	0.48	0.42	0.25	0.28	0.41	0	0	0	0
Average rate of return per transaction	2.73	31.07	33.93	-3.09	31.04	29.17	-1.84	29.27	29.43	-2.09	25.61	25.28	-4.70	27.91	32.07	-2.41	-1336	2697	3454	-259
Value of evaluation function (y)	0.62	0.24	0.41	0.29	0.41	0.37	0.41	0.32	0.39	0.43	0.38	0.36	0.30	0.46	0.42	0.39	0.12	0.19	0.21	0.04

6   Conclusions

We presented four different heuristic algorithms for the problem, built on top of a variety of well-known tools. Some results of the experimental comparison were presented. It should be noted, that in the case of High-Frequency Trading the computational complexity of agents' selection algorithms should as low as possible. The algorithms considered in this paper allow for achieving better performance than the consensus determined without agents' selection. The total rate of return gathered by some of our algorithms is higher than the benchmark strategy. However, they generate a larger number of transactions, which may entail a higher transaction cost. Therefore, the agents' evaluation based only on historical data is not sufficient.

The problem of the evaluation function definition is still open. The linear function is not the best solution from a scientific point of view, but it is easy to use from the investors' point of view. We want to address this issue in our upcoming research.

References

1. Barbosa, R.P., Belo, O.: Multi-agent forex trading system. In: Hãkansson, A., Hartung, R., Nguyen, N.T. (eds.) Agent and Multi-agent Technology for Internet and Enterprise Systems, pp 91–118. Studies in Computational Intelligence, vol. 289. Springer, Heidelberg (2010). https://doi.org/10.1007/978-3-642-13526-2_5
2. Bohm, V., Wenzelburger, J.: On the performance of efficient portfolios. J. Econ. Dyn. Control **29**(4), 721–740 (2005)
3. Ivanović, M., Vidaković, M., Budimac, Z., Mitrović, D.: A scalable distributed architecture for client and server-side software agents. Vietnam J. Comput. Sci. **4**(2), 127–137 (2017)
4. Khosravi, H., Shiri, M.E., Khosravi, H., Iranmanesh, E., Davoodi, A.: TACtic-a multi behavioral agent for trading agent competition. In: Sarbazi-Azad, H., Parhami, B., Miremadi, S.-G., Hessabi, S. (eds.) CSICC 2008. CCIS, vol. 6, pp. 811–815. Springer, Heidelberg (2008). https://doi.org/10.1007/978-3-540-89985-3_109
5. Korczak, J., Hernes, M., Bac, M.: Risk avoiding strategy in multi-agent trading system. In: Proceedings of Federated Conference Computer Science and Information Systems (FedCSIS), Kraków, pp. 1131–1138 (2013)
6. Korczak, J., Hernes, M., Bac, M.: Collective intelligence supporting trading decisions on FOREX market. In: Nguyen, N.T., Papadopoulos, G.A., Jędrzejowicz, P., Trawiński, B., Vossen, G. (eds.) ICCCI 2017. LNCS (LNAI), vol. 10448, pp. 113–122. Springer, Cham (2017). https://doi.org/10.1007/978-3-319-67074-4_12
7. Kozierkiewicz-Hetmańska, A., Pietranik, M.: The knowledge increase estimation framework for ontology integration on the concept level. J. Intell. Fuzzy Syst. **32**, 1–12 (2016)
8. Sycara, K.P., Decker, K., Zeng, D.: Intelligent agents in portfolio management. In: Jennings, N., Wooldridge, M. (eds.) Agent Technology, pp. 267–282. Springer, Heidelberg (2002). https://doi.org/10.1007/978-3-662-03678-5_14
9. Tatikunta, R., Rahimi, S., Shrestha, P., Bjursel, J.: TrAgent: a multi-agent system for stock exchange. In: Proceedings of the 2006 IEEE/WIC/ACM International Conference on Web Intelligence and Intelligent Agent Technology (WI-IATW 2006), pp. 505–509. IEEE Computer Society, Washington, DC (2006)

Ratings vs. Reviews in Recommender Systems: A Case Study on the Amazon Movies Dataset

Maria Stratigi, Xiaozhou Li, Kostas Stefanidis[✉], and Zheying Zhang

Tampere University, Tampere, Finland
{maria.stratigi,xiaozhou.li,konstantinos.stefanidis,
zheying.zhang}@tuni.fi

Abstract. Together with the prevalence of e-commerce and online shopping, recommender systems have been playing an increasingly important role in people's daily lives in terms of discovering their potential preferences. Therein, users' preferences are mostly reflected by their online behaviors, specially their evaluation towards particular items, e.g., numeric ratings and textual reviews. Many existing recommender systems focus on using item ratings to determine users' preferences, while others provide approaches using textual reviews instead. In this work, via a case study on the Amazon movies data, we compare the recommendation results when using ratings or reviews, as well as that of combining both.

Keywords: Recommender systems · Ratings · Reviews

1 Introduction

Recommender systems facilitate the selection of data items by users by issuing recommendations for items they might like. In particular, they aim at providing suggestions to users by estimating their item preferences and recommending those items featuring the maximal predicted preference. Typically, recommendation approaches are classified as content-based and collaborative filtering approaches. In content-based approaches, information about the features/content of the items is processed, and the system recommends items with features similar to items a user likes. In collaborative filtering approaches, we produce interesting suggestions for a user by exploiting the taste of other similar users.

Nowadays, recommendations have more broad applications, beyond products, like news recommendations [9], links recommendations [13,15], and more innovative ones like query recommendations [3], health recommendations [14], open source software recommendations [7] and diverse venue recommendations [4]. For achieving efficiency, there are approaches that build user models for computing recommendations. For example, [10] applies subspace clustering to organize users into clusters and employs these clusters, instead of a linear scan of the database, for making predictions.

© Springer Nature Switzerland AG 2019
T. Welzer et al. (Eds.): ADBIS 2019, CCIS 1064, pp. 68–76, 2019.
https://doi.org/10.1007/978-3-030-30278-8_9

In this study, we investigate the effectiveness of using the sentiment analysis of users' textual reviews as the rating of the users for the target items [8]. Comparing to the traditional collaborative filtering approach, when calculating the similarity between users and the relevance of them towards items, we use the ratings obtained based on the sentiment scores of their textual reviews. The effectiveness of this textual review sentiment-based recommender system is evaluated via a case study on the Amazon movie review dataset. According to the findings, using sentiment score-based rating mechanism can provide more reasonable numeric score for the target items and therefore a more intuitive view of the item quality. In addition, the effectiveness of the recommender system based on sentiment rating is as high as that with regular numeric ratings.

The remainder of the article is organized as follows. Section 2 introduces the recommendation model used in this study. Section 3 presents the method of processing textual reviews towards producing sentiment scores. Section 4 provides the general information regarding the data used in the case study. Section 5 evaluate the recommender system of sentiment score by comparing to that with regular numeric ratings based on the previously introduced dataset. Section 6 concludes the article with a summary of our contributions.

2   The Recommendation Model

Assume a recommender system, where I is the set of data items to be rated and U is the set of users in the system. A user $u \in U$ might rate an item $i \in I$ with a score $p(u, i)$, i.e., $p(u, i)$ reflects the numeric rating (range in $[1, 5]$) u gave directly to i. Comparatively, $p(u, i)$ can also reflect the user's evaluation for an item via a textual review (more in Sect. 3). Let R be the set of all ratings recorded in the system. Typically, the cardinality of I is high and users rate only a few items. The subset of users that rated an item $i \in I$ is denoted by $U(i)$, whereas the subset of items rated by a user $u \in U$ is denoted by $I(u)$.

For the items unrated by the users, recommender systems estimate a relevance score, denoted as $r(u, i)$, $u \in U$, $i \in I$. There are different ways to estimate the relevance score of an item for a user. In the content-based approach (e.g., [11]), the estimation of the rating of an item is based on the ratings that the user has assigned to similar items, whereas in collaborative filtering systems (e.g., [12]), this rating is predicted using previous ratings of the item by similar users. In this work, we follow the collaborative filtering approach. First, similar users are located via a similarity function that evaluates the proximity between two users. Then, items relevance scores are computed for users taking into account their most similar users. For computing the similarities between user u and u', denoted as $s(u, u')$, we use the Pearson Correlation, that is defined as follows:

$$s(u, u') = \frac{\sum_{i \in X}(p(u, i) - \mu_u)(p(u', i) - \mu_{u'})}{\sqrt{\sum_{i \in X}(p(u, i) - \mu_u)^2}\sqrt{\sum_{i \in X}(p(u', i) - \mu_{u'})^2}} \quad (1)$$

where $X = I(u) \cap I(u')$ and μ_u is the mean of the ratings in $I(u)$, i.e., the mean of the ratings for user u. Given a user u and the group of users that are considered similar to him/her, P_u, if u has expressed no preference for an item i, the relevance of i for u, denoted as $r(u,i)$, is estimated as:

$$r(u,i) = \frac{\sum_{u' \in (P_u \cap U(i))} s(u,u')p(u',i)}{\sum_{u' \in (P_u \cap U(i))} s(u,u')} \tag{2}$$

After estimating the relevance scores of all unrated user items for a user u, namely A_u, the items A_u^k with the top-k relevance scores are suggested to u.

3 Processing Textual Reviews

3.1 Sentiment Analysis with VADER

To assign a sentiment score to each textual review, we adopt a robust tool for sentiment strength detection on social web data, namely the Valence Aware Dictionary for sEntiment Reasoning (VADER) approach [5]. Compared with other sentiment analysis tools, VADER has a number of advantages. Firstly, the classification accuracy of VADER on sentiment towards positive, negative and neutral classes is even higher than individual human raters in social media. In addition, its overall classification accuracy on product reviews from Amazon, movie reviews, and editorials from NYTimes also outperform other sentiment analysis approaches, such as SenticNet [2], SentiWordNet [1], and Word-Sense Disambiguation [6], and run closely with the accuracy of individual human.

As any text can be seen as a list of words, the approach first selects a lexicon that will determine the sentiment score of each word in the given list. The lexicon for sentiment analysis is a list of words used in English language, each of which is assigned with a sentiment value in terms of its sentiment valence (intensity) and polarity (positive/negative). To determine the sentiment of words, we assign a rational value within a range to a word. For example, if the word "okay" has a positive valence value of 0.9, the word "good" must have a higher positive value, e.g., 1.9, and the word "great" has even higher value, e.g., 3.1. Furthermore, the lexicon set shall include social media terms, such as Western-style emoticons (e.g., :-)), sentiment-related acronyms and initialisms (e.g., LOL, WTF), and commonly used slang with sentiment value (e.g., nah, meh).

With the well-established lexicon, and a selected set of proper grammatical and syntactical heuristics, we to determine the overall sentiment score of a review, which is in the range of $[-1, 1]$ with VADER. The grammatical and syntactical heuristics are seen as the cues to change the sentiment of word sets. Therein, punctuation, capitalization, degree modifier, and contrastive conjunctions are taken into account. For example, the sentiment of "The book is EXTREMELY AWESOME!!!" is stronger than "The book is extremely awesome", which is stronger than "The book is very good.". With both the lexicon value for each word of the review, and the calculation based on the grammatical and syntactical heuristics, we can then assign unique sentiment values to each review.

3.2 Sentiment-Based Ratings

In order to analyze the relation between the users' rating and their text review sentiments, we use the VADER approach to calculate the corresponding sentiment score for each user review text. Due to the fact that VADER is a lexicon-based method which is more suitable for sentence-level sentiment analysis [5], we firstly divide each review text r_i (i.e., the review of user u to item i) into k sentences $s_{i,1}, s_{i,2}, ... s_{i,k}$ using the tokenize function of NLTK. Then the sentiment score of each sentence (i.e., $sent(s_{i,j})$) is calculated accordingly. Therefore, the overall sentiment of review text r_i from user u (i.e., $sentO(u, r_i)$) is calculated by the mean of the sentiment scores of all its review sentences.

$$sentO(u, r_i) = \frac{\sum_{j=1}^{k} sent(s_{i,j})}{k}, \quad where \quad r_i = \{s_{i,1}, s_{i,2}, ... s_{i,j} ... s_{i,k}\} \quad (3)$$

To ease the comparison, we further transform the obtained sentiment score $sentO(s_{i,j})$ into quintiles (i.e., sentiment-based ratings), corresponding to the original numeric 5-star ratings with the following equation.

$$p(u, i) = \begin{cases} 1, & \text{if } sentO(u, r_i) \geq -1 \text{ and } sentO(u, r_i) < -0.6 \\ 2, & \text{if } sentO(u, r_i) \geq -0.6 \text{ and } sentO(u, r_i) < -0,2 \\ 3, & \text{if } sentO(u, r_i) \geq -0.2 \text{ and } sentO(u, r_i) < 0,2 \\ 4, & \text{if } sentO(u, r_i) \geq 0.2 \text{ and } sentO(u, r_i) < 0,6 \\ 5, & \text{if } sentO(u, r_i) \geq 0.6 \end{cases} \quad (4)$$

By doing so, for each textual review given by a user u, we obtain the new ratings based on the sentiment score of the texts, namely, $p(u, i)$.

4 The Amazon Movies Reviews Dataset

In this case study, we use the Amazon movies reviews dataset[1]. The dataset spans a period between August of 1997 and October of 2012. It consists of 889.176 users and 253.059 movies. The users have given 7.911.684 reviews, and their median word count is 101. Each entry in the dataset consists of a user identification number, the movie id that the user reviews, the plain text review and the rating that the user ultimately gave to the product.

To better determine the sentiment score from a user review, we eliminate all entries that the reviews were composed with less than 101 words. After this step the dataset is reduced to 487.134 users, 211.903 movies and 4.086.968 reviews. Also, we eliminate all movies with less than 20 reviews. This resulted in our final dataset that contained 320.451 users, 46.421 movies and 3.301.125 reviews. The distribution of the ratings per user follows a power law distribution. To better demonstrate that, in Fig. 1, we show the 600 users with the most ratings given.

---

[1] https://snap.stanford.edu/data/web-Movies.html.

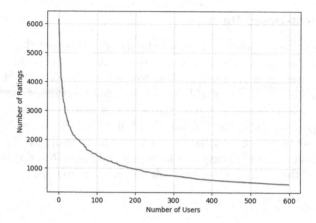

Fig. 1. The distribution of ratings for the 600 users with the most ratings given.

All other users belong in the long tail of the power law distribution with less than 1000 ratings per user.

Figure 2(a) presents the distribution of the original numeric ratings, while Fig. 2(b) presents that of the sentiment-based ratings. Figure 2(a) shows that the majority of the users give 5 stars to the movies on Amazon.com, while the number of 4-star ratings or below is much lower. Comparatively, from Fig. 2(b), we could observe that the majority of the users mean to give 3 or 4 star ratings according to the sentiment of their textual reviews. It is rare that users choose to give completely positive or negative reviews. Figure 3 visualizes the number of user reviews along with their difference between numerical ratings and sentiment scores.

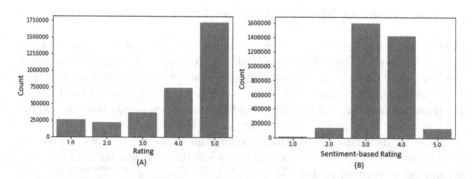

Fig. 2. The distribution of (a) ratings, and (b) sentiment-based ratings.

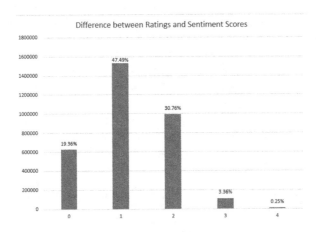

Fig. 3. The difference between the ratings and sentiment scores.

5   Evaluation

To calculate the MAE and RMSE errors, we hidden k items from $I(u)$ for 20 different users. Then applied the recommender algorithm twice and tried to predict them. The fist time we utilized the ratings and the second the sentiment scores. Finally, for each different input dataset we average the errors over all users. Figure 4 shows the results for $k = 10, 20, 30, 40, 50$.

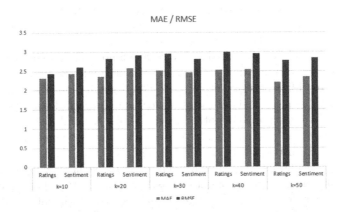

Fig. 4. MAE and RMSE errors for the rating and sentiment scores, calculated for different numbers of hidden items.

To calculate the distance between the recommendation lists produced by the ratings and the sentiment scores, we applied the recommendation algorithm twice on 20 different users. Afterwards, we calculated the distance between these lists for each user and finally we averaged them.

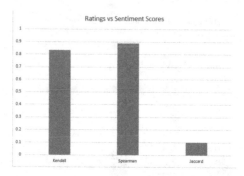

Fig. 5. Kendall, Spearman and Jaccard for the 20 top users. Comparisons between ratings and sentiment scores.

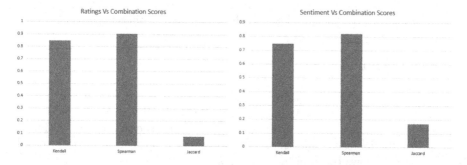

Fig. 6. Kendall, Spearman and Jaccard for the 20 top users. Comparisons between ratings and combination scores (left), and sentiment and combination scores (right).

Figures 5 and 6 shows the calculated distances among the obtained sentiment-based ratings, the combination ratings and the original user-provided ratings. We apply three different ways of distance-based evaluation, including Spearman's footrule distance, Kendall tau distance, and Jaccard distance. From the shown figures, we observe that both Spearman and Kendall distances, for any two recommendation lists, are significant while the Jaccard similarity of such is relevantly low. Such result interestingly demostrates that the recommendation results for a particular user based on the sentiment of his/her textual reviews can be largely different from that based on his/her numeric ratings. Furthermore, averaging the results from both numeric ratings and sentiment-based ratings shall lead to new recommendation lists which are different from either of the results from the two. Without proper evaluation, this study does not indicate any recommendation result is more accurate and preferred than the other two. It instead presents that the preference shown from users' textual reviews towards particular items can be different from the ratings they give to them, which results in largely different recommendation results.

6   Summary

In this study[2], we provide a pilot investigation on the recommendation results using sentiment analysis on users textual reviews. Compared with the recommendation based on users' numeric ratings, as well as that of the combination of both, we find that the similarity among such results are very low. The results of the case study on Amazon users' ratings and reviews for movie items suggests the possibility of users' potential inconsistency in showing preferences with numeric ratings and textual reviews. In our future work, we will evaluate the obtained recommendation results and further investigate the usefulness of applying review sentiment analysis towards recommendation systems.

References

1. Baccianella, S., Esuli, A., Sebastiani, F.: Sentiwordnet 3.0: an enhanced lexical resource for sentiment analysis and opinion mining. In: LREC (2010)
2. Cambria, E., Speer, R., Havasi, C., Hussain, A.: Senticnet: a publicly available semantic resource for opinion mining. In: Commonsense Knowledge (2010)
3. Eirinaki, M., Abraham, S., Polyzotis, N., Shaikh, N.: Querie: collaborative database exploration. IEEE Trans. Knowl. Data Eng. **26**(7), 1778–1790 (2014)
4. Ge, X., Chrysanthis, P.K., Pelechrinis, K.: MPG: not so random exploration of a city. In: MDM (2016)
5. Hutto, C.J., Gilbert, E.: VADER: a parsimonious rule-based model for sentiment analysis of social media text. In: ICWSM (2014)
6. Júnior, E.A.C., de Andrade Lopes, A., Amancio, D.R.: Word sense disambiguation: a complex network approach. Inf. Sci. **442–443**, 103–113 (2018)
7. Koskela, M., Simola, I., Stefanidis, K.: Open source software recommendations using github. In: Méndez, E., Crestani, F., Ribeiro, C., David, G., Lopes, J.C. (eds.) TPDL 2018. LNCS, vol. 11057, pp. 279–285. Springer, Cham (2018). https://doi.org/10.1007/978-3-030-00066-0_24
8. Li, X., Zhang, Z., Stefanidis, K.: Mobile app evolution analysis based on user reviews. In: SoMeT (2018)
9. Liu, J., Dolan, P., Pedersen, E.R.: Personalized news recommendation based on click behavior. In: IUI (2010)
10. Ntoutsi, E., Stefanidis, K., Rausch, K., Kriegel, H.: Strength lies in differences: diversifying friends for recommendations through subspace clustering. In: CIKM (2014)
11. Pazzani, M.J., Billsus, D.: Content-based recommendation systems. In: Brusilovsky, P., Kobsa, A., Nejdl, W. (eds.) The Adaptive Web. LNCS, vol. 4321, pp. 325–341. Springer, Heidelberg (2007). https://doi.org/10.1007/978-3-540-72079-9_10
12. Sandvig, J.J., Mobasher, B., Burke, R.D.: A survey of collaborative recommendation and the robustness of model-based algorithms. IEEE Data Eng. Bull. **31**(2), 3–13 (2008)

---

[2] This work has been partially supported by the Virpa D project funded by Business Finland.

13. Stefanidis, K., Ntoutsi, E., Kondylakis, H., Velegrakis, Y.: Social-based collaborative filtering. In: Encyclopedia of Social Network Analysis and Mining, 2nd edn (2018)
14. Stratigi, M., Kondylakis, H., Stefanidis, K.: FairGRecs: fair group recommendations by exploiting personal health information. In: Hartmann, S., Ma, H., Hameurlain, A., Pernul, G., Wagner, R.R. (eds.) DEXA 2018. LNCS, vol. 11030, pp. 147–155. Springer, Cham (2018). https://doi.org/10.1007/978-3-319-98812-2_11
15. Yin, Z., Gupta, M., Weninger, T., Han, J.: LINKREC: a unified framework for link recommendation with user attributes and graph structure. In: WWW (2010)

Solution Pattern for Anomaly Detection in Financial Data Streams

Maciej Zakrzewicz[1]([⊠]), Marek Wojciechowski[1],
and Paweł Gławiński[2]

[1] Poznan University of Technology, Poznan, Poland
{maciej.zakrzewicz,
marek.wojciechowski}@cs.put.poznan.pl
[2] Softman SA, Piaseczno, Poland
pawel.glawinski@softman.pl

Abstract. Anomaly detection in versatile financial data streams is a vital business problem. Existing IT solutions for business anomaly detection usually rely on explicit Complex Event Processing or near-real time Business Activity Monitoring. In this paper we argue that business anomaly detection should be considered an implicit infrastructural BPM service and we propose a corresponding Solution Pattern. We describe how a Business Anomaly Detector can be architectured and designed in order to handle fast dynamic streams of business objects in BPM environments. The presented solution has been practically verified in Oracle SOA/BPM Suite environment which handled real-life financial controlling business processes.

Keywords: Service oriented systems · Design patterns · Risk management

1 Introduction

Business anomaly detection is a technique used to identify business items or events which do not conform to valid data patterns. Typical business anomalies include credit card frauds, purchase card frauds, telecommunication subscription fraud, phone call fraud, financial reporting fraud, insurance fraud, fraudulent claims for health care, credit applications fraud, credit transactional fraud, etc. Since fraud is a multi-million dollar business, efficient IT solutions are demanded by the business sector to timely detect anomalous/fraudulent activities attempted by performers of business processes fed with streams of complex data. Anomalies in business process execution may exhibit themselves in different manners, e.g., as an unusual process object state, as an unusual process execution path, as an unusual performer-to-activity assignment.

For a long time, SOA-based Business Process Management Systems (BPMS) have been used to design and execute services (typically SOAP Web Services) orchestrated into complex business processes performed by humans and applications. However, despite the maturity of business process modeling techniques and efficiency of execution platforms, effective monitoring of the flow of such business processes still lacks usability and flexibility, especially in the area of detecting anomalous business behavior, e.g., representing fraudulent actions. Existing solutions rely either on explicit

© Springer Nature Switzerland AG 2019
T. Welzer et al. (Eds.): ADBIS 2019, CCIS 1064, pp. 77–84, 2019.
https://doi.org/10.1007/978-3-030-30278-8_10

calls to rule evaluation systems (Complex Event Processing) in order to validate business data, engage external tools for near-real time business reporting (Business Activity Monitoring) or simply assume that anomaly detection is outside functional requirements. We argue that monitoring capabilities of BPMSs should be expanded with functions to automatically monitor every single business process instance to detect anomalous activities and report their findings to other components/processes.

Architectural, Design, and Solution Patterns are known as generalized, formalized descriptions of reusable solutions to common problem classes within a given context, supposed to transfer knowledge about successful designs and implementations. Examples of patterns that may partially support business anomaly detection include Complex Event Processing (CEP) Design Pattern and Business Activity Monitoring (BAM) Design Pattern. Unfortunately, the diverse nature of business processes and their business data objects make it challenging to develop a universal CEP or BAM framework for business anomaly detection.

In this paper we describe our proposal for the Business Anomaly Detection Solution Pattern and we present its successful implementation in the form of an asynchronous Java EE service which can be easily injected into existing Business Process Management (BPM) environments, allowing business processes to benefit from automated detection of anomalous behavior.

The remainder of this paper is organized as follows. Section 2 contains related work. In Sect. 3 our Business Anomaly Detector Solution Pattern is characterized. Section 4 describes architecture and design of a real-life Business Anomaly Detector. Section 5 summarizes results of selected experimental validation tests that we have conducted to justify our design decisions. Conclusions are presented in Sect. 6.

2 Related Work

General anomaly detection methods have been covered by numerous papers and surveys. Statistical outlier detection techniques have been described in [3]. Machine learning anomaly detection methods have been surveyed in [9]. A review of anomaly detection techniques for numerical and symbolic data has been provided in [2].

SOA best practices and technologies have been covered in [7]. Surveys on Web Service composition methods can be found in [8]. Implementation best practices and messaging patterns for SOA have been described in [5]. In [6] the authors discussed Business Activity Monitoring functional requirements and applications. Patterns for business processes have been extensively studied in [1]. In [10] the motivation for design patterns for Complex Event Processing has been presented and a foundation for them has been provided.

3 Problem Definition

In this section we will follow a usual three-parted scheme [4] to describe our Solution Pattern.

3.1   Context: SOA, BPM, Dynamic Data Streams

The background is a SOA-based Business Process Management environment, where complex business processes orchestrate both software services and user tasks. The business processes are fed with data entered by users as well as by data stream sources like sensors, POS terminals, automatic document feeder scanners, IoT devices, etc. Software services are invoked either directly or through an Enterprise Service Bus (ESB).

3.2   Problem: Anomaly Detection Based on Rules and Learning Models

Automatic identification of business data objects or events which do not conform to patterns. The patterns can be static in nature (defined by an operator) or dynamically discovered (by using machine learning methods). The business data objects/events to be analyzed can be any business process objects created, retrieved or transmitted by business process tasks. The business data objects/events will be delivered explicitly or implicitly intercepted in-flow during a business process execution. The solution should be platform-agnostic to integrate with various BPM environments.

3.3   Solution: Infrastructural Service

A new architectural component of BPMS, a form of an infrastructural service which performs on-line monitoring of business data objects being transmitted between activities of a business process. The captured business data objects are validated against the static and discovered patterns in order to detect anomalies. When an anomaly is detected, a BPM message or signal is generated to notify other processes or applications about the finding (Fig. 1).

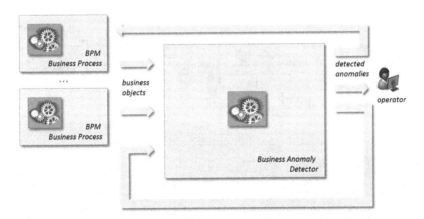

Fig. 1. Business Anomaly Detector as an infrastructural service

We have successfully developed a prototype of a Business Anomaly Detector based on the Java EE platform, using SOAP Web Service messaging, JMS/Kafka message buffering, Drools Rule Management System, Weka Machine Learning, and ESB message interception. The prototype has been validated in an Oracle SOA/BPM Suite environment which handled real-life financial controlling business processes.

4 Software Architecture and Design

4.1 Overview

An overview of the architecture of the Business Anomaly Detector is shown in Fig. 2. Business objects are delivered to SOAP Web Service interfaces as XML documents. JAXB converts the XML documents into Java objects, which are sent to a throttling JMS queue or Kafka topic. The objects in the queue/topic are periodically propagated by the Controller to the embedded JBoss Drools rule engine, which then executes business anomaly detection rules on the received objects. The business anomaly detection rules are designed by a business user using a visual rule editor (part of our solution). The rules can be based on the expert's knowledge or rely on statistical models obtained through machine learning. The models are learnt using Weka algorithms integrated into the Business Anomaly Detector. When anomalies are detected, new business objects are created and delivered to an output JMS queue or Kafka topic. The Dispatcher splits the queued objects into classes and delivers them to external consumers (SOAP Web Services) based on defined allocation schemes. All management tasks are handled by a Web-based administration console. A database repository is used as a persistence store to protect the state of the Business Activity Detector in case of failures or planned unavailability. MongoDB has been selected for that purpose in our implementation due to its permissive license, simple but adequate data model, and small write overhead.

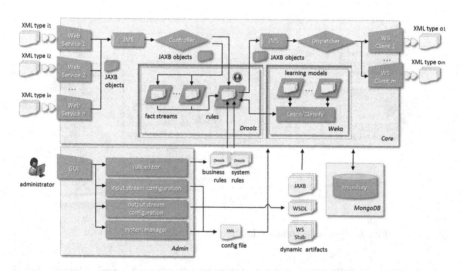

Fig. 2. Business Anomaly Detector software architecture

Business objects need to be delivered to the Business Anomaly Detector in order to be processed using static or dynamically discovered anomaly detection rules. Several scenarios can be considered for this action: (1) a business process can explicitly invoke the Business Anomaly Detector, providing objects to be analyzed, (2) a service call can be intercepted on ESB level and its object can be delivered to the Business Anomaly Detector, (3) an external application (e.g., a database table trigger, a network firewall) can invoke the Business Anomaly Detector and pass business object data.

Whenever an anomaly has been detected, a new business object is generated asynchronously. The object can be then consumed by an event-based business process to perform standard anomaly-related activities (e.g., notifications and alerting) or it can be processed by an existing Business Activity Monitoring tool in order to visualize the findings to the operator.

4.2 Input and Output Configuration

Business Anomaly Detector receives input data in the form of streams of XML documents. Each input stream corresponds to a particular class of analyzed business objects, e.g., invoices, bank transfers, credit card payments, etc. In order to define a new input stream a user has to upload an XML Schema document describing the structure of XML documents that will form the stream. Based on the uploaded XML Schema, the system will automatically complete the process of input stream definition by performing the following steps: (1) generation and compilation of JAXB classes, (2) generation of a SOAP Web Service with one operation, called "Receive", (3) deployment of the created Web Service. The automatically generated code of the Web Services' "Receive" operation forwards the JAXB object corresponding to the received XML document to the input JMS queue or Kafka topic.

The output of Business Anomaly Detector is a collection of streams of XML documents representing discovered anomalies, sent to registered external consumers. It is assumed that a consumer provides a SOAP Web Service to receive XML messages from Business Anomaly Detector. When defining an output stream, a user has to upload an XML Schema describing the outgoing XML documents and provide parameters identifying the consuming SOAP Web Service and its operation. As part of the output stream definition process, the system will automatically perform the following steps: (1) generation and compilation of JAXB classes, (2) adding a record to the system configuration associating the generated JAXB class with the external Web Service.

4.3 Rule-Based Anomaly Detection

The heart of the Business Anomaly Detector is its rule engine. As this crucial component we incorporated Drools since it was the only open source rule engine with a permissive license that supported all types of rules included in our functional requirements. Drools is implemented using Java, has a well-defined API conforming to the JSR-94 specification, and therefore seamlessly integrates with Java EE applications.

Business rules in Drools are expressed in a textual format called DRL and have a WHEN-THEN syntax. Coding business rules directly in DRL can certainly be beneficial for advanced users due to its flexibility and expressive power. However, relying on free-text rule editing only could make the system too difficult to use for novice users, especially those without a technical background. Based on this observation, we developed a GUI to define rules of low or moderate complexity in the form of an interactive dialog window, still allowing advanced users to switch into the direct DRL editing mode.

Four types of rules are supported by the Business Anomaly Detector and handled by its Drools rule engine component:

1. Simple rules based on the current business object only. Example: When a credit card payment exceeds 1500 EUR, then raise an alert.
2. Aggregation rules based on moving window aggregates calculated from collections of business objects received in the past. Example: For each new bank transfer calculate the number of bank transfers within the last 24 h for the given customer and when this number is greater by at least 50% than the average daily number of bank transfers for this customer from the last 30 days, then raise an alert.
3. Calendar rules based on schedules to validate business objects received recently. Example: Every day at 23:59 calculate a balance for each customer for the time window of the last 30 days and when the balance is below -10000 EUR, then raise an alert.
4. Learning model rules based on patterns learnt from business objects received recently. Example: When the actual decision for a loan application is different than predicted by the statistical model, then raise an alert.

For the rules of type 2 we had to extend the out-of-the box Drools engine with aggregate materialization functionality to achieve satisfactory performance of the system on large amounts of data (see Sect. 5). For the rule of type 4 we integrated the Weka data mining library to build prediction models and implemented a solution for keeping the prediction models up to date (see Sect. 4.4).

4.4 Discovery-Based Anomaly Detection

Data mining subsystem within our Business Anomaly Detector serves two purposes: (1) generates statistical models based on event history from a defined time window, (2) based on the built statistical model, discovers anomalies in incoming new business events. The system reacts to discovered anomalies (e.g., by raising an alert) according to defined business rules.

Despite the fact that the input to our system can be regarded as a data stream, we decided against using a dedicated data stream mining library due to immaturity of available solutions. Instead, we incorporated Weka, which is a well-established and flexible library. From the collection of algorithms offered by Weka for anomaly prediction we selected two classification algorithms: J48 decision trees and Naïve Bayes. To keep the anomaly detection models up to date we periodically rerun the prediction algorithms on recent data according to a user-defined time window. We also enable operators to define derived or calculated facts/attributes in order to provide for anomaly detection in correlated parallel data streams.

As building prediction model takes significant time, our system performs this task in the background. Each statistical model is learnt in one separated execution thread. When a new prediction model is ready, it is put into production, and immediately afterwards the new model is being learnt from the most recent time window including the data that came during the previous model learning cycle. Thus, each statistical model exists in two versions: the completely learnt Foreground Model which is used for scoring incoming data and the Background Model being learnt that will become the new Foreground Model when completed. Expired statistical models (ones replaced by new Foreground Models) can be archived (kept in the MongoDB repository) for the sake of future analytics or auditing.

The detailed procedure of machine learning and prediction within Business Anomaly Detector consists of the following steps: (1) A new business event object is retrieved from the input JMS queue or Kafka topic. (2) The business event object is placed in two Drools streams. The first stream is temporary, i.e., the object is removed from it after being processed by all defined business rules. The second stream maintains an event time window on which statistical models are to be built. Objects are removed from that stream when they fall out of the time window. (3) Drools business rules are called. (4) Business rules apply predictive functions based on statistical models to the business object.

5 Testing and Evaluation

In order to validate the proposed solution pattern and to verify its architectural and design approaches, we have implemented the Business Anomaly Detector in Oracle SOA/BPM Suite environment which handled real-life financial controlling business processes. The prototype implementation served as a proof of concept and was used as a testbed to verify the effect of aggregate materialization on the performance of aggregation rules processing.

The impact of our aggregate materialization techniques has been experimentally verified using a 4-CPU, 16 GB RAM Linux machine. Our test business anomaly detection rule alerted when an invoice exceeded the 30-day moving average for a customer. The number of customers was set to 100, the daily number of invoices per a customer varied between 1 and 1000 (uniform distribution). The performance metric was the number of invoices processed per second. Throughout the experiments aggregate materialization resulted in almost constant throughput of 10000 invoices processed per second. With materialization turned off, the performance degraded to an unacceptable level (i.e., the system unable to process invoices in real time) already around 100 invoices per a customer, which clearly showed that aggregate materialization is a must for aggregation-based rules.

6 Summary

We have proposed a new solution pattern called Business Anomaly Detector and provided good practices for its implementation. The Business Anomaly Detector can be perceived as an infrastructural service, intercepting (explicitly or implicitly) business objects from SOA BPM business process flows in order to detect anomalous behavior. The good practices include asynchronous architecture, four types of anomaly detection rules, aggregate materialization, and off-line learning of discovery-based business rules. A prototype system has been implemented according to the proposed solution pattern and has been validated in a real-life environment.

The current deployment architecture is based on Java EE, JMS/Kafka, SOAP technology stack. Our future work will focus on development of alternative integration interfaces and runtime platforms, including application containerization and RESTful Web Services (with business events represented in JSON).

References

1. Van der Aalst, W., Hofstede, A., Kiepuszewski, B., Barros, A.: Workflow patterns. Distrib. Parallel Databases **14**(1), 5–51 (2003)
2. Agyemang, M., Barker, K., Alhajj, R.: A comprehensive survey of numeric and symbolic outlier mining techniques. Intell. Data Anal. **10**(6), 521–538 (2006)
3. Barnett, V., Lewis, T.: Outliers in Statistical Data, 3rd edn. Wiley, Hoboken (1994)
4. Buschmann, F., Meunier, R., Rohnert, H., Sommerlad, P., Stal, M.: Pattern-Oriented Software Architecture Volume 1: A System of Patterns. Wiley, Hoboken (1996)
5. Chaterjee, S.: Messaging patterns in service-oriented architecture (parts 1 and 2). Microsoft Archit. J. (2, 3) (2004)
6. DeFee, J., Harmon, P.: Business activity monitoring and simulation. In: Fischer, L. (ed.) Workflow Handbook, pp. 53–74. Future Strategies Inc., Lighthouse Point (2005)
7. Dodani, M.: Where's the SOA Beef? J. Object Technol. **3**(10), 41–46 (2004)
8. Dustdar, S., Schreiner, W.: A survey on web services composition. Int. J. Web Grid Serv. **1**(1), 1–30 (2005)
9. Hodge, V., Austin, J.: A survey of outlier detection methodologies. Artif. Intell. Rev. **22**(2), 85–126 (2004)
10. Paschke, A.: Design patterns for complex event processing. In: Proceedings from Distributed Event-Based Systems Symposium (2008)

Exploring Pattern Mining for Solving the Ontology Matching Problem

Hiba Belhadi[1(✉)], Karima Akli-Astouati[1], Youcef Djenouri[2], and Jerry Chun-Wei Lin[3]

[1] Department of Computer Science, USTHB, Algiers, Algeria
{hbelhadi,kakli}@usthb.dz
[2] Department of Computer and Information Sciences, NTNU, Trondheim, Norway
youcef.djenouri@ntnu.no
[3] Western Norway University of Applied Sciences, Bergen, Norway
jerrylin@ieee.org

Abstract. This paper deals with the ontology matching problem, and proposes a pattern mining approach that exploits the different correlation and dependencies between the different properties to select the most appropriate features for the matching process. The proposed method first discovers the frequent patterns from the ontology database, and then find out the most relevant features using the patterns derived. To demonstrate the usefulness of the suggested method, several experiments have been carried out on the DBpedia ontology databases. The results show that our proposal outperforms the state-of-the-art ontology matching approaches in terms of both execution time and quality of the matching process.

Keywords: Semantic web · Ontology matching · Matching instances · Frequent patterns

1 Introduction

The ontology matching is the process to build a bridge between different instances that represent the same real world, by the set of instances, where each instance is characterized by different properties. It is applied in diverse fields such as biomedical data [15], and Natural Language Processing [9]. Trivial solutions for ontology matching compare each instance of the first ontology with each instance of the second ontology by taking into account all the properties of both ontologies. $n \times n' \times m \times m'$ comparisons are required to find the alignment, where n and n' are the numbers of instances, and m and m' are the corresponding numbers of the data properties of the first ontology and the second ontology, respectively. These solutions have polynomial complexity. However, for some high dimensional data like DBpedia ontology[1] containing 4,233,000 instances,

---

[1] http://wiki.dbpedia.org/Datasets.

© Springer Nature Switzerland AG 2019
T. Welzer et al. (Eds.): ADBIS 2019, CCIS 1064, pp. 85–93, 2019.
https://doi.org/10.1007/978-3-030-30278-8_11

and 2,795 different properties, 144×10^{18} possible comparisons are required for such solutions. As a result, the matching process became a high time consuming, and also it can be reduced on the alignment's quality performance.

Emergent solutions to the ontology matching problem attempt to improve the quality of the overall matching process, by exploiting the enumeration search space using partitioning algorithms [3], evolutionary algorithms [18] and using high performance computing [16]. However, the overall performance of these algorithms is still low when dealing with high dimensional data. Pattern mining is a data mining technique that finds frequently co-occurring items in a database, and accordingly provides relevant patterns useful in the decision making process. Pattern mining largely applies as preprocessing step for solving complex problems [4,5]. Motivated by the success of the pattern mining discovery process, and in order to improve the overall performance of the ontology matching problem on high dimensional data, this paper investigates the use of pattern mining in selecting the most relevant features for solving the ontology matching based instance problem. In the pattern mining related literature, several algorithms have been proposed such as the Apriori [1], Fpgrowth [8], and many others. These approaches are both time-consuming and memory-consuming, especially when dealing with a low minimum support value. Recently, SSFIM [7] was proposed to extract the frequent itemsets using only a single pass, and it was proven to be non-sensitive to the minimum support value. The experimental study reported in [6] reveals that the SSFIM outperforms the state-of-the-art pattern mining algorithms. Therefore, in this work, the SSFIM is adopted to study the correlations of the properties of the given ontologies. The main contributions of our work are threefold: (i) Propose a new framework called PMOM (Pattern Mining for Ontology Matching) which adopts the pattern mining techniques to solve the ontology matching problem. In this context, SSFIM [7] is applied to discover the frequent patterns of the given ontologies. (ii) Develop a new strategy based on the extracted patterns for selecting the most relevant properties of the given ontologies. This is realized by computing the probability of each property in the set of the derived patterns. (iii) An intensive experiments have been performed to demonstrate the usefulness of the PMOM framework. The results reveal that PMOM outperforms the state-of-the art ontology matching algorithms.

The rest of this paper is organized as follows. Section 2 reviews some existing works related to the ontology matching problem. Section 3 introduces the ontology matching based instance problem. Section 4 presents a detail explanation of the PMOM framework. The evaluation performance is sketched in Sect. 5. Section 6, concludes this paper with some perspectives for a future work.

2   Related Work

In the last decade, several works have been proposed for solving the ontology matching problem [12,14]. Wang et al. [10] developed a generic VMI approach, which reduces the number of similarity computations by introducing multiple indexes, and candidate selection rules. Li et al. [17] first combines multiple lexical matching strategies using a novel voting-based aggregation method, and then utilizes the structural information and the correspondences already found to discover additional ones. Hu et al. [13] proposed RiMOM, an iterative matching framework where the distinctive information is based on a blocking strategy to reduce the number of candidate instance pairs. It uses predicates and their distinctive object feature as a key of the index for the instances. It also employs a weighted exponential function based similarity aggregation approach to assure the high accuracy of instance matching. Niu et al. [11] developed EIFPS, a semi-supervised learning algorithm, to recursively refine the matching process by using rules extracted by the association rule mining approach. A small number of existing properties are used as seeds and the matching rules are treated as parameters for maximizing the precision. Cerón-Figueroa et al. [2] proposed LOM by presenting the power of homogeneity resources in e-learning context. The application of an original associative classifier to the problem of ontology matching is investigated, in order to extend and improve the available tools for online learning in the semantic way. Although, the ontology matching-based approaches perform well on small and medium ontology databases, they are inefficient, in terms of runtime performance and solution's quality, for large ontologies (i,e high number of instances), and high dimensional data (instances with high number of properties). To deal with these two challenging issues, in this work, we present a pattern mining-based approach that explores the discovered patterns to select the most relevant features for solving the ontology matching process. Before detailing on our proposal, the next section presents the ontology matching based instance problem.

3 Ontology Matching Based Instance Problem

The goal of the ontology matching problem is to determine the common features between two ontologies. The result of this process is to represent the alignment between these ontologies. For that, an ontology O described by a set of instances $I = \{I_1....I_m\}$. Each of which is defined by a set of attributes (data properties) $P = \{p_1....p_n\}$. The properties may have more than one value.

It exists many variants of ontology matching problem. We are interesting in this work to ontology matching based instance. This variant of matching considers the common instances between two ontologies, with considering that the instance could have the same values for some properties and could also have missing values for other properties. Note that the name of properties for each ontology can be different, this issue causes the alignment more difficult. Thus, all the values of the instances of the first ontology should be scanned and compared to all the values of the instances of the second ontology. Consequently, the aim of the ontology matching based instance problem is to find the same information represented differently.

The alignment's result depends to the method used in the matching process, for that each matching can result a different number of instances in alignment. For that, each resulted alignment is evaluated and compared to an alignment reference. Alignment reference represents an alignment that is suggested by the domain expert. The alignment reference contains all the common instances between the ontologies.

4　PMOM: Pattern Mining for Ontology Matching Based Instances

4.1　Overall Framework

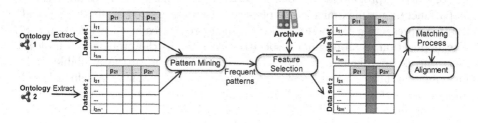

Fig. 1. PMOM framework

In this part, we present the main components of the proposed framework called PMOM (Pattern Mining for Ontology Matching based instances). The aim of PMOM is to improve the ontology matching based instance problem by taking into account the relevant features of the two ontologies to be aligned. This reducing allows on the one hand to boost the matching process for finding the common instances between two ontologies, on the other hand, it aims to improve the quality of the resulted alignment. PMOM is mainly composed into two steps: feature selection and matching process steps (See Fig. 1 for more details). The feature selection step is first performed to the set of attributes for each ontology, which results in an optimal subset of attributes that represents perfectly the two ontologies. This step is considered as a pre-processing step, (it will be executed only one time). To do so, an archive folder will be constructed for each ontology in the ontology base system. The matching process is then applied between the instances of the ontologies by taking only the attributes selected of the above step. The K-cross-validation model is used here, where at each pass of the algorithm, the training and the test alignments are performed. For the training matching process, the proposed model is learned to fix the best parameters. If the alignment rate exceeds the given threshold, then the test alignment is started. In this work, we are interested in the feature selection part, by proposing the pattern mining strategy for selecting the relevant attributes for the input ontologies. Any existing methods could be used for the ontology matching process.

Feature Selection. This strategy studies the correlation between the data properties of the ontologies to select the best features of the matching process. It is inspired by the pattern mining process [1] which is used to extract the most relevant data properties that cover the maximum number of possible instances. PMOM denotes the extraction of the relevant patterns that satisfies the minimum support constraint ($minsup$) from the transactions. Motivated by the success of SSFIM [6] in discovering the pattern mining problem, this work proposes the use of SSFIM in PMOM framework. The mining process is performed through two main steps: generation and extraction. In the generation step, starting from the first instance I_1, we refer $Pattern(I_1)$ by the set of all possible combinations of the literals of this instance. The result is added to the hash table H by creating an entry for each itemset in $Pattern(I_1)$. The frequency of each pattern is initialized by one in the hash table H. Then, the patterns of the second instance I_2 are generated for each pattern in $Pattern(I_2)$; if this pattern exists in H, then, its frequency is incremented by one; otherwise, a new entry is created with the frequency set to 1. This process is repeated until all instances I are processed. The second step aims to extract the frequent patterns (frequent literals in our case) from the hash table H. For that purpose, the support of each pattern t is computed as $\frac{h(t).freq}{|I|}$. If the support of t is greater than $minsup$, then t is called the frequent literal and added to the set of frequent literals S. Based on the frequent literals S discovered above, the appropriate set SP is selected. The probability $P(i, S)$ denotes the probability of the apparition of the ith property in the set of frequent literals S. A threshold μ which is between [0-1], is used to select the appropriate data properties. Indeed, for each property, if its probability value is greater than μ then, it is added to SP set.

Matching Process. After the selection of the appropriate properties, it is time to compare the instances of the basic ontology to the second one. In this part, we consider two ontologies, the basic ontology BO that will be matched with the second ontology O. $<P, I>$ is the set of property P and the instances I of the basic ontology. $<P', I'>$ is the set of property P' and the instances I' of the second ontology. P and P' are the set resulted by the feature selection described above. The iterative matching consists to scan the whole set of instances I of the basic ontology, and compares it with each instance in the set of instances I' of the second ontology. The comparison between two instances is established by checking each value in the ith instance in BO with all the values in the jth instance in O.

5 Performance and Evaluation

To validate the usefulness of the PMOM framework, extensive experiments have been carried out using well-known ontology matching databases. The approach has been implemented in Java[2] and experiments have been run on a desktop machine equipped with an Intel $I7$ processor and $16GB$ memory. DBpedia (See footnote 1), a well-known ontology database, is used in this experiment. It is a hub data that represent the Wikipedia knowledge and make this structured

---

[2] Code available at https://github.com/YousIA/PMOM.

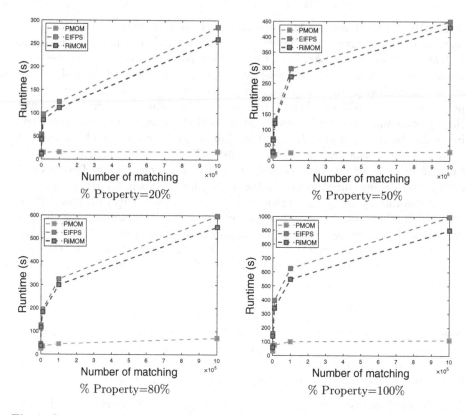

Fig. 2. Comparison of the runtime performance (in seconds) of the PMOM, the EIFPS, and the RiMOM using the DBpedia by varying the percentage of the data properties (%P) from 20% to 100%

information available on the Web. This database ontology contains 4,233,000 instances, and 2,795 different properties. The quality of the returned solutions is evaluated using recall, precision, and fmeasure, which are common measures for the evaluation of ontology matching methods. The experiment aims to compare PMOM with the state-of-the-art algorithms (EIFPS [11], and RiMOM [13]) using the DBPedia ontology database. Figure 2 shows the runtime of the three approaches at the percentage of data properties from 20% to 100%, considering all instances. When the number of matching varied from 100 to 1,000,000, PMOM outperforms the two other approaches. Moreover, the runtime of PMOM stabilized at a high number of data properties, where the two baseline approaches were highly time-consuming for a large number of instances and a large number of matchings. Thus, EIFPS and RiMOM need more than 900 seconds for dealing 1,000,000 matching in the whole DBpedia ontology database, whereas, it took only 111 seconds for PMOM. These results were obtained using the pre-processing step, where only the most relevant features were selected using the pattern mining approach. Table 1 compares the quality of matching of PMOM

Table 1. Comparison of Recall, Precision, and Fmeasure of PMOM, the EIFPS, and the RiMOM using the DBpedia by varying both the percentage of instances (%I) and the percentage of the data properties (%P) from 20% to 100%

%I	%P	PMOM			EIFPS			RiMOM		
		Rec.	Prec.	Fmeas.	Rec.	Prec.	Fmeas.	Rec.	Prec.	Fmeas.
20	20	0.97	0.95	0.96	0.97	0.94	0.95	**0.98**	**0.95**	**0.96**
	50	**0.97**	**0.95**	**0.96**	0.92	0.92	0.92	0.95	0.93	0.94
	80	**0.97**	**0.95**	**0.96**	0.90	0.91	0.90	0.93	0.92	0.92
	100	**0.97**	**0.95**	**0.96**	0.88	0.90	0.89	0.89	0.90	0.89
	20	**0.96**	**0.94**	**0.95**	0.93	0.92	0.92	0.95	0.92	0.93
	50	**0.96**	**0.94**	**0.95**	0.89	0.87	0.88	0.91	0.89	0.90
	80	**0.96**	**0.94**	**0.95**	0.87	0.84	0.85	0.89	0.86	0.87
	100	**0.96**	**0.94**	**0.95**	0.85	0.82	0.83	0.87	0.83	0.85
80	20	**0.95**	**0.92**	**0.93**	0.90	0.89	0.89	0.91	0.90	0.90
	50	**0.95**	**0.92**	**0.93**	0.88	0.86	0.87	0.89	0.88	0.88
	80	**0.95**	**0.92**	**0.93**	0.82	0.80	0.81	0.83	0.81	0.82
	100	**0.95**	**0.92**	**0.93**	0.78	0.75	0.76	0.80	0.79	0.79
100	20	**0.94**	**0.90**	**0.92**	0.85	0.82	0.83	0.87	0.86	0.86
	50	**0.94**	**0.90**	**0.92**	0.83	0.81	0.82	0.84	0.82	0.83
	80	**0.94**	**0.90**	**0.92**	0.78	0.75	0.76	0.80	0.77	0.78
	100	**0.94**	**0.90**	**0.92**	0.74	0.70	0.72	0.73	0.72	0.72

framework and the baseline algorithms (EIFPS and RiMOM) using the DBpedia ontology database. By varying the percentage of data properties and the percentage of instances from 20% to 100%, PMOM outperforms the two other algorithms regarding the quality (recall, precision, and fmeasure) in all cases, except in the first case that containing 20% of data properties and instances. Moreover, the results showed that the quality of PMOM is non-sensitive to the number of data properties and the number of instances. Thus, the quality of PMOM exceeds 92%, whereas the quality of the EIFPS and RiMOM does not reach 70% and 72%, respectively. These results were obtained thanks to the feature selection procedure, which find out the most relevant data properties of the given ontologies.

6   Conclusions

This paper presents PMOM (Pattern Mining for Ontology Matching based instances) framework, for ontology matching problem. The approach explores different correlations between the data properties and selects the most frequent data properties describing the overall instances of the given ontology. To evaluate PMOM framework, intensive experiments have been carried on DBpedia database. The results show that PMOM outperforms the baseline methods (EIFPS, and RiMOM) in terms of execution time and solution's quality. As future work, we plan to explore other data mining techniques for ontology matching problem. Dealing with big ontology databases is also in our future agenda.

Acknowledgment. Youcef Djenouri's work was carried out at the Norwegian University of Science and Technology (NTNU), funded by a postdoctoral fellowship from the European Research Consortium for Informatics and Mathematics (ERCIM).

References

1. Agrawal, R., Imieliński, T., Swami, A.: Mining association rules between sets of items in large databases. In: ACM SIGMOD Record, vol. 22, pp. 207–216. ACM (1993)
2. Cerón-Figueroa, S., et al.: Instance-based ontology matching for e-learning material using an associative pattern classifier. Comput. Hum. Behav. **69**, 218–225 (2017)
3. Del Vescovo, C., Parsia, B., Sattler, U., Schneider, T.: The modular structure of an ontology: atomic decomposition. In: IJCAI Proceedings-International Joint Conference on Artificial Intelligence, vol. 22, p. 2232 (2011)
4. Djenouri, Y., Djamel, D., Djenoouri, Z.: Data-mining-based decomposition for solving maxsat problem: owards a new approach. IEEE Intell. Syst. **32**(4), 48–58 (2017)
5. Djenouri, Y., Belhadi, A., Fournier-Viger, P., Lin, J.C.W.: Fast and effective cluster-based information retrieval using frequent closed itemsets. Inf. Sci. **453**, 154–167 (2018)
6. Djenouri, Y., Comuzzi, M., Djenouri, D.: SS-FIM: single scan for frequent itemsets mining in transactional databases. In: Kim, J., Shim, K., Cao, L., Lee, J.-G., Lin, X., Moon, Y.-S. (eds.) PAKDD 2017. LNCS (LNAI), vol. 10235, pp. 644–654. Springer, Cham (2017). https://doi.org/10.1007/978-3-319-57529-2_50
7. Djenouri, Y., Djenouri, D., Lin, J.C.W., Belhadi, A.: Frequent itemset mining in big data with effective single scan algorithms. IEEE Access **6**, 68013–68026 (2018)
8. Han, J., Pei, J., Yin, Y.: Mining frequent patterns without candidate generation. In: ACM SIGMOD Record, vol. 29, pp. 1–12. ACM (2000)
9. Iwata, T., Kanagawa, M., Hirao, T., Fukumizu, K.: Unsupervised group matching with application to cross-lingual topic matching without alignment information. Data Min. Knowl. Disc. **31**(2), 350–370 (2017)
10. Li, J., Wang, Z., Zhang, X., Tang, J.: Large scale instance matching via multiple indexes and candidate selection. Knowl.-Based Syst. **50**, 112–120 (2013)
11. Niu, X., Rong, S., Wang, H., Yu, Y.: An effective rule miner for instance matching in a web of data. In: Proceedings of the 21st ACM International Conference on Information and Knowledge Management, pp. 1085–1094. ACM (2012)
12. Otero-Cerdeira, L., Rodríguez-Martínez, F.J., Gómez-Rodríguez, A.: Ontology matching: a literature review. Expert Syst. Appl. **42**(2), 949–971 (2015)
13. Shao, C., Hu, L.M., Li, J.Z., Wang, Z.C., Chung, T., Xia, J.B.: Rimom-im: a novel iterative framework for instance matching. J. Comput. Sci. Technol. **31**(1), 185–197 (2016)
14. Shvaiko, P., Euzenat, J.: Ontology matching: state of the art and future challenges. IEEE Trans. Knowl. Data Eng. **25**(1), 158–176 (2013)

15. Smith, B., et al.: The obo foundry: coordinated evolution of ontologies to support biomedical data integration. Nat. Biotechnol. **25**(11), 1251 (2007)
16. Thayasivam, U., Doshi, P.: Speeding up batch alignment of large ontologies using mapreduce. In: 2013 IEEE Seventh International Conference on Semantic Computing (ICSC), pp. 110–113. IEEE (2013)
17. Wang, Z., Li, J., Zhao, Y., Setchi, R., Tang, J.: A unified approach to matching semantic data on the web. Knowl.-Based Syst. **39**, 173–184 (2013)
18. Xue, X., Liu, J.: Collaborative ontology matching based on compact interactive evolutionary algorithm. Knowl.-Based Syst. **137**, 94–103 (2017)

Systematic Creation of Cumulative Design Science Research Knowledge with a Case Study in the Field of Automatic Speech Recognition

Udo Bub[(⊠)]

Faculty of Informatics, Eötvös Loránd University (ELTE), Budapest, Hungary
udobub@inf.elte.hu

Abstract. This paper proposes a method for the systematic build-up of incremental Information Systems Design Science Research knowledge using tight linkage with a process model for digital innovation. This results in the enhanced presentation of a new design science approach to the engineering of digital innovation that systematically builds up information systems design theories. We evaluate this combined new method artifact in a case study for an innovative conversational interface of an interactive voice response system that was developed in a university-industry innovation lab, trained on a gender and age specific speech database and offering automatically tailored services for each identified user group. Our work targets the goal of providing the methodical basis that enables researchers to systematically build up and re-use, and transfer Design Science Research knowledge.

Keywords: Digital innovation · Information systems design science · Cumulative design science knowledge · Automatic speech recognition

1 Introduction

The design of novel information systems (IS) and its related artifacts that follows scientific guidelines and processes is generally accepted as research [5, 7, 9]. Within the field of IS this paper aims at contributing to the systematic build-up and extension of the knowledge base of IS by incrementally adding to information systems design theory (ISDT). ISDT are theories of design and action (Gregor [4]), where the word *theory* encompasses a body of knowledge with conjectures, models, and frameworks in the scientific discipline. This view is based on Walls et al. [18] who claim that "the purpose of a theory is prediction and/or explanation of a phenomenon". Although these foundations were introduced over a decade and more ago and individual ISDTs were extensively built in numerous scientific IS publications, Schuster et al. [16] state an "alarming paucity of follow-up research" building on existing ISDTs and thus a failure to leverage the powerful concept in order to build cumulative and interlinked reusable Design Science Research (DSR) knowledge on a broad scale.

© Springer Nature Switzerland AG 2019
T. Welzer et al. (Eds.): ADBIS 2019, CCIS 1064, pp. 94–101, 2019.
https://doi.org/10.1007/978-3-030-30278-8_12

On the other hand, Offermann, Blom and Bub [14] have developed a concept of for the generalization, reuse, and transfer of DSR knowledge by means of distinct design strategies building on ISDT. We argue that the above mentioned paucity can be remedied by installing a so far missing framework to ensure intensive and systematic use of this concept. Consequently, in order to overcome the deficit of a not yet satisfactory build-up of cumulative DSR knowledge, this paper aims at combining above concept with the processual engineering of Digital Innovation (DI). We follow both the definitions of DI by Hevner, vom Brocke, Maedche [6] as "the appropriation of digital technologies in the process of and as the result of innovation" and Yoo [19] as "carrying out of new combinations of digital and physical components to produce novel products". It is an emerging topic of research within the field of DSR in IS.

The paper is structured as follows: Sect. 2 presents the chosen research approach, Sect. 3 presents prior work and literature. Section 4 presents the new artifact: the extension of the method from Bub [1] by the use of design strategies of [14] to overcome the shortcomings described above and by combining above concepts to a new "factory" approach to create, re-use, and transfer DSR knowledge. Subsequently, the proposal is evaluated in a case study in Sect. 5 for the establishment of a novel Interactive Voice Response (IVR) system as innovation output that was developed in the setting of a joint academic and industrial innovation lab. Section 6 adds discussion and conclusion on the lessons learned and reflects about future work.

2 Research Approach

The new artifact is an extended method for the Engineering of Digital Innovation that allows for the cumulative build-up of DSR knowledge. This article focuses on the extended process model as part of the method. The way to achieve this is to combine existing artifacts, more precisely a method described by Bub [1] with work by Offermann, Blom and Bub [14] and Offermann et al. [15]. As a matter of fact, the improved method itself is a scientific design artifact according to the guidelines as exposed by the seminal paper of Hevner et al. [7] itself and is first published here.

The project of the case study "Tailored Call Center Process" was already finished by the time of the publication of this paper, but it was carried out within the Innovation Process described in Bub [1]. The scientific output by the described combined Innovation Process that is re-used in this case study has been well published in other domains of engineering [11, 12], but not analyzed and labeled as DSR although its engineering work is similar to tasks of design and action and thus is eligible for augmenting ISDT. This present paper retells the well-documented story from a DSR perspective in the case study. As a matter of fact, the research is restructured from a new DSR point of view while applying the newly enriched method for the build-up of cumulative DSR knowledge. Consequently, the proposed new artifact is validated using past documentation. This kind of causal approach classifies as Reverse Engineering and Design Recovery in the sense of Chikofsky and Cross [2].

3 Prior Work

Because the proposed work is a combination of an approach to create cumulative DSR knowledge and a process for Digital Innovation and DSR we briefly present the state-of-the-art of these two lines of action.

3.1 Cumulative DSR Knowledge

We generate cumulative knowledge for the body of knowledge of information systems by incrementally adding to ISDTs in the sense of Gregor [4]. Schuster et al. [16] give a comprehensive overview on the state of affairs in the area of cumulative DSR knowledge and state a deficit of follow-up research that builds on ISDTs. Offermann et al. [15] have identified 12 strategies enabling systematic reuse, generalization, and transfer of design knowledge from previous designs. They adopted the concept of *ranges* from the field of social sciences [10], implying that there are different levels of theory relative to their distance to empirical observations: a design for a specific setting is classified as *short-range design*, design for a specific type or class of setting is classified as *mid-range design*, whereas general insights about a type of design approach are classified as *long-range design* [14]. The idea of different design ranges are very important for modelling DSR knowledge, as mobility between these ranges, such as generalization, derivations, and validations play an important role in research and can be modeled explicitly this way and thus is preferred more coarse concepts presented e.g. by Gregor and Hevner [5] that focus solely on mid-range designs.

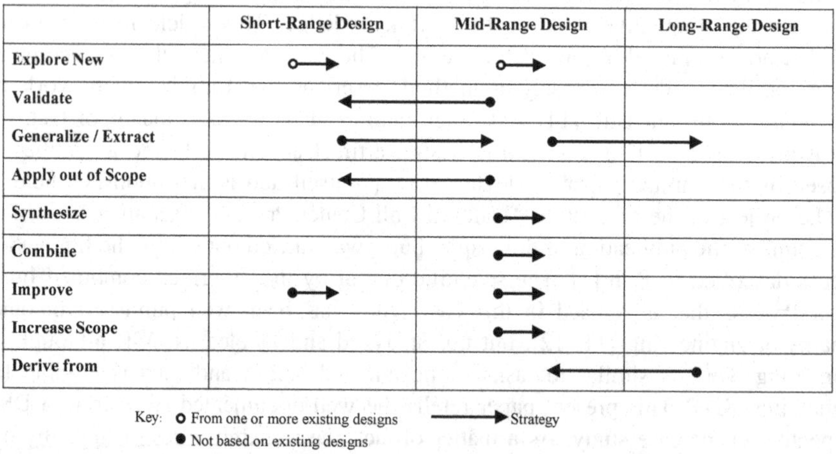

	Short-Range Design	Mid-Range Design	Long-Range Design
Explore New	o—▶	o—▶	
Validate	◀———————●		
Generalize / Extract	●————————▶	●————————▶	
Apply out of Scope	◀———————●		
Synthesize		●—▶	
Combine		●—▶	
Improve	●—▶	●—▶	
Increase Scope		●—▶	
Derive from		◀————————●	

Key: o From one or more existing designs ———————▶ Strategy
      ● Not based on existing designs

Fig. 1. The 12 design strategies as proposed by Offermann et al. [15].

In follow-up work Offermann et al. [15] conceptualized cumulative knowledge development by identifying 12 distinct design strategies, building on previous ISDTs, that have been evaluated in an extensive literature research of DSR publications. However, a roll-out of the concept at a greater scale has not taken place, yet. These 12 design strategies as exhibited in Fig. 1 will be applied to these artifacts in the innovation design of our approach to the engineering of innovation.

3.2 Digital Innovation and Information Systems

Bub [1] has introduced a combination of state-of-the-art, practice-driven stage-gate-innovation processes such as Cooper [3] with Design Science Research Processes as presented e.g. in [13]. This combination provides a systematic way of combining practical relevance and scientific rigor as demanded e.g. by Hevner et al. [7] – provided that the degree of innovation allows for work classifiable as research.

4   Artifact Description: Methodical DSR Approach to Digital Innovation

As mentioned before, the artifact is constructed through a combination of the concept of *generalization* and *transfer* presented in [14] and the combined DSR and Innovation Management approach presented in [1]. Interestingly, this design itself follows one of the presented design strategies from Fig. 1: *combine*, which results in a new mid-range design.

Stage-gate oriented processes offer the benefit of providing synchronization points at the gates. Criticism arose against stage-gate oriented idea-to-launch processes because of their emphasis on convergent thinking and possible censorship at the gates and thus suppressing creativity (e.g. Kressel, Winarsky [8]). As a basis for present work we agree with Bub [1] about the advantages of offering a bigger umbrella framework with synchronization points with project stakeholders ensuring practical relevance and transferability of the results into production. At the same time, within these synchronization points the integration of more frequent iterative search processes like a DSR research process (as described e.g. in [13]) can be integrated, ensuring scientific rigor. In order to address the criticism we adjust the process from [1] by attributing the individual stages the status of maturity levels that are reached when passing the individual gates. This way it is emphasized better that the main function of the gate-reviews is not the filtering, but rather the focus point of input by stakeholders that lead to a higher maturity of the innovation project. This is reflected by usage of the process model depicted in Fig. 2. It is important to note that the Innovation Process runs only once per project, whereas the DSR process runs several times - considering DSR is an iterative search problem [7]. However, because DSR processes also have stage-gate shape with the stages *Problem Identification*, *Solution Design*, and *Evaluation* [13], it is possible to synchronize the iterations with the gates at each maturity level in the overarching innovation process in Fig. 2.

In the targeted combined design, the design strategies from Fig. 1 are applied during the phase *Solution Design*. Because design is a search process, the design strategy can be iteratively developed. However, due to its importance for following decisions the design strategy should be ready at Gate 2 and further changes are rather refinements or smaller adjustments.

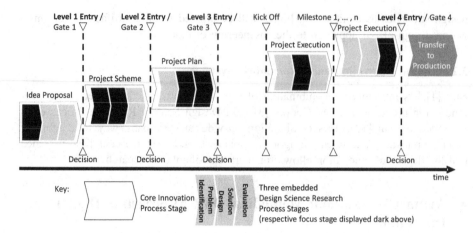

Fig. 2. Innovation and design science research process with maturity stages

5 Case Study: Tailored Call Center Process

We analyze the design in the context of an innovation project that has already taken place, but with a different scientific embedding in engineering sciences [11, 12]. The project has been run through the stage-gate innovation process as described in Fig. 2 with further process explanations in Bub [1]. This resulting case study is structured as proposed by vom Brocke and Mendling [17] in the following sections: Situation Faced, Action Taken, Results Achieved, and Lessons Learned.

Situation Faced

An incumbent telecommunications company faced the problem of too high operational costs in the call center when compared to the competition. The decision was taken to set up an innovation project to introduce cost savings to automate part of the work of call center agents by means of automatic speech recognition resulting in an interactive voice response system (IVR). The company uses commercial off-the-shelf modules, but also has the skills to differentiate from the competition by developing additional new own modules. A project team was set up lead by the innovation process manager of the company internal innovation lab and business stakeholders. During innovation workshops, it turned out that a main pain from the stakeholders is the high costs per call due to tedious manual interaction of live call center agents. As a matter of fact, an automation of manual process steps would yield a high benefit (efficiency business case). Likewise, new value added services for target groups (gender and age dependent) would offer additional marketing and business opportunities at the call center customer front end.

Recognition of non-verbal features like age and gender beyond speech-to-text from a speech signal has been a topic that had only just emerged at the time of the project with no commercial off-the-shelf recognizers available. It was goal of this project to use classification of such non-verbal features in parallel to recognizers that convert speech to text in a call centers to make skill-based routing and market analyses in call centers possible. These features had to be developed individually in an innovation project.

Action Taken

In this context emerged the idea to tailor the IVR call process flow according to age and gender of the caller. This would enable the call center agents to save time for pre-classifying the caller and automate part of the dialog script. This would result in saving time and effort and thus reducing the costs per call as the human interaction is the most cost intensive.

The researchers identified the research design problem of Age and Gender recognition in the domain of speech recognition. Knowledge of literature yielded several approaches, but none of them ready for immediate use on the problem. The second research question was how to best tailor the IVR dialog given knowledge of age and gender in order to achieve the goals.

Thus, the research design strategy followed in two steps. First, the team contributed to science with the design and comparison of four different approaches for age and gender recognition and the subsequent comparative empirical evaluation in a laboratory experiment [11] on the same speech database. The recognition task was to differentiate 7 groups for age and gender: children of 13 years and younger, young people between 14 and 19 years (male/female), adults between 20 and 64 years (male/female) and seniors of 65 years and older (male/female).

The best performing method was an adapted design based on an existing Parallel Phone Recognizer (PPR) [20]. For easy tasks its precision is comparable to human performance [11]. PPR was originally developed to recognize languages (like English, German, Hungarian, etc.), not gender and age. According to the classification from Fig. 1 the design adhered to the strategy *Increase Scope* as the scope was increased from language identification by recognition of age and gender of the speakers. For more details of the recognizer please refer to System A in [11].

In the second design step the team combined commercially available recognizers that are used for speech-to-text tasks with our own PPR classifier for non-verbal speech used in parallel on the same speech signal. In the classification exhibited in Table 1 the chosen innovation design is *Combine* because the innovation is based on a novel combination of existing designs.

For the Gate 2 presentation the design strategy was stable according to Table 1.

Table 1. Research design for case study, applying the design strategies from Fig. 1.

Design contributions	Applied design strategy	Description of implementation
Design Step 1: Create innovative module for recognition of age and gender	*Increase Scope*	Use PPR recognizers from Language ID scope and apply them to age and gender task
Design Step 2: Create new tailored IVR dialog	*Combine*	Use existing Commercial Off-The-Shelf automatic speech recognizers/IVR tools and combine with new module for age and gender recognition

Researchers also participated in preparing Gate 4 as their innovative artifact was built- into a live system and they needed to evaluate it also scientifically. This evaluation is published in [12].

Results Achieved

The empirical evaluation showed improvements of mean opinion scores of live users and average ratings of users when compared to a conventional routing [12]. Yearly efficiency gains could be quantified with 42 Million Euros.

Although the new artifact was originally not built in a formal DSR process it was possible to align all its contributions in detail to the phases Problem Identification, Solution Design, and Evaluation [13]. All of the activities of the researchers could be mapped to the phases of the DSR process according to its embedding in Fig. 2.

Lessons Learned

The outcome of the case study is that the application of the research strategies is very well applicable and it smoothly integrates into the combined process artifact. On the one hand it works well in order classify the actual design work and thus document better the actual innovation steps. It could be used for the analysis of a wealth of past engineering projects and make their results available for the DSR body of knowledge.

6 Conclusion and Future Work

The value of these proposed design strategies lies in consciously documenting, communicating, and educating about the scientific design, as well as in evaluating critical parts in a new or unfamiliar design. They make creative steps of the designer explicit and document them in a common structure. Moreover, they help researchers in identifying and performing design science and engineering research projects as they offer criteria to categorize their design. They help to make re-use of design knowledge more efficient. The producer of knowledge has a reference against which the work can be described and the consumer of knowledge can describe the information need in a more standardized way. It enhances transparency and maturity of the design process and its subsequent communication – be it for scientific publications, project documentation, strategic reflection or educational purposes.

At the same time, applying these design strategy successfully addresses the deficit of cumulative knowledge building in design science. On the one hand, we could demonstrate that also past knowledge can be categorized and thus be made useful for use within the body of knowledge of IS DSR. On the other hand, the proposed framework is intended to be rolled out in combination with existing innovation processes and thus deliver design knowledge at a greater scale. We realize the incremental build-up of design knowledge by combining Research and Innovation processes with Research Strategy templates by means a combination of these concepts and contributing an enhanced artifact.

In total, this results in the new design proposal and evaluation that opens the way for building a "Design Science Research Factory". The integration of this present work into the formulation of a complete methodical framework as started by Bub [1] will be realized in a next step.

References

1. Bub, U.: Towards an integrated method for the engineering of digital innovation and design science research. In: Benczúr, A., et al. (eds.) ADBIS 2018. CCIS, vol. 909, pp. 327–338. Springer, Cham (2018). https://doi.org/10.1007/978-3-030-00063-9_31
2. Chikofsky, E., Cross, J.: Reverse engineering and design recovery – a taxonomy. IEEE Softw. **7**(1), 13–17 (1990)
3. Cooper, R.: Stage-gate systems - a new tool for managing new products. Bus. Horiz. **33**(3), 44–54 (1990)
4. Gregor, S.: The nature of theory in information systems. MIS Q. **30**(3), 611–642 (2006)
5. Gregor, S., Hevner, A.: Positioning and presenting design science research for maximum impact. Int. J. MIS Q. **37**(2), 337–355 (2013)
6. Hevner, A., vom Brocke, J., Maedche, A.: Roles of digital innovation in design science research. Int. J. Bus. Inf. Syst. Eng. **11**(1), 3–8 (2019)
7. Hevner, A., March, S., Park, J., Ram, S.: Design science in information systems research. MIS Q. **28**(1), 75–105 (2004)
8. Kressel, H., Winarsky, N.: If You Really Want to Change the World. Harvard Business Review Press (2015)
9. March, S., Smith, G.: Design and natural science research on information technology. Decis. Support Syst. **15**(4), 251–266 (2015)
10. Merton, R.: Social Theory and Social Structure. The Free Press (1968)
11. Metze, F., et al.: Comparison of four approaches to age and gender recognition for telephone applications. In: Proceedings of the IEEE International Conference on Acoustic, Speech, and Signal Processing (ICASSP), Hololulu, HI (2007)
12. Metze, F., Englert, R., Bub, U., Burkhardt, F., Stegmann, J.: Getting closer: tailored human-computer speech dialog. J. Univ. Access Inf. Soc. **8**(2), 97–108 (2008)
13. Offermann, P., Levina, O., Schönherr, M., Bub, U.: Outline of a design science research process. In: Proceedings of the 4th International Conference on Design Science Research in Information Systems and Technology (DESRIST), Malvern, PA (2009)
14. Offermann, P., Blom, S., Bub, U.: Strategies for creating, generalising and transferring design science knowledge – methodological discussion and case analysis. In: Proceedings of the 10th International Conference on Wirtschaftsinformatik, Zürich (2011)
15. Offermann, P., Blom, S., Schönherr, M., Bub, U.: Design range and research strategies in design science publications. In: Jain, H., Sinha, A.P., Vitharana, P. (eds.) DESRIST 2011. LNCS, vol. 6629, pp. 77–91. Springer, Heidelberg (2011). https://doi.org/10.1007/978-3-642-20633-7_6
16. Schuster, R., Wagner, G., Schryen, G.: Information systems design science and cumulative knowledge development: an exploratory study. In: Proceedings International Conference on Information Systems, San Francisco, CA (2018)
17. vom Brocke, J., Mendling, J.: Frameworks for business process management: a taxonomy for business process management cases. In: vom Brocke, J., Mendling, J. (eds.) Business Process Management Cases. MP, pp. 1–17. Springer, Cham (2018). https://doi.org/10.1007/978-3-319-58307-5_1
18. Walls, J., Widmeyer, G., Sawy, O.: Building an information systems design theory for vigilant EIS. Inf. Syst. Res. **3**(2), 36–59 (1992)
19. Yoo, Y., Henfridsson, O., Lyytinen, K.: Research commentary – the new organizing logic of digital innovation: an agenda for information systems research. Inf. Syst. Res. **21**(4), 724–735 (2010)
20. Zissman, M.: Comparison of four approaches to language identification of telephone speech. IEEE Trans. Speech Audio Process. **4**(1), 31 (1996)

SLFTD: A Subjective Logic Based Framework for Truth Discovery

Danchen Zhang[1(✉)], Vladimir I. Zadorozhny[1], and Vladimir A. Oleshchuk[2]

[1] School of Computing and Information, University of Pittsburgh, Pittsburgh, USA
{daz45,viz}@pitt.edu
[2] Department of Information and Communication Technology, University of Agder,
Kristiansand, Norway
vladimir.oleshchuk@uia.no

Abstract. Finding truth from various conflicting candidate values provided by different data sources is called truth discovery, which is of vital importance in data integration. Several algorithms have been proposed in this area, which usually have similar procedure: iteratively inferring the truth and provider's reliability on providing truth until converge. Therefore, an accurate provider's reliability evaluation is essential. However, no work pays attention to "how reliable this provider continuously providing truth". Therefore, we introduce subjective logic, which can record both (1) the provider's reliability of generating truth, and (2) reliability of provider continuously doing so. Our proposed methods provides a better evaluation for data providers, and based on which, truth are discovered more accurately. Our framework can handle both categorical and numerical data, and can identify truth in either a generative or discriminative way. Experiments on two popular real world datasets, Book and Population, validates that our proposed subjective logic based framework can discover truth much more accurately than state-of-art methods.

Keywords: Data fusion · Truth discovery · Subjective logic

1 Introduction

Data conflict is a common problem in data management area. For example, for a given flight, different websites may report different departing time. Figuring out the (most likely) truth from conflicting values provided by different sources is an important and challenging task. Naive methods, such as voting, do not consider the data provider's reliability, and hence may fail in particular cases. Therefore, many methods [1–11] paying attention to accurately evaluate trustworthiness of data provider are proposed. With provider reliability considered, these methods then identify the truth usually by selecting the value with the maximum averaged provider reliability.

However, in past studies, provider's reliability is usually evaluated in a probabilistic logic, which uses an evidence based probability (ranging from 0 to 1) to

T. Welzer et al. (Eds.): ADBIS 2019, CCIS 1064, pp. 102–110, 2019.
https://doi.org/10.1007/978-3-030-30278-8_13

represent people's opinion. For example, after observing the flipping the coin for hundreds of times, people believe the probability of "head" is 0.5, and believe the probability of "tail" is 0.5, too. However, when sample size is too small, the probability is unreliable. In such a situation, Subjective Logic (SL), proposed by Jøsang [12], can provide more information for this situation. With SL, an opinion from a person p towards a statement s can be represented by a triple $\omega_s^p = \{t, d, u\}$, with $t, d, u \in [0, 1]^3$, and $t + d + u = 1$, where t means trust, d means distrust, and u means uncertainty. With a too small sample, we may use $\{0.3, 0.3, 0.4\}$ to describe our uncertainty and impression towards the coin than a simple 0.5. In terms of truth discovery, SL allows us to record our trust (of provider providing truth) and certainty (of provider continuously doing so) towards each provider. In turn, we can identify truth more accurately.

To summarize, our paper has following major contributions: (1) our study is the first to pay attention to "how reliable the provider is able to continuously provide truth"; (2) SL is first introduced to truth discovery area, and it can perfectly records above mentioned two kinds of reliability; (3) The experiments on two popular real world dataset show that, compared with state-of-art methods, our framework can improve the truth discovery performance by a large degree.

2   Related Works

In truth discovery area, the simplest mechanism is voting, which does not consider the provider's reliability. Many studies show that a good evaluation of provider reliability can improve the performance largely. In [1], Dong et al. proposed to use Accuracy, which is calculated as the probability of each value being correct, and average the confidence of facets provided by the source as the provider trustworthiness. After that, they proposed the concept of AccuracySimilarity, which further considers the similarity of two values. In [2], authors proposed POPAccuarcy, which differs from Accuracy by releasing the assumption that false value probability is uniformly distributed. Another popular method is the TruthFinder, proposed by Yin et al. [5], which differs from Accuracy by not normalizing the confidence score of each entity. In [8], Pasternack et al., proposed three methods: (1) AverageLog is a transformation of Hub-Authority algorithm, with source trustworthiness being the averaged confidence score of provided values multiplying the log of provided value count; (2) Investment, where the confidence score of the value grows exponentially with the accumulated providers' trustworthiness. (3) PooledInvestment, where the confidence score of the value grows linearly. In [4] authors proposed a semi-supervised reliability assessment method, SSTF. It is basically a PageRank method assuming that there is a set of entities having the true value, which will affect the result in the PageRank iteration. [3], proposes 2-Estimates, which is a transformation of Hub-Authority algorithm, whose provider trustworthiness is the average instead of the sum of the vote count. They further proposed 3-Estimates, which additionally considers the value's trustworthiness. Another group, involving CRH [9], CATD [10] and GTM [11], can generate the truth for data with numerical values, and they can also be adapted to categorical dataset with slight modification. They share a similar idea, trying to generate/select the true value of each entity to minimize the difference between "estimated true value" and the "observed input value". Additionally, CATD is designed to smoothly predict truth on the long tail data with chi-squared distribution.

SL is a powerful decision making tool extending the probabilistic logic by including uncertainty and subjective belief ownership [12]. It is widely used in trust network analysis, conditional inference, information provider reliability assessment, trust management in sensor networks, etc. SL uses subjective opinions to express subjective beliefs about the truth of propositions with degrees of uncertainty.To the best of our knowledge, our work is the first one applying it to area of reliable truth discovery.

3   SL Based Framework for Truth Discovery (SLFTD)

Consider a dataset that contains a set of entities $E = \{e_1, e_2, ..., e_n\}$, and a set of data providers $P = \{p_1, p_2, ..., p_m\}$, the value of entity e_i provided by provider p_j is named as v_{ij}, constructing the value set V. Different providers may provide different values for same entity, and truth discovery aims to find the true value for each entity.

In the procedure of identifying the truth, our proposed framework involves three steps: (1)evaluate the provider's reliability score and entity's discrimination score in an iterative way, then (2) for each provider, SL based opinions are constructed based on the converged scores, and (3) the true value are inferred based on the fused opinions in either a discriminative manner or a generative manner. In the framework, two SL operations are utilized, and more detail can be referenced in [12].

- **Recommendation.** Assume two persons, A and B: A has an opinion towards B, and B has an opinion towards a statement s. Then according to B's recommendation, A can generate an opinion towards this statement s, described as $\omega_s^{AB} = \omega_B^A \otimes \omega_s^B = \{t_s^{AB}, d_s^{AB}, u_s^{AB}\}$.
- **Consensus.** If two persons A and B have opinions towards one statement s, then consensus operator $\oplus$ can be used to combine their opinions, described as $\omega_s^{A,B} = \omega_s^A \oplus \omega_s^B = \{t_s^{A,B}, d_s^{A,B}, u_s^{A,B}\}$.

3.1   Accurately Infer Providers' Reliability

Our first step is to iteratively evaluate the provider's reliability score and entity's discrimination score. In this study, the degree, to which the algorithm can infer the true value of the entity in an undisputed convincing manner, is defined as the **entity discrimination ability**. The entity whose majority candidate values are from reliable providers and are very similar/close to each other should be given a higher score. Thus discrimination score of entity E_i is defined as:

$$Disc(E_i) = \frac{\sum_{P_j,P_l \in P_{E_i}} Rel(P_j)Rel(P_l)Imp(V_{il} \to V_{ij})}{\sum_{P_j,P_l \in P_{E_i}} Rel(P_j)Rel(P_l)}, \tag{1}$$

where P_{E_i} is the set of providers that gives value on entity E_i; $Rel(P_j)$ is the reliability score of provider P_j, which will be described later. Please notice that $Imp(V_{il} \to V_{ij})$ reflects the implication from $V_{i.}$ to V_{ij}, introduced from [5]. It is a value (ranging from 0 to 1) reflecting to what degree V_{ij} is (partially) true if V_{il} is correct. At the first iteration, all provider have equal weights; and after each round, normalization is conducted with weights summing up to 1.

When measure the **providers' reliability**, more attention is paid to providers' performance on entities with higher discrimination score. Given such entities, if the values from this provider obtain lots of implications from other values of same entity, this provider reliability should be boosted; otherwise, should be lower down. Such impact from entities with low discrimination score should be relatively discounted. Thus reliability score of provider P_j is defined as:

$$Rel(P_j) = \frac{\sum_{E_i \in E_{P_j}} Disc(E_i)Imp(V_{i.} \to V_{ij})}{\sum_{E_i \in E_{P_j}} Disc(E_i)}, \tag{2}$$

where E_{P_j} is the set of entities to whom P_j gives value; and $V_{i.}$ consists of all candidate values of entity E_i. Also, normalization is conducted in the end of each iteration. This iterative procedure will continue until all scores converge.

3.2   Construct SL Opinions

Then SL opinions of each provider is computed based on converged scores. Two kinds of provider reliability should be considered: (1) reliability of generating the true value; (2) reliability of continuously doing so. Provider's reliability in last subsection describes the first kind of reliability, and we propose a new concept, **certainty**, to describe the second one. Certainty of provider P_j is defined as:

$$Certainty(P_j) = \frac{\sum_{E_i \in E_{P_j}} Disc(E_i)}{|E_{P_j}|}. \tag{3}$$

With SL, we proposed to record the provider's reliability of generating the true value in **trust**, and record the reliability of continuously doing so with **uncertainty**. In this way, the algorithm's opinion towards the P_j is defined as $\omega_{P_j}^{Algo} = \{t_{P_j}^{Algo}, d_{P_j}^{Algo}, u_{P_j}^{Algo}\}$:

$$t_{P_j}^{Algo} = (1 - u_{P_j}^{Algo})Rel(P_j) \tag{4}$$

$$d_{P_j}^{Algo} = 1 - t_{P_j}^{Algo} - u_{P_j}^{Algo} \tag{5}$$

$$u_{P_j}^{Algo} = \gamma(1 - Certainty(P_j)) + \alpha. \tag{6}$$

where $Algo$ is short for "algorithm". α describe people's fundamental uncertainty, since even given by enough evidence, people can still be skeptical. γ is a parameter to limit the certainty to a certain range. Both parameters range from 0 to 1. In this way, provider's reliability can be accurately described.

3.3 Infer True Value in Generative Manner

This manner only fits numerical data, i.e., $V_{ij} \in R$. For each entity, we define a statement "true value of entity is the largest candidate value", and generate $Algo$'s opinion towards them. Higher trust means truth is close to the max candidate value; otherwise, truth is close to the min candidate value. First, on each entity E_i, we normalize all the candidate values in the following manner:

$$V'_{ij} = \frac{V_{ij} - min(V_{i.})}{max(V_{i.}) - min(V_{i.})}, \tag{7}$$

so that $V'_{ij} \in [0,1]$. Then, the original statement is mapped to "true value of entity in the normalized space is 1". Thereby, given provider E_i, the provider P_j's opinion towards the statement can be defined as:

$$\omega^{P_j}_{truth(E_i)=1} = \{(1-\beta)V'_{ij}, 1 - (1-\beta)V'_{ij} - \beta, \beta\}, \tag{8}$$

where β also describes people's fundamental uncertainty, similar to α. Second, the provider can recommend his opinion of the entity's truth to $Algo$. Thus, $Algo$'s opinion towards truth of E_i by P_j's recommendation is defined as:

$$\omega^{Algo,P_j}_{truth(E_i)=1} = \omega^{Algo}_{P_j} \otimes \omega^{P_j}_{truth(E_i)=1}. \tag{9}$$

Entity E_i has a set of candidate values from several providers $\{P_j, ..., P_k\}$, and $Algo$ should have a summarized opinion based on all recommendations with Consensus operation. The algorithm's final opinion towards truth of E_i is defined as:

$$\omega^{Algo,P_j,...,P_k}_{truth(E_i)=1} = \omega^{Algo,P_j}_{truth(E_i)=1} \oplus ... \oplus \omega^{Algo,P_k}_{truth(E_i)=1}. \tag{10}$$

In the fused opinion, the trust reflects the true value of E_i in the normalized space, and final step is to map it to the original numerical space by:

$$V^{true}_{ij} = t^{Algo,P_j,...,P_k}_{truth(E_i)=1}(max(V_{i.}) - min(V_{i.})) + min(V_{i.}). \tag{11}$$

3.4 Infer True Value in Discriminative Manner

In this model, we generate a SL opinion for each candidate value, and then for each entity, select the value with highest trust as the truth. Given a provider P_j, the algorithm's opinion towards a value V_{ij} is defined as:

$$\omega^{Algo,P_j}_{V_{ij}} = \{t^{Algo}_{P_j}, d^{Algo}_{P_j}, u^{Algo}_{P_j}\}. \tag{12}$$

If a value is provided by several providers $\{P_j, ..., P_k\}$, consensus operation is used to fuse opinions together. Thus we have algorithm's final opinion towards a value V_{ij}:

$$\omega_{V_{ij}}^{Algo,P_j,...,P_k} = \omega_{V_{ij}}^{Algo,P_j} \oplus ... \oplus \omega_{V_{ij}}^{Algo,P_k}. \tag{13}$$

4 Experiments

In this section, we evaluate our proposed framework on two popular real word datasets, **Book** and **Population**, one being categorical another being numerical. In the experiment, we name our proposed method SLFTD generating the true value in a generative way, as **SLFTD-Gen**; and name SLFTD selecting the true value from existing candidates in a discriminative manner, as **SLFTD-Dis**. Naive baselines include **Voting, Median, Average**. State-of-art methods include **TruthFinder** [5], **Accuracy** [6], **AccuracySim** [6], **Sums, Investment, PooledInvestment, Average.Log** [8], **CRH** [9], **CATD** [10] and **GTM** [11].

4.1 Finding True Book Author List

Dataset: Book. It is a popular categorical dataset in truth discovery area. Its data describes that for each book, online bookstores post author list in their web pages, but some data is wrong. It contains the information on ISBN, book name, authors, online bookstore name for 1265 books. Totally, there are 894 bookstores and they generate 26,494 author lists. In this study, we use two testing data: (1) the gold testing dataset in the original dataset consisting of 100 books, (2) a new silver testing dataset composed of 161 book, containing the first 100 books and other 61 books. The 61 books are selected because different methods appearing in our experiments gives different true data. Thus it is more challenging than the first one. The true author list of both testing data are manually assigned by people reading the cover page of the book. All the codes and preprocessed datasets in this paper are posted online[1].

Settings. The implication appeared in Eq. 1, is defined as $Imp(V_{il} \rightarrow V_{ij}) = \frac{\#|V_{il} \cap V_{ij}|}{\#|V_{ij}|}$, where $\#|V_{il}|$ is the amount of elements in V_{il}; the implication appeared in Eq. 2 is defined as $Imp(V_{i.} \rightarrow V_{ij})$ is defined as $\frac{\sum_{P_l,P_j \in P_{E_i}} Imp(V_{il} \rightarrow V_{ij})}{\sum_{P_l,P_j \in P_{E_i}} 1}$. Following past studies, the parameters of all methods are set with optimal performance on the testing data. In TruthFinder, $\{\gamma = 0.1, \rho = 0.7\}$. In AccuracySim, $\{\lambda = 0.9\}$. In SLFTD-Dis, $\{\gamma = 0.2, \alpha = 0.2\}$.

---

[1] https://github.com/daz45.

Table 1. Precision of all methods on dataset BOOK. Best results are in bold.

Method	Golden Testing	Silver Testing
SLFTD-Dis	**94%**	**77.6%**
TruthFinder	93%	**77.6%**
AccuracySim	91%	68.9%
Accuracy	89%	68.9%
PooledInvestment	87%	72.75%
Average.log	82%	62.1%
Voting	80%	62.1%
Investment	79%	63.4%
Sums	74%	55.3%

Results. From Table 1, we can see that our proposed method SLFTD-Dis has the best performance on both testing data. TruthFinder provides a nearly same good result. Also, methods (SLFTD-Dis, TruthFinder, and AccuracySim) that use value similarity, shows a better performance than those do not. Please note that our preprocessed dataset is cleaner than the data used in prior works [6], as the voting results is 82%, while past studies showed only 71%.

4.2   Finding True Population of the City

Dataset: Population. It is a sample of Wikipedia edit history of city population, proposed in [8], and algorithms need to identify the true population for each city of each year. This data is picked to test the system's performance on numerical data. It is preprocessed in the same way as that in [11], except σ_0 is set to be 0.91 instead of 0.9. The final data consists of 4,183 tuples on 1,172 city-year from 1,926 providers, and methods are evaluated on 277 city-year.

Settings. The implication appeared in Eq. 1, is defined as: $Imp(V_{il} \rightarrow V_{ij}) = 1 - \frac{|V_{ij} - V_{il}|}{max(V_{i.}) - min(V_{i.})}$; the implication appeared in Eq. 2 is defined as $Imp(V_{i.} \rightarrow V_{ij}) = 1 - \frac{|V_{ij} - avg(V_{i.})|}{max(V_{i.}) - min(V_{i.})}$. Following past studies, the parameters of all methods are set based on optimal performance on the testing data. In TruthFinder, $\{\gamma = 0.3, \rho = 0.01\}$. In terms of GTM, we have two set of parameters, $(\alpha = 10, \beta = 10, \mu_0 = 0, \sigma_0^2 = 1)$ suggested by [11], and $(\alpha = 1, \beta = 1, \mu_0 = 0, \sigma_0^2 = 1)$ suggested by our experiment. For CATD, significance level α is set to be 0.03. For SLFTD-Gen, $(\alpha = 0, \beta = 0.01, \gamma = 0.001)$; for SLFTD-Dis, $(\alpha = 0.01, \gamma = 0.001)$.

Results. Experiment results are shown in Table 2. We can see that SLFTD-Dis gives best performance on three metrics, then TruthFinder provides second best on MAE and Error Rate, while SLFTD-Gen gives second best on RMSE.

Comparing two groups, Error Rate shows that discriminative methods generally makes less errors than generative models. Naive methods, especially Average, gives a much worse performance. Also, SLFTD-Dis, SLFTD-Gen, TruthFinder, CATD can find the true value with a smaller error.

Table 2. Experiment results on dataset Population. First group are discriminative models; second group are generative models. Best results are in bold; second best is labeled with *. Error Rate is based on mismatch 10%.

Methods	MAE	RMSE	Error Rate
SLFTD-Dis	**1489.57**	**5819.00**	14.44%
TruthFinder	1744.05*	8942.86	16.97*%
Voting	2511.04	11328.71	23.10%
Investment	2614.21	11378.42	25.99%
CRH - weighted median	3030.23	12696.96	25.99%
Median	2475.05	9759.71	33.57%
CATD	1796.67	8765.81	21.30%
SLFTD-Gen	2132.35	7070.54*	55.60%
GTM - parameters by us	2424.10	8659.36	57.04%
GTM - parameters in [11]	2710.30	9290.32	58.12%
Average	3336.49	9799.66	58.48%
CRH - weighted average	3805.10	11898.04	58.48%

5   Conclusion

In this study, we proposed a SL based framework for the truth discovery, which can predict truth either in a discriminative way or a generative manner. SL is introduced to more accurately describe the provider's reliability. Experiments on two real world datasets validates the effectiveness of our proposed methods.

References

1. Dong, X.L., Berti-Equille, L., Srivastava, D.: Integrating conflicting data: the role of source dependence. In: Proceedings of the VLDB Endowment 2.1, pp. 550–561 (2009)
2. Dong, X.L., Saha, B., Srivastava, D.: Less is more: selecting sources wisely for integration. In: Proceedings of the VLDB Endowment, vol. 6. no. 2. VLDB Endowment (2012)
3. Galland, A., et al.: Corroborating information from disagreeing views. In: Proceedings of the Third ACM International Conference on Web Search and Data Mining. ACM (2010)
4. Yin, X., Tan, W.: Semi-supervised truth discovery. In: Proceedings of the 20th International Conference on World Wide Web. ACM (2011)

5. Yin, X., Han, J., Philip, S.Y.: Truth discovery with multiple conflicting information providers on the web. IEEE Trans. Knowl. Data Eng. **20**(6), 796–808 (2008)
6. Dong, X.L., Srivastava, D.: Big data integration. In: 2013 IEEE 29th International Conference on Data Engineering (ICDE), p. 2. IEEE (2013)
7. Pelechrinis, K., et al.: Automatic evaluation of information provider reliability and expertise. World Wide Web **18**(1), 33–72 (2015)
8. Pasternack, J., Roth, D.: Knowing what to believe (when you already know something). In: Proceedings of the 23rd International Conference on Computational Linguistics. Association for Computational Linguistics (2010)
9. Li, Q., et al.: Resolving conflicts in heterogeneous data by truth discovery and source reliability estimation. In: Proceedings of the 2014 ACM SIGMOD International Conference on Management of Data. ACM (2014)
10. Li, Q., et al.: A confidence-aware approach for truth discovery on long-tail data. Proc. VLDB Endow. **8**(4), 425–436 (2014)
11. Zhao, B., Han, J.: A probabilistic model for estimating real-valued truth from conflicting sources. In: Proceedings of QDB (2012)
12. Jøsang, A.: Subjective Logic. Springer, Heidelberg (2016)

A Cellular Network Database
for Fingerprint Positioning Systems

Donatella Gubiani[1]([⊠]), Paolo Gallo[2], Andrea Viel[2,3], Andrea Dalla Torre[3],
and Angelo Montanari[2]

[1] University of Nova Gorica, Nova Gorica, Slovenia
donatella.gubiani@gmail.com
[2] University of Udine, Udine, Italy
{paolo.gallo,angelo.montanari}@uniud.it
[3] u-blox Italia SpA, Trieste, Italy
{andrea.viel,andrea.dallatorre}@ublox.com

Abstract. Besides being a fundamental infrastructure for communication, cellular networks are increasingly exploited for positioning via signal fingerprinting. Here, we focus on cellular signal fingerprinting, where an accurate and comprehensive knowledge of the network is fundamental. We propose an original multilevel database for cellular networks, which can be automatically updated with new fingerprint measurements and makes it possible to execute a number of meaningful analyses. In particular, it allows one to monitor the distribution of cellular networks over countries, to determine the density of cells in different areas, and to detect inconsistencies in fingerprint observations.

Keywords: Cellular network · Signal fingerprinting ·
Multilevel database · Data analysis

1 Introduction

Nowadays, our society is characterized by a pervasive use of mobile devices. The most common example of a mobile device is the smartphone, which combines voice communication with data services, Wi-Fi connection, and localization services to support advanced activities. As a matter of fact, many commonly employed applications make use of the current position of the user.

In order to compute the current location of a device, the most widely known solution is the Global Positioning System (GPS). Despite its widespread use, GPS has some significant drawbacks. On the one hand, the GPS signal cannot be received in certain conditions; on the other hand, energy consumption of GPS modules can be a problem with battery-powered devices [8,15].

Cellular signal fingerprinting offers a viable alternative to GPS solutions [1,2,7]: an estimation of the current position of a device can be obtained by comparing the signals received at that position with those recorded in a database of

T. Welzer et al. (Eds.): ADBIS 2019, CCIS 1064, pp. 111–119, 2019.
https://doi.org/10.1007/978-3-030-30278-8_14

observations taken at known positions. The use of signals coming from cellular networks in positioning systems has a number of advantages. The most significant ones are high coverage and low cost. Moreover, monitored devices need to be equipped with a cellular module and a simple software component only.

The most critical aspect is the need of collecting and maintaining a large set of fingerprints with their position. As a matter of fact, there are some ready-to-use repositories. Some of them are free of charge, others require a subscription fee. The most famous community dataset is OpenCellID [13], which is the result of a crowd source effort. We will make use of such a dataset. Unfortunately, as it happens with the other publicly-available datasets, collected data are poorly structured: information is recorded in a raw table (in csv format) of cellular signal readings paired with their position in a global reference system. In particular, no data structuring reflecting the organization of the cellular networks is present. The lack of an organization and of a user-friendly presentation of data complicates and limits their utilization. To overcome these weaknesses, we designed and implemented a cellular network relational database for fingerprint positioning systems, which integrates a large set of relevant data about cellular network, at different levels of granularity, in a coherent and systematic way.

It is well known that signal fingerprinting heavily relies on a comprehensive and accurate knowledge of cellular network configurations. In view of that, we start with an in-depth analysis of cellular networks in order to define a conceptual schema able to capture all meaningful aspects of their organization (Sect. 2). Then, we develop (Sect. 3) and populate (Sect. 4) the database, and show, by means of some representative examples, how useful information about the configuration of the network can be easily obtained from it (Sect. 5).

2   Basics of Cellular Networks

Cellular networks support wireless communication between mobile devices (both voice and data transmissions), and allow for seamless nation or even worldwide roaming with the same mobile phone. Different cellular technologies have been proposed over the years, each one with its own distinctive features [6,9,11].

Cellular radio networks are based on the deployment of a large number of low-powered base stations for signal transmission, each one with a limited transmission area, covering the surroundings with typically more than one cell. Cells are grouped into clusters to avoid adjacent cells to use the same frequency. Usually, a cell overlaps one or more other ones; a mobile device can distinguish among them by making use of their frequencies and scrambling codes (in the case of UMTS and LTE). Cells in a mobile network are put together into administrative areas, known as Location Areas (LA) in 2G/3G voice services, Routing Areas (RA) in 2G/3G data services, and Tracking Areas (TA) in 4G networks. These administrative areas are used to determine in a rough way the current location of a mobile device in the idle mode, that is, when it is switched on, but it is not using the network for any call or data exchange.

Independently of the adopted technology, there is a Public Land Mobile Network (PLMN), which can be identified by the Mobile Country Code (MCC), which indicates the country where the network is located, and the Mobile Network Code (MNC), which identifies the network in the country. On the basis of the administrative organisation, a Global Cell Identifier (GCI) can be used to globally identify every cell. Even though the properties of each element depend on the specific technology, the GCI can be viewed as the concatenation of the PLMN identifier, the LAC or the TAC identifier, and the Cell Identifier (CI).

3   A Multilevel Database for Cellular Networks

Following the consolidated methodology for database design, we start with the conceptual schema, which has been developed by using the ChronoGeoGraph (CGG) model [3,4], a spatiotemporal extension of the Entity-Relationship model.

Conceptual Design. The analysis of cellular networks reported in Sect. 2 makes it clear that they are organized in a hierarchical way (Fig. 1 - left side). Each network (entity *PLMN*) can be univocally identified by the values of the attributes *mcc* and *mmc*, and it consists of a number a distinct components ("subnetwork" entity *subPLMN*). Such a decomposition depends on the specific cellular technologies. According to the administrative perspective, each subnetwork consists of a number of cells (entity *CELL*), grouped into administrative areas (entity *ADMINISTRATIVE AREA*), on the basis of their registration and/or routing services. The properties of each level of the network organization depend on the specific technology. The differences among technologies are modeled by means of a suitable specialization of the entity *subPLMN*. The most significant one occurs at the level of the administrative areas (Fig. 1 - right side): the first two generations (2G GSM and 3G UMTS) distinguish two kinds of area, namely, Location and Routing Areas (entities *LA* and *RA*, respectively), which are identified by specific *lac* and *rac* codes. In the fourth generation, Location and Routing Areas are replaced by the Tracking Area (entity *TA*), which is characterized by a *tac* code. In all cases, a single code, combined with the *ci*, is used to identify the cells: the *lac*, for 2G and 3G, and the *tac*, for 4G.

Logical Design and Implementation. The CGG schema is turned into a relational one by applying the standard rules for the ER-to-relational schema mapping paired with dedicated rules for the encoding of the CGG spatial features [4]. The resulting schema is then implemented in the DBMS PostgreSQL with its spatial extension PostGIS. PL/SQL triggers are used for the automatic population of the tables. More precisely, when a new observation is received, it is checked and, if valid, the corresponding cell is inserted (or updated) in the table of cells. When a cell is inserted (or updated), the corresponding administrative area is inserted (or updated) in the table of the location/tracking areas, and so on. To obtain a comprehensive description of cellular networks via observations, the logical schema has been extended with some derived attributes whose value can be automatically computed, such as timestamp attributes (`firstview` and `lastview` attributes), that model the lifespan of the instances, as inferred from

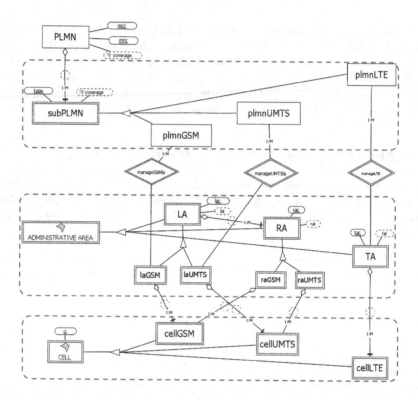

Fig. 1. A hierarchical schema for cellular networks: the administrative organization.

observations, and a counter attribute (attribute **numobs**), which keeps track of the number of observations of a single component to evaluate its reliability.

4   Data Ingestion and Filtering

In order to test our proposal, we used the OpenCellID dataset [13], which is probably the most popular, publicly available, crowd source project in the field. Basically, users collect observations about signals from cell towers and locations using a specific mobile phone application. On 2017, the project was acquired by the Unwired Labs company, a geolocation service provider enterprise that affected both the privacy policies and the level of detail of published data.

We downloaded the OpenCellID dataset on April 2017, just before the change to the data and privacy policies. Data is organized in a tabular format (csv format), where each observation is encoded by means of a number of attributes: *mcc, net, area, cell, lon, lat, signal, measured, created, rating, speed, direction, radio, ta, rnc, cid, psc, tac, pci, sid, nid,* and *bid.* As a matter of fact, not all these attributes are available for all technologies, and even their meaning slightly changes from one technology to the other. Moreover, since several devices contributed to the dataset, there are significant differences in terms of attribute structure and attribute subsets. The dataset covers an interval of about three years (from 2014-01-01 to 2017-03-17). It includes 42,951,377 observations distributed among three different cellular technologies. On the basis of a preliminary cleaning step, we restricted ourselves to 26,840,87 GSM observations, 6,177,024 UMTS, and 9,848,455 LTE (total 42,865,566) over the entire globe.

Fig. 2. Frequent errors: oversized cell over country.

Fig. 3. Frequent errors: oversized cell within country.

Data from a real scenario is generally affected by different types of error. As noticed in [12], there are at least four phenomena leading to incorrect readings: erroneous Cell IDs, antenna dragging, outliers, and unrealistic cell sizes. As a consequence, one of the most important operation in data acquisition is cleaning. To this end, we developed two filters to detect two relevant error situations. The first one deals with erroneous Cell IDs, and it consists of a domain check for each cellular parameter. The study of the cellular network standards allowed us to list a complete set of domains for each technology, e.g., ci ranges from 0 to 65535 for 2G and 3G, and from 65535 to 268435455 for 4G. If an observation exhibits one or more attribute values out of range, this is a valid reason to discard the entire observation, as some error may have occurred. The second filter focuses on the quality of the GPS position associated with each observation. First, to avoid inaccurate GPS positions, we excluded observations with less than 3 visible satellites. Later, a more interesting check has been done by comparing the GPS location of observations with borders of the country corresponding to the associated mcc. This spatial filter allowed us to avoid errors as the outlier depicted in Fig. 2. It is clear that a single cell belonging to a country cannot be received from such a long distance, in the middle of another country.

One of the main advantages of the proposed database is that, by keeping the geometry of cells constantly updated, it allows one to easily integrate additional filters. As an example, we may think of a filter that excludes observations which are not coherent with the physical characteristics of the cellular network. Consider the scenario in Fig. 3, where the coverage of the shown cell is clearly too wide with respect to the transmitting range. This can be interpreted as an error, which may be caused by various phenomena (e.g., a device may have submitted discrepant positions and signals due to a failure or a switch on after a flight).

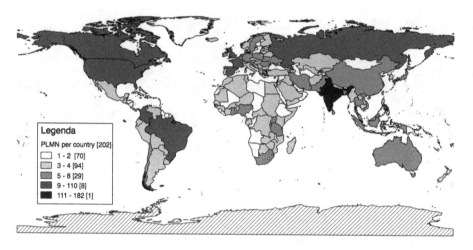

Fig. 4. Cellular networks per country.

5   Data Analysis

Once the data ingestion and filtering phases have been completed, the hierarchical structure of the schema allows us to execute some meaningful analysis tasks on the filtered data. It is worth pointing out that all the analyses rely upon a reconstruction of the actual network as perceived trough the observations recorded in the OpenCellID dataset. Moreover, despite the application of a couple of filters, that ruled out some inconsistent data, most probably data are still affected by errors. However, our main goal is to validate the proposed data model, and the above limitations have a little impact on it.

Let us first focus on the PLMNs occurring at the coarsest level. In Fig. 4, we give a graphical account of the result of a query that computes the number of cellular networks available in any single country, ignoring the specific cellular technology. India turns out to be the country with the maximum number of PLMNs (182), followed by USA (110) and Brazil (17). If we consider density (number of PLMNs divided by the area of the country), the first positions are occupied by small countries like Monaco, Gibraltar, and Macau.

Let us consider now the finer levels of the schema where administrative areas and cells come into play. In [10], the authors make it evident the existence of a correlation between the size of administrative areas and the density of population in the area. Other experiments showed that the number of cells in a given administrative area can also be used to distinguish between rural and urban areas. This is the case, for instance, with the results reported in [16], where the density of base stations, and thus of cells, is taken as a good criterion to classify an area as rural or not. By exploiting the relationships between the various spatial entities, we can easily determine the internal composition of the administrative areas, and compute the density of the cells.

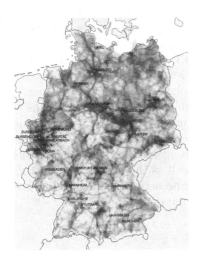

Fig. 5. Density of cells.

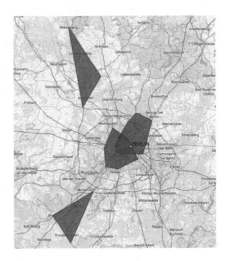

Fig. 6. Urban (red) and rural (blue) areas. (Color figure online)

In Fig. 5, we give a graphical account of the density of cells, in relation to the localization of urban areas, as it emerges from the pieces of information recorded in the considered dataset. It is clear from the picture that areas with a high density of cells (dark purple) are located where urban areas are, in particular where the cities with the highest population, such as Berlin, Hamburg, and the area of Koln, are (the largest cities are labeled with their name). As another example, working with GSM location areas with an extension between $50\,Km^2$ and $500\,m^2$ (avoiding areas which are not significant for the lack of a sufficient number of observations or for the presence of errors), it is possible to select the two areas with the highest (reps. lowest) density. As already pointed out, we cannot assume the dataset to be complete, and thus we may expect new cells to be added in the areas under consideration. However, the overall result confirms the original idea, as shown in Fig. 6: the two areas with the highest density (red areas) are located in the urban area of Berlin; the two areas with the lowest density (blue areas) are located in the rural area near Berlin.

6   Conclusions and Further Work

Thanks to their global coverage, cellular networks play a relevant role in a number of different contexts, including positioning systems based on cellular fingerprint observations. A comprehensive and accurate knowledge of their configuration is thus extremely important to optimise their usage. This paper is a first step towards the creation of a data store providing carefully structured information coming from observations of cellular networks.

We focused on the administrative organization of the networks (pieces of information usually available on the mobile device side), and we proposed a multilevel database that can be automatically updated with new cellular fingerprint measurements. We implemented the database in PostgreSQL, taking advantage of its spatial extension PostGIS, and populated it with an open data collection. Then, we demonstrated by some representative examples that it allows one to filter out inconsistent data and to perform a number of meaningful analyses.

We are currently exploring various possible improvements to the work done. One is the integration of other (public and proprietary) data sources in the database. We are also investigating an extension of the database schema with information about the network architecture as well as the physical parameters. Finally, to systematically deal with changes of the network configuration over time [14], we are thinking of adding some temporal dimensions [3,5].

References

1. Benikovsky, J., Brida, P., Machaj, J.: Localization in real GSM network with fingerprinting utilization. In: Chatzimisios, P., Verikoukis, C., Santamaría, I., Laddomada, M., Hoffmann, O. (eds.) Mobilight 2010. LNICST, vol. 45, pp. 699–709. Springer, Heidelberg (2010). https://doi.org/10.1007/978-3-642-16644-0_60
2. Chen, M.Y., et al.: Practical metropolitan-scale positioning for GSM phones. In: Dourish, P., Friday, A. (eds.) UbiComp 2006. LNCS, vol. 4206, pp. 225–242. Springer, Heidelberg (2006). https://doi.org/10.1007/11853565_14
3. Gubiani, D., Montanari, A.: ChronoGeoGraph: an expressive spatio-temporal conceptual model. In: Proceedings of the 15th SEBD, pp. 160–171 (2007)
4. Gubiani, D., Montanari, A.: A tool for the visual synthesis and the logical translation of spatio-temporal conceptual schemas. In: Proceedings of the 15th SEBD, pp. 495–498 (2007)
5. Gubiani, D., Montanari, A.: A relational encoding of a conceptual model with multiple temporal dimensions. In: Bhowmick, S.S., Küng, J., Wagner, R. (eds.) DEXA 2009. LNCS, vol. 5690, pp. 792–806. Springer, Heidelberg (2009). https://doi.org/10.1007/978-3-642-03573-9_67
6. Hoy, J.: Forensic Radio Survey for Cell Site Analysis. Wiley, New York (2013)
7. Paek, J., Kim, K.-H., Singh, J.P., Govindan, R.: Energy-efficient positioning for smartphones using cell-id sequence matching. In: Proceedings of the 9th MobiSys, pp. 293–306 (2011)
8. Li, X., Zhang, X., Chen, K., Feng, S.: Measurement and analysis of energy consumption on android smartphones. In: Proceedings of the 4th ICIST, pp. 242–245 (2014)

9. Pahlavan, K., Krishnaumurty, P.: Principles of Wireless Access and Localization. Wiley, New York (2013)
10. Ricciato, F., Widhalm, P., Craglia, M., Pantisano, F.: Estimating population density distribution from network-based mobile phone data (2015)
11. Sauter, M.: From GSM to LTE: An Introduction to Mobile Networks and Mobile-Broadband. Wiley, New York (2011)
12. Ulm, M., Widhalm, P., Brändle, N.: Characterization of mobile phone localization errors with OpenCelliD data. In: Proceedings of the 4th ICALT, pp. 100–104 (2015)
13. Unwired Labs: OpenCell ID (2017). http://www.opencellid.org. Accessed 28 Feb 2019
14. Viel, A., et al.: Dealing with network changes in cellular fingerprint positioning systems. In: Proceedings of the 7th ICL-GNSS, pp. 1–6 (2017)
15. Zhuang, Z., Kim, K.-H., Singh, J.P.: Improving energy efficiency of location sensing on smartphones. In: Proceedings of the 8th MobiSys, pp. 315–330 (2010)
16. Zhou, Y., et al.: Large-scale spatial distribution identification of base stations in cellular networks. IEEE Access **3**, 2987–2999 (2015)

Automatically Configuring Parallelism for Hybrid Layouts

Rana Faisal Munir[1,2]($\boxtimes$), Alberto Abelló[1], Oscar Romero[1], Maik Thiele[2], and Wolfgang Lehner[2]

[1] Universitat Politècnica de Catalunya, Barcelona, Spain
{fmunir,aabello,oromero}@essi.upc.edu
[2] Technische Universität Dresden, Dresden, Germany
{maik.thiele,wolfgang.lehner}@tu-dresden.de

Abstract. Distributed processing frameworks process data in parallel by dividing it into multiple partitions and each partition is processed in a separate task. The number of tasks is always created based on the total file size. However, this can lead to launch more tasks than needed in the case of hybrid layouts, because they help to read less data for certain operations (i.e., projection, selection). The over-provisioning of tasks may increase the job execution time and induce significant waste of computing resources. The latter due to the fact that each task introduces extra overhead (e.g., initialization, garbage collection, etc.).

To allow a more efficient use of resources and reduce the job execution time, we propose a cost-based approach that decides the number of tasks based on the data being read. The proposed cost-model can be utilized in a multi-objective approach to decide both the number of tasks and number of machines for execution.

Keywords: Big data · Hybrid storage layouts · Parallelism · Parquet · Spark

1 Introduction

The competition in businesses demands quick insights from data, which is exponentially growing from petabytes to zettabytes [15]. Researchers have proposed distributed processing frameworks (e.g., Hadoop ecosystem[1] and Spark[2]) for quickly processing such large volumes of data to meet the business demands. These frameworks provide distributed storage (e.g., HDFS[3]) and distributed processing [6]. In addition, for more efficient analysis, very wide tables [3,10] are being used to store non-normalized data in hybrid layouts [2,11]. Through their built-in operations (e.g., projection, selection), these layouts read data more efficiently from the disk. Hybrid layouts allow to read less data from the disk. This

---

[1] https://hadoop.apache.org.
[2] https://spark.apache.org.
[3] https://hadoop.apache.org/docs/r1.2.1/hdfs_design.html.

© Springer Nature Switzerland AG 2019
T. Welzer et al. (Eds.): ADBIS 2019, CCIS 1064, pp. 120–125, 2019.
https://doi.org/10.1007/978-3-030-30278-8_15

is not thoroughly exploited by distributed frameworks when deciding the number of tasks for processing the data. They always decide the number of tasks based on the total table size and not on the portion of the table being read. This leads to the over-provisioning of tasks, where many tasks remain idle—without any data to process, still present extra overhead (e.g., initialization time, garbage collection). Furthermore, the idle tasks also waste the computational resources which are assigned to them. The latter is not considered even in the area of cloud computing [9,16,18], where computational resources are decided based on the total data size. This leads to wastage of resources and money.

As argued above, we need to decide the number of tasks based on the actual data read from the disk. To do that, we first need to estimate the read size, which can be done by utilizing our cost model presented in [12], which estimates the scan, projection, and selection sizes for hybrid layouts.

In this paper, we propose to extend it further to estimate the makespan of the job implementing a query based on the estimated reading size. Thus, we design a framework which takes a user query and data statistics as inputs to estimate the reading size, and then through a multi-objective optimization method decides the number of *tasks* and *executors*. After configuring the number of tasks and executors, the query would be automatically submitted to a distributed processing framework.

The main contribution of this work is to discuss the main variables to be considered in a multi-objective optimization method to configure the number of tasks and executors of a given query.

The remainder of this paper is organized as follows: In Sect. 2, we discuss the related work. In Sect. 3, we present our approach for configuring the number of tasks and executors for a given query. Finally, in Sect. 4, we conclude the paper.

2 Related Work

Estimating Number of Tasks. There are research works [13,17] for Hadoop, which estimate the number of mappers and reducers tasks. Moreover, these approaches do not consider the amount of data read, while estimating the number of tasks. These works only estimate the tasks based on the available number of machines and some objectives (such as deadline). As previously argued, the amount of data read is an important factor in deciding the number of tasks.

Resource Provisioning in Cloud. There have been extensive research works [9, 16,18] by cloud community on resource provisioning. These works focus more on deciding the number of machines to process an application. They aim at saving energy and computational resources, which indirectly leads to cost savings. However, they make these decisions without considering the reading size. Our approach could help them to decide resource provisioning in more granular level and overall, it can help these works to achieve their goals more efficiently.

Tuning Configurable Parameters. There are research works [8,14] to tune the configurable parameters of distributed processing frameworks. In [5], the shuffle performance in Spark is improved by controlling the total number of shuffle files. These works do not explicitly consider the degree of parallelism. Their main aim is to fine tune a distributed processing framework. Our approach can be complementary to these works.

[1] presents a cost model for Spark SQL to evaluate different query plans. However, it does not configure the number of tasks and executors. We can use this work as complementary to ours, as well.

3   Our Approach

In this section, we discuss our proposed approach. It is based on a cost model which can be utilized in a multi-objective optimization method for configuring the number of tasks and executors.

For cost model, we propose to extend our previous work [12], that estimates the reading size for hybrid layouts. The reading size can be further used in estimating the number of tasks and executors. The number of tasks always depends on the size of partition (also known as *input split*), which we need to consider in the extended cost model.

Moreover, we focus on read-only analytical jobs, to estimate the amount of data read for their first operation and based on that, we try to find the best partition size to control the number of tasks. Given the simplicity of a file system (far from that of a DBMS), only three operations need to be considered: scan, projection, and selection. These three operations can be generalized to *selection sorted* and *selection unsorted*, because scan and projection operations are just the extreme cases of selection unsorted with selectivity factor of 1 (i.e., they read all Row Groups - *RGs*), and when you can choose the attributes in the output.

3.1   Estimating Number of Tasks

Modern distributed processing frameworks decide the number of tasks based on the total file size (which is the actual size of data without metadata) and the partition size. Moreover, all tasks cannot be executed at once, if the number of executors is less than the total number of tasks. Thus, we need multiple rounds/waves to finish the job.

3.2   Types of Partitions

As discussed earlier, data is processed by dividing into multiple partitions and each partition is processed in a separate task. These tasks process different amount of data presented in each partition, based on the number of referred attributes and the selection predicate. For instance, *selection unsorted* always reads all RGs, thus every task processes a full partition except the last one, whose partition might not be completely full, as shown in Fig. 1a.

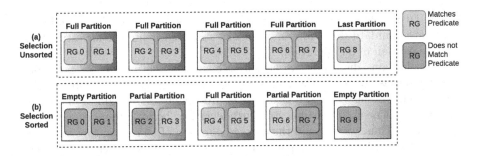

Fig. 1. Type of partitions in selection sorted and unsorted

On the other hand, *selection sorted* has high probability of skipping some RGs, thus, it can have empty partitions, which only read metadata. Additionally, it has full partitions that contain all matching RGs and two partial partitions, the first (from where selection starts) and last one (where selection ends), because requested data will not start just at the beginning and finish just at the end of a partition, as exemplified in Fig. 1b.

3.3 Task's Cost Estimation

The total cost of a task depends on four factors: *initialization cost*, *I/O cost*, *CPU cost*, and *networking cost*. The *initialization cost* is constant and can be determined according to the execution environment. The *I/O cost* depends on the amount of data read within a task and the disk bandwidth. We do not consider *CPU cost* due to its negligible impact compared to *I/O cost* (existing works [2,11] already proved that this is enough to capture the execution trend). Finally, we focus on the first operation loading data, thus *networking cost* for shuffling is also considered to be zero [2]. However, there is still a networking cost for metadata, because current solutions require to sequentially transfer metadata to all other executors before start processing the data. Typically, it is read and transferred by the master or driver executor.

Each partition has an initialization cost, which is a constant, and I/O cost (which depends on the amount of metadata and data read inside the partition). As shown in Fig. 1, *full partitions* read all RGs inside and *partial partitions* only read the matching RGs. Whereas, *last partitions* read the remaining data and *empty partitions* only read metadata. These costs help to estimate the total cost of each task, which can help to estimate the average cost.

3.4 Estimating Makespan

As discussed earlier, each task processes different amounts of data and thus, some tasks can finish earlier compared to others. Likewise, each executor can finish its assigned tasks in different times. Thus, we should estimate makespan based on the executor that is processing largest stack of tasks, which can be estimated using the number of executors active in the last wave. This would help to estimate standard deviation among tasks and used it further for estimating overall makespan of an operation.

For makespan, there are two scenarios based on the number of executors active in the last wave. In the first scenario, there is only one executor in the largest stack. In this case, the last task is processing the remaining data and then, we do not need to take any standard deviation, because there is one single largest stack. Thus, we just add the average duration of all task in that stack. In the second scenario, the makespan depends on metadata transfer, the average cost of a task, the number of executors running in the last wave, and their standard deviation. Thus, we need to estimate expected maximum [4] of those, which accounts for the standard deviation of the addition of tasks, as well as the maximum among executors in the last wave.

3.5 Multi-objective Optimization

As presented earlier, we would like to optimize two objectives (i.e., makespan and resource usage), which are mutually contradicting, i.e., if we want to reduce makespan, we require more computational resources and vice versa. Thus, we need to find a trade-off between them that satisfies user requirements and constraints. Additionally, to avoid unfavorable or even impossible configurations, we also need to consider three constraints. Firstly, the partition size must always be greater than or equal to the RG size. Secondly, we must have enough partitions to utilize all assigned executors. Finally, it must enforce the maximum limit on the number of executors.

We propose to use an existing multi-objective optimization approach, namely NSGA-II [7], implementing genetic algorithms. It takes objective functions along with constrains as input, and produces the Pareto front as an output. Typically, there is no single optimum in a multi-objective optimization problem, but a Pareto front which contains many potentially optimal solutions depending on user prioritization of one objective or another. Our framework[4] facilitates the user choice by reducing the many possible configurations to very few (belonging or close to the Pareto front), so helping her to select one according to her preferences.

4 Conclusions

Big Data systems process data on a cluster by creating multiple tasks. Typically, they create tasks based on the total size of the table, rather than based on the reading size of the query. Thus, we propose a multi-objective approach based on our extended cost model to configure the number of tasks and executors for a given query based on the reading size. The proposed approach will be implemented as a framework, that automatically configures the number of tasks and executors for a given query.

---

[4] http://www.essi.upc.edu/dtim/tools/adbis2019.

Acknowledgement. This research has been funded by the European Commission through the Erasmus Mundus Joint Doctorate "Information Technologies for Business Intelligence - Doctoral College" (IT4BI-DC).

References

1. Baldacci, L., Golfarelli, M.: A cost model for Spark SQL. TKDE **31**(5), 819–832 (2019)
2. Bian, H., Tao, Y., Jin, G., Chen, Y., Qin, X., Du, X.: Rainbow: adaptive layout optimization for wide tables. In: ICDE, pp. 1657–1660 (2018)
3. Bian, H., et al.: Wide table layout optimization based on column ordering and duplication. In: SIGMOD (2017)
4. Dasarathy, G.: A simple probability trick for bounding the expected maximum of n random variables. Technical report, Arizona State University (2011)
5. Davidson, A., Or, A.: Optimizing shuffle performance in Spark. Technical report, UC Berkeley (2013)
6. Dean, J., Ghemawat, S.: MapReduce: simplified data processing on large clusters. Commun. ACM **51**(1), 107–113 (2008)
7. Deb, K., Agrawal, S., Pratap, A., Meyarivan, T.: A fast and elitist multiobjective genetic algorithm: NSGA-II. IEEE Trans. Evol. Comput. **6**(2), 182–197 (2002)
8. Gounaris, A., Torres, J.: A methodology for Spark parameter tuning. Big Data Res. **11**, 22–32 (2018)
9. Islam, M.T., Karunasekera, S., Buyya, R.: dSpark: deadline-based resource allocation for big data applications in Apache Spark. In: e-Science, pp. 89–98 (2017)
10. Li, Y., Patel, J.M.: WideTable: an accelerator for analytical data processing. PVLDB **7**(10), 907–918 (2014)
11. Munir, R.F., Abelló, A., Romero, O., Thiele, M., Lehner, W.: ATUN-HL: auto tuning of hybrid layouts using workload and data characteristics. In: Benczúr, A., Thalheim, B., Horváth, T. (eds.) ADBIS 2018. LNCS, vol. 11019, pp. 200–215. Springer, Cham (2018). https://doi.org/10.1007/978-3-319-98398-1_14
12. Munir, R.F., Abelló, A., Romero, O., Thiele, M., Lehner, W.: A cost-based storage format selector for materialization in big data frameworks. In: Distributed and Parallel Databases (2019)
13. Nghiem, P.P., Figueira, S.M.: Towards efficient resource provisioning in MapReduce. JPDC **95**, 29–41 (2016)
14. Petridis, P., Gounaris, A., Torres, J.: Spark parameter tuning via trial-and-error. In: Angelov, P., Manolopoulos, Y., Iliadis, L., Roy, A., Vellasco, M. (eds.) INNS 2016. AISC, vol. 529, pp. 226–237. Springer, Cham (2017). https://doi.org/10.1007/978-3-319-47898-2_24
15. Shvachko, K.V.: HDFS scalability: the limits to growth. Login **35**(2), 6–16 (2010)
16. Sidhanta, S., Golab, W.M., Mukhopadhyay, S.: Optex: a deadline-aware cost optimization model for Spark. In: CCGrid, pp. 193–202 (2016)
17. Verma, A., Cherkasova, L., Campbell, R.H.: Resource provisioning framework for MapReduce jobs with performance goals. In: Kon, F., Kermarrec, A.-M. (eds.) Middleware 2011. LNCS, vol. 7049, pp. 165–186. Springer, Heidelberg (2011). https://doi.org/10.1007/978-3-642-25821-3_9
18. Wu, W., Lin, W., Hsu, C., He, L.: Energy-efficient Hadoop for big data analytics and computing: a systematic review and research insights. Future Gener. Comput. Syst. **86**, 1351–1367 (2018)

Automated Vertical Partitioning
with Deep Reinforcement Learning

Gabriel Campero Durand[✉], Rufat Piriyev, Marcus Pinnecke,
David Broneske, Balasubramanian Gurumurthy, and Gunter Saake

Otto-von-Guericke-Universität, Magdeburg, Germany
{campero,piriyev,pinnecke,david.broneske,gurumurthy,saake}@ovgu.de

Abstract. Finding the right vertical partitioning scheme to match a
workload is one of the essential database optimization problems. With
the proper partitioning, queries and management tasks can skip unnecessary data, improving their performance. Algorithmic approaches are
common for determining a partitioning scheme, with solutions being
shaped by their choice of cost models and pruning heuristics. In spite
of their advantages, these can be inefficient since they don't improve
with experience (e.g., learning from errors in cost estimates or heuristics
employed). In this paper we consider the feasibility of a general machine
learning solution to overcome such drawbacks. Specifically, we extend
the work in GridFormation, mapping the partitioning task to a reinforcement learning (RL) task. We validate our proposal experimentally
using a TPC-H database and workload, HDD cost models and the Google
Dopamine framework for deep RL. We report early evaluations using 3
standard DQN agents, establishing that agents can match the results of
state-of-the-art algorithms. We find that convergence is easily achievable
for single table-workload pairs, but that generalizing to random workloads requires further work. We also report competitive runtimes for our
agents on both GPU and CPU inference, outperforming some state-of-
the-art algorithms, as the number of attributes in a table increases.

Keywords: Vertical partitioning · Deep reinforcement learning

1 Introduction

The efficiency of data management tools is predicated on how well their configuration (e.g. physical design) matches the workload that they process. Database
administrators are commonly responsible for defining such configuration, but
even for the most experienced practitioners finding the optimal remains challenging given: (a) the high number of configurable knobs & possible settings,
(b) rapidly changing workloads, and finally (c) the uncertainty in predicting

We thank Dr. Christoph Steup, Milena Malysheva and Ivan Prymak for valuable feedback. This work was partially funded by the DFG (grant no.: SA 465/50-1).

T. Welzer et al. (Eds.): ADBIS 2019, CCIS 1064, pp. 126–134, 2019.
https://doi.org/10.1007/978-3-030-30278-8_16

the impact of choices when based on cost models or assumptions (e.g. knob independence) that might not match real-world systems. To alleviate these challenges either fully or partially automated tools are used (e.g. physical design advisors [1]). Specially relevant in these tools is the incorporation of machine learning models, helping tools to learn from experience, relying less on initial assumptions. In recent years, due to reinforcement learning methods (RL) outperforming humans in highly complex game scenarios (including Atari games, Go, Poker and real-time strategy games), teams from academia [3,7,9,10,12] and industry[1] propose data management solutions that learn from real-world signals using RL or deep RL (DRL, i.e., the combination of RL methods, with neural networks for function approximation). In this context, DRL enables models to have a limited memory footprint, a competitive inference process that can use massively-parallel processors and models can generalize from past experiences to unknown states. In this work we study how vertical partitioning can be supported with a novel DRL solution, in order to learn from experience and accelerate the decision making process on real-world signals. Vertical partitioning is a core physical design task, responsible for dividing an existing logical relation into optimally-defined physical partitions, where each partition is a group of attributes from the original relation. The main purpose of this operation is to reduce I/O related costs, by keeping in memory only data that is relevant to an expected workload. The right vertical partitioning improves query performance and other database physical design decisions [8].

Our contributions in this study are: (1) Building upon previous work [3], we design necessary components to make vertical partitioning a problem amiable to be learned with DRL models. (2) We offer a prototypical implementation of our solution, using OpenAI Gym and the Google Dopamine framework for DRL [4], adapting 3 standard value-based agents: DQN, distributional DQN with prioritized experience replay [2,6] and distributional DQN with implicit quantiles [5]. (3) We contribute a study on the learnability of vertical partitioning for a TPC-H workload, showing that DRL can indeed learn to produce the same solutions as algorithms. We find single table-workloads simple to learn, whereas generalizing to random workloads requires more effort. (4) We show that the out-of-the-box inference of DRL is competitive with the performance of state-of-the-art algorithms, specially as table sizes increase.

We structure our work as follows: We start with a brief background (Sect. 2) and the design of our solution (Sect. 3), covering the rewarding scheme, the observation and action spaces. Our evaluation and results follow (Sect. 4). We conclude by summarizing our findings and outlining future work (Sect. 5).

---

[1] https://blogs.oracle.com/oracle-database/oracle-database-19c-now-available-on-oracle-exadata.

2 Background

Automated Vertical Partitioning: The problem of algorithmically finding the best vertical partitioning is not new – specially for complementing DBA's manual partitioning [8]. Since this problem is NP complete [1], most solutions seek a trade-off between finding the optimal and the runtime required to produce the result. This trade-off is managed with the adoption of heuristics that prune the search process. Among some solutions, as covered by Jindal et al. [8], we can mention the following: Navathe, one of the earliest algorithms, consisting of keeping an affinity matrix and clustering as a means to create partitions. Hill-Climb, a simple approach that makes the process iterative, with each iteration exploring all possible combinations into two, of the existing partitions, passing to the next iteration the most promising candidate. This process ends when there is no cost improvement with respect to former iterations. AutoPart, a variant of HillClimb, including categorical partitioning and pruning the search space based on query coverage. O2P is a variant of Navathe, which seeks to perform faster, making more concessions on optimality. Apart from algorithmic solutions, the task has also been studied with unsupervised learning and genetic algorithms.

Deep Reinforcement Learning: RL is a class of machine learning solution, where agents are set to learn how to act by interacting with an environment through actions, observing the states of the environment and the rewards obtained. Formally, the scenario is modeled as a Markov decision process. In developing RL solutions, the ability to scale to cases where the space of possible states is quite large poses challenges to the exploration process and to the learned model. Rather than storing all observations, a function approximation solution is required to limit the memory employed, and also to generalize knowledge from visited states to unvisited ones. DRL adopts neural networks for the function approximation task. There are 3 general families of RL and DRL algorithms. Namely, value-based solutions, policy gradient solutions and model-based solutions. The latter form a model of the transitions and expected rewards in the environment, reducing the learning task to a planning problem. From the former approaches, they either learn the long-term value of performing all actions in a state, or simply the policy to pick the best action. In our study we employ 3 value-based methods: DQN, a deep approach to Q-learning relying on experience replay and fixed-Q targets; Distributional DQN with prioritized experience replay (in Dopamine, this model is called Rainbow), which mainly learns to approximate the complete distribution rather than the approximate expectation of each Q-value [2] (in our study we use a categorical distribution, as considered for C51), in addition to using a priority sampling strategy; and Distributional DQN with implicit quantiles, this model extends the previous, using a deterministic parametric function to reparameterize samples from a base distribution to the quantile values of a return distribution [5].

3   Design

Observation Space: Our state space is represented as a 2D array. For the case of our study, it is of size 16×23. Each column represents one partition. If there is a state with only two partitions, these should take the two leftmost positions of the state array. The remaining columns are set to zero. Hence, this observation space can represent tables of up to 16 columns. This choice fits well the TPC-H database, since the largest table consists of 16 columns (LINEITEM). The first row in the observation space collects the size of each partition, which is the sum of the attribute sizes in the partition, multiplied by the row count of the table. The remaining rows represent 22 queries of a workload. At each position the query holds a 1 if the attribute is accessed by the query, or a 0 otherwise.

Actions: Actions can be either to merge or to split partitions, forming bottom-up or top-down processes. In our case we focus on merging. Given the number of columns in our implementation, we have a maximum of 120 actions (which is the number of combinations of 16 items into groups of 2, without repetitions or order). The semantic of actions is given by their number, with action 0 being joining the columns *[0, 1]*, and action 119 being joining *[14,15]*. Once joined, the sizes of the columns are added (first row), and the entries in the remaining rows are resolved by a logical OR between the two entries. Finally, the rightmost of the two columns is removed from the observation space, and the result of the operation is stored in the other column, shifting all non-zero columns to the left.

Rewarding Scheme: Before defining a reward, we calculate the value (V) of a state as the inverse of the cost (in our case the HDD cost). The reward itself is calculated with (1).

$$Reward_{current\_state} = \begin{cases} 0, & \text{if not end of the game} \\ 100 * \frac{V_{current\_state} - V_0}{\Delta_{best}}, & \text{otherwise} \end{cases} \tag{1}$$

Here V_0 - is the value at the beginning of the episode (with all the attributes in different partitions), $\Delta_{best} = V_{beststate} - V_0$ - where $V_{beststate}$ is the value of the state reached when following all the expert steps (getting this value from the expert). By expressing the final reward in terms of the percentage compared to the expert, we are able to normalize the rewards obtained, solving the problem of having rewards at different magnitudes based on table and workload characteristics. As an expert we employ HillClimb. Alternatively, the expert could also be a fixed heuristic (e.g. the cost of supporting the workload with a columnar partitioning), as considered in related work [12]. The game reaches a terminal state (game over) when one of these conditions occur: if the number of columns $= 1$ (only one partition left), if $V_{state} < V_{state-1}$, if the selected action is not valid (e.g. it refers to a column that does not exist in the current table).

4   Evaluation

4.1   Experiment Configuration

The main components of our prototype are shown in Fig. 1. We created an OpenAI Gym environment, encapsulating the logic of our solution. The *Runner* class from Dopamine was adapted to select the table-workload pair for each episode. This class, in its role of organizing the learning process, also launches an experiment decomposed into iterations, in turn

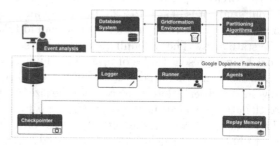

Fig. 1. Architecture of our implemented solution, based on the Google Dopamine Framework [4]

composed of test and evaluate phases. During the experiment the *Runner* fulfills the task of keeping up-to-date the states of the agent and environment, communicating between them, while logging and checkpointing the experimental data. In our implementation, for initializing the environment, it interacts with the vertical partitioning algorithms (e.g. HillClimb, AutoPart, etc.), such that the cost and actions performed by the algorithms/expert can be used to normalize the rewards obtained by the agents during an episode, across diverse cases.

Rewarding and Cost Model: For our experiments we implemented in our environment the rewarding scheme described previously (Sect. 3). This is based on the value of a state, which in turn depends on a cost model applied to each partition in the given state. For evaluating the cost of each partition we used an HDD cost model employed in previous work [8]. Though we use a cost model to calculate the rewards (i.e., cost-model boot-strapping, as considered in other research [11]) given by our environment to the agents, it should be possible to replace this component, either with other cost models, or with real-world signals, without affecting to a large extent the overall solution.

Agents: We adopted the provided agents from the Google Dopamine Framework (DQN, distributional DQN with prioritized experience replay/Rainbow, and implicit quantiles), using provided hyper-parameters. Agents differ in the last layers, which are crucial in determining the predicted values. In our adoption of the agents we changed the last layer, to match the number of actions, and we introduced a hidden layer preceding it, with 512 neurons. As activation function for this layer we used ReLu. When appropriate, we used multi-step variants of the agents.

Benchmark Data: We report results on a TPC-H database of scale factor 10 (SF10), using all tables and the complete TPC-H workload. For our observation space we create "views" for how the workload is seen from the perspective of a table. We use state-of-the-art open source Java implementations of the algorithms [8]. For normalization, in our results we use the cost obtained by HillClimb. We could have also used a simple heuristic.

Hardware: We used a commodity multi-core machine running Ubuntu 16.04, with an Intel® Core™i7-6700HQ CPU @ 2.60 GHz (8 cores in total), an NVIDIA® GeForce GTX 960M graphic processing unit and 15.5 GiB of memory.

4.2   Training

TPC-H Workloads - Single Table-Workload Pair, or Set of Table-Workload Pairs: In order to give an optimal partitioning for the tables in the TPC-H workload at SF10 cases generally require few actions, with 4 being the highest (CUSTOMER), and some cases requiring no action (PART). Considering the simplicity of each case, we started training by evaluating the convergence of the agents in finding a solution for each table-workload pair. The results from our evaluation of convergence are shown in Fig. 2. For conciseness, we omit the results for other tables, since they all converge in less than 10 iteration (with 500 training steps per iteration, each iteration corresponding to an average over 100 evaluation steps). Figure 3 considers the training on the set of 8 TPC-H table-workload pairs, using an update horizon of 3 (i.e., multi-step of 3). From these results we see, at first, that when going beyond 1 table-workload pair, convergence requires up to 300 iterations, overall. In this experiment we compare agents with regards to pruning or not pruning actions. This refers to the heuristic of pruning the Q-values obtained from the neural network, by artificially removing those that correspond to actions that would be invalid (i.e., we hard-code it before choosing an action, rather than letting the agent explore bad actions). Results show that by not pruning actions, convergence on very simple cases takes more time.

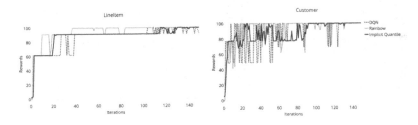

Fig. 2. Training for single table-workload pairs

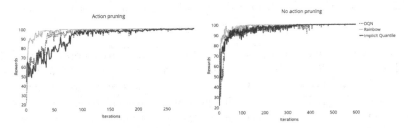

Fig. 3. Pruning or non-pruning invalid actions

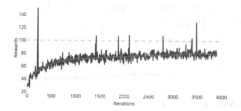

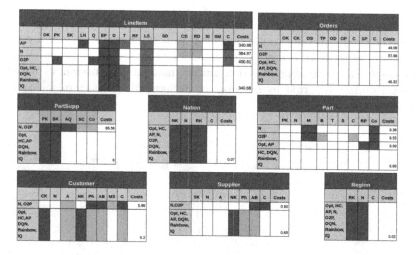

Generalizing to Random Workloads: We evaluate the learning process with a fixed table (CUSTOMER), and a series of random workloads. Figure 4 portrays the results of an Implicit Quantile agent. At 4k iterations there's only convergence to 80% of expertise. To master such challenging scenario, extra work is required. In practice, however, workloads are not entirely random.

Fig. 4. Implicit quantile agent with update horizon 3 (multi-step of 3), training on CUSTOMER table, random workloads

Hence, scoped stochastic workloads might be better to evaluate generalization. Figure 4 also shows cases where the agent outperforms HillClimb (HC). This is understandable as HC does not guarantee the optima.

4.3   Inference

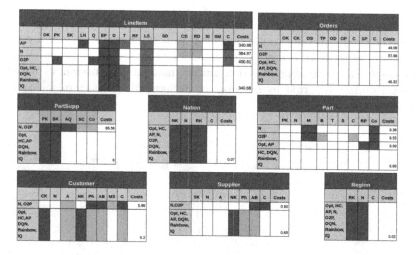

Fig. 5. Resulting partitions (in same color, except when white) and costs obtained, for TPC-H tables. Agents predict the cost optimal partitioning. Algorithms: AutoPart (AP), Brute force (Opt), HillClimb (HC), Navathe (N), O2P. (Color figure online)

Figure 5, presents the exact costs and partitions of alternative approaches. DRL, trained on the 8 TPC-H tables-workloads, reach for this case the exact costs of the optimal (from brute force). Concerning the time the agents require, after training, to make predictions (and change state based on actions), Fig. 6 shows average runtimes across 100 repetitions for the given tables. Out-of-the-box inference is able to outperform brute-force approaches for tables with increasing number of attributes. This inference also remains in the same order of magnitude, being competitive with solutions like HillClimb, as the number of attributes in a table increases. Vanilla DQN provides a faster inference, as expected, given its simpler end layers. Still, more studies are required, on a tuned inference process.

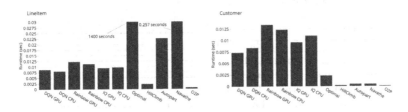

Fig. 6. Inference time of the agents, compared with partitioning algorithms.

5   Conclusion

In this paper we present a novel DRL solution for vertical partitioning, and early evaluations using a cost model and a TPC-H database & workload. We report that the partitioning for individual table-workload pairs or for a small set of pairs is simple to learn. However, generalizing to entirely random workloads given a table, or to random tables, remains challenging, requiring further work. In terms of inference, we validate that our agents are able to learn the same behavior of algorithmic approaches. We show that on cases (when training for generalizing) it is possible to spot instances where our approach outperforms algorithmic solutions. We find that the brute force algorithm can be outperformed (in optimization time) by our off-the-shelf solution, and that our solution is competitive with algorithms, remaining in the same order of magnitude, and becoming more competitive as the number of attributes increases. Future work will consider more tuned model serving procedures (inference and observation space transitions) and training with a database. Further aspects relevant to the production-readiness of our solution should be studied in future work: more challenging workloads, extensions to observation and action (bottom-up and top-down actions) spaces, model improvements (specialized architectures and designs enabling the model to scale to a larger number of columns), different forms of partitioning (horizontal, hybrid), learning from demonstrations and, finally, interface design such that DRL solutions can be offered as useful plugins to data management systems.

References

1. Agrawal, S., Narasayya, V., Yang, B.: Integrating vertical and horizontal partitioning into automated physical database design. In: SIGMOD, pp. 359–370. ACM (2004)
2. Bellemare, M.G., Dabney, W., Munos, R.: A distributional perspective on reinforcement learning. In: ICML, pp. 449–458 (2017)

3. Durand, G.C., Pinnecke, M., Piriyev, R., et al.: GridFormation: towards self-driven online data partitioning using reinforcement learning. In: AIDM@SIGMOD, p. 1. ACM (2018)
4. Castro, P.S., Moitra, S., Gelada, C., et al.: Dopamine: a research framework for deep reinforcement learning (2018). arXiv preprint arXiv:1812.06110
5. Dabney, W., Ostrovski, G., Silver, D., et al.: Implicit quantile networks for distributional reinforcement learning (2018). arXiv preprint arXiv:1806.06923
6. Hessel, M., Modayil, J., Van Hasselt, H., et al.: Rainbow: combining improvements in deep reinforcement learning. In: AAAI (2018)
7. Hilprecht, B., Binnig, C., Röhm, U.: Towards learning a partitioning advisor with deep reinforcement learning. In: AIDM@SIGMOD (2019)
8. Jindal, A., Palatinus, E., Pavlov, V., et al.: A comparison of knives for bread slicing. Proc. VLDB Endow. **6**(6), 361–372 (2013)
9. Krishnan, S., Yang, Z., Goldberg, K., et al.: Learning to optimize join queries with deep reinforcement learning (2018). arXiv preprint arXiv:1808.03196
10. Marcus, R., Papaemmanouil, O.: Deep reinforcement learning for join order enumeration. In: AIDM@SIGMOD, p. 3. ACM (2018)
11. Marcus, R., Papaemmanouil, O.: Towards a hands-free query optimizer through deep learning (2018). arXiv preprint arXiv:1809.10212
12. Sharma, A., Schuhknecht, F.M., Dittrich, J.: The case for automatic database administration using deep reinforcement learning (2018). arXiv preprint arXiv:1801.05643

BrainFlux: An Integrated Data Warehousing Infrastructure for Dynamic Health Data

Jonathan Elmer[2], Quan Zhou[1], Yichi Zhang[1], Fan Yang[1],
and Vladimir I. Zadorozhny[1(✉)]

[1] Department of Informatics and Networked Systems,
University of Pittsburgh, Pittsburgh, PA, USA
{quz3,yiz141,fan_yang,viz}@pitt.edu
[2] Departments of Emergency Medicine, Critical Care Medicine and Neurology,
University of Pittsburgh, Pittsburgh, PA, USA
elmerjp@upmc.edu

Abstract. Intensively sampled longitudinal data are ubiquitous in modern medicine and pose unique challenges. Future progress in medicine will exploit the fact that multiple physiological measurements can be recorded electronically at arbitrarily high temporal resolution, but novel information systems are needed to create knowledge from these data. In this paper, we present BrainFlux - a novel data warehousing technology that implements a holistic paradigm for storage, summarization and discovery of trends in dynamic data from continuously measured physiological processes. We focus on one specific example of high-resolution longitudinal data: electroencephalographic (EEG) recordings obtained in a large cohort of comatose survivors of cardiac arrest. Post-arrest EEG conveys important clinical and prognostic information to clinicians, but the rigor of prior analyses has been limited by a lack of proper data processing infrastructure. Our EEG data are complemented by a complete set of contemporaneously recorded electronic medical record data, clinical characteristics, and patient outcomes. In this paper, we describe the architecture and performance characteristics of BrainFlux's scalable data warehousing infrastructure that efficiently stores these data and optimizes information-preserving summarization.

Keywords: Medical data streams · Large-scale data warehousing · Time series data aggregation

1 Introduction

Does a given critically ill patient have sufficient potential to recover to justify continued aggressive care? How should we store, monitor, and analyze health data to help answer such questions in a most efficient way? Can we rescue the huge amounts of valuable and continuously accumulating health data that are

© Springer Nature Switzerland AG 2019
T. Welzer et al. (Eds.): ADBIS 2019, CCIS 1064, pp. 135–143, 2019.
https://doi.org/10.1007/978-3-030-30278-8_17

currently discarded because of the limitations of data storage and processing infrastructures? How can we use these data to efficiently summarize patients' current clinical conditions and provide timely warnings when deterioration appears imminent? In this paper we present BrainFlux - a novel data warehousing technology that allows us to address above questions in a systematic and efficient way. BrainFlux implements a holistic paradigm for preservation, aggregation and discovery of trends in large amounts of data from continuously measured physiological processes.

Individual health changes dynamically over time. Proper understanding of a patient's current status, anticipated clinical course and likely outcome is a fundamental requirement in precision medicine. Hospitalized patients often have multiple physiological parameters monitored continuously with innumerable other measures sampled repeatedly over time. Despite the wealth of available information, clinical providers and biomedical researchers typically rely on massively down-sampled data. For example, the heart's electrical activity may be observed continuously using a telemetry monitor. Waveform telemetry data are typically stored for a short time to allow qualitative review by clinical providers then discarded. Only instantaneously estimated heart rate is documented and retained, typically no more than once every hour. This convention for handling complex biomedical data has persisted for two reasons: alternative methods are intensive in terms of computational and storage needs, and human providers interacting with these data are limited in their capacity to integrate complex data into clinical decision making.

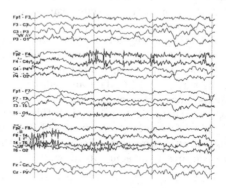

Fig. 1. 15 seconds of EEG activity

While expedient, oversimplification of complex, continuous physiological processes lacks a sound biological underpinning. A wealth of important data may be lost. Returning to our example of telemetry data, change in beat-to-beat heart rate variability ("RR-variability") is a sensitive predictor of impending deterioration not captured by either qualitative review of heart rhythm or measures of absolute heart rate [4]. Even subtler information that can be derived from these signals, such as the correlation structure between several simultaneously measured biological processes [3], may also be informative but is lost in current practice. Future progress in medicine will exploit the fact that multiple physiological measurements can be recorded electronically at arbitrarily high temporal resolution (thousands of samples per second).

A unique feature of our BrainFlux architecture is its efficient and information-preserving summarization of huge time series at different aggregation levels, which makes it highly scalable and applicable for a wide array of health monitoring tasks. In particular, we focus on a well-defined and challenging patient population: comatose survivors of cardiac arrest. Comatose post-arrest patients typically undergo intensive monitoring for hours to days after hospital arrival resulting in multiple data streams. One of the richest streams of available information that might inform post-arrest care is electroencephalography (EEG). EEG measures brain activity and is typically recorded at 256 Hz from 22 electrodes adhered to standard positions on a patient's scalp (Fig. 1). EEG can guide clinical care. For example, seizures are common after cardiac arrest, worsen brain injury, but may be detected on EEG prompting treated [5]. EEG can also inform neurological prognostication: certain quantitative and qualitative EEG characteristics strongly predict severity of brain injury and potential for recovery [6,7]. A single day's EEG recording on one patient results in approximately 4.5×108 data points. Current standard of care is for a specially trained physician to interpret daily EEG recordings subjectively, generally providing no more than 10 to 20 impressions of various features (for example "no seizures were observed"). Other work has explored quantitative EEG features like amplitude or fast Fourier Transform measures of band-pass filtered spectral power. These are informative, but again are typically have been considered as single summary measures during a given time epoch without consideration of their change over time. Contemporaneous exogenous factors (e.g. medication administration) and endogenous factors (e.g. patient temperature or blood pressure) affect the EEG signal and may modify the relationship between EEG and outcomes. There is an urgent need for quick and clinically effective processing and summarization of these complex data.

In this work, we focus on the following primary contributions: (a) an efficient data warehousing architecture integrating massive amounts of dynamic health data that utilizes advanced time-series database processing; (b) development and evaluation of advanced data aggregation strategies for a continuously growing health data repository; (c) implementation of the proposed solutions in a large-scale integrated framework using the computational and data handling resources of Pittsburgh Supercomputing Center (PSC). Because BrainFlux scales to extremely high data loads, endogenous and exogenous clinical factors can be considered as covariates or effect modifiers and serve to further refine clinical decision support systems. Our system constitutes a model for productive collaboration of medical and data scientists who seek to advance precision medicine built on solid data.

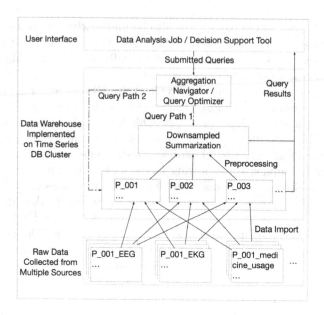

Fig. 2. General BrainFlux architecture

2   BrainFlux Architecture

A generalized BrainFlux architecture is shown in Fig. 2. It has three major layers. At the bottom is a data collection and storage layer, where multiple clinically derived data streams are transferred for future aggregation and processing. Some data are EEG-based while others are patient-level metadata, time-stamped medication administration data, etc. Note that data are heterogeneous and distributed in multiple files for each patient.

The BrainFlux imports raw data from the bottom layer and organizes them based on record timestamps. While imported data are theoretically suitable for data analytics, the huge amount of data make any meaningful analysis of these raw data challenging. Instead, time-varying summary measures of aggregated data are often of greater clinical and research interest. It is not uncommon for BrainFlux to handle hundreds of terabytes of raw data on regular basis, which risks creating a barrier between the data and end user while the data are imported, organized and aggregated. With a data transfer rate of 12 MB/s, for example, a user would have to wait a full day simply for the import process on 1 TB of data to be completed. To address this problem, BrainFlux takes advantage of system idle time to prepare aggregated data for future analyses based on user-defined parameters. For example, BrainFlux may generate 10-s, 60-s, and 120-s summaries (e.g. mean, median, variance and between-electrode correlations) from the original imported data. One set of such summary measures may require a full week of processing time, but is then added to the data warehouse as a new sub-layer. BrainFlux's top layer is an aggregation navigator that intelligently select the most appropriate data aggregation sub-layer in the warehouse to which to direct new user-defined queries. The navigator is designed to maximize new query efficiency while preserving information accuracy for data-driven decision making. For user defined or ML-based data exploration, the aggregation navigator directs queries to the most appropriate sub-layer to maximize efficiency while preserving information. The performance boost is considerable. For example, if we wished to explore 1-h mean values of two qEEG trends averaged across all electrodes, the system would direct this aggregation task to the lowest resolution data that support this query without information loss. Using 120-s means instead of 1 Hz data provides a linear (120-fold) boost to data processing time.

3   Implementation

BrainFlux uses on demand, distributed computing resources of Pittsburgh Supercomputing Center [10] for the main data warehouse. We also support a test version of the system with the same functionality but a fraction of the raw data. The test version runs on a single node and can be used to test the performance of our algorithms and to identify system flaws without using PSC resources. Both versions of the system fully utilize the multi-threading functionality of powerful computers to increase system performance. To handle large scale medical time-series BrainFlux uses InfluxDB [2] distributed time-series database with high-speed read/write capabilities. Non-EEG data, including metadata of patients, user queries, warehouse log, etc., are stored in a separate MySQL database to facilitate the metadata management. The web-based user interface is written in Java with Spring Framework. We use Grafana analytics and monitoring platform [1] for visualization of the data on every level of the data warehouse.

3.1   Importing BrainFlux Data

It is standard medical practice at many hospitals to monitor all comatose post-arrest patients with continuous EEG. After EEG acquisition, we deidentify and export raw EEG waveforms (in open source European Data Format, or .EDF+). We then use Persyst software (Persyst Development Corp, Prescott, AZ) and MATLAB (MathWorks Inc., Natick, MA) to generate over 5,000 standard and custom qEEG features a frequency of 1 Hz across individual electrodes for the duration of monitoring. We then transfer these data to BrainFlux for aggregation and analysis. If there are embedded errors within input files, BrainFlux automatically flags the file as problematic during the import process.

3.2   Querying BrainFlux Data

After data are imported, researchers may wish to explore specific hypotheses that require only a subset of the patients contained in BrainFlux. For example, patients with a EEG feature above or below a threshold value, patients receiving a particular medication, or those younger than 50 years of age could be of particular interest.

3.3   Summarizing and Exporting BrainFlux Data

A key feature of BrainFlux is the ability to efficiently summarize qEEG features over space (e.g. EEG electrodes distributed across anatomic regions of the scalp, or adjacent frequency bands of a fast Fourier transformed frequency (FFT) spectrogram) and time (i.e. consecutive seconds of data). For example, a user may be interested in the FFT spectral power within the alpha frequency band (8 Hz to 13 Hz, which in BrainFlux is stored in 0.5 Hz bins or 10 separate columns), across all anatomical regions calculated against a common average reference (18 columns per frequency bin) summarized as hourly averages (3,600 consecutive seconds).

3.4   Data-Driven Decision Making

BrainFlux can be efficiently used for patient outcome prediction based on statistical models of EEG features. For instance, group-based trajectory modeling (GBTM) is a specialized example of a finite mixture model that identifies clusters of individuals that follow a similar trajectory of one or more repeated measures over time [9]. We built a 5-group model describing the evolution of suppression ratio (SR), a single EEG feature, over time [6]. The model strongly predicts patients outcomes after cardiac arrest, and discriminates between those who may recover and those who cannot earlier than previously possible. Model estimation required that we aggregate SR values across anatomic regions of the brain and down sampled to various time intervals. We describe the performance of this summarization strategy below, and demonstrate how BrainFlux balanced efficiency with information loss.

4   BrainFlux Performance

For the BrainFlux performance evaluation, we used data recorded for 2,331 patients that had cardiac arrest. Each patient also had varied total recording length, from hours to months depending on how long the patient stayed in the ICU resulting in 95,000 h or 4,000 days of data. We used this data along with the metadata of each patient totaling approximately 15 TB.

4.1   Data Loading

For data loading, we utilize 28 cores on each node with every core performing data loading at 800 KB/s rate. We found that BrainFlux loads up to 22.5 MB, or approximately 2.6 million data points per second. Since the patients' brain signal is recorded every second, BrainFlux could perform real-time monitoring of more than 450 patients on a single node system. We also used a remote loading mechanism, where source data on a local machine are loaded into the remote database on a regular memory node with 28 cores and 112 GB of RAM hosted by PSC. This achieved loading data rate of up to 17.3 MB/s. Overall, multi-threaded data loading gives close to linear increase in the processing speed.

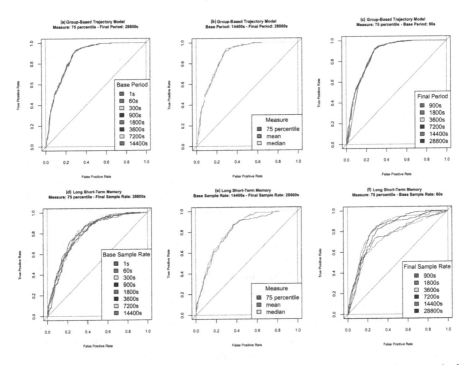

Fig. 3. ROC curves demonstrating sustainable accuracy in predicting patient survival probability on progressively more aggregated data.

4.2 Impact of Data Aggregation

To explore performance of data-driven decision making for different data aggregation strategies we conducted multiple queries on suppression ratio values aggregated at different levels. We summarized mean, median and 75-percentile values over 36 h of observation at time intervals of 1 min, 5 min, 15 min, 30 min, 1 h, 2 h, 4 h and 8 h, in a sequential manner. For each coarser interval, the query was performed iteratively from each possible finer data granularity. We found that the query execution time significantly decreases (from a few hours to a few seconds) as the down-sampling interval of the base data increases. As a more solid proof to the robustness of our optimized query output, we fed SR data into two of statistical learning models widely used for clinic data: GBTM [9] and long short-term memory (LSTM) [8]. We used these to predict individual survival probability from SR. Figure 3 shows receiver operating characteristic (ROC) curves for both models with varying aggregation granularities. We find that overall discriminatory power of GBTM was minimally affected by data granularity. LSTM outperformed GBTM analyzing high-resolution data but was more sensitive to data aggregation with significant decrement in performance at lower data resolutions.

execution time of analytical queries and maintains overall high accuracy of the
data-driven decision making, in particular, predicting the survival probability of
patients.

5 Conclusion

We presented BrainFlux - a novel data warehousing technology for storage, summarization and discovery of trends in dynamic data from continuously measured
physiological processes. In particular, the BrainFlux data efficiently exploits data
aggregation to improve query execution time while maintaining accuracy of the
data-driven prediction. The BrainFlux system helps medical experts build complex queries for data analysis and support decision-making. The complex data
summarization and aggregation navigation functionality is transparent for the
end user. The BrainFlux architecture allows data analysts and medical experts
to collaborate more efficiently. Note, that the BrainFlux can be used both for retrospective exploration of existing data for the purpose of research and discovery,
and real-time data exploration for the purpose of clinical decision making.

While the BrainFlux data aggregation significantly improves the execution
time of analytical queries and maintains overall high accuracy of the data-driven
decision-making, different modeling tools might have different susceptibilities to
aggregation and our navigator needs to be tuned to specific model performance
characteristics. In future work we will devise various data summarization techniques to capture the significant information for different kind of data that can be
used to build an efficient and effective decision-making system. We will explore
their applicability to the task of medical data monitoring and develop specific
aggregation/disaggregation methods.

Acknowledgement. This work was supported by a grant from UPMC Enterprise, a
not for profit entity, to the University of Pittsburgh.

References

1. Open platform for analytics, G.T., monitoring: (2013). https://grafana.com/
2. InfluxData - Time Series Data Products Analytics (2013). https://www.influxdata.com/
3. Bose, E., et al.: Risk for cardiorespiratory instability following transfer to a monitored step-down unit. Respir. Care **62**, 415–422 (2017). https://doi.org/10.4187/respcare.05001
4. Chen, W.L., Kuo, C.D.: Characteristics of heart rate variability can predict
   impending septic shock in emergency department patients with sepsis. Acad.
   Emerg. Med. **14**(5), 392–397 (2007)
5. Elmer, J., Callaway, C.W.: The brain after cardiac arrest. Semin. Neurol. **37**, 19–24 (2017)
6. Elmer, J., et al.: Group-based trajectory modeling of suppression ratio after cardiac
   arrest. Neurocrit. Care **25**(3), 415–423 (2016)

7. Elmer, J., et al.: Clinically distinct electroencephalographic phenotypes of early myoclonus after cardiac arrest. Ann. Neurol. **80**(2), 175–184 (2016)
8. Gers, F.A., Schmidhuber, J., Cummins, F.: Learning to forget: continual prediction with LSTM (1999)
9. Nagin, D.S., Odgers, C.L.: Group-based trajectory modeling in clinical research. Annu. Rev. Clin. Psychol. **6**, 109–138 (2010)
10. Nystrom, N.A., Levine, M.J., Roskies, R.Z., Scott, J.: Bridges: a uniquely flexible HPC resource for new communities and data analytics. In: Proceedings of the 2015 XSEDE Conference: Scientific Advancements Enabled by Enhanced Cyberinfrastructure, p. 30. ACM (2015)

Modelling and Querying Star and Snowflake Warehouses Using Graph Databases

Alejandro Vaisman$^{(\boxtimes)}$, Florencia Besteiro, and Maximiliano Valverde

Instituto Tecnológico de Buenos Aires, Buenos Aires, Argentina
{avaisman,mabestei,mvalverd}@itba.edu.ar

Abstract. In current "Big Data" scenarios, graph databases are increasingly being used. Online Analytical Processing (OLAP) operations can expand the possibilities of graph analysis beyond the traditional graph-based computation. This paper studies graph databases as an alternative to implement star and snowflake schemas, the typical choices for data warehouse design. For this, the MusicBrainz database is used. A data warehouse for this database is designed, and implemented over a Postgres relational database. This warehouse is also represented as a graph, and implemented over the Neo4j graph database. A collection of typical OLAP queries is used to compare both implementations. The results reported here show that in ten out of thirteen queries tested, the graph implementation outperforms the relational one, in ratios that go from 1.3 to 26 times faster, and performs similarly to the relational implementation in the three remaining cases.

1 Introduction

Online Analytical Processing(OLAP) [6,8] comprises a set of tools and algorithms that allow querying large data repositories called data warehouses (DW). At the conceptual level, these DWs are modelled as *data cubes*, following the multidimensional (MD) model. In such model, each cell of the cube contains one or more *measures* of interest, that quantify *facts*. Measure values can be aggregated along *dimensions*, organized as sets of hierarchies. The most popular OLAP operations on cube data are aggregation and disaggregation of measure values along the dimensions (called roll-up and drill-down, respectively); selection of a portion of the cube (dice); or projection of the data cube over a subset of its dimensions (slice operation).

Property Graphs [5,7] underlie the most popular graph database engines [1]. In addition to traditional graph analytics, it is also of interest of the data scientist to have the possibility of performing OLAP on graphs. This paper explores graph databases as an alternative for storing DWs or data marts (that is, DWs oriented to analyse focused particular problems, generally at departmental level). Relational OLAP models are typically of two kinds: star schema-based and snowflake

© Springer Nature Switzerland AG 2019
T. Welzer et al. (Eds.): ADBIS 2019, CCIS 1064, pp. 144–152, 2019.
https://doi.org/10.1007/978-3-030-30278-8_18

schema-based. Both of them are composed of a collection of fact and dimension tables. Fact tables contain the foreign keys of the dimension tables, and a set of measures quantifying the facts. Dimension tables contain dimensional data. In the star model, dimension tables are denomalized, whereas in the snowflake model, dimension tables are normalized. This paper studies graph databases as an alternative to implement the star and snowflake DW schemas. The main hypothesis here is that graph databases can be more efficient than relational ones to address the typical kinds of OLAP queries, which consist, basically, in sequences of selection-projection-join-aggregation operations (SPJA).

A case study is discussed throughout the paper, based on the MusicBrainz database (http://musicbrainz.org/), an open music encyclopedia that collects music metadata and makes it available to the public. In this work, MusicBrainz is considered the OLTP (Online Transactional Processing) database, from which a DW is designed following the snowflake schema. This design is implemented on a PostgreSQL database, populated with data from the OLTP database through an ETL process (not reported in this paper). The same logical design is performed following the property graph data model, and implemented over a Neo4j database, which is then populated from the relational data. A collection of SPJA queries is defined and executed on both implementations, and the results are reported and discussed. *The experiments showed (with some exceptions) a clear advantage of the graph alternative over the relational one.*

The remainder of this paper is organized as follows: In Sect. 2 related work is discussed. Section 3 presents the case study, and the corresponding relational and graph models. Section 4 presents the queries implemented in the paper, and Sect. 5 describes the experiments, and reports and discusses the results. Section 6 concludes the paper and suggest future research directions.

2   Related Work

There is an extensive bibliography on graph database models, comprehensively studied in [1,3]. The interested reader is referred to this corpus of work for details. In real-world practice, two graph database models are used, namely (a) Models based on RDF (https://www.w3.org/RDF/) oriented to the Semantic Web; and (b) Models based on Property Graphs. Models of type (a) represent data as sets of triples where each triple consists of three elements that are referred to as the subject, the predicate, and the object of the triple. Informally, a collection of RDF triples is an RDF graph. In the *property graph* data model, [2] nodes and edges are labelled with a sequence of (attribute, value)-pairs. Extending traditional graph database models, Property Graphs are the usual choice in modern graph databases used in real-world practice.

The present paper works with models based on Property Graphs, and it is based on the work of Gómez et al. [4], who propose a model for performing OLAP on hypergraphs, based on the notion of graphoids, which are defined as graphs aggregated at different levels of granularity, using a collection of OLAP dimension hierarchies. The proposal supports *heterogeneous graphs*, which is required for typical multidimensional problems. The authors show that this model captures the semantics of the star and snowflake schemas.

3   Data Model

The model adopted in this paper will be presented through a running example based, as mentioned above, on the MusicBrainz database. For clarity of presentation, and to make the analysis more comprehensive, only the portion of the database containing information about music track releases and musical events, is tackled here. The core data in the original database includes, for example: *Artists* (with, e.g., name, aliases, type, begin and end dates); *Releases* (title, artist credit, type, status, language, date, country, label, etc.); *Recordings* (title, artist credit, duration, etc.); *Labels* (name, aliases, country, type, code, begin and end dates). Based on the original MusicBrainz database, a relational DW is defined at a conceptual level. Then, an equivalent graph DW is created, following [4]. In the next two sections, the relational and graph representations of the DW are described and discussed.

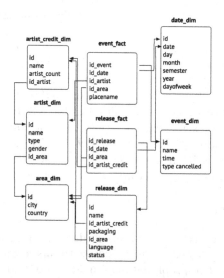

Fig. 1. ROLAP design for the musicbrainz database.

The MusicBrainz Relational Data Warehouse. Figure 1 depicts the snowflake schema for the MusicBrainz DW. There are two fact tables, namely release_fact and event_fact, representing the occurrence of the release of a music piece, or of an event, respectively. Each fact table refers to a dimension table called release_dim and event_dim. Artists in the release_fact fact table are described at the artist credit level (using the id_artist_credit attribute, which is a foreign key of the artist_credit_dim dimension table). An artist credit indicates a reunion of artists for a release, like, for example, David Bowie and Queen for the "Under Pressure" track. For the events, artists are described in the artist_dim dimension table, referred through the id_artist attribute. There is also an area_dim dimension table

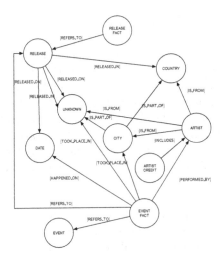

Fig. 2. Graph design for the musicbrainz database.

which includes the city → country dimension hierarchy. Thus, actually the design is a mixture of a star and snowflake schemas, as it is usual in data warehousing design practice. Finally, there is a time dimension represented by a dimension table denoted date_dim. This DW was implemented on a PostgreSQL database, and part of the data exported using an ETL process (not described here).

The MusicBrainz Graph Datawarehouse. Figure 2 depicts the schema of the graph database model for the MusicBrainz DW represented in Fig. 1. Attributes are not included in the graph representation, for the sake of clarity, and only the node types are depicted. There is a node for each event fact and a node for each release fact. In addition, there is an "Unknown" node type, which indicates missing data. The references to the dimensions are materialized through links to nodes which represent the members of the dimensions. Therefore, there is one node for each dimension member linked to a fact, such that the descriptive attributes are properties of these nodes (e.g., a Date node, with month and year properties). Some dimension hierarchies are made explicit. For example, the area_dim dimension is transformed into a node hierarchy of the kind city → country. Event and release facts are associated with nodes representing event and release data (e.g., the date of the event). In summary, with respect of the model introduced in [4], Fig. 2 represents the *base graphoid* (the graphoid at the finest level of granularity), and the background dimension hierarchies, one for each dimension in the relational schema, plus an Unknown dimension, with one level. This DW was implemented on a Neo4j graph database, which was populated using a script containing a sequence of Cypher statements.

4   Case Study and Discussion

This section shows how four kinds of OLAP queries can be expressed over both representations described in Sect. 3: (a) PJA queries, that means, queries that perform a projection after join operations, and finally aggregate the result; (b) SPJA queries, analogous to (a), but including a selection (a filter using a Boolean condition); (c) FPJA queries, that is, PJA queries that involve self joins of fact tables (in the snowflake model) of self references to the same kind of node (in the graph model); (d) PJA-DA queries, that is, PJA queries referring to different kinds of facts. For the sake of space, only some of Cypher expressions are shown in the queries below.

(a) PJA Queries. These typical OLAP queries, generally link facts and dimensions, and finally perform some kind of aggregation. The first query is a simple join, representing the climbing along the artist hierarchy starting from artist_credit, aggregating the releases.

Query 1. *Compute the number of releases per artist.*

In SQL, this is a join between the fact table containing the releases (release_fact, and the dimension tables artist_credit_dim and artist_dim. In Cypher:

```
MATCH (r:ReleaseFact)-[]->(a:ArtistCredit)-[]->(a1:Artist)
RETURN a1.name, count(r) ORDER BY a1.name ASC
```

Queries in Cypher are evaluated by means of pattern matching. In this case, the join is implemented through the matching of the ReleaseFact → ArtistCredit → Artist path, against the MusicBrainz graph. In this case, the relationship names of the path are not needed since the edges between nodes can be inferred from the node types. In the next query, aggregation is performed climbing along two dimensions, namely Artist and Time.

Query 2. *Compute the number of releases per artist and per year.*

```
MATCH (r:ReleaseFact)-[r1:RELEASED_BY]->(a:ArtistCredit)-[]->(a1:Artist),
(d:Date)<-[rd:RELEASED_ON]-(r)
RETURN a1.name, d.year, count(r) ORDER BY a1.name ASC,d.year ASC
```

Note that in Cypher, the aggregation a1.name, d.year, count(r) is a concise way to express a GROUP BY clause in SQL. Query 3 below, is similar to Query 1, but uses the event facts rather than the release facts.

Query 3. *Compute the number of events per artist.*

In the next query, aggregation of event facts is performed along the Event and Artist dimensions.

Query 4. *Compute the number of times the artist performed in each event.*

```
MATCH (e1:Event)<-[:REFERS_TO]-(e:EventFact)-[:PERFORMED_BY]->(a:Artist)
RETURN e1.name,a.name, count(*) ORDER BY    e1.name ASC, a.name ASC
```

It can be seen that the kind of event is given in the event type of node (of the event_dim dimension in the relational model). The last PJA query aggregates data along three dimensions (Event, Artist, and Time).

Query 5. *For each (event, artist, year) triple, compute the number of times the artist performed in an event on an year.*

(b) SPJA Queries. These queries add, to the join condition between facts and dimensions, a selection (Boolean) condition. The first query to be analysed operates over event facts.

Query 6. *Same as Query 5, for artists in the United Kingdom and events occurred after year 2006.*

```
MATCH (e1:Event)<-[r1:REFERS_TO]-(e:EventFact)-[r:HAPPENED_ON]->(d:Date)
WHERE d.year > 2006
WITH   e,d,e1
MATCH (e)-[p:PERFORMED_BY]->(a:Artist)-
[IS_FROM]->(c:Country{name:'United Kingdom'})
RETURN e1.name,   d.year,a.name, count(*)
ORDER BY   e1.name asc, d.year asc, a.name asc
```

In the query above, the WITH expression passes the variables on to the next step of the computation. The next query operates over releases.

Query 7. *Compute the number of releases, per language, in the UK.*

The next query requires joining the event facts with themselves, once. Queries 9 and 10, require two and three self joins, respectively.

Query 8. *Compute, for each pair of artists, the number of times they have performed together at least twice in an event.*

```
MATCH (a1:Artist)<-[]-(e:EventFact)-[]->(a2:Artist) WHERE a1.id < a2.id
WITH a1, a2, COLLECT(e) AS events WHERE SIZE(events)   > 1
RETURN a1.name, a2.name, SIZE(events) ORDER BY SIZE(events) desc
```

The COLLECT statement builds a list with all the events for each pair of artists. The SIZE function computes the length of this list.

(c) FPJA Queries But Involving More Than One Self Fact Table Joins. This kind of queries joins several fact nodes with other ones of the same type. The next query requires two of such joins.

Query 9. *Compute the triples of artists, and the number of times they have performed together in an event, if this number is at least 3.*

```
MATCH (a1:Artist)<-[]-(e:EventFact)-[]->(a2:Artist) WHERE a1.id < a2.id
WITH a1,a2,COLLECT(e) AS events WHERE SIZE(events) > 2
MATCH (a1:Artist)<-[]-(e1:EventFact)-[]->(a2:Artist)
MATCH (a3:Artist)<-[]-(e1) WHERE a2.id < a3.id
WITH a1.name as name1, a2.name as name2,a3.name as name3 ,
COUNT(e1.idEvent) as nbrTimes WHERE nbrTimes > 2
RETURN name1,name2,name3,  nbrTimes  ORDER BY nbrTimes DESC
```

Query 10. *Compute the quadruples of artists, and the number of times they have performed together in an event, if this number is at least 3.*

(d) PJA-DA Queries. Finally, queries involving events and releases are evaluated. In OLAP this is called a drill-across operation between event and release facts.

Query 11. *Compute the pairs of artists that have performed together in at least two events and that have worked together in at least one release, returning the number of events and releases together.*

Query 12. *List the artists who released a record and performed in at least an event, and the year(s) this happened.*

Table 1. Dataset sizes for the relational representation (left); Dataset sizes for the graph representation (center); Results of the experiments (right).

Table	# tuples
date_dim	90,033
artist_dim	1,151,920
artist_credit_dim	1,871,875
area_dim	74,211
release_dim	1,715,636
event_dim	19,441
event_fact	55,281
release_fact	1,724,365

Element(node/edge type)	# nodes	# edges
Artist	1,151,920	
ArtistCredit	1,871,875	
City	73,955	
Country	255	
Event	19,441	
Release	1,715,636	
ReleaseFact	1,724,365	
EventFact	19,457	
Unknown	1	
HAPPENED_ON		19,348
IS_FROM		1,151,920
INCLUDES		4,098,689
IS_PART_OF		73,955
PERFORMED_BY		52,599
RELEASED_BY		1,913,494
REFERS_TO		1,743,822
RELEASED_IN		1,755,598
RELEASED_BY		1,913,494
RELEASED_ON		1,407,504
TOOK_PLACE_IN		19,475

Query	PostgreSQL	Neo4j	Pg/Neo4j
query 1	22	11	2
query 2	30	12.5	2.4
query 3	0.8	0.12	6.66
query 4	1.5	0.2	7.5
query 5	2.4	0.4	6
query 6	0.2	0.2	1
query 7	0.2	0.2	1
query 8	2	1.5	1.33
query 9	24	5	4.8
query 10	1110	43	25.8
query 11	9	1.8	5
query 12	3	3	1
query 13	2	1	2

In Cypher, the query reads:

```
MATCH (a:Artist)<-[PERFORMED_BY]-(e:EventFact)-[:HAPPENED_ON]->(d:Date)
WITH distinct a,d.year as year
MATCH (r:ReleaseFact)-[r1:RELEASED_BY]->(a2:ArtistCredit)-[]->(a),
(r)-[r2:RELEASED_ON]->(d1:Date) WHERE d1.year=year
RETURN DISTINCT a.name, year
```

Query 13. *Artists who released a record and performed in at least an event, and the year(s) this happened, for events and releases occurred since 2007.*

5   Experiments

The thirteen queries in Sect. 4 where run over relational and graph databases designed following the models described in Sect. 3. The solution based on the graph model is compared against the relational alternative containing exactly the same data. For the relational representation, the DW was implemented as a snowflake schema and stored in a PostgreSQL database. All tables were fully indexed for the workload described in the previous section. The left-hand side of Table 1 shows the sizes of the tables. For the graph representation, the number of nodes and edges in the graph, stored in a Neo4j database, community version 3.5.3, are depicted in the center part of the figure. As for the relational alternative, all the required indexes were defined. The right-hand side of Table 1 shows the results of the experiments. The queries introduced in Sect. 4 were run on machine with a i7-6700 processor and 32 GB of RAM, a 1 TB hard disk, and a 256 MB SSD disk. The execution times are depicted as the averages of five runs of each experiment, expressed in seconds. The best time for each query is highlighted in bold font. The ratio between the execution times in PostgreSQL and Neo4j are indicated in the fourth column.

The results of the experiments show that only in three out of thirteen queries the execution times were the same. In the remaining eleven queries, Neo4j clearly outperformed the relational alternative, ranging from 1.33 to almost 26 times faster. For all *PJA queries* (Queries 1 through 5), the graph alternative clearly outperforms the relational one. Note that this occurs for queries addressing both, events (the smaller fact table, Queries 1 and 2) and releases (Queries 3, 4, and 5). The probable reason for this is that neighbours are found very fast in Neo4j, due to its internal graph representation mechanism. Once the neighbours are found, aggregation is performed very efficiently. In the case of SQL, the join must be computed, and this turns out, in general, to be more expensive than finding the neighbours of a node. For *SPJA queries* (Queries 6 through 8), execution times are similar for both alternatives, except for Query 8, which requires a self join of the event facts, and the graph implementation behaves better than the relational one. The reason here is that selections in RDBMSs are very efficiently performed thanks to the indexing of the filtering attributes, and this compensates the cost of the joins. Probably the most surprising results are obtained for self-join queries (the *FPJA queries*), Queries 9 and 10, where the differences are clearly in favour of the graph alternative. Finally, for the *PJA-DA (drill-across) queries* (Queries 12 and 13), execution times seem to depend highly on the selectivity of the filtering attributes, but nevertheless, Neo4j behaves at least the same than SQL (like in Query 12, which does not include a selection).

6   Conclusion and Open Problems

This paper studied the plausibility of graph databases, and the property graph data model, for representing and implementing data warehouses modelled as star and snowflake schemas, using the MusicBrainz database. The results in most situations, the graph representation was clearly faster, up to one order of magnitude. Building on the results reported in this paper, future work includes looking for new case studies, that would lead to building a benchmark to evaluate graph databases for OLAP queries.

Acknowledgments. Alejandro Vaisman was partially supported by project PICT-2017-1054, from the Argentinian Scientific Agency.

References

1. Angles, R.: A comparison of current graph database models. In: Proceedings of ICDE Workshops, Arlington, VA, USA, pp. 171–177 (2012)
2. Angles, R., Arenas, M., Barceló, P., Hogan, A., Reutter, J.L., Vrgoc, D.: Foundations of modern query languages for graph databases. ACM Comput. Surv. **50**(5), 68:1–68:40 (2017)
3. Angles, R., Gutierrez, C.: Survey of graph database models. ACM Comput. Surv. **40**(1), 1:1–1:39 (2008)
4. Gómez, L.I., Kuijpers, B., Vaisman, A.A.: Performing OLAP over graph data: query language, implementation, and a case study. In: Proceedings of BIRTE, Munich, Germany, 28 August 2017, pp. 6:1–6:8 (2017)
5. Hartig, O.: Reconciliation of RDF* and property graphs. CoRR, abs/1409.3288 (2014)
6. Kimball, R.: The Data Warehouse Toolkit. Wiley, New York (1996)
7. Robinson, I., Webber, J., Eifrém, E.: Graph Databases. O'Reilly Media, Sebastopol (2013)
8. Vaisman, A., Zimányi, E.: Data Warehouse Systems: Design and Implementation. Springer, Heidelberg (2014). https://doi.org/10.1007/978-3-642-54655-6

Towards Integrating Collaborative Filtering in Visual Data Exploration Systems

Houssem Ben Lahmar[⊠] and Melanie Herschel

IPVS, University of Stuttgart, Stuttgart, Germany
{houssem.ben-lahmar,melanie.herschel}@ipvs.uni-stuttgart.de

Abstract. Visual data exploration assists users in investigating data by providing recommendations. These recommendations take the form of queries that retrieve these data for the next exploration step, paired with suited visualizations. This paper extends the content-based recommendation techniques adopted so far in EVLIN for query recommendations with collaborative-filtering recommendation techniques. For that, we propose a merge approach that fuses evolution provenance graphs representative of individual user's exploration session into a global multi-user graph. This merged graph is then used by our collaborative recommendation approach that searches similar queries for a given user query in the multi-user graph to then recommend queries that were previously explored in exploration steps adjacent to these similar queries.

Keywords: Visual data exploration · Provenance · Recommendations

1 Introduction

Visual data exploration tools support users in understanding and investigating data to reveal interesting information they are a priori unaware of. We recently presented EVLIN [1,2], a visual data exploration tool that leverages provenance to recommend queries and their visualizations during the interactive exploration of data stored in a data warehouse.

EVLIN provides recommendations by computing a set of potentially interesting queries whose results are to be explored at the next step of a user's exploration session. Each query is associated with a score that quantifies its potential interestingness. The recommended queries with associated scores are presented as an impact matrix to the user, as illustrated in Fig. 1. Each cell presents a recommended query and its color encodes the interestingness. Essentially, queries are recommended for typical data warehouse operations such as drill-down, slice, etc. (columns) that are parameterized by attributes or attribute values (rows) that are considered interesting.

In [1], we have adopted content-based filtering techniques to recommend queries and associated visualizations. Throughout extensive experimentation

© Springer Nature Switzerland AG 2019
T. Welzer et al. (Eds.): ADBIS 2019, CCIS 1064, pp. 153–160, 2019.
https://doi.org/10.1007/978-3-030-30278-8_19

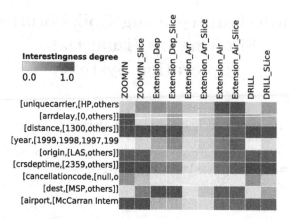

Fig. 1. Impact matrix visualizing ranked query recommendations (Color figure online)

with our system, we have observed that while the recommendations generally succeed in distinguishing interesting from non-interesting next exploration steps, users are often still left with a large number of equally interesting options to choose from. For instance, the impact matrix of Fig. 1 shows around 17 (out of 90) dark red, i.e., most interesting queries, whose scores are indistinguishable (visually and also when looking at the actual recommendation scores). Consequently, the choice of the next exploration step remains difficult.

Contributions. This paper incorporates collaborative-filtering techniques in the recommendation process of EVLIN. The goal is to diversify the recommendations that users have previously obtained, by taking into account globally interesting trends in addition to (limited) locally obtained insights. More precisely, our contributions are:

- A method to incrementally merge evolution provenance graphs that describe user's exploration sessions, into a global multi-user graph based on a similarity matching between the local session graph and the global graph.
- A collaborative-filtering query recommendation approach that first searches exploration steps similar to a user's current exploration in the multi-user graph. Then, it recommends queries that have been previously explored in exploration steps succeeding those similar exploration steps.
- A preliminary experimental evaluation that measures the performance of our merge and recommendation approaches.

Structure. Section 2 introduces EVLIN and relevant concepts. Merging evolution provenance graphs is discussed in Sect. 3. We cover our collaborative-filtering recommendation approach in Sect. 4. Section 5 presents a preliminary evaluation, while we discuss related work in Sect. 6 and conclude in Sect. 7.

2   System Overview and Preliminaries

EVLIN [1] is a provenance-based recommendation system for interactive visual data exploration of data warehouses. Its general processing is demonstrated in [2].

Initially, a user issues an exploration query Q to visually explore a sub-region in a data warehouse D.

This exploration query takes the general form

SELECT $f(m)$, A FROM $rel(D)$ WHERE cond GROUP BY A

where m is a measure in the fact table, $A = \{a_1, \ldots, a_n\}$ is a set of attributes, f is an aggregation function, $rel(Q)$ refers to one or more relations of D, and cond is a conjunction of predicates.

The user's query result $Q(D)$ is visualized and rendered to the user. This forms an exploration step that is under investigation. More formally, we define an exploration step as follows.

Definition 1 (Exploration step). *An exploration step over data in D is defined as $X_D = \{Q, V\}$ where Q is the query whose result $Q(D)$ over database D is visualized with an interactive visualization described by V.*

The user can interact with an exploration step X_D, selecting a sub-result of interest, denoted r. This interaction triggers the content-based recommendation process that identifies, as explained in [1], the set of attributes and values that may interest the user. For each identified pair, we determine variations Q' of the query Q that correspond to typical data warehouse operations such as drill down, slice, etc. This results in a set of recommended queries, visualized as an impact matrix as shown in Fig. 1 where cells refer to recommendations, rows map to identified attribute values and columns refer to operations types.

Later, the recommended query selected by the user undergoes a visualization recommendation process where we use data previously seen during the session, captured as *evolution provenance*, to recommend a visualization that consistently renders similar concepts in a similar way through a complete exploration session that spans multiple exploration steps. Essentially, we model the evolution provenance as an exploration session graph defined as follows.

Definition 2 (Exploration session graph). *An exploration session graph over a database D is a labeled directed acyclic graph (DAG) $G_{XS,D}(N, E)$ where N is a set of nodes and E a set of labeled edges. Each node $n \in N$ corresponds to an exploration step X. An edge $e = (n, n', L)$ represents the transition from one exploration step $X_D = \{Q, V\}$ to the next exploration step $X'_D = \{Q', V'\}$ whose query Q' is a recommendation directly derived from Q over the same D. L is a 3-tuple $\langle op, a, s \rangle$ where op is an identifier of the query type of Q' wrt Q, a is the relevant attribute used to construct Q' based on Q, and s an impact score, reflecting the number of subsequent exploration steps triggered by the navigation from X_D to X'_D computed as $s(e) = 1 + \sum_{e_c = (n', n_i, L') \in E} s(e_c)$.*

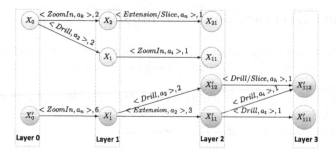

Fig. 2. Two exploration session graphs

Figure 2 shows two sample exploration session graphs. Note that when D is clear from the context, we omit the subscript D from X_D and $G_{XS,D}$.

Intuitively, the score labeling edges reflects how many exploration steps have been pursued upon reaching a given exploration step, giving an indication of how many potentially interesting steps can be reached from a session.

To this end, we leverage in this paper the evolution provenance recording exploration session graphs in order to incorporate collaborative-filtering query recommendations. For that, we propose firstly a new approach to merge multiple exploration sessions graphs in one global graph defined as follows.

Definition 3 (Multi-user exploration graph). *A multi-user exploration graph $G_{MU}(N_{MU}, E_{MU})$ is the result of aggregating many individual exploration session graphs $\{G_{XS1}, \ldots, G_{XSn}\}$.*

Accordingly, we propose a collaborative-based query recommendation approach that considers G_{MU} to recommend a set of potentially interesting queries for a given (current) exploration step $X_D = \{Q, V\}$.

Finally, the set of recommendations computed by the *collaborative-filtering* and the *content-based* recommendation techniques undergo a quantification process where we employ various utility scores depending on the types of recommendations (collaborative or content). Indeed, we adopt dissimilarity metric [10] to quantify content-based recommendations while we leverage impact scores of edges in G_{MU} to quantify collaborative recommended exploration steps. These scores are combined and rendered as a visualization of an impact matrix.

3   Merging of Evolution Provenance Graphs

Merging evolution provenance graphs takes as input G_{XS}, G_{MU}, and a similarity threshold θ_{sim}. Note that for simplicity, we slightly abuse the notation and consider that nodes of both graphs represent queries only. More formally, during the matching phase, given the set of vertices $N \in G_{XS}$ and $N_{MU} \in G_{MU}$, we determine a one-to-one matching $\mathcal{M}$ between nodes $n_i \in N$ and $n_j \in N_{MU}$ such that for each $m = (n_i, n_j) \in \mathcal{M}$, $sim(n_i, n_j) \geq \theta_{sim}$. For that, we adopt in this paper the Jaccard coefficient to implement the similarity function sim.

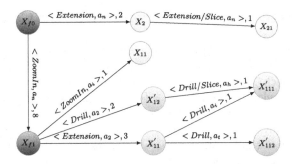

Fig. 3. Fused mutli-user graph

The merge phase then produces a multi-user graph G'_{MU} with $N'_{MU} = N_{MU} \cup_{\approx} N$ and $E'_{MU} = E_{MU} \cup_{\approx} E$. The symbol $\cup_{\approx}$ denotes that the usual equality considered by set union is extended to consider similar objects as "equal".

To perform this merge phase, we propose an approach that we call root-layer-merge (RLM). Our approach is inspired by a match-merging algorithm that merges unordered trees [9]. The problem is transformed to a variant of the stable marriage problem [4] where sub-trees are compared and merged recursively. The match-merging strategy relies on the hierarchy of trees to identify sub-trees prone to merge, recursively descending through the tree. Given that exploration session graphs are DAGs, we can transform them into layered graphs [5], as illustrated in Fig. 2. Similarly to [9], we only match and merge nodes at layer i in N with nodes at layer i in N_{MU}. More precisely, given an exploration session graph, a multi-user graph, and a similarity threshold, we take initially the set of roots of the two graphs. Using the stable marriage algorithm [4], we determine a stable matching $\mathcal{M}$ between these two sets. For each match, we replace the individual matching nodes by a merged node in the merged graph. This leads to the update of edges pointing to (and going from) replaced nodes. Thus, they are "rerouted" to point at the merged node. Finally, we recursively proceed with matching and merging children of merged nodes.

For illustration, consider the two layered exploration graphs shown in Fig. 2. We assume that both root nodes are matched using our RLM approach. These are then combined into a single node, which becomes the common root node at layer 0 of the merged graph (depicted in Fig. 3). Then, the algorithm proceeds to layer 1, where the exploration steps X_1 and X'_1 are matched. This entails a merge of both these nodes at layer 1. As a consequence, at layer 2, further possible matches among pairs in $\{X_{11}\} \times \{X'_{11}, X'_{12}, X'_{13}\}$ are searched. Assuming no further matches are found, the RLM approach stops at this layer, resulting in the merged multi-user graph depicted in Fig. 3.

Given that our RLM approach requires that all input exploration session graphs begin with a similar exploration step, we expect that the merging process will produce a moderate number of matchings. To counter this effect, further merge strategies may be explored in the future.

4   Collaborative-Filtering Recommendation Computation

The multi-user graph G_{MU} obtained using our RLM merge method, is accessed during the recommendations computation by our collaborative-filtering recommendation approach. This latter searches top-k similar exploration steps in G_{MU} given a current exploration step $X_D = \{Q, V\}$ and considers children of those as interesting exploration steps to be recommended.

More specifically, our collaborative recommendation approach performs a brute-force search of top-k similar exploration steps in G_{MU}. Essentially, it computes distances between all nodes in G_{MU} and the current exploration step X_D. Hence, nodes in G_{MU} having distances bigger than a distance threshold θ_{dist}, are immediately filtered out. Remaining candidate nodes are stored in a priority queue of size k in a descending order in a way that the priority queue ultimately contains the k most similar queries to X_D. Subsequently, we determine the children of each top-k similar node present in the priority queue and add them to the set of recommended queries Rec. Recall that each edge is labeled by (operator, att, score) as explained in Definition 2 where scores reflect the impact beyond the selection of underlying navigation. That is, we leverage this information to quantify Rec. Subsequently, we combine scores available in Rec with the (previously computed) dissimilarity scores computed using our previous approach [1,2]. This is done by multiplying the two normalized query-recommendation scores.

Finally, we point out that the complexity of our collaborative recommendation approach is in $O(N_{MU})$ as it is dominated by the top-k similar exploration steps search. Investigating further improvement to speed up the recommendation process (especially when processing large G_{MU}) is left for future work.

5   Preliminary Evaluation

This section presents the preliminary evaluation of our methods implementing the evolution provenance merge and the collaborative recommendation.

For that, we consider exploration sessions over a data warehouse about US domestic flights. It contains a fact table recording information about 1 million flights as well as three dimension tables that contain further information about

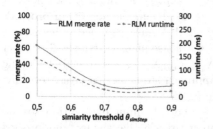

Fig. 4. θ_{sim} vs. merge rate (solid lines) and runtime (dashed lines)

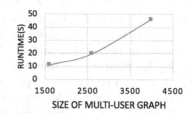

Fig. 5. Recommendation runtime for varying G_{MU} sizes

departure and destination airports, airlines, and plane types. To evaluate the performance of our proposed approaches, we implemented an exploration session generator that generates synthetic exploration sessions based on a set of real exploration sessions. All experiments were run on a single machine with a 2.2 GHz quad-core Intel processor and 16 GB RAM.

Our first experiment studies the runtime and the merge rates (computed as $\frac{size\_multi\_user\_graph}{\sum size\_users\_exploration\_sessions}$) of our merge approach RLM for varying similarity threshold θ_{sim} (set to 0.5, 0.7, and 0.9). Thus, we used five exploration workloads, each comprising 1000 exploration steps. Individual exploration sessions, present in each workload have sizes ranging between 10 and 20 exploration steps.

Our RLM approach iterates over the exploration sessions of each workload, incrementally merging each session into the multi-user graph. Figure 4 reports results as averages over the five randomly generated workloads. We observe the merge rate (the same for the runtime) decreases significantly when using more strict similarity threshold values. This is mainly related to the restrictive merge policy adopted by RLM that requires that root of graphs are similar to be able to perform further merges recursively. This requirement does not hold for exploration sessions that may start with different roots and converge later.

Consequently, we plan in the future to improve our merge approach to perform more robust merge rates. This is important to generate a compact multi-user graph, fast to browse during the collaborative recommendation process.

In a second experiment, we evaluate our collaborative recommendation approach proposed in Sect. 4. As our collaborative recommendation approach relies on a current exploration step within an exploration session G_{XS}, we randomly generate four queries that simulate the current exploration step. For each query, we compute top-3 collaborative recommendations using our approach applied on the three multi-user graphs of sizes 1600, 2600 and 4000 exploration steps, respectively with $\theta_{dist} = 0.2$. Figure 5 reports the average runtime over the four queries with increasing multi-user graph size. Clearly, runtime of our recommendation approach increases with the size of the multi-user graph.

6   Related Work

Prominent works related to our collaborative-filtering recommendation approach belong to two classes (i) systems that assist SQL query formulation with limited visual interaction facilities and (ii) visual data exploration systems that provide recommendations to assist users in browsing visually through interesting data.

In relation to the first class, we find several approaches e.g., [3,6] that employ collaborative recommendation to assist users (lacking SQL knowledge) in writing queries. Yet, these works rely on a basic form of history presented as a serial trace, whereas we focus on interactive visual exploration whose history contains various users' navigations and forms thereby a direct graph. Accordingly, we benefit from our rich history model to infer ratings of exploration steps (via edges' scores), used later in our collaborative recommendation framework.

For the second class of collaborative recommendation work, we find REACT [8] that uses users' prior explorations made over various data sets to provide collaborative recommendations. While REACT leverages the users' history collected when exploring different datasets to mitigate the problem of cold-start e.g., absence of history about the explored data set, our work remedies the cold-start situation by providing both content and collaborative recommendations.

Further related work comes from other areas relevant to our proposal such as the graph matching e.g., [11] and the graph summary e.g., [7].

7 Conclusion

In this paper, we discuss the extension of our visual interactive data exploration system EVLIN by a novel, collaborative-filtering recommendation framework. For that, we propose an approach to merge evolution provenance collected from many previous users' exploration sessions into a global graph. This latter is used to compute and rank recommended queries.

Acknowledgements. This research is funded by the Deutsche Forschungsgemeinschaft (DFG, German Research Foundation) – Projektnummer 251654672 – TRR 161.

References

1. Ben Lahmar, H., Herschel, M.: Provenance-based recommendations for visual data exploration. In: TaPP (2017)
2. Ben Lahmar, H., Herschel, M., Blumenschein, M., Keim, D.A.: Provenance-based visual data exploration with EVLIN. In: EDBT (2018)
3. Eirinaki, M., Abraham, S., Polyzotis, N., Shaikh, N.: Querie: collaborative database exploration. TKDE **26**(7), 1778–1790 (2014)
4. Gale, D., Shapley, L.S.: College admissions and the stability of marriage. Am. Math. Mon. **69**(1), 9–15 (1962)
5. Healy, P., Nikolov, N.S.: How to layer a directed acyclic graph. In: Mutzel, P., Jünger, M., Leipert, S. (eds.) GD 2001. LNCS, vol. 2265, pp. 16–30. Springer, Heidelberg (2002). https://doi.org/10.1007/3-540-45848-4_2
6. Khoussainova, N., Kwon, Y., Balazinska, M., Suciu, D.: SnipSuggest: context-aware autocompletion for SQL. VLDB **4**(1), 22–33 (2010)
7. Liu, Y., Safavi, T., Dighe, A., Koutra, D.: Graph summarization methods and applications: a survey. ACM Comput. Surv. **51**(3), 1–34 (2018)
8. Milo, T., Somech, A.: Next-step suggestions for modern interactive data analysis platforms. In: KDD (2018)
9. Schulz, C., Zeyfang, A., van Garderen, M., Ben Lahmar, H., Herschel, M., Weiskopf, D.: Simultaneous visual analysis of multiple software hierarchies. In: VISSOFT (2018)
10. Vartak, M., Rahman, S., Madden, S., Parameswaran, A., Polyzotis, N.: SeeDB: efficient data-driven visualization recommendations to support visual analytics. VLDB **8**, 2182–2193 (2015)
11. Yan, J., Yin, X., Lin, W., Deng, C., Zha, H., Yang, X.: A short survey of recent advances in graph matching. In: ICMR, pp. 167–174 (2016)

ADBIS 2019 Workshop: Modelling is Going to Become Programming – M2P

Usage Models Mapped to Programs

András J. Molnár[1,2]([⊠]) [ID] and Bernhard Thalheim[1] [ID]

[1] Computer Science Institute,
Christian-Albrechts-University Kiel, 24098 Kiel, Germany
{ajm,thalheim}@is.informatik.uni-kiel.de
[2] MTA-SZTAKI Institute for Computer Science and Control,
Hungarian Academy of Sciences, Budapest 1111, Hungary
modras@ilab.sztaki.hu

Abstract. Model-based programming can replace classical programming based on compilation and systematic development of models as well on explicit consideration of all model components without hiding intrinsic details and assumptions. A key element of model-based programming is the proper definition and management of model suites, by which multiple, interrelated models can be transformed from one another and their consistency is ensured after modifications. A usage model is based on the specification of user roles and types, together with an interaction space described in a form of a storyboard, showing which activities are supported, in which order, by which actors. A workflow model is an extended, well-formed declaration of how specific processes should be carried out. It can directly be translated to program code, using a proper workflow or process engine. A novel way of programming is being opened up by usage modeling, which is being investigated in this paper: given a storyboard with supported usage scenarios, it is possible to derive a workflow model from it. We present our two translation methods using a working example, identifying guidelines as requirements for model refinement and normalization, rules for model translation, and propose considerations towards improved methods and model specifications.

Keywords: Model-centered programming · Model to program · Model suite · Model transformation · Storyboard · Process model

1 Introduction

1.1 Programming by Modeling

Programming is nowadays a socio-technical practice in most disciplines of science and engineering. Software systems are often developed by non-programmers or non-computer scientists, without background knowledge and skills, or insight into the culture of computer science, without plans for systematic development. Maintenance, extension, porting, integration, evolution, migration, and modernisation become an obstacle and are already causing problems similar to the

© Springer Nature Switzerland AG 2019
T. Welzer et al. (Eds.): ADBIS 2019, CCIS 1064, pp. 163–175, 2019.
https://doi.org/10.1007/978-3-030-30278-8_20

software crisis 1.0, since such systems often have a poor structure, architecture, documentation, with a lost insight of specific solutions. Programs of the future must be understandable by all involved parties and must support reasoning and controlled realisation and evolution at all levels of abstraction.

Our envisioned *true fifth generation programming* [13] is a new programming paradigm where models are essentially programs of next generation and models are translated to code in various third or fourth generation languages. Programming is done by model development, relying on the compilation of these models into the most appropriate environment.

Application engineers and scientists are going to develop and use models instead of old-style programming, supported by templates from their application area. They can thus concentrate on how to find a correct solution to their problems, managing the complexity of software intensive systems. The process will be supported by model-backed reasoning techniques, as developers will appreciate and properly evaluate the model suite at the desired level of abstraction.

1.2   Usage Models and Workflow Models

In our study we are considering the case of web information system development.

A *usage model* of a web-is consists of specification of user roles and types, their associated goals and tasks, and an interaction space. The latter can be expressed as a graph, called a *storyboard*, describing what activities are supported and in which possible order, by which actors [9]. Supported interaction playouts can be formulated as *scenarios* (exact graph paths), *story algebra expressions* (path scemata), or more generally, subgraphs of the storyboard, including actor-specific views. The usage model is developed by a global-as-design approach.

A *workflow model* is an extended, well-formed declaration of how specific processes should be carried out, in a notation that is readily understandable by all stakeholders, including business analysts, technical developers and people who manage and monitor those processes [7]. A de facto standard is *BPMN* [7], but it is possible to use another workflow description language. The workflow model can directly be translated to software process components, using a proper workflow or process engine (e.g. [3]). This opens up a novel way of programming by usage modeling, via intermediate translation to a workflow model.

1.3   Related Work

Our current contribution can be related – amongst others – to the following previous works. Notions of *models* are discussed in [12]. [11] introduces *model suites* consisting of multiple, explicitly associated models, where the association uses maintenance modes, similar to integrity support in databases [15]. Amongst others, MetaCASE tools [1] were developed to support the definition of metamodel packages and the creation and customization of CASE tools based on them. The *models as programs – true fifth generation programming* agenda is proposed in [13]. For data structuring, translation of entity-relationship models to relational database schemata is well-known [4]. We are proposing a similar approach for the dynamics of functionality, motivated by compilers [8]: phases of preprocessing, parsing and syntax checking is followed by semantic analysis resulting an intermediate structure, and finally a possible optimization phase of the resulting, translated model. [10] discusses *proceses-driven applications* and *model-driven execution* in terms of BPMN [7] diagrams. [2] elaborates a generative approach to the functionality of interactive information systems. [14] introduces dynamically combinable *mini-stories* to handle workflow cases with large flexibility. Although these latter works consider steps and ideas we can apply here, our currently addressed problem of usage model translation to workflow model is not explicitly discussed in any of the publications known to us.

1.4   Goal and Outline of the Paper

Our general vision is to generate running program code based on a usage model specification. We investigate on a particular sub-case in this paper: given a usage model as a storyboard with supported scenarios [9], is there a formalizable method to derive a workflow model in BPMN [7] from it. We present our proposed path and general framework for modeling as next generation programming in Sect. 2, based on [13]. Section 3 introduces our target case of workflow model elicitation from a usage model, illustrated by a working example, with general guidelines for model refinement and enhancement, rules and two different methods for translation. We conclude and close with future issues in Sect. 4.

2   Modeling and Programming Based on Model Suites and Layering

Models are universal instruments for communication and other human activities. Ideas and thought chunks can be presented to those who share a similar culture and understanding without the pressure to be scientifically grounded. A model is an adequate (i.e. analogous, focused, purposeful) and dependable (i.e. justified, sufficient in quality) instrument that represents origins and performs functions in some deployment scenario [12]. As an instrument, the model has its own background (i.e. grounding, basis) and should be well-formed. Models are more abstract than programs, but can be as precise and appropriate as programs. They support understanding, construction of system components, communication, reflection, analysis, quality management, exploration, explanation, etc. Models can be translated to programs to a certain extent, therefore, models can be used as higher-level, abstract, and effective programs. They are, however, independent of programming languages and environments. Models encapsulate, represent and formulate ideas both as of something comprehended and as a plan. Models declare what exactly to build and can be understandable by all stakeholders involved in software system development. They become general and accurate enough, and can be calibrated to the degree of precision that is necessary for high quality [13].

A *model suite* [11] consists of a coherent collection of explicitly associated models. A model in the model suite is used for different purposes such as communication, documentation, conceptualisation, construction, analysis, design, explanation, and modernisation. The model suite can be used as a program of next generation and will be mapped to programs in host languages of fourth or third generation. Models delivered include informative and representation models as well as the compilation of the model suite to programs in host languages. Consistency can be ensured similarly to relational databases [15]. Models will thus become executable while being as precise and accurate as appropriate for the given problem case, explainable and understandable to developers and users within their tasks and focus, changeable and adaptable at different layers, validatable and verifiable, and maintainable.

Similarly to database modeling, *layering* has already often and successfully been used, including most program language realisations and application development methodologies. We assume a general layered approach as the universal basis for treatment of models as programs [13]. Layering has also been the guiding paradigm of the TeX and LaTeX text processing realisations [5,6].

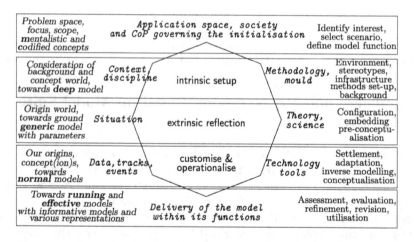

Fig. 1. The layered approach to model suite development and program generation

Model suite development and deployment will be based on separation of concern into *extrinsic* and *intrinsic* parts of models. Models typically consist on the one side of a *normal model* that displays all obviously relevant and important aspects of a model and on the other side of a *deep model* that intrinsically reflects commonly accepted intentions, the accepted understanding, the context, the background that is commonly accepted, and restrictions for the model. The model suite will be layered into models as shown in Fig. 1. Taking it as basis, we can formulate our proposed agenda for usage and workflow models.

The *initialisation layer* is given by the application and the scenarios in which models are used, by the problem characterisation, by background elements of the CoP and especially commonly accepted concepts in this community, and additionally by interest, intensions, and the value. In our case, it consists of a declaration that a website is needed for a specific application, analogously to selecting a *documentclass* in LATEX. It determines the possible syntax and semantics of the underlying layers.

The *enabling strategic setup layer* defines the opportunity space and especially the hidden background for the model. Its main result is the deep model that is typically assumed to be given (normal models are not entirely developed from scratch). In our case, it will correspond to what a website means, what are the side conditions and underlying infrastructure of it and the selected application domain. It gives an opportunity space and can impose requirements or proposals for the way of system development.

The *tactic definition layer* starts with some generalisation, i.e. select a ground generic model that will be customised and adapted to become the normal model. It can be, for example, a generalization of a previous storyboard development, or a configurable storyboard composed of best-practice patterns. Decision of the modeling framework or language (here, the use of storyboarding, with or without story algebra usage, in which format) must have been taken. Generic modeling must be supported by meta-models assumed to be available as (re)usable packages. Further model contents are interpreted based on the selected packages.

The *operational customisation layer* fits, calibrates and prunes the model suite to the problem space. This is where the actual design is made, forming a normal model (here: the generic storyboard is customized as needed or allowed by the generic model: missing parameters are set up, defaults can be overridden). Requirements for an acceptable normal model must have been given in the generic model or the metamodel, in order to ensure well-formedness and consistency, and to allow proper model transformations possible on the delivery layer. The normal model(s) must be validated according to these requirements.

Finally, the *model is delivered* in various variants depending on the interest and the viewpoints of the CoP members. It is elicitated from the normal model using a model translation, extraction or enhancement method. The target model language (here, BPMN) must be given with the selection and customization of the available translation methods. Interrelations and consistency management between the normal and the delivered model can be further declared.

The complete model suite thus becomes the source for the code of the problem solution, and for the system to be built [13].

3   Elicitation of Workflow Models from Usage Models

One of the main challenges for model translation is the existence of different intentions behind the two modeling languages. BPMN is stricter than storyboarding, while a proper translation needs to make use of the inherent flexibility of the storyboard. Therefore, besides a direct and full translation option, we are proposing a way for constructing BPMN workflows by formulating story path schemata, based on selected parts of the storyboard.

3.1   An Application Case and Its Usage Model

We are assuming the development an information system for a touristic and recreational trail network, providing guidance for visitors, as well as facility management of the trails and related field assets. A map interface is being provided with planning and navigation features along the designated trails, connected to an issue tracking system for reporting and managing trail and asset defects.

The high-level usage model is given as a storyboard on Fig. 2. Abstract usage locations represented by graph nodes are called *scenes*. Users can navigate between these scenes, by performing actions associated to *directed transition links*. Some indicative *action names* are given for the reflexive links (which are in fact, denote multiple links, one by named action). The *entry point* is marked with a filled black circle. An *end point* may also be added as a double circle (by default, each scene is assumed to be a potential end point).

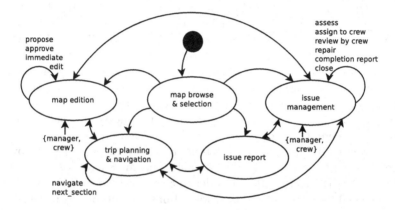

Fig. 2. High-level storyboard graph for a sample trail management system

We declare three *actor roles*: *visitor*, trail *manager* and trail *crew*. A set of authorized actors are pointed to the bottom of restricted scecnes by vertical arrows [9, p. 410]. By default, all actors are allowed to enter a scene.

The storyboard is about to represent supported *normal* scenarios, as specific means the users can accomplish given tasks. Context-loosing random navigations (e.g. back to the main page at any time) can be treated as breaking or canceling the started scenario, and starting a new scenario with a new context. These moves are not explicitly modeled so the focus can be kept on meaningful issues.

We are limiting our current discourse for one-session, one-actor scenarios.

The storyboard can be enhanced with input-output content specifications for each scene. We use the notation of [9, p. 410] so that input and output content for a scene is displayed using a short horizontal arrow on the left and the right side, respectively. Input-output content is named and an output content is assumed to be delivered as an input content to the next scene along each transition link, where the content names are equal. Square brackets denote optional input or output. Content names can be prefixed by generic database operation names.

3.2   Refinement and Normalization of the Storyboard

The top-level storyboard (Fig. 2) has to be refined and enhanced, so that actual scenarios as paths in the graph will be self-descriptive and consistent, and the graph is formally sound and contains enough details for a working and meaningful translation into workflow model(s). We state the following semantical considerations and guidelines for developing the refined usage model. If all these criteria are met, and guidelines have considered, we call the storyboard *normalized*. This is only partially verifiable formally – for the items marked with (*) – and refers to a quality and stage of model development:

- Complex scenes must be decomposed into atomic sub-scenes, each having a single, well-defined action, task or activity which is fully authorized by a given set of actor roles. The interaction paths must be modeled by directed links between the sub-scenes and directly connected to outside (sub)scenes.
- Each transition link with active actor participation (action) must be replaced by a link-scene-link combination, where the action or activity is performed at the scene and the new links are only for navigation. This new scene can be handled and parametrized together with other scenes in a unified way.
- No parallel links between two scenes are allowed (*). They must either be translated using separate scenes (see above), or merged into one link, or their source or target scenes must be decomposed to separate sub-scenes.
- The routing decision (which link to follow after a scene) is assumed to be taken as part of the activity inside a scene, by default. If it is not intended, then only one outgoing link is allowed and an extra routing decision scene must be explicitly introduced after the original scene as necessary (this may be later optimized out).
- Unique names are assumed for all scenes and links (except that two or more links pointing to the same target scene can have the same name) (*).
- There must be a unique start node (entry point) with a single link to an initial scene and either a unique end node or a default rule declaring which scenes can be places for story completion (*).
- Each scene must be enhanced with a set of authorized actor roles. Without that, a default rule must be supplied. There must be no (normal) links between scenes without at least one common authorized actor role. (*)
- Input and output content is to be specified by symbolic names for each scene wherever applicable. Content names will be matched along the links (*): For each input content of a scene s there must be an output content with the same name provided by the source scene of each link directed to s. Optional content is written in square brackets.
- Input and output content names can be prefixed by database operations: SELECT is allowed for input, while INSERT, UPDATE, and DELETE are allowed for output content. Without detailed semantics of these operations, a single central application database is assumed by default.

3.3 View Generation by Actor Roles

Given a selected actor role, a specific storyboard view can be generated for it as a basis of role-specific workflow models, by removing unauthorized scenes for a selected role with their links, resulting a cut-out of the storyboard, with reachable scenes by actors of the chosen role. An enhanced, normalized version of the visitors' storyboard view is shown on Fig. 3, with multiple sub-scenes. Links are denoted by italic numbers. An explicit end node is placed additionally, reachable from chosen scenes.

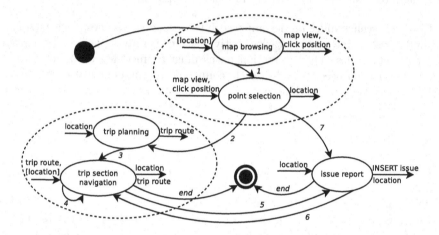

Fig. 3. Visitors' view of the storyboard after refinement and normalization

3.4 Graph-Based, Direct Translation Method

At this point, a default translation algorithm we have developed, can be applied to generate a BPMN process flow diagram, based on the graph connectivity of the storyboard. Details of the algorithm are omitted due to space limitations, but the result of the translation of Fig. 3 is shown on Fig. 4 as a demonstrative example. The translation process can continue with enhancements of Sect. 3.8.

3.5 Modeling Supported Scenarios by Story Algebra Expressions

Alternatively to the previous method, a more sophisticated and targeted method is developed, if specific scenarios, which are intended to be supported by the system, are collected and expressed as patterns in a story algebra.

A particular playout of system usage becomes a path in the storyboard and is called a *scenario*. A set of possible scenarios can be modeled as using the *story algebra SiteLang* [9, p. 76], similar to regular expressions. Such a *scenario schema* can be a pattern for generating a workflow model. The original notation uses link names for description. We found that using scene names in the story algebra more naturally supports the translation to workflow models.

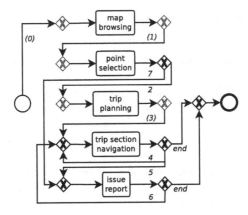

Fig. 4. Direct translation of visitor usage to BPMN, based on the storyboard graph. The full connectivity of the usage model is represented as possible process flow paths. Scenes become tasks. Numbers denote choices based on scene transition edges. Bracketed numbers are only for information, referring to original transition edges without alternatives. The grey-colored gates can be removed by merging their connections. Further refinements and optimizations are possible.

For example, a scenario schema of a visitor can be modeled out of the following variations: a visitor looks at the map, selects a destination point. The scenario may continue by reporting an issue for the selected point, or by planning a trip, navigating along it, and maybe at certain points, reporting an issue on-site. Each of these variants correspond to different scenarios the system should support and can be summarized as one or more scenario schemata.

Using abbreviated scene names (by first letters of words, e.g. *mb* stands for *map browse*), the above mentioned visitor scenarios can be modeled by the following story algebra expression (semicolon is used for denoting sequential steps, plus sign for at-least-once iteration, square brackets for optionality and box for expressing alternatives):

$$mb; ps; (ir\Box(tp; (tsn; [ir])^{+})) \tag{1}$$

Given a storyboard (view) specification, a scenario schema must be *compatible* with the given scene transitions, which means the following: Atoms of the story algebra expression must match to authorized scenes of the storyboard. The defined scenarios must correspond to valid directed paths within the storyboard (view). The defined scenarios must start with the marked initial scene and finish at the defined (or default) end scene(s).

Expression (1) is compatible with the visitors' storyboard view (Fig. 3). Consistency of the input-output content declarations can also be checked along the possible playouts. Note the link *6* will not be available if the visitor is coming from link *7* (there is no navigated route to go back to).

3.6   Decomposition into Mini-Stories

A scenario or story schema might contain semantically meaningful, reusable patterns of scene transition playouts, which can be combined with each other flexibly. Story algebra expressions, however, may be too complex and hard to handle by human modelers, and such semantical information remains hidden. A possible solution is to take the union of the relevant scenarios and decompose them into *mini stories* [14] (or, at least, extract some mini-stories from it).

A *mini-story* is a semantically meaningful, self-contained unit, which can be used flexibly in different scenarios, sometimes by possibly different actors. It can be modeled explicitly and translated as a reusable subprocess in the workflow model. Syntactic hints or heuristics can reveal possible mini-story candidates, but at the end the modeler has to explicitly define or verify them.

In our case, given the story algebra expression (1), candidate mini-stories can be recognized by maximal, non-atomic subexpressions with none of its nontrivial parts appearing elsewhere. Based on modeler decision taking into account semantics as well, we define the following two mini-stories, and substitute them in the story algebra expression (in a real case, with more scenarios, their reusability could be better verified): 1. *Select location from map*: $Slfm ::= mb; ps$ and 2. *Navigate along trip (with reporting issues)*: $Nat ::= (tsn; [ir])^+$. We keep referring to scenes ir and tp as atomic mini-stories. The resulting story algebra expression with the above mini-story substitutions of (1) becomes:

$$Slfm; (ir \square (tp; Nat)) \tag{2}$$

3.7   The Story-Based Translation Method

After the storyboard (viewed by an actor role, refined and normalized) and the desired story schemata (story algebra expressions) are given as above, with the mini-stories modeled, the workflow model in BPMN for each story schema can be elicitated the following, inductive way:

- Translate atomic mini-stories,
- Translate compound mini-stories based on their story algebra expressions (which are not translated yet),
- Compose the complex workflow based on the full story algebra expression.

Translation can be hierarchically carried over using structural recursion along the story algebra atoms and connectives, as displayed on Fig. 5. A choice for rule alternatives is proposed, with given defaults. The modeler can either leave the defaults as they are, or utilize the alternatives by applying pragma-like declarations to the usage model, or stereotypes associated to story algebra elements or subexpressions. For example, compound mini-stories can be translated as subprocesses, or connected using the link event notation. Conditionals for process flow gates match the names of corresponding storyboard edges (based on user choice) or their associated conditions or triggers (if such conditions are given for links of the storyboard).

3.8 Enhancement of the Translated Model

Transition link names (here, numbers) can be added to the workflow model, as well as input-output content as data objects and database connections associated to the workflow tasks and subprocesses (see the additional rules of Fig. 5).

Figure 6 displays a result of the refined, normalized visitors' storyboard view (Fig. 3) being translated to BPMN, based on story algebra expression (2) and mini-stories of Sect. 3.6, using rules of Fig. 5.

Model translation may be guided by additional information in forms of scene or link stereotypes. BPMN provides a variety of assets and some of them could be directly elicitated. Stereotypes offer more semantic information such as data or user-driven navigation, cancellation or rollback of started transactions, etc., to be mapped to native BPMN constructs.

The modeling process is based on laying out default values for model formats, start/end scenes, authorized actor roles, context objects containing scenario history, handling of exceptions and invalid routing, stereotypes and other semantical or transformative guidance (e.g. how to connect mini-stories together, how to translate iterated sub-processes). Defaults should work for conventional modeling cases. For customized, more sophisticated modeling, defaults can be overwritten. A possible post-translation optimization phase can improve the workflow model in each case. Most of these issues are left for future investigation.

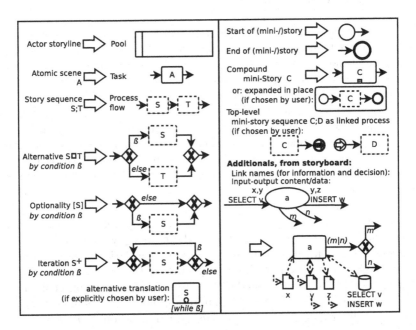

Fig. 5. Translation rules for story algebra expressions and additional assets based on storyboard. Dotted-lined rectangles denote arbitrary workflow model parts already translated from story subexpressions. There is a default translation for each construct, with possible alternatives that can explicitly be chosen by the modeler.

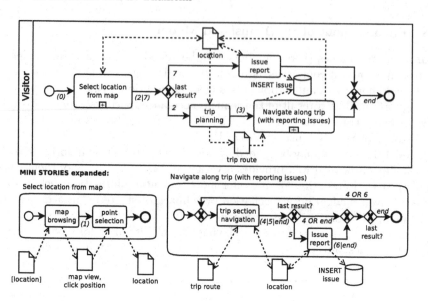

Fig. 6. BPMN workflow translation of visitors' view usage model, based on story algebra expression (2)

4   Conclusion and Future Work

Model-based programming can be the true fifth generation programming, supported by sound foundation and appropriate tools, based on model suites of explicitly interrelated models. Models have their specific functions, viewpoints and can be given in various levels of details. Ensuring coherence, consistence and translatability among them is a crucial issue. In this paper, we have presented a general, layered modeling framework as a basis, and showed its feasibility by giving methods and guidelines for model development and translation between two specific types of models: the usage model (expressed by storyboard graphs and story algebra expressions) and the workflow model (expressed by BPMN).

The workflow model is claimed to be directly translatable to program code [3,10]. We have introduced the concept of user view and the normalization of the storyboard, providing guidelines to the modeler to refine an initial, top-level usage model. We gave two methods for translating the refined usage model to workflow models, and successfully applied the mini-story concept for semantically structured and flexible workflow elicitation. Translation is based on default rules, while alternatives can be chosen explicitly by the modeler.

The method is ready to be tested with more examples or prototype implementations. Future issues include actor collaboration modeling, defining stereotypes and pragmas determining model semantics and translations. The metamodeler has to implement packages of generic models and add-ons, enrich generic models with pre-defined patterns and templates. The actual application modeler can choose among them or let the modeling system decide on which defaults it uses for which cases. It points towards a generic model-suite framework, which is, in our view, essential for truly working general model-based programming.

References

1. Alderson, A.: Meta-case technology. In: Endres, A., Weber, H. (eds.) SDE 1991. LNCS, vol. 509, pp. 81–91. Springer, Heidelberg (1991). https://doi.org/10.1007/3-540-54194-2_27
2. Bienemann, A.: A generative approach to functionality of interactive information systems. Ph.D. thesis, CAU Kiel, Department of Computer Science (2008)
3. Camunda: The Camunda BPM manual. https://docs.camunda.org/manual/7.10/. Accessed 17 May 2019
4. Chen, P.: Entity-relationship modeling: historical events, future trends, and lessons learned. In: Broy, M., Denert, E. (eds.) Software Pioneers, pp. 296–310. Springer, Heidelberg (2002). https://doi.org/10.1007/978-3-642-59412-0_17
5. Knuth, D.E.: The METAFONTbook. Addison-Wesley, Boston (1986)
6. Lamport, L.: LaTeX: a document preparation system. Addison-Wesley, Boston (1994)
7. OMG: Business process model and notation (BPMN) version 2.0 (2010)
8. Pittman, T., Peters, J.: The Art of Compiler Design: Theory and Practice. Prentice Hall, Upper Saddle River (1992)
9. Schewe, K., Thalheim, B.: Design and Development of Web Information Systems. Springer, Heidelberg (2019). https://doi.org/10.1007/978-3-662-58824-6
10. Stiehl, V.: Process-Driven Applications with BPMN. Springer, Switzerland (2014). https://doi.org/10.1007/978-3-319-07218-0
11. Thalheim, B.: Model suites for multi-layered database modelling. In: Information Modelling and Knowledge Bases XXI, volume 206 of Frontiers in Artificial Intelligence and Applications, pp. 116–134. IOS Press (2010)
12. Thalheim, B.: Normal models and their modelling matrix. In: Models: Concepts, Theory, Logic, Reasoning, and Semantics, Tributes, pp. 44–72. College Publications (2018)
13. Thalheim, B., Jaakkola, H.: Models as programs: the envisioned and principal key to true fifth generation programming. In: 29th International Conference on Information Modelling and Knowledge Bases. IOS Press (2019)
14. Tropmann, M., Thalheim, B.: Mini story composition for generic workflows in support of disaster management. In: DEXA 2013, pp. 36–40. IEEE Computer Society (2013)
15. Türker, C., Gertz, M.: Semantic integrity support in SQL:1999 and commercial (object-)relational database management systems. The VLDB J. **10**(4), 241–269 (2001)

Phenomenological Framework for Model Enabled Enterprise Information Systems

Tomas Jonsson[1(✉)] and Håkan Enquist[2]

[1] Genicore AB, Gothenburg, Sweden
tomas@genicore.se
[2] University of Gothenburg, Gothenburg, Sweden
hakan.enquist@gu.se
http://www.genicore.se, http://ait.gu.se

Abstract. Models are fundamental to enterprise information systems design. In recent years information systems which are generated from models, such as conceptual models, has emerged among researchers as well as in practice. We present Phenomenological Foundational Ontology for ontology driven conceptual modeling, with the purpose of improving information quality and manageability of information systems generated from models. The ontology is outlined and its application is exemplified using an easy-to-use web based tool which generates runtime systems including user interface, with the capability to process and communicate data.

Keywords: Information systems · Conceptual models · Ontology · Phenomenology · Semantic consistency

1 Introduction

We consider Model Enabled Information Systems (MEIS) to be Information Systems (IS) which are fully defined by and consistent with a model, either through model execution or by generating executable code from the model. Model driven development and execution of IS has a long history [1–4]. Automated generation of IS software, from requirements and conceptual models, is available from some independent suppliers and has proven to work for full-scale IS. Documented benefits include improvements in orders of magnitude concerning lead-time, cost and quality for comparable runtime IS.

However, model driven development and execution of IS is still not a mainstream approach and essentially it means a paradigm shift for the software community. In order to facilitate this paradigm shift, methods and tools have to be simplified, refined and knowledge about them has to be disseminated beyond the academic sphere.

T. Welzer et al. (Eds.): ADBIS 2019, CCIS 1064, pp. 176–187, 2019.
https://doi.org/10.1007/978-3-030-30278-8_21

1.1   From Enterprise Actors via Conceptual Model to Information System

Enterprises are *evolving socio-technical systems*, with a purpose of creating some value. Enterprises consist of people and artifacts and exist in context of society together with other enterprises. Enterprise ACTors (EACT) are people and other enterprises which act in relationship to the enterprise in question. EACTs have a need to communicate, share, manage and process information, thus there is a need for Enterprise Information Systems (EIS).

Our research and development focus on seamless life cycle management of EIS, over the life cycle of an enterprise. Systems which are semantically consistent with concepts of EACT and adapted to the traditions and culture as well as evolution of the enterprise. For this purpose, a method for agile and incremental, enterprise lifecycle information management is needed where MEIS is an essential mean. A method where EACT concepts are modeled into conceptual models, which can be transformed into IS, retaining the conceptual structure of the conceptual model, for semantic consistency between IS and EACT (Fig. 1).

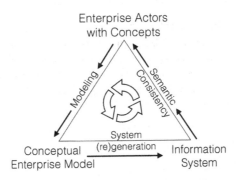

Fig. 1. Life cycle management of MEIS in evolving enterprises. A continuous process of modeling, (re)generating information system and deliver information semantically consistent with concepts of enterprise actors

We consider the three following phases of development of EIS.

1. Finding the phenomenon essence of the enterprise i.e. fundamental enterprise object types, their relations, properties and rules, forming a structure upon which information system support can be built and maintained, over the life cycle of an enterprise.
2. Implementing information system support starting with a limited number of actor roles (one or two) and a few essential activities, amending the fundamental object types with properties, relations, rules and possibly additional object types.
3. Continuously adding and refining support for actor roles and activities, leading to additional phenomena types, attributes and rules but, if 1 and 2 is properly performed, no restructuring of phenomena types and relations.

1.2 Requirements on Information Systems, Models and Tools

Methods and tools for MEIS should be designed based on the answers to two fundamental questions.

- Which characteristics of information systems are desirable?
- Which kind of models will yield desired characteristics?

In the context of EIS design, we consider two EIS characteristics being of most importance.

- EIS should convey concepts and data relevant to EACT
- EIS should be managed over the lifecycle of an evolving enterprise

In this paper we focus on a kind of model, based on phenomenological philosophy, which will result in EIS corresponding to these characteristics and we present an example of model and EIS in an easy to use web environment.

1.3 Relevant Structure of Concepts

To convey concepts and data relevant to EACT, there need to be a semantic consistency between the IS and EACT as stated by Langefors infological Eq. 1 [5]. The equation expresses the infological perspective that data alone is not information, but can give rise to information in the minds of people if data is presented within a frame of reference or perception of reality in the minds of people.

$$I = i(D, S, t) \tag{1}$$

I = conveyed information in specific actors mind
i = information function
D = set of data received by actor
S = interpreting structure (actors knowledge, in a wider sense, perception of reality)
t = time available to interpret data

In general IS need to be consistent with S, so design of IS needs to be guided by some model of S. This kind of model is generally referred to as a conceptual model. To make a conceptual model of S, some general understanding of the structure of S and some modeling guidance, such as an ontology, is needed.

The infological equation is related to individuals interpretation of data, however enterprises have many actors with different S, of which only fragments are related to perception of enterprise. Further, in order for EACT to collaborate with each other, there need to be some shared S elements, i.e., structures of concepts which has been agreed on between groups of actors. For EIS it is these shared elements of S which are of interest to find, model and implement consistently.

1.4   Sustainable Life Cycle IS Management in Evolving Enterprises

A key to sustainable life cycle IS management is to capture, in a model, a stable structure of concepts, so that changes in the evolving enterprise results in amending the model, rather than changing concept definitions and structures. Changing concept definitions and structures is problematic for two reasons. It means that defined concepts disappear or change meaning, which from a user perspective is confusing, it might break system integrations and there is also a data conversion problem, data which exists related to one definition either is lost or need to be converted into new concept definitions.

We suggest that a phenomenological ontology can assist in finding stable structures of concepts. An example of life cycle manageability of a MEIS ERP system [6], based on the described modeling principle, has been developed and managed over a period of 25 years.

2   Enterprise Information Models, Philosophy and Ontology

A systematic literature review on Ontology Driven Conceptual Models [7] indicate a lack of an ontology for executable conceptual models, with the purpose of modeling in such a way that information systems both will communicate relevant structures of concepts and being maintainable over the life cycle of an enterprise. A part of our current effort is therefore to focus on a new foundational ontology to assist in finding and defining the stable essence upon which information system support can be built and maintained, over the life cycle of an enterprise.

A phenomenological approach for conceptual models has been applied for successful and proven model driven EIS life cycle management, in large scale information systems (1500+ users) [6] over a period of 25 years. Lately, it has also been applied for successful design and maintenance of EIS for the social network enterprise Project Lazarus, a fantasy live action role play organization as demonstrated at ER 2018 [8].

In these implementations, the system model is composed of three layers. Phenomenon layer define semantic structure and data dependencies of the EIS, organization layer define a structure of roles in the enterprise and the perspective layer define fragments of information which should be available to specific roles in specific situations. In this paper we focus on the Phenomenological Foundational Ontology (PFO) for guiding semantic content of phenomenon layer.

2.1   Phenomenological Philosophy

Phenomena are defined as pre-conceptual mind entities [9], the fundamental entities from which we can build understanding and reasoning about the world, either from perceptions of reality or from creative thoughts. In Logical Investigations [10], Husserl lays out the foundation for phenomenological philosophy. Woodruff-Smith [11] summarize the Unified Theory of Husserlian System as depicted in Fig. 2.

Our focus of interest is the right side of the figure, the Formal Ontology and the Material Ontology Regions, indicating the existence of two orthogonal ontologies. Formal Ontology, the structure of (mental) existence as objects and Material Ontology Regions as mental (perceptual) regions to which these objects belong.

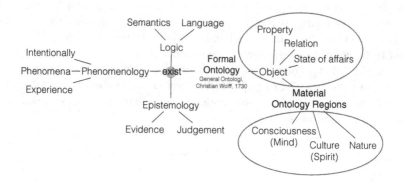

Fig. 2. The unified theory of the Husserlian system

In this Husserlian System there is no explicit notion of time, however Heidegger's work Being and Time [12] extends phenomenological philosophy with aspects of time as a fundamental ingredient. Starting from Husserlian System we add the time aspect, both to formal ontology and to ontology regions. In the formal ontology a state graph is added representing state concepts and transitions between these concepts. In the ontology regions, the region of Nature is further specialized into static physical world and temporal physical world.

Some other additions and modifications were made to transform Husserl's formal ontology into an ontology for conceptual modeling languages, such as the notion of object type and formal expressions to augment definitions of property and relation concepts. The elements object type, property, relation and state represent concepts which can be labeled and defined in a structure given by the language ontology. The PFO system is shown in Fig. 3.

Phenomenology of Husserl and others was from the beginning focusing on perception of reality of individuals (as first person view), not in social contexts such as societies and enterprises. The work of Schutz [13] and others, put phenomenology into the context of social worlds, adding the notion of intersubjectivity, a shared worldview of phenomena among groups of individuals. When applying PFO in the context of modeling EIS, it is specifically this shared worldview that is of interest, modeling the types of phenomena that are shared and communicated about, between actors related to an enterprise.

3   Phenomenological Foundational Ontology (PFO)

PFO is based on the notion of object as the superior structure for concepts, where objects represent mind objects, named phenomena. Phenomena are the

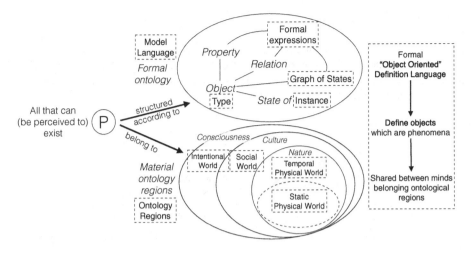

Fig. 3. PFO System, combining an object ontology with an ontology of mind objects. Additions to Husserlian system with blue labels. (Color figure online)

foundation for perception of reality in a metaphysical sense, expressed as Life World (Lebenswelt) in the philosophical school of phenomenology. Phenomenological Foundational Ontology (PFO) is thus different from the most referred-to ontologies [14] Bunge Wand Weber (BWW) and Unified Foundational Ontology (UFO).

BWW is an *object-type* ontology stipulating that object types are a superior structure of concepts (meta model) to be used when modeling a domain. BWW is designed with the purpose of defining data structures and transformations of an information system. However, BWW does not express a relation to S in the infological equation.

UFO is a concept classification ontology based on philosophical and linguistic theories. A model developed with UFO represent a structure of concepts, related to that which is used in a community, e.g. actors in an enterprise. UFO intent is to assist in creating models related to S in the infological equation, but does however not adhere to the object principle. When applied using the domain language OntoUML, the resulting model is an entity relation ship model where labeled concepts are entities and relations are relations between these concepts. I.e. all concepts are related at one level, without a superior structure. However, recent work on OntoUML demonstrate rules of how to group concepts into object like structure [15].

PFO combines an object ontology with an ontology of mind objects, as the domain to be modeled. PFO assist modelers to find the essence of enterprise and creating mind harmonizing models, reflecting on how the mind distinguish different kinds of phenomena and organizes concepts.

3.1 Phenomenon Kinds

Phenomenon kinds are grouped in four areas, Fig. 4, representing world-view awareness domains, building outwards from the most concrete and primitive awareness of existence. We use the term *physical object* to refer to that which exist outside the mind, observable through perception.

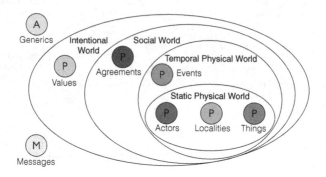

Fig. 4. Phenomenon kinds, their world-view awareness domains, combined with a coloring scheme used in modeling tool and runtime environment. (Color figure online)

Static Physical World: Phenomena as mind items, representing that which is considered to exist in a static physical world. Physical here means, that which could be observed. However, could be observed, does not mean that something has to exist for observation, it can be just thought of or imagined to exist as observed.

- Things: Predominantly inanimate physical objects, objects which are not considered to act nor have their own will.
- Actors: Predominantly animate physical objects and groups of such objects. Also automata such as automated machines, robots and information systems, if they seem to act i.e. take actions.
- Localities: Spacial relations, concepts of location, position, area or volume, coordinates in a coordinate system defining position, area or volume.

Temporal Physical World: A physical world which changes over time. All phenomena have a past, a present and a future, but when we want to understand and reason about change in the world of phenomena, change is the phenomenon.

- Events: Phenomena of change. Events can be observed only when they happen, are in progress, but as mind items they can exist as plans before and memories after they are observed.

Social World: The social world is regulated, driven, by agreements between actors in relation to the temporal physical world.

– Agreements: Phenomena of social relationships between actors, often also related to other kinds of phenomena. Agreements are either informal, undocumented, subconscious, or formalized and documented i.e. legal systems and written contracts.

Intentional World: A world of intentionality, that which motivates, drives, gives purpose to, individuals and organizations.

– Value: In the perspective of an actor, related to social and physical world, describing that which is value and possibly its measures. For actors as individuals, Maslow's hierarchy of needs indicates a starting point for possible values. For profit making enterprises, monetary value is fundamental but also other values such as customer values are considered. For non-profit enterprises such as social clubs, value could be emotional satisfaction for its members. For government organizations, value for citizens such as security, physical wellbeing as well as continuity of society, education, legal system, etc., could be considered as values.

Additional Phenomenon Kinds: The following two kinds of phenomena are not considered to be pre-conceptional mind items from the point of view of phenomenological philosophy. They are conceptual abstractions, which however may play an important role as entities in conceptual modeling for information systems.

– Message: As in the widest sense, generally unstructured (non modeled) information, usually as textual or image data. E.g. email, documents, pictures, books in their non physical sense i.e. in the digital universe.
– Generic: Should not be used when modeling according to PFO, except when a phenomenon type represents an abstraction of two different phenomenon kinds, e.g., phenomenon sales item can be either something physical (thing) or service (event).

3.2   Phenomenon Abstractions

Phenomena (P) of the kinds in PFO as shown in Fig. 4, except generic kind, can in turn be categorized into 5 levels of abstractions, see Fig. 5.

– **Instance** P: An individual identifiable phenomenon with properties and property values, e.g., Person, Car.
– **Type** P: A set of Instance P, which are thought of, in a community, as having the same kind of properties, e.g., People, Cars.
– **Amount type** P: A set of instances representing amounts of a phenomenon, which are thought of, in a community, as having the same kind of properties, e.g., amount of unidentified individuals or bulk handled phenomena, such as Nails, Sand, Liquids.
– **Meta type** P: A set of instances which represent types of P, e.g., Product Types, Car Models.

Kind \ Abstraction	Instance	Type	Amount Type	Meta Type	Abstract Type
Thing		2			1
Locality	1				
Actor		2			1
Event	1				
Agreement		2			1
Value					
Message					
General					

Fig. 5. Phenomenon classification system including both kinds and abstraction. Quantities refer to the shipping model in Fig. 7

- **Abstract type** P: A type P which has properties common to a set of other P types (Type, Amount type, Meta type, Abstract type) and represent the union of instance sets of those types, e.g., Vehicles (Car, Bus, Motorcycle), Primates (Humans, Simia, Lemur, Vespertilio).

Notes: Type P, Amount Type P and Meta Type P are called concrete types, as they represent the most concrete type representations in a PFO model. If two concrete types have the same properties, they should be considered as same type P (and joined into one type). Abstract type P, allows for introducing generalized type concepts and thus can enrich and simplify model as well as runtime system.

4   Example – Shipping Model with UFO and PFO

Core Model Language (CoreML) is a language concept for declarative object oriented executable models, with a structure as shown in Fig. 3. CoreML has been implemented in various tools and used in various projects over 30 years. CoreML has been used both for modeling CoreML tools (meta modeling) and information systems modeling. In CoreWEB, a recent implementation of CoreML, support for PFO is added both for modeling and system executions.

As an illustration of interpretability of models for enterprise actors we present two examples of the same enterprise case, one made with UFO and one with PFO. UFO in combination with the language OntoUML, published in [16], is shown in Fig. 6. The UFO model contains 25 concepts as 25 entities related to each other with 29 relations.

We then apply PFO with the language CoreML for the same case and the result is shown in Fig. 7. This model contains 11 entities (8 concrete, 3 abstract) using five of the eight phenomena kinds of PFO, shown in Fig. 5, and 8 relationships.

Graphical syntax of CoreML phenomena map is related to graphical representation of sets i.e. in this model the type People represents a set of person instances and type Organization represents a set of organization instances. Accordingly,

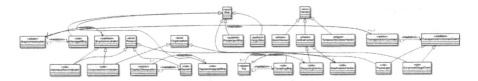

Fig. 6. Shipping model with UFO and OntoML. Image provided for an understanding of complexity of the non object conceptual approach, not for specific details.

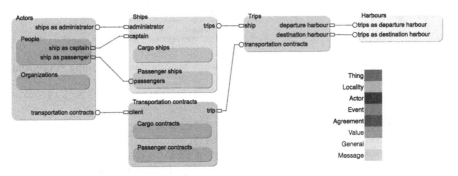

Fig. 7. Shipping model according to PFO in CoreML, represented as a phenomenon map

the abstract type phenomenon Actors, represent the set of all People and Organization instances. Relationship properties have multiplicity of $[0, 1]$ represented by rectangle or $[0, \infty]$ represented by a circle.

Phenomenon map in CoreML graphically represents entities and relationships, thus some concepts from UFO ship model are not shown. Those concepts are included as properties in CoreML modeling tool and textual representation of the model. For instance "harbor phases" will in CoreML be, *harbor* property *state* with possible values *Extinct, Active* and *Temporarily closed*.

To make the model meaningful for execution, as an IS, some additional properties are assumed and added, such as name of actors and ships and departure time, arrival time for trips. A relationship between transportation contracts and trips is also added. From this model a runtime system is generated in CoreWEB.

Figure 8 show the default user interface of the generated runtime system, from the phenomenological shipping model, in CoreWEB environment. For further examination, the model and runtime system are provided as a samples in CoreWEB website, and can be accessed by registering at Association for Model Enabled Systems (AMEIS) https://www.ameis.se/cml.

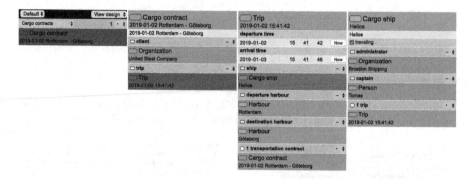

Fig. 8. The generated IS for shipping, with some sample data, retaining conceptual structure of the model

5   Concluding Remarks

The paper present PFO and an application of PFO to demonstrate the impact on EIS design. We show that PFO for Ontology Driven Conceptual Modeling have potential to contribute to increased information quality as well as to semantic consistency between user and information systems, in model driven EIS life cycles.

Significant research and design work remain to fully explore and implement complete support for PFO. Continued work on PFO relates to categories of phenomenon properties, such as identity, measurement, state and categories of phenomenon relations.

Researchers are encouraged to use the free CoreWEB tool to model, execute and evaluate models as well as the ontology. We urge researchers and practitioners with interest in EIS to join in a triple helix effort to research and spread MEIS and its practice in digitization of society.

References

1. Reenskaug, T.: PROKON/PLAN - a modelling tool for project planning and control. In: IFIP Proceedings, Toronto, Canada, pp. 717–722 (1977)
2. Pastor, O., González, A.: OO-method: an object-oriented methodology for software production. In: Tjoa, A.M., Ramos, I. (eds.) Database and Expert Systems Applications, pp. 121–126. Springer Vienna, Vienna (1992). https://doi.org/10.1007/978-3-7091-7557-6_21
3. Pawson, R.: Naked Objects. University of Dublin, Trinity College (2004)
4. Jonsson, T., Enquist, H.: Semantic consistency in enterprise models-through seamless modelling and execution support. In: CEUR Workshop Proceedings, Valencia, pp. 356–359 (2017)
5. Langefors, B.: Infological models and information user views. Inf. Syst. **5**, 17–32 (1980)
6. Jonsson, T., Enquist, H.: CoreEAF - a model driven approach to information systems. In: CEUR Workshop Proceedings, Stockholm, pp. 137–144 (2015)

7. Verdonck, M., Gailly, F., de Cesare, S., Poels, G.: Ontology-driven conceptual modeling: a systematic literature mapping and review. Appl. Ontol. **10**, 197–227 (2015)
8. Jonsson, T., Enquist, H.: Phenomenological ontology guided conceptual modeling for enterprise information systems. In: Woo, C., Lu, J., Li, Z., Ling, T.W., Li, G., Lee, M.L. (eds.) ER 2018. LNCS, vol. 11158, pp. 31–34. Springer, Cham (2018). https://doi.org/10.1007/978-3-030-01391-2_7
9. Zahavi, D. (ed.): The Oxford Handbook of Contemporary Phenomenology. Oxford University Press, Oxford (2012)
10. Husserl, E.: Logische Untersuchungen. Second edition 1913 and 1921 (1921)
11. Woodruff Smith, D.: "Pure" logic, ontology, and phenomenology. Revue internationale de philosophie **2**, 21–44 (2003)
12. Heidegger, M.: Being and Time. Trans. by J. Macquarrie, E. Robinson. Original published 1927. Harper and Row, New York (1962)
13. Schutz, A.: The Phenomenology of the Social World. Trans. by G. Walsh, F. Lehnert. Original published in 1932. Northwestern University Press, Evanston (1967)
14. Verdonck, M., Gailly, F.: Insights on the use and application of ontology and conceptual modeling languages in ontology-driven conceptual modeling. In: Comyn-Wattiau, I., Tanaka, K., Song, I.-Y., Yamamoto, S., Saeki, M. (eds.) ER 2016. LNCS, vol. 9974, pp. 83–97. Springer, Cham (2016). https://doi.org/10.1007/978-3-319-46397-1_7
15. Guizzardi, G., Figueiredo, G., Hedblom, M.M., Poels, G.: Ontology-based model abstraction. Presented at the IEEE Thirteen International Conference on Research Challenges in Information Science, Brussels (2019)
16. Figueiredo, G., Duchardt, A., Hedblom, M.M., Guizzardi, G.: Breaking into pieces: an ontological approach to conceptual model complexity management. In: 2018 12th International Conference on Research Challenges in Information Science (RCIS), Nantes, pp. 1–10. IEEE (2018)

Query-Based Reverse Engineering of Graph Databases – From Program to Model

Isabelle Comyn-Wattiau[1]([⊠]) and Jacky Akoka[2]

[1] ESSEC Business School, Cergy, France
wattiau@essec.edu
[2] CEDRIC-CNAM & IMT-BS, Paris, France
jacky.akoka@lecnam.net

Abstract. Graph databases have been developed to meet data persistence requirements, notably from social networks. They are, like the other NoSQL databases, often schemaless. This paper describes an incremental approach deriving a conceptual model from a graph database by analyzing a Cypher flow of queries. This reverse engineering approach embeds three main contributions: (1) a set of transformation rules of Cypher queries into chunks of conceptual schemas, (2) an incremental approach based on these rules, (3) an illustration on an example. This contribution enables, from a Cypher code, to generate a conceptual model that will facilitate the evolution of the existing graph database. This research is part of a project aiming at building an environment enabling round-trip engineering of relational and NoSQL databases.

Keywords: Conceptual model · Graph database · Cypher query · Reverse engineering · Schema integration

1 Introduction

NoSQL databases are considered a relevant choice when dealing with performance aspects. In addition, real time analytics is better suited to a NoSQL setting. Finally, in cases where data come from many different sources, NoSQL appears to be the only solution. However, most NoSQL databases use non-relational data models and, therefore, are schemaless. Although the absence of a schema offers a certain flexibility, it does not allow to fully benefit from its advantages. As pointed out by Klettke et al. [1], "most NoSQL data stores do not enforce any interesting structural constraints on the data stored. Instead, the persisted data often has only implicit structural information". Without a conceptual representation, a database is difficult to understand, to query, to migrate, and to transform. The lack of a conceptual schema may generate some overhead and complexity. Moreover, a logical schema facilitates query optimization and data integrity. Finally both logical and conceptual models remain useful, even when dealing with NoSQL databases.

This paper seeks to present and illustrate a reverse engineering approach, allowing us to transform queries expressed in graph database query language into a conceptual model. We propose a systematic and rule-based mechanism to transform a set of Neo4 J graph queries expressed in Cypher into a conceptual Extended Entity-Relationship

© Springer Nature Switzerland AG 2019
T. Welzer et al. (Eds.): ADBIS 2019, CCIS 1064, pp. 188–197, 2019.
https://doi.org/10.1007/978-3-030-30278-8_22

schema. The proposed transformation rules can handle a flexible set of graph database queries, enabling us to consider users' requirements. The approach consists of three major steps. The first step is based on a set of transformation rules enabling the development of conceptual model chunks. The second step queries the graph database in order to enrich the set of chunks. In the third step, we perform an integration mechanism leading to a global Extended Entity-Relationship schema. We present an illustrative scenario of this Program to Model (P2M) reverse engineering approach.

The rest of the paper is organized as follows. In Sect. 2 we present a state of the art on modeling NoSQL databases using both forward, reverse, and roundtrip engineering, especially for graph databases. We describe in Sect. 3 our reverse engineering approach. The latter and the associated transformations rules are illustrated in Sect. 4. Finally, Sect. 5 presents some conclusions as well as some perspectives in terms of future research.

2   State of the Art

Model Driven Engineering (MDE) is considered as a methodology providing several benefits such as an improved code quality and a better traceability. It consists of the application of models to increase the level of abstraction required to develop and evolve software products. Its aim is to offer software development approaches in which abstract models of software systems are created and transformed facilitating their implementations. MDE is based on model transformation which takes one or more source models and transform them into one or more target models. Several authors have proposed transformations using forward, reverse and roundtrip engineering approaches.

2.1   NoSQL Forward Engineering

Forward engineering of databases consists in going from an abstract conceptual representation of data to an implementable physical data model. Even if many NoSQL databases are schemaless, several authors have described forward engineering approaches [2]. An approach dedicated to data migration from a relational database into the document-oriented database is proposed by Banerjee et al. [3]. Vera et al. proposed a methodology dedicated to the design of NoSQL document databases [2]. As for graph databases, the approach of Aggarwal et al. enables the transformation of relational models as well as RDF models to property graphs [4]. Daniel et al. define transformation rules mapping conceptual UML class diagrams and OCL constraints, into logical graph databases [5]. Finally, for other NoSQL databases, let us mention the contribution of Abdelhedi et al. [6]. They propose to map a UML class diagram into a logical model of column-family databases. Li et al. propose transformations that map a logical relational model into a physical HBase database (an example of column-family DBMS) [7]. Although the mapping of conceptual or logical relational models into a NoSQL family or into a specific NoSQL DBMS has received significant attention, few works have investigated the specific case of mapping conceptual multidimensional (i.e. OLAP) models into NoSQL [8].

2.2 NoSQL Reverse Engineering

Database reverse engineering consists in deriving a conceptual model from the source code of an application. Since forward engineering leads to loss of semantics, the reverse process allows the elicitation of the missing semantic concepts. It provides a relevant high level description of data independent from physical characteristics. Although many authors have shown how to generate rich conceptual models from instances, few are related to NoSQL databases. Klettke et al. propose an algorithm for schema extraction of JSON data. Ruiz et al. have proposed a reverse engineering approach to derive the implicit schema of aggregate-oriented NoSQL databases [9]. Lamhaddab et al. present an MDE-based reverse engineering approach for inferring different graph models from mobile applications [10]. In [11], we developed a method deriving conceptual models from property graph databases.

2.3 NoSQL Roundtrip Engineering

Roundtrip engineering (RTE) represents one facet of Model Driven Engineering (MDE). Since code and model are interrelated, changing code will change the model and vice versa. RTE can be considered as a way to improve software engineering process. It consists mainly of forward engineering and reverse engineering. Demeyer et al. define RTE as the "seamless integration between design diagrams and source code, between modeling and implementation" [12]. Code generation, described as a push method, is obtained using forward engineering. The transformation of the source code into a conceptual model is obtained by a reverse engineering process based on a pull method. RTE is a solution enabling the synchronization of models by keeping them consistent, thus maintaining conceptual-implementation mappings under evolution. The main advantage of RTE is that the design and implementation artifacts are automatically synchronized all the time [13]. RTE has been first used with UML. It has been extended to other technologies such as graphical user interface design, database design, and to other software modeling artifacts. In [14], we have presented a framework describing a roundtrip engineering process for NoSQL database systems.

3   Our Approach

In [11], we described an approach taking into account Cypher CREATE statements and generating an EER conceptual model, using a graph meta-model as an intermediate step. In this paper, we go beyond Cypher's definition language dedicated to database generation by considering the query manipulation language (MATCH statements).

As mentioned by Angles et al., in graph databases, a graph structure contains nodes, edges, and properties storing the data [15]. Graph data models provide index-independent adjacency, avoiding lookup for indexing. Among graph data models, Neo4 J is a very popular one. Neo4j supports ACID transactions. It is schemaless system where the implicit data model is based on nodes representing entities and edges corresponding to relationships between entities. Its declarative query language, called Cypher, enables the creation and the update of nodes, edges, and properties of the graph. It considers the main functionalities of graph query languages such as: subgraph matching corresponding to conjunctive queries, nodes finding where nodes are connected by regular paths and/or conjunctive regular paths, paths comparison and returning, aggregation, node creation for nodes that were not part of the input, and approximate matching and ranking [16]. Its basic query syntax is as follows:

```
MATCH: pattern matching
WHERE: filter of the outcome
RETURN: output
```

We propose to analyze all MATCH queries and to elicit the underlying semantics in order to build a conceptual schema. The main steps of our reverse approach are: (Step 1) collect all Cypher code related to queries being performed; (Step 2) parse the code to deduce the conceptual EER schemas using transformation rules; (Step 3) complete and enrich the conceptual schema by querying the graph database; (Step 4) integrate all the schemas obtained in steps 2 and 3 into a global EER schema. We describe these steps below.

STEP 1. Collect all Cypher queries (manipulation language) related to the graph database under consideration. This step is relatively easy to achieve and therefore not subject to specific development.

STEP 2. Parse the resulting code and generate conceptual schema chunks using transformation rules

We present below the main transformation rules used in our reverse engineering approach and generating the conceptual model from Cypher manipulation language (Table 1). **Rule R1** generates an entity each time a Cypher query selects a node of the graph. The filter and projection properties are all inserted as attributes of this entity. **Rule R2** handles the case of queries that look for nodes with multiple labels. It generates a generalization hierarchy of entities and assigns to the generic entity all the properties present in the Cypher query. **Rule R3** considers queries looking for an arc of the graph knowing the type of the source and target nodes as well as the arc's label. It generates a binary relationship between the two entities corresponding to the source and target nodes. This rule handles oriented arcs and maintains arc orientation using the Source and Target roles attached to the lines connecting the relationship with the involved entities. **Rule R3.1** deals with the syntax where the arc is in the opposite direction. The Source and Target roles are thus reversed in the conceptual model. **Rule R4** handles multi-label arcs and generates, in this case, as many relationships between the entities describing the corresponding nodes. **Rule R4bis** is an alternative to Rule 4 which uses, where appropriate, the concept of hierarchy of relationships. **Rule R5** exploits Cypher queries that look for a sequence of two labeled arcs connecting three nodes. It generates a chain of two relationships. **Rule R6** translates queries with a WHERE clause filtering segments based on the properties of their nodes and/or their arcs. For readability reasons, the rule considers a reduced number of properties. **Rule R7** handles similar queries in which the WHERE clause filters based on node labels instead of property values. **Rule R8** uses the WITH DISTINCT clause. When the Label1 origin node has an already known identifying property, the presence of WITH DISTINCT followed by a label node Label2 guarantees the cardinality N of the binary relationship.

Table 1. Transformation rules

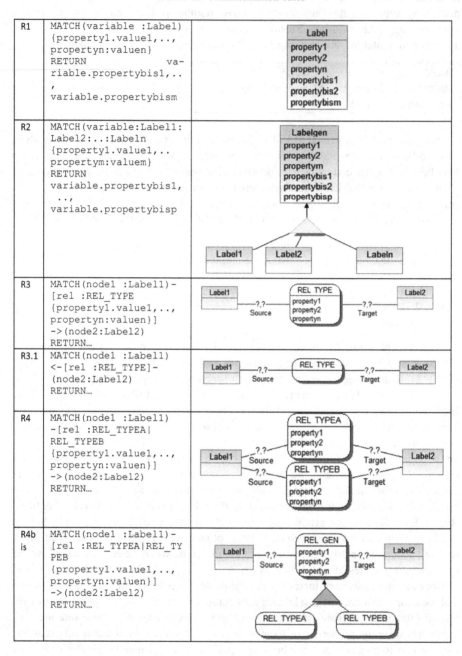

R1	`MATCH(variable :Label){property1.value1,..,propertyn:valuen} RETURN variable.propertybis1,.. , variable.propertybism`	
R2	`MATCH(variable:Label1:Label2:..:Labeln {property1.value1,.. propertym:valuem} RETURN variable.propertybis1, .., variable.propertybisp`	
R3	`MATCH(node1 :Label1)-[rel :REL_TYPE {property1.value1,.., propertyn:valuen}]->(node2:Label2) RETURN...`	
R3.1	`MATCH(node1 :Label1)<-[rel :REL_TYPE]-(node2:Label2) RETURN...`	
R4	`MATCH(node1 :Label1)-[rel :REL_TYPEA|REL_TYPEB {property1.value1,.., propertyn:valuen}]->(node2:Label2) RETURN...`	
R4b is	`MATCH(node1 :Label1)-[rel :REL_TYPEA|REL_TYPEB {property1.value1,.., propertyn:valuen}]->(node2:Label2) RETURN...`	

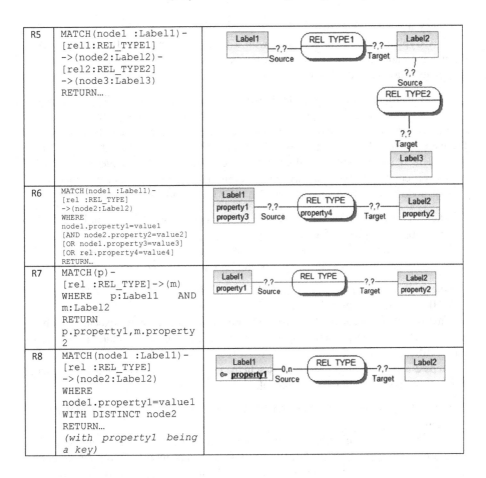

R5	`MATCH(node1 :Label1)-`
	`[rel1:REL_TYPE1]`
	`->(node2:Label2)-`
	`[rel2:REL_TYPE2]`
	`->(node3:Label3)`
	`RETURN...`	

R6	`MATCH(node1 :Label1)-`
	`[rel :REL_TYPE]`
	`->(node2:Label2)`
	`WHERE`
	`node1.property1=value1`
	`[AND node2.property2=value2]`
	`[OR node1.property3=value3]`
	`[OR rel.property4=value4]`
	`RETURN...`	

R7	`MATCH(p) -`
	`[rel :REL_TYPE]->(m)`
	`WHERE    p:Label1    AND`
	`m:Label2`
	`RETURN`
	`p.property1,m.property`
	`2`	

R8	`MATCH(node1 :Label1)-`
	`[rel :REL_TYPE]`
	`->(node2:Label2)`
	`WHERE`
	`node1.property1=value1`
	`WITH DISTINCT node2`
	`RETURN...`
	(with property1 being
	a key)	

Step 2 produces an EER schema chunk for each firing of a rule. The resulting schema chunks require to be integrated. This is the aim of step 4. But, beforehand we will enrich it if the database is available.

STEP 3. Schema enrichment

This optional step may be fired when the graph database is available and can be queried. We provide below a set of four Cypher queries that can be executed on the graph database to extract more semantics and complete the resulting conceptual schemain. The objective is to enrich the conceptual schema obtained at the end of the previous step. It is particularly useful since the set of queries parsed in step 2 can ignore certain parts of the graph.

Query a: Call db.indexes

This Cypher query returns a table with a Label column and a Properties column. We use it to add the corresponding entities with the related properties if they do not already exist as a result of step 2.

Query b: MATCH (node1: Label1) - [rel] -> (node2: Label2) RETURN type (rel)

The Cypher primitive type(rel) returns all labels, carried by arcs from nodes labelled Label1 to nodes labelled Label2. If it returns a single label value that was not discovered in step 2, we create a new relationship whose name is this label value. In the case where it returns several values, not yet elicited in step 2, a hierarchy of relationships is created in a similar way to rule R4bis. This query b must be executed for all pairs of node labels discovered in step 2 and/or in the indexes.

Query c: MATCH (node: Label) WHERE node.property1 = value1 RETURN node.property1, COUNT (*)

This query makes it possible to search the possible identifying properties of the nodes of the graph whose labels have been previously elicited.

Query d: MATCH (node1: Label1) - [rel: REL_TYPE] -> (node2: Label2) WHERE node1.property1 = value1 RETURN node2

This query deduces the 1 maximum cardinalities. It must be executed with all possible values of the property1 attribute before being able to conclude to a 1 maximum cardinality.

Steps 2 and 3 produce a set of EER schema chunks. The aim of Step 4 is to merge all these chunks in a unique EER model.

STEP 4. Integrate all the schema chunks obtained in step 2 into a global EER schema

This step uses a simplified schema integration process. As an example, if two schema chunks contain an entity called Label1, they are merged into a single entity, considering that the naming conflicts are solved. In the same vein, if the two corresponding entities have distinct properties, the resulting entity will contain the mathematical union of all properties. For space reasons, and given the fact that the main focus of this paper is on the transformation rules, we do not describe in more details the integration algorithm. The latter produces a single EER conceptual schema abstracting the graph database.

4   Illustrative Scenario

Let's consider the following set of Cypher queries extracted from Neo4j website. We apply steps 2, 3 and 4 of our approach.

STEP 2. Application of transformation rules

```
MATCH    (offer:PromotionalOffer)-[:USED_TO_PROMOTE]->(product:Product)
RETURN offer, product;
```
Applying rule R3 leads to the following schema chunk:

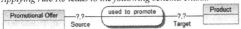

```
MATCH      (:Customer)-[:ADDED_TO_WISHLIST|:VIEWED]->(notebook:Product)-
[:IS_IN]->(:Category {title: 'Notebooks'}) RETURN notebook;
```
Applying rules R4 and R5 leads to:

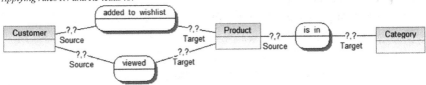

```
MATCH      (offer:PromotionalOffer     {type:      'discount_offer'})-
[:USED_TO_PROMOTE]->(product:Product)<-[:ADDED_TO_WISHLIST|:VIEWED]-
(customer:Customer)
RETURN offer, product, customer;
```
Applying rules R5 and R4 leads to:

```
MATCH (alex:Customer {name: 'Alex McGyver'})
MATCH (free_product:Product) WHERE NOT ((alex)-->(free_product))
MATCH (product:Product) WHERE ((alex)-->(product))
MATCH (free_product)-[:IS_IN]->()<-[:IS_IN]-(product) WHERE ((prod-
uct.price - product.price * 0.20) >= free_product.price <= (prod-
uct.price + product.price * 0.20)) RETURN free_product;
```
Applying rules R1 and R5 leads to:

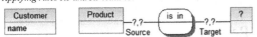

```
MATCH (p:Product)-[:IS_IN]->(m:Category:Family) RETURN p;
```
Applying rule R2 leads to:

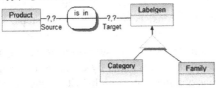

```
MATCH (a:Customer {fname:alex})-[r:BOUGHT]  ->(p:Product {availabil-
ity:true}) WHERE p.price<1000 RETURN p;
```
Applying rule R6 leads to:

```
MATCH  (c:Customer)-[:BOUGHT]->(p:Product)  WHERE  c.name='Alex'  WITH
DISTINCT p RETURN
```
Applying rule R8 leads to:

STEP 3. Enrichment

Firing "**Query a**" generates the entity Promotional Offer with property type as an additional entity. "**Query b**" applied to Customers and Products returns the following type(rel) values: added_to_wishlist, bought, viewed and dislike. The last value leads to a new relationship not found in our subset of Cypher queries. "**Query c**" applied to the customer allows us to elicit the identifying property *name*. "**Query d**" did not generate new information.

STEP 4. Schema integration

The integration process leads to the following global EER schema:

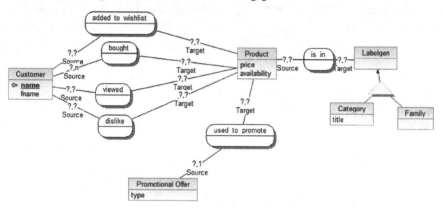

This simple example aimed at illustrating our reverse engineering process. We plan to test it on a real Neo4j database in order to refine the rules and check the completeness and the correctness of the resulting conceptual schema

5   Conclusion and Further Research

This paper presents a Program-to-Model approach enabling reverse engineering of NoSQL property graph databases manipulation language. We illustrate our approach using Neo4j environment. Our reverse approach consists of the following steps. We start by collecting all Cypher code related to queries being performed. We then parse the code to deduce the conceptual EER schemas using transformation rules. In the following step, we complete and enrich the conceptual schema by querying the graph database. Finally, we integrate all the schemas obtained into a global EER schema. We present an illustrative scenario.

We plan to extend this approach to reverse engineering of other graph database systems, such as OrientDB. Further work should target other NoSQL databases including key-value or document databases. Beyond the basic graphs, we will study the case of hypergraphs. This will necessitate the extension of our physical and logical models. Thus, we will be able to generate N-ary relationships. More experiments will be conducted on large databases enabling a comprehensive validation process. To this end, we are implementing the transformation rules in a prototype. Finally, experimenting with this approach will allow us to evaluate its robustness, as well as its completeness.

References

1. Klettke, M., Störl, U., Scherzinger, S.: Schema extraction and structural outlier detection for JSON-based NoSQL data stores. Datenbanksysteme für Business, Technologie und Web (BTW 2015) (2015)
2. Vera, H., Wagner Boaventura, M.H., Holanda, M., Guimaraes, V., Hondo, F.: Data modeling for NoSQL document oriented databases. In: CEUR Workshop Proceeding, pp. 129–135 (2015)
3. Banerjee, S., Sarkar, A.: Logical level design of NoSQL databases. In: IEEE Region 10 Conference (TENCON) (2016)
4. Aggarwal, D., Davis, K.C.: Employing graph databases as a standardization model towards addressing heterogeneity. In: 17th International Conference on Information Reuse and Integration. IEEE (2016)
5. Daniel, G., Sunyé, G., Cabot, J.: UMLtoGraphDB: mapping conceptual schemas to graph databases. In: Comyn-Wattiau, I., Tanaka, K., Song, I.-Y., Yamamoto, S., Saeki, M. (eds.) ER 2016. LNCS, vol. 9974, pp. 430–444. Springer, Cham (2016). https://doi.org/10.1007/978-3-319-46397-1_33
6. Abdelhedi, F., Ait Brahim, A., Atigui, F., Zurfluh, G.: Logical unified modeling for NoSQL DataBases. In: 19th International Conference on Enterprise Information Systems (ICEIS 2017), Porto, Portugal, pp. 249–256, April 2017
7. Li, Y., Gu, P., Zhang, C.: Transforming UML class diagrams into HBase based on meta-model. In: International Conference on Information Science, Electronics & Electrical Engineering, pp. 720–724 (2014)
8. Chevalier, M., El Malki, M., Kopliku, A., Teste, O., Tournier, R.: Implementation of multidimensional databases with document-oriented NoSQL. In: Madria, S., Hara, T. (eds.) DaWaK 2015. LNCS, vol. 9263, pp. 379–390. Springer, Cham (2015). https://doi.org/10.1007/978-3-319-22729-0_29
9. Sevilla Ruiz, D., Morales, S.F., García Molina, J.: Inferring versioned schemas from NoSQL databases and its applications. In: Johannesson, P., Lee, M.L., Liddle, S.W., Opdahl, A.L., Pastor López, Ó. (eds.) ER 2015. LNCS, vol. 9381, pp. 467–480. Springer, Cham (2015). https://doi.org/10.1007/978-3-319-25264-3_35
10. Lamhaddab, K., Elbaamrani, K.: Model driven reverse engineering: graph modeling for mobiles platforms. In: 15th International Conference on Intelligent Systems DeSign and Applications (ISDA) (2015)
11. Comyn-Wattiau, I., Akoka, J.: Model driven reverse engineering of NoSQL property graph databases: the case of Neo4j. In: 2017 IEEE International Conference on Big Data (Big Data), Boston, MA, pp. 453–458 (2017)
12. Demeyer, S., Ducasse, S., Tichelaar, S.: Why unified is not universal. In: France, R., Rumpe, B. (eds.) UML 1999. LNCS, vol. 1723, pp. 630–644. Springer, Heidelberg (1999). https://doi.org/10.1007/3-540-46852-8_44
13. Kellokoski, P.: Round-trip engineering. MA thesis, University of Tampere, Finland ((2000))
14. Akoka, J., Comyn-Wattiau, I.: Roundtrip engineering of NoSQL databases. Enterp. Model. Inf. Syst. Archit. (EMISAJ) **13**, 281–292 (2018)
15. Angles, R., Gutierrez, C.: Survey of graph database models. ACM Comput. Surv. (CSUR) **40**(1), 1 (2008)
16. Wood, P.T.: Query languages for graph databases. ACM SIGMOD Rec. **41**(1), 50–60 (2012)

A Model-Driven Needs Based Augmented Reality: From Model to Program

Manal A. Yahya[(✉)] and Ajantha Dahanayake

Lappeenranta University of Technology, 53851 Lappeenranta, Finland
manal.yahya@student.lut.fi

Abstract. The process of creating augmented reality experiences has developed greatly to break the lab boundaries and become available to every person with a smartphone. AR experiences for the most part focus on the object for which they are developed: the product to be advertised, the process to be taught, or the history to be shown for examples. Little focus is given to the person experiencing AR. It is similar to stepping into an enhanced world with information at the user's fingertips that was dictated by the experience developer rather than what the user actually needs to have from this experience. This controversy originated the idea for this research. In this research we aim to explore the concept for Model Driven Development to create augmented reality based on human needs. Towards achieving this concept, we present a conceptual model, analyze its components and study the application of Model driven concepts to this application idea.

Keywords: Model driven development · Human needs · Conceptual model · Augmented reality

1 Introduction

The use of Augmented Reality (AR) has grown over the years. It started from highly complex systems at labs available to only few researchers [1] to applications available for use in daily life in smartphones and tablets. Augmented reality experiences are usually pre-developed at the design and implementation phase. In a similar manner, the recognition of a human need in software systems is usually done during the requirements gathering phase. However, developing a software to recognize a need in Real-time is a novel concept. These conditions inspired us to develop an augmented reality application that responds to human needs.

The main question in this research is: *How to apply a model driven development (MDD) method to create a needs-based augmented reality system?*

The paper is structured as follows: Sect. 2 presents background information about the main concepts in this study, Sect. 3 presents our conceptual model with description of its components, Sect. 4 explains our application of MDD on the proposed system, and finally Sect. 5 offers the conclusion.

© Springer Nature Switzerland AG 2019
T. Welzer et al. (Eds.): ADBIS 2019, CCIS 1064, pp. 198–209, 2019.
https://doi.org/10.1007/978-3-030-30278-8_23

2   Background

To study human needs is mainly a field in psychology. Linking that concept to the computer world and attempting to understand human needs using computing systems is an effort that might draw near the concept of ubiquitous computing. This section provides basic definitions of the main concepts in this research.

An **Augmented reality** system as defined by [2] embodies the following properties:

- "combines real and virtual objects in a real environment;
- Runs interactively, and in real time;
- Registers (aligns) real and virtual objects with each other".

Human Needs differ from the concept of wants in the sense that needs are finite and common to all human beings while wants are changing and specific to each person. Human needs are an established field of study in psychology. Many models are developed throughout the years including the renowned Maslow's pyramid of needs [3], Max-Neef's matrix of needs [4], Doyal & Gough's intermediate needs [5], and Ryan & Deci's self-determination theory [6].

In all of these models, a basic rule applies: the human body tends to keep all systems in a *balanced state*: homeostasis. And to maintain homeostasis, several mechanisms are in action such as: osmoregulation, thermoregulation and chemical regulation [7]. These mechanisms aid in restoring the state of homeostasis.

The homeostasis control process starts with the balanced state, when a stimulus triggers imbalance, the receptors detect it, and send the information to the control system (the brain), which enables the effectors to perform required action to return the balance.

Needs create a drive to reduce them and satisfy the need. Needs may be recognized by first recognizing the homeostasis state and detecting the change in that state.

In order to develop methods to capture human needs, we need to understand the triggers of need. According to Myers [8], three main triggers elicit needs. These are: deviation from homeostasis, incentives, and stimulation, Table 1 describes the triggers.

Table 1. Needs triggers

Concept	Description	Example
Homeostasis imbalance	A change in the steady internal state that triggers action to retain it	Thirst
Incentive	"A positive or negative environmental stimulus that motivates behavior"	The sight of a cold lemonade in a hot day
Stimulation	A thing, activity or event that evokes a reaction	Excitement about mountain climbing

A Need Lifecycle

Originating from the idea of the homeostasis process through which the human body resolves and satisfies basic biological needs, we develop the concept of Need Lifecycle. The life of a human need starts when a trigger elicits the need, it then moves to possible methods to satisfy those needs, the action taken, and finally the satisfaction of the need. The cycle continues while the need persists, and terminates when it is either satisfied or disregarded (Fig. 1).

Fig. 1. Need life cycle

2.1 Model Driven Development

First, we start with Forrester's Definition of Model driven development: "An iterative approach to software development where models are the source of program execution with or without code generation" [9]. In this section we examine available literature to find best practices and implementation tools of MDD in software development.

Modelling for an augmented reality application requires a different set of skills than most regular applications. The number of studies on the specific topic of model-driven development for augmented reality applications is limited. Some of these studies are shown in Table 2 and discussed below.

Table 2. Studies on model driven development for augmented reality

Authors	Study description	Tools, languages, approaches	Findings/Conclusion
Bucciero, Mainetti [10]	Proposed a model driven method for easy content generation in collaborative virtual environments	WebTalk, Mobile WebTalk, A concept map for CVE	The gap between domain experts and system designers must be reduced by means of MDD
Feuerstack, de Oliveira, Anjo, Araujo, Pizzolato [11]	Presented an interaction focused extension for an existing MDD design for AR	State chart XML (SCXML), scxmlgui editor	It is possible to create change on the abstract level by combining modeling with flow chart notations

(*continued*)

Table 2. (*continued*)

Authors	Study description	Tools, languages, approaches	Findings/Conclusion
Fiore, Mainetta, Manco, Marra [12]	Presented the Cultural Compass application which allows spatial, semantics, and time navigation in cultural sites	Wikitude SDK	Cultural experience is enhanced
Raso, Cucerca, Werth, Loos [13]	Proposed a paradigm that enables automatic AR content creation by defining domain ontology	Java EE	Provided a new definition of hybrid media and means to utilize it in business
Ghandorh, Mackenzie, Eagleson, Ribaupierre [14]	Developed two AR simulator systems for neurosurgical training using Model driven engineering	Unity 3D engine, Vuforia SDK, 3D meshes	Further testing with the prototypes is needed for user performance
Swain, Mohan, Choppella, Reddy [15]	Used model driven approach to develop a virtual Lab Authoring Kit	Builder Design Pattern, Javascript Node.js, interactjs library	"The approach reduced the conversion of Flash Labs from weeks to days"
Vaupel, Taentzer, Gerlach, Guckert [16]	Presented a modelling language for iOS and Android role-based app development	Domain specific modelling language, CRUD functionality	The presented approach is best suited for data management apps

The goal of using model driven development is to increase abstraction and reduce complexity. In order to shorten the bridge between domain experts and engineers in the development of collaborative virtual environments (CVEs), the researchers in [10] created an extensive *concepts map* describing all the elements involved in the development of CVEs and used the map to create the WebTalk and Mobile WebTalk engines.

Another approach to add abstraction away from the source code level is the use of state charts [11]. The researchers applied a "Model-driven design of user interfaces (MDDUI) process" and proposed the Multimodal Interaction Framework (MINT). They used state chart XML in scxmlgui editor to develop the interactor models.

The authors in [13] recognized the lacking of automatic content creation in mixed media. They proposed a system that allows for automatic AR content creation by defining domain ontology and using a recommender system. This research is similar to our study in the sense that AR content is to be generated according to individual user needs rather than having a general predefined AR content.

Model driven engineering for augmented reality is also used for more critical application such as neurosurgical simulator that aids in training for brain procedures [14]. The researchers used Hierarchical Task Analysis (HTA) to develop the project schemas.

As a means to enable students worldwide easy access to science labs, researchers in [15] developed a model-based virtual lab system named the Lab Authoring Kit. This system allows easy creation of science experiments away from the detailed programming work.

Studying the literature, we find that the main goal in using Model driven or Model based development is to reduce the gap between two ends in a development project such as: between the domain experts and system designers [10], and between authoring environments and source code or framework [11]. Figure 2 shows the levels of abstraction in AR development.

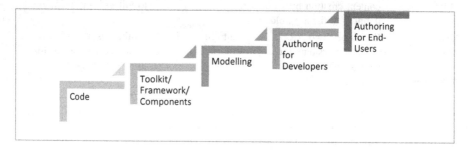

Fig. 2. Levels of abstraction in AR development [11]

3   Conceptual Model for a Needs-Based Augmented Reality

3.1   Model Description

The model we propose in this research is based on a previously developed framework: The Needs-Context-Technology (NCT) Framework [17] (Table 3), The NCT Framework describes the technology needed to detect needs based on the four existential categories of Max-Neef's matrix of needs [4]: Being, Having, Doing, and Interacting. These categories span across all the nine axiological categories: Subsistence, Protection, Affection, Understanding, Participation, Leisure, Creation, Identity, and Freedom.

Table 3. Needs context technology framework

Fundamental human needs existential categories	Being (qualities)	Having (things)	Doing (actions)	Interacting (settings)
Context aware categorization	User, who (Identity)	Things	What (activity)	Where (location), weather, social, networking
	When (time)			
Sensors and technology	Emotion sensors Body sensors	IoT systems and sensors	Activity recognition through motion sensors	Location awareness, nearby user device (for proximity with other users)

The **Being** existential category corresponds to the internal state and the defining characteristics of the human. To recognize change or a lacking (imbalance) in the Being category, if we consider for instance the subsistence need (Table 4), it requires:

- Being: Healthy physically and mentally
- Having: Food, shelter, work
- Doing: Feed, procreate, rest, work
- Interacting: Living environment, social setting

The secret is in the balance, once all states (internal state and external state) are balanced, a need may be considered satisfied or dormant. The occurrence of imbalance in either state (internal or external) requires attention from the system to move towards needs satisfaction.

Table 4. Subsistence need according to Max-Neef [4]

Needs according to axiological categories	Needs according to existential categories			
	Being	Having	Doing	Interacting
Subsistence	1/Physical health, mental health	2/Food, shelter, work	3/Feed, procreate, rest, work	4/Living environment, social setting

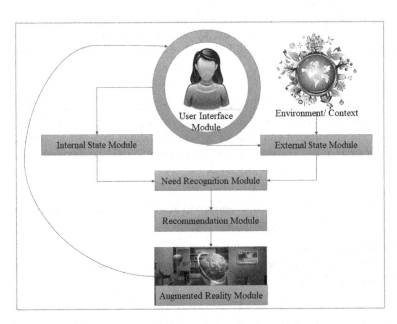

Fig. 3. Conceptual model of a needs based augmented reality system

We propose the conceptual model in Fig. 3 to represent our system. The application of this model results in a scenario similar to this: "It is noon in the weekend, the user is relaxing at home. Given the same context information, the user was interested to check some food services in the past. The probable satisfier for this **need** for food is determined to be going out to a restaurant, based on user preference from profile information and current context. Then, an AR view displays information about a nearby restaurant and allows the user to view directions and more information about it" [18]. In this example scenario we notice that AR can be used as a means to need satisfaction rather than being the satisfier.

3.2 Model Components and Description

The proposed model Fig. 3 consists of a number of modules, each of which fulfills a certain function and collectively they achieve the overall goal of the system. The modules are as follows:

- User Interface module
- Internal state module
- External state module
- Need recognition module
- Recommendation module
- Augmented Reality module

User Interface Module
The user interface module is concerned with the input and output devices with which the user interacts. For the user interface we must consider the physical sensors, and devices that the user must use to fulfill the goal of the system: mainly, input and output devices.

Internal State Module
This module checks the internal state of the human being using the system. It requires sensors to read the various physiology and psychology states. The main goal is to monitor health information and changes in sensor readings to detect a state of imbalance for a certain time. The occurrence of imbalance should be reported to the need recognition module to recognize the need.

External State Module
The external state module requires sensors to read the information surrounding the human being: context, environment, and objects.

Need Recognition Module
The need recognition module receives input from the previous two modules and decides which need is active.

Recommendation Module

The recommendation module matches the need with appropriate satisfier based on relation, or preferences.

Augmented Reality Module

Finally, the augmented reality module displays the AR experience associated with the satisfier.

4 Model to Programming

The purpose of this research is to study the transfer from model to program code in the case of augmented reality applications. Therefore, we chose to focus on the **Need Recognition** and **Augmented Reality modules**. The recommendation module is left out on purpose since it is mainly concerned with machine learning algorithms and certain parameters.

We follow the method presented by Vaupel et al. [16] in their work aimed at model-driven development for mobile applications. While the authors' work focused on the development of a domain-specific language, our work focuses on model-development and code-generation to match the requirements of the application presented in this paper. Our resultant method provides the following models:

1. A Data Model
2. A Graphical User Interface Model
3. A Behavior Model

4.1 Data Model

A Data model defines the classes and class structure of the application. This model affects the data presentation in user interface [16]. It also affects the processing of needs recognition and satisfier recommendations. To support the application in the correct manner, our proposed data model is defined along the axes of the existential needs' categories of Max-Neef Matrix of needs: *Being, Having, Doing,* and *Interacting.* Figure 4 shows the Ecore data model based on the Eclipse modeling framework (EMF) [19]. The model is developed using the GenMyModel platform, and its goal is to establish a data structure with all data required for needs recognition. Figure 5 depicts the Metamodel template that generates MySQL code from the Ecore model (on the left) and the automatically generated MySQL code (on the right side). The goal of applying MDD in this case is to abstract model development from code generation.

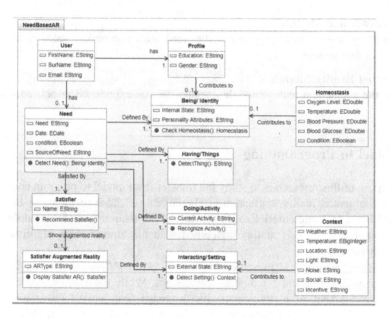

Fig. 4. Data model of the needs based augmented reality application

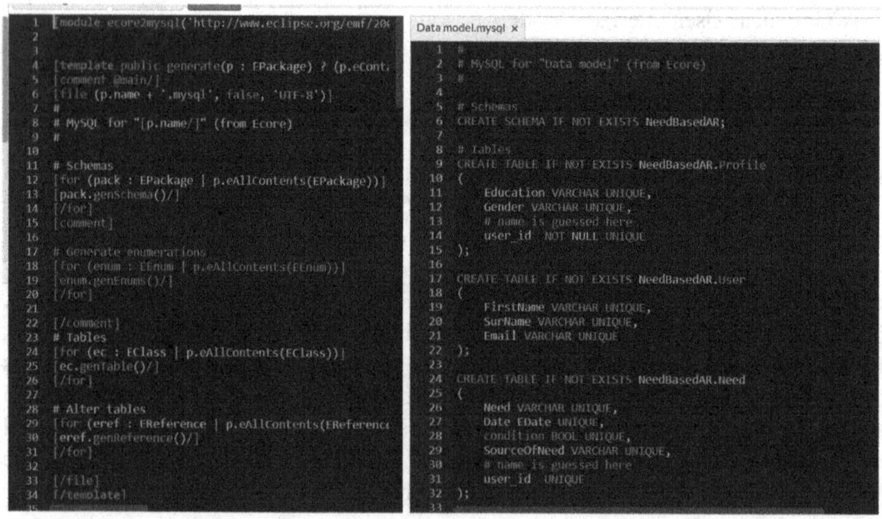

Fig. 5. Code generator from Ecore to MySQL (Left), Generated MySQL Schema code (Right)

4.2   Graphical User Interface (GUI) Model

It contains the main pages for interaction with the application user. In our application, the GUI is composed of two elements: user registration and information, and satisfier AR display. For the Augmented Reality display, we demonstrate the output of the proposed application using the ROAR platform (Fig. 6), it provides drag and drop features to create AR experiences. Such a user interface should be connected to the recommendation module and user profile to provide the most suitable suggestions to the user.

Fig. 6. ROAR editor (Left) Example of AR output (Right)

4.3 Behavior Model

For the behavior model, we developed an activity diagram that depicts the flow of control, the dynamics between the elements and their relations (Fig. 7). The main interacting parties are the user, the application, and sensors.

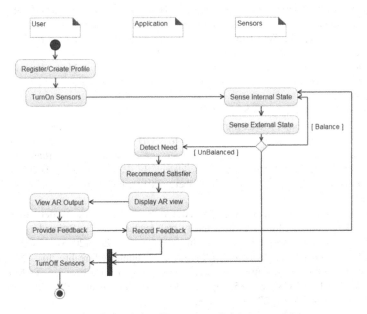

Fig. 7. Activity diagram as a behavior model

5 Conclusion

In this paper we presented a Need-Based Augmented Reality application, provided the conceptual model and component analysis of the application. We also described our approach to a model-driven development of the presented application. Model-driven development provides an excellent base to ease the application development process, improve the quality and performance due to standardization of code generators [20]. Nevertheless, the process requires much thought and planning to enable a complete application generation especially in the case of Augmented reality where 3D objects, markers, locations, and many elements play a vital role. In this research, we presented two methods for MDD: first, an automatically generated code for the Ecore data model; and second, a drag and drop AR creation example. For future direction, we aim to test the presented model and develop a complete application with real life participants. While the presented models provide a necessary representation of the application, more refinement is required to depict all elements related to the development of such a time dependent system, as well as developing the integration process for the various models.

References

1. Caudell, T.P., Mizell, D.W.: Augmented reality: an application of heads-up display technology to manual manufacturing processes. In: Proceedings of the Twenty-Fifth Hawaii International Conference on System Sciences, Kauai, USA. IEEE (1992)
2. Azuma, R., Baillot, Y., Behringer, R., Feiner, S., Julier, S., MacIntyre, B.: Recent advances in augmented reality. IEEE Comput. Graphics Appl. **21**(6), 34–47 (2001)
3. Maslow, A.H.: A theory of human motivation. Psychol. Rev. **50**(4), 370 (1943)
4. Max-Neef, M.A.: Human Scale Development. The Apex Press, New York (1991)
5. Doyal, L., Gough, I.: A Theory of Human Need. Palgrave Macmillan (1991)
6. Ryan, R.M., Deci, E.L.: Self-determination theory and the facilitation of intrinsic motivation, social development, and well-being. Am. Psychol. **55**(1), 68 (2000)
7. Sarvani, P.: Role of homeostasis in human physiology: a review. J. Med. Physiol. Ther. **1**(2) (2017). https://www.omicsonline.org/open-access/role-of-homeostasis-in-human-physiology-a-review.pdf
8. Myers, D.G.: Psychology, 13th edn. Worth Publishers, New York (2013)
9. Nare, A.: Introduction to Model Introduction to Model-Driven Development (MDD). https://www.utdallas.edu/~chung/SYSM6309/Intro-to-MDD-UTD.pdf. Accessed 07 July 2019
10. Bucciero, A., Mainetti, L.: Model-driven generation of collaborative virtual environments for cultural heritage. In: Petrosino, A., Maddalena, L., Pala, P. (eds.) ICIAP 2013. LNCS, vol. 8158, pp. 268–277. Springer, Heidelberg (2013). https://doi.org/10.1007/978-3-642-41190-8_29
11. Feuerstack, S., de Oliveira, A., Anjo, M., Araujo, R., Pizzolato, E.: Model-based design of multimodal interaction for augmented reality web applications. In: Proceedings of the 20th International Conference on 3D Web Technology, pp. 259–267 (2015)
12. Fiore, A., Mainetti, L., Manco, L., Marra, P.: Augmented reality for allowing time navigation in cultural tourism experiences: a case study. In: De Paolis, L.T., Mongelli, A. (eds.) AVR 2014. LNCS, vol. 8853, pp. 296–301. Springer, Cham (2014). https://doi.org/10.1007/978-3-319-13969-2_22

13. Raso, R., Cucerca, S., Werth, D., Loos, P.: Automated augmented reality content creation for print media. In: Lugmayr, A., Stojmenova, E., Stanoevska, K., Wellington, R. (eds.) Information Systems and Management in Media and Entertainment Industries. ISCEMT, pp. 245–261. Springer, Cham (2016). https://doi.org/10.1007/978-3-319-49407-4_12

14. Ghandorh, H., Mackenzie, J., Eagleson, R., de Ribaupierre, S.: Development of augmented reality training simulator systems for neurosurgery using model-driven software engineering. In: 2017 IEEE 30th Canadian Conference on Electrical and Computer Engineering (CCECE) (2017)

15. Swain, S., Mohan, L., Choppella, V., Reddy, Y.: Model driven approach for virtual lab authoring - chemical sciences labs. In: 2018 IEEE 18th International Conference on Advanced Learning Technologies (ICALT), pp. 241–243 (2018)

16. Vaupel, S., Taentzer, G., Gerlach, R., Guckert, M.: Model-driven development of mobile applications for Android and iOS supporting role-based app variability. Softw. Syst. Model. **17**(1), 35–63 (2018)

17. Yahya, M., Dahanayake, A.: A needs-based personalization model for context aware applications. Front. Artif. Intell. Appl. **292**, 63–82 (2016)

18. Yahya, M.A.: A Context-Aware Personalization Model for Augmented Reality Applications. Prince Sultan University (2016)

19. Steinberg, D., Budinsky, F., Paternostro, M., Merks, E.: EMF: Eclipse Modeling Framework 2.0, 2nd edn. Addison-Wesley Professional (2009)

20. Küster, J.: Model-Driven Software Engineering - Code Generation. IBM Research (2011)

Graphical E-Commerce Values Filtering Model in Spatial Database Framework

Michal Kopecky$^{(\boxtimes)}$ [iD] and Peter Vojtas [iD]

Department of Software Engineering, Faculty of Mathematics and Physics,
Charles University, Prague, Czech Republic
{kopecky,vojtas}@ksi.mff.cuni.cz

Abstract. Our customer preference model is based on aggregation of partly linear relaxations of value filters often used in e-commerce applications. Relaxation is motivated by the Analytic Hierarchy Processing method. In low dimensions our method is well suited also for data visualization.

The process of translating models to programs is formalized by Challenge-Response Framework CRF. CRF resembles remote process call. In our case, the model is automatically translated to a program using spatial database features. This enables us to define new metrics with spatial motivation.

We provide experiments with simulated data (items) and users.

Keywords: E-commerce values filtering · Spatial database ·
Recommender systems · User preference learning ·
Pivot based indexing · Experiments · Spatial evaluation measures

1 Introduction, Motivation, Contributions

Our main motivation are recommender systems so far they point us to interesting items on e-commerce sites. Such a system has to be personalized to each user/customer preferences separately.

In representation of customer preferences we restrict to Fagin-Lotem-Naor-class of models (FLN models). Fagin, Lotem and Naor in their paper [5] described a (middleware) top-k query system where each object in a database has m scores (somewhere out in the web), one for each of m attributes that represent relevance degrees. To each object is then (on the middleware) assigned an overall score that is obtained by combining the attribute scores using a fixed monotone combining rule. This approach enables multi-criterial ordering.

We work on idea to use these types of models for e-commerce value filtering. Our goal is to present intuitiveness of visual features of these models and automated translation to programs. It is also suitable for implementation of "best match" in case, when system has to respond "we weren't able to find any matching results, but we found these similar listings for you".

Supported by Czech grants Progres Q48.

In [6] we have studied a subclass of FLN models - Linear combination of Triangular attribute modes - LT-models. LT-models are motivated by softening/relaxing value filtering. Degree of relaxation is motivated by the Analytic Hierarchy Processing method [9] and has to be done by a domain expert - this is out of the scope of this paper. We develop further this idea.

We consider pivot based learning from [6] with some stochastic noise. Results are evaluated through new metrics calculating spatial data characteristics of LT-models.

The process of translating models to programs is formalized by Challenge-Response Framework CRF. CRF resembles remote process call.

To have our models intuitive, we visualize them - the price we have to pay is we can depict only two or three dimensions. The idea is to use most important attributes and/or some aggregated ones. In our case, the model is automatically translated to a program using spatial database features. This enables us to define new metrics with spatial motivation.

We provide offline experiments with simulated data (items) and users.

Main contributions of this paper are:

- Spatial representation of linear triangular model for most important attributes
- Challenge-Response Framework for translating models to programs
- Visual aspects of our model making it intuitive for decision maker
- New spatial metrics
- Prototype and Experiments

2 Models

First we describe Fagin-Lotem-Naor class of models. In [5] they assume, that each object o has assigned m-many attribute scores $x_i^o \in [0;1]$. Combination function $t : [0;1]^m \longrightarrow [0;1]$ is assumed to fulfill: $t(0,\dots,0) = 0$, $t(1,\dots,1) = 1$ and t preserves ordering, i.e. if $x_j \leq_{[0,1]} y_j$ for all $1 \leq j \leq m$ then

$$t(x_1,\dots,x_j,\dots,x_m) \leq_{[0,1]} t(y_1,\dots,y_j,\dots,y_m) \tag{1}$$

Because of this inequality we call this function monotone. The overall score of object o is $t(o) = t(x_1^o,\dots,x_j^o,\dots,x_m^o)$.

In [6] we have described a class of LT-linear triangular models which is a subclass of FLN models. Especially, such a model can be generated by domain preference functions $f_i : D_i \longrightarrow [0,1]$ and $x_i^o = f_i(o.A_i)$ and a combination function c. Special case of such domain preference function are triangular (or trapezoidal) functions which can be considered as softening / relaxing of value filters in e-commerce.

In Fig. 1 there are two such preference models - the green model m_c (given by f_1, f_2 and c_{Pc}) and the red model m_l (given by g_1, g_2 and c_{Pl}). We can think of m_c as being the correct model and of m_l as being the computed (learned) model. This should illustrate a decision maker (e-shop owner, customer) situation when comparing two decision alternatives. In this figure, attribute preferences are triangular - this is a softening of one element value filter. There are two aggregations $\frac{2*x_1+x_2}{3}$ and $\frac{2*x_2+x_1}{3}$ represented by 2/3 contour line in the preference cube.

Main feature of our model is visualization of these contour lines in the data cube. Each contour line corresponds to the polygon in the data cube. Using attribute preferences can be endpoints of preference cube contour lines traced back (by respective horizontal and vertical lines and their intersections with attribute preferences) to respective attribute values in domains (in our figure there are always two of them). This should be intuitive for user/customer. Here we see areas in data cube (with preference bigger or equal to 2/3) and area of their intersection.

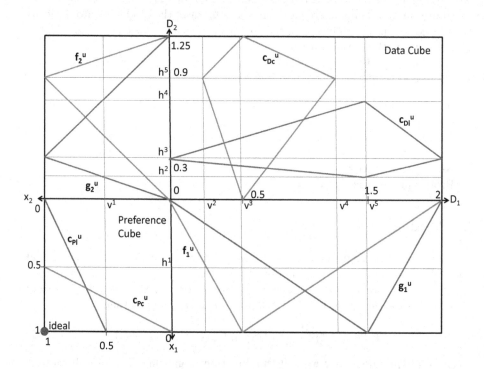

Fig. 1. Illustration of our model

We will consider three metrics motivated by this spatial representation of relaxed value filters - the area metrics, number of data points metric and metric calculating with average distribution of produced items given by a measure. Having the "correct" model m_c and computed (learned) model m_l, it is natural to ask for precision and recall of such models at different levels (here is depicted the 2/3 and higher).

For $p \in c, l$ let us denote $A_{m_p^h}$ the area above the h-contour line in data cube of the m_p model, the area-precision P_a^h is defined as follows

$$P_a^h = \frac{A_{m_c^h} \cap A_{m_l^h}}{A_{m_l^h}} \tag{2}$$

and area recall R_a^h

$$R_a^h = \frac{A_{m_c^h} \cap A_{m_l^h}}{A_{m_c^h}} \tag{3}$$

When having some data D, instead of area metrics we can calculate these fractions by number of data points in each area and we get P_D^h and R_D^h.

Sometimes we know the measure μ of distribution of production of items (data points). In this case instead of area or number of data points we can use

$$\int_{A_{m_p^h}} x d\mu \tag{4}$$

3   Challenge Response Framework

Motivated by an old mathematical idea from [11] we use the terminology of [3] and define the Challenge Response Framework CRF. Our goal is to use CRF as a formal framework for description of translation of models to programs.

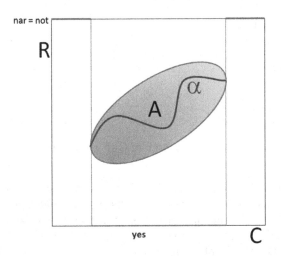

Fig. 2. Illustration of CRF situation

Challenge Response Situation $S = (C, R, A)$ consists of a set C of challenges, set of responses R and an acceptability relation $A \subseteq C \times R$ (can be preferential).

For an $c \in C, r \in R$ we read $A(c,r)$ as "r is an acceptable response (in some degree) for challenge c". We assume, that each set R contains also a special element nar representing "there is no acceptable response". See Fig. 2.

We assume: $A(c, nar)$ is equivalent to $(\forall r \in R \setminus \{nar\})(\neg A(c,r))$. The set $R \setminus \{nar\}$ are meaningful responses, nar is like logical "not" in combinatorial decision problems.

Challenge Response Reduction of a situation $S_1 = (C_1, R_1, A_1)$ to a situation $S_2 = (C_2, R_2, A_2)$ consists of a pair of functions (f^-, f^+) such that $f^- : C_1 \longrightarrow C_2$, $f^+ : R_2 \longrightarrow R_1$, such that $f^+(nar_2) = nar_1$, $f^+(r_2) = nar_1$ implies $r_2 = nar_2$ and following holds:

$$(\forall c_1 \in C_1)(\forall r_2 \in R_2)(A_2(f^-(c_1), r_2) \longrightarrow A_1(c_1, f^+(r_2)))(*) \tag{5}$$

Note that $A_2(f^-(c_1), nar_2) \longrightarrow A_1(c_1, nar_1)$ is equivalent to $\neg A_1(c_1, nar_1) \longrightarrow \neg A_2(f^-(c_1), nar_2)$ and this to $(\forall c_1 \in C_1)(\exists r_1 \in R_1 \setminus \{nar_1\})(A_1(c_1, r_1)) \longrightarrow (\exists r_2 \in R_2 \setminus \{nar_2\})(A_2(f^-(c_1), r_2))$. See Fig. 3.

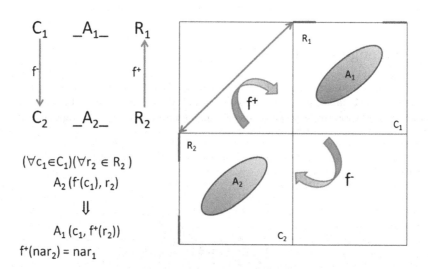

Fig. 3. Illustration of CRF reduction.

This CRF reduction can be used to formally represent transformations of models to programs, see Fig. 4. In our model situation challenges can come from user interaction with visual interface (so far not implemented) were the user can slide ideal points, descent of relaxation, combination function etc. Transformation f^-_{m2p} has to be fully automated and sends these actions to program situation challenges (inputs). In what follows we describe implementation of f^-_{m2p} in Oracle Spatial. Challenges of the program situation are input and responses are output. So far we have implemented calculation of polygons of contour lines (after possible user changes), visualization and spatial metrics. f^+_{m2p} sends this back to model response - here user visual interface.

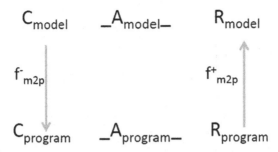

Fig. 4. Model to program transformation as CRF reduction.

4   Spatial Representation and Area Based Experiments

In this section we would like to describe our acquaintances with model to program transformation.

4.1   Practical Challenge-Response Construction

Our experiments are using user and item data from [6] together with sparse preference matrix $M = \{<0;1> \cup null\}^{|U| \times |I|}$ where U is set of users and I is set of items.

Training data for each user $u \in U$ contain only the corresponding row of preference matrix, i.e. set of ranked items with their preferences.

In our simulated environment, each user u, is fully represented by the triple $<i_1, i_2, w_1>$, that can be understood as quadruple $<i_1, i_2, w_1, w_2 = 1 - w_1>$, where

- i_1 represents the ideal point in first data dimension.
- i_2 represents the ideal point in second data dimension.
- w_1 represents the weight (importance) of first data dimension.
- w_2 represents the weight (importance) of second data dimension.

This way we can know all preferences for all items.

Challenges C_{model} in model situation equal to rows of preference matrix M, i.e. known preferences of single user for all items. R_{model} contain contour lines - polygons P_p^u for chosen levels of preference p (see Fig. 1). Mappings f_{m2p}^- and f_{m2p}^+ are identities on respective domains. $A_{program}$- finds an approximation of the user by choosing closest pivot m in the dataset [6] and computes polygons $P_{m,p}^u$ for this approximation. Evaluation of implication 5 is done by computing precision and recall on polygons obtained by A_{model} and $A_{program}$.

For given level of preference p, we need to compute corresponding polygon $P_{m,p}^u = [K, L, M, N]$ in the data cube, that represents the contour line $[A, B]$ at the level p in the preference cube (see Fig. 5). First we compute the contour line itself.

$$A = [1; p - w_1/w_2 * (1 - p)] \tag{6}$$

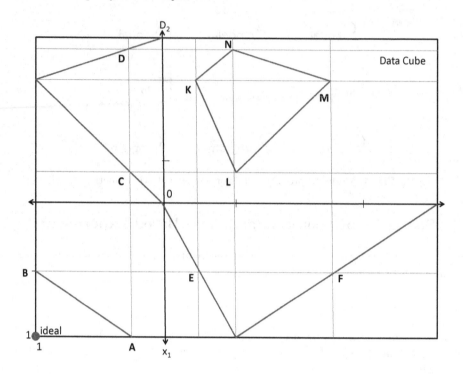

Fig. 5. Illustration of polygon computation for given level of contour line

$$B = [p - w2/w1 * (1 - p); 1] \tag{7}$$

Next we can compute intersections C, D, E, F with (triangular-shaped) preference functions in both dimensions.

$$C = [A_2 * i_2; A_2] \tag{8}$$

$$D = [i_2 + (1 - i_2) * (1 - A_2); A_2] \tag{9}$$

$$E = [B_1; B_1 * i_1] \tag{10}$$

$$F = [B_1; i_1 + (1 - i_1) * (1 - B_1)); \tag{11}$$

Finally, we can compute boundary of the polygon as follows:

$$K = [E_2; i_2]; \tag{12}$$

$$L = [i_1; C_1] \tag{13}$$

$$M = [F_2; i_2] \tag{14}$$

$$N = [i_1; D_1] \tag{15}$$

Both ideal point $[i_1; i_2]$ and polygons $P_{m,p}^u = [K, L, M, N]$ were stored in Oracle database using Oracle Spatial extension using MDSYS.SDO_GEOMETRY point and polygon ring.

Precision at level 0.7 – histogram

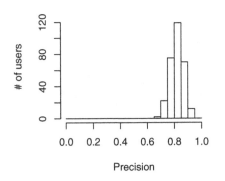

Precision at level 0.8 – histogram

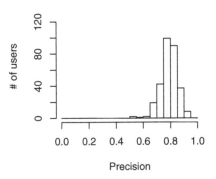

Precision at level 0.9 – histogram

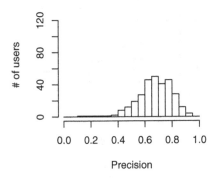

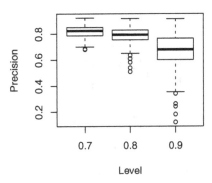

Fig. 6. Histograms of area based precision at level 0.7 (top-left), 0.8 (top-right), 0.9 (bottom-left) and corresponding box-plots (bottom-right).

4.2   Experiments

This allowed us to effectively compute areas of both user's and model's polygons $Poly_u$, $Poly_m$ and their intersections.

From this we can for given level of preference p compute both area based precision P_p and recall R_p for each user estimation.

$$P_p^u = \frac{Area(P_p^u \cap P_{m,p}^u)}{Area(P_{m,p}^u)} \qquad (16)$$

$$R_p^u = \frac{Area(P_p^u \cap P_{m,p}^u)}{Area(P_p^u)} \qquad (17)$$

Figure 6 represents distribution of area based precision for levels 0.7, 0.8 and 0.9 over 300 randomly generated points with different rating frequencies.

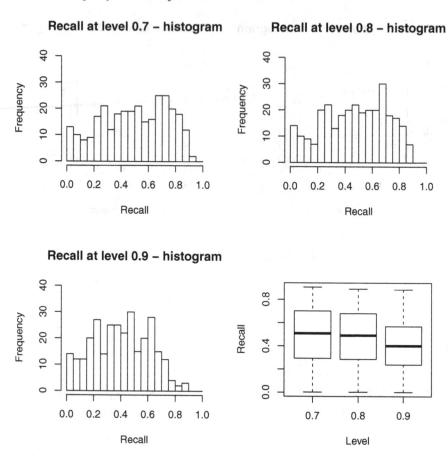

Fig. 7. Histograms of area based recall at level 0.7 (top-left), 0.8 (top-right), 0.9 (bottom-left) and corresponding box-plots (bottom-right)

Figure 7 represents distribution of area based recall for levels 0.7, 0.8 and 0.9 over 300 randomly generated points with different rating frequencies.

Figure 8 then presents comparisons of area sizes of user's contour polygon at level 0.7 to size of its intersection with preference polygon of corresponding model. Every circle represents one user.

5   Related Research

Value filtering is well known and studied topic in HCI, see e.g. [4]. Although it is more involved in areas where attributes are not so easy to describe (e.g. clothing), for our motivation value filtering in numeric (ordered) domains is sufficient. They consider also interactive graphical filtering through (double) slider, or various ways of specifying intervals of values.

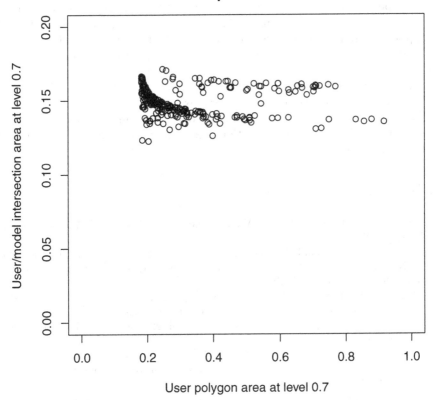

Fig. 8. User and user/model intersection areas at level 0.7

Our research was also motivated by original idea of QBE - querying by example - the untrained user should be able to specify query (without any knowledge of programming). It induces also a form of tableau mode (see [1]).

Further motivation is relaxing selects. Acceptable violation of ideal values is considered in AHP - analytic hierarchy process method of Saaty, [9].

Another way of representing relaxation are fuzzy systems, initiated by paper [2]. Surprisingly, we found fuzzy motivation in the paper [5] of Fagin, Lotem and Naor, which uses combination functions for modeling multicriterial ordering to find top-k objects (see also [10]). We used fuzzy sets to model preferences in [7]. We considered learning user preferences for recommender systems in [8].

Challenge response framework was originally motivated in set theoretic study in real analysis in [11]. Later Blass [3] showed that this idea appeared in different setting and can be used also in computer science.

6   Conclusions, Future Work

In this paper we described spatial representation of low dimensional value filtering for e-commerce. We introduced Challenge-Response Framework for formal description of translating models to programs. We hope that visual aspects of our model makes it intuitive for decision makers and makes them usable without any programming.

For evaluation of our experiments we introduced new spatial metrics

With our prototype we provided several experiments.

For future work we plan to use real world data, study impact of choosing two, three attributes for visual representation and learning user preferences restricted to these attributes.

References

1. Abiteboul, S., Hull, R., Vianu, V.: Foundations of Databases. Addison-Wesley, Boston (1995)
2. Bellman, R.E., Kalaba, R.E., Zadeh, L.A.: Abstraction and Pattern Classification, RAND memorandum RM-4307-PR, October 1964. https://www.rand.org/pubs/research_memoranda/RM4307.html
3. Blass, A.: Questions and answers - a category arising in linear logic, complexity theory, and set theory. In: Girard, J.-Y., et al. (eds.) Advances in Linear Logic. London Mathematical Society Lecture Notes, vol. 222, pp. 61–81. London Mathematical Society (1995)
4. Holst, Ch.: The Current State of E-Commerce Filtering. https://www.smashingmagazine.com/2015/04/the-current-state-of-e-commerce-filtering/. Accessed 8 May 2019
5. Fagin, R., Lotem, A., Naor, M.: Optimal aggregation algorithms for middleware. JCSS **66**(4), 614–656 (2003)
6. Kopecky, M., Vomlelova, M., Vojtas, P.: Basis functions as pivots in space of users preferences. In: Ivanović, M., et al. (eds.) ADBIS 2016. CCIS, vol. 637, pp. 45–53. Springer, Cham (2016). https://doi.org/10.1007/978-3-319-44066-8_5
7. Peska, L., Eckhardt, A., Vojtas, P.: Preferential interpretation of fuzzy sets in recommendation with real e-shop data experiments. Archives for the Philosophy and History of Soft Computing No 2 (2015). http://aphsc.org/index.php/aphsc/article/view/32/2
8. Peska, L., Vojtas, P.: Using implicit preference relations to improve recommender systems. J. Data Semant. **6**(1), 15–30 (2017)
9. Saaty, T.L.: Decision making with the analytic hierarchy process. Int. J. Serv. Sci. **1**(1), 83–98 (2008)
10. Hudec, M., Sudzina, F.: Construction of fuzzy sets and applying aggregation operators for fuzzy queries. In: 14th International Conference on Enterprise Information Systems (ICEIS 2012), Wroclav, Poland, pp. 253–258. SCITEPRESS (2012)
11. Vojtas, P.: Generalized Galois-Tukey connections between explicit relations on classical objects of real analysis. In: Israel Mathematical Conference Proceedings, vol. 6, pp. 619–643 (1993)

Transforming Object-Oriented Model to a Web Interface Using XSLT

Jiri Musto$^{(\boxtimes)}$ and Ajantha Dahanayake

Lappeenranta-Lahti University of Technology, 53850 Lappeenranta, Finland
{Jiri.Musto,Ajantha.Dahanayake}@lut.fi

Abstract. Citizen science projects are based on gathering data from citizens. Citizen science projects start from understanding what type of data is needed and from whom. This can lead to creating a conceptual model for the project. Citizen science projects often use web interfaces and creating the interfaces require specific tools or programming knowledge. Having a method for creating the interface from a conceptual model would increase project development speed. This paper tackles the presented idea by using simple steps to create a web interface from an object-oriented conceptual model using XML transformation language XSLT. The model is transformed into a text-based representation using XSD that is then turned into an XML template. Finally, XSLT is used to generate a web interface by transforming the XML file. Results show that the process from object-oriented model to XML can be automated and the programming required for a working web interface is minimal but without other programming languages, the functionality is lacking. At the end of this article, some design suggestions are given to increase the functionalities while maintaining a low amount of programming.

Keywords: Object-oriented model · XML · Web interface · Citizen science · Model to program

1 Introduction

Graphical models are commonly used when designing a system or for process modeling. There are many different types of models such as class diagrams, business process models, activity diagrams among others. Models are an easy way to present ideas and they give some form of understanding of the problem or solution at hand. Models can be abstract or they can be extremely detailed depending on the context where the models are used. When implementing a graphical model, it needs to be transformed into code, events, activities, or programs. Transforming models into programs is the main goal in model-driven engineering (MDE) [6] and conceptual model programming (CMP) [3].

Creating working programs with models alone requires automation and different programs. The tools can create unoptimized code and non-commercial options are more limited and can misinterpret input models. The benefit of models is that they do not require extensive knowledge and skills in programming,

© Springer Nature Switzerland AG 2019
T. Welzer et al. (Eds.): ADBIS 2019, CCIS 1064, pp. 221–231, 2019.
https://doi.org/10.1007/978-3-030-30278-8_25

thus allowing fast designing and changes with less effort compared to making changes into systems or software [4].

Citizen science projects often use web interfaces and creating the interfaces require specific tools or programming knowledge. Having a method for creating the interface from a conceptual model would increase project development speed.

In this research, the main objective is to transform a conceptual object-oriented (OO) model for citizen science applications into a web interface. This is done by transforming the OO model into text-based representation using extensible markup language (XML) [21] with the help of XML schema definition (XSD) [20]. XML is then transformed into a web interface using extensible stylesheet language transformation (XSLT) [22]. The conceptual model is simplified and used as a test case for transforming an OO model into a web interface. This research aims to find how well a web interface can be created from a conceptual model using minimal amount of programming.

The paper is structured as follows. Section 2 introduces related work. Section 3 presents the conceptual model used in this research and in Sect. 4, the conceptual model is transformed into XSD. Section 5 introduces the web interface and discussion and Sect. 6 draws the conclusions and possible future work.

2   Related Work

Using models to create programs has been done countless of times. With the progression of knowledge and expertise, new tools have been created to automate the program generation from models as much as possible.

A Swedish company Genicore has created a platform for creating information systems purely from a model that is given to it. This does require for the model to be as detailed as possible to avoid misinterpretations [5,7].

Using business process model to generate and update user interface components is researched in [8]. Different user interface components are derived from the model using elementary transformation patterns. The business process model includes processes that can be transformed into functionalities of the user interface.

In Java, there are tools to generate blocks of code from UML class diagrams [12]. The more detailed the design is, more code can be generated. When generating programs from models, the models need a way to express functionalities as well. Class diagrams give names for methods, but they do not have expressions for the functionalities, so they require another type of diagram such as a sequence or an activity diagram.

Creating applications and interfaces from models using Java programming language has been a topic of many researches. Using three different types of models proves to be effective in capturing system requirements. Models are then transformed into text using a formal object-oriented specification. The specification is used to create an execution model for the application to be implemented [13]. Java has been used to create a tool for web interface generation using class diagram and basic CRUD operations. The interface can be modified in more detail during any step of the development process if necessary [9]. Java can also be used to generate Rich Internet Application interfaces using different modeling tools and frameworks [2].

WYSIWYG editors can be thought of as tools that automatically generate programs based on a model. The user drags and drops elements on the canvas and the application writes the code automatically without the user's knowledge. These types of editors are common in amateur web design, such as WordPress [23]. WYSIWYG editors transform low-level models to programs as the programs are built from the actual building blocks rather than abstract models.

As there are many types of models, there are also debates on what is the best model. People often forget that not all types of models work in all cases. Different models are better in some situations than others. Having a collection of different models provide a broader spectrum and can lead to better results than just relying on one type of model [17].

3   Conceptual Model of Citizen Science Applications

Most citizen science applications have the same basic functionality where a *citizen* observes nature or environment, submits the *observation* through a web page or a mobile app and other *citizens* can see the information related to the observation. Observations have *location* and *date/time* and citizens can edit their observations with new information, creating new *versions*. Another common point in most citizen science applications is the use of community validation, such as voting, on observations. This act of voting often affects the *reputation* of citizens and observations. The reputation changes are tracked to be able to identify possible harmful citizens [11].

Using the earlier concepts, a conceptual model can be created using OO modeling as seen in Fig. 1. This model considers only the most common information that is found in citizen science applications. The model can be extended for specific usage if necessary by adding new objects or additional variables. For testing purposes, the model in Fig. 1 is enough.

4   XSD Design

The OO model in Fig. 1 works as a base for a text-based representation of the model. XML is a multipurpose language that can be used for database management and website creation among other things. Using XML, we can represent concrete objects based on the OO-model but there needs to be a high-level text-based representation as well. XSD is a formal representation of how specific XML elements are formed and it is possible to represent different relations between objects and represent the same level of abstraction that the OO model presents. Using XSD, we transform the OO model into text-based representation while maintaining a high-level of abstraction. There are commercial tools available that can generate XSD files from UML [14,19]. There are also tools,

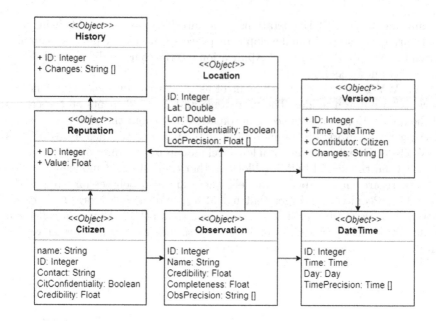

Fig. 1. Simplified object-oriented model for citizen science applications.

such as XMLSpy from Altova [1], that can generate template XML documents from XSD documents automatically or vice-versa. In this research, the XSD file is created by hand because it is less error prone without a commercial tool but it is possible to automate the process of generating XSD and XML files from a UML diagram.

Figure 2 shows a part of Observation-object as complexType. ComplexTypes are used as templates in XSD to model classes (object definition). The templates can have elements, attributes and other relevant structures within them. Elements are used to represent concrete objects (e.g. strings, other complexTypes) and attributes are used for primitive data types (e.g. integers). Elements and attributes can have additional information or restrictions, such as number of occurrences, that can be used to represent restrictions on variables or relations between objects. ComplexTypes are implemented as elements where they can have specific information added similar to a constructor receiving parameters.

There are two ways of representing relations between complexTypes in XSD. One way is to have elements as parts of other elements and another way is to use referencing. XSD allows referencing by using key/keyref in the XSD file. These are similar to primary/foreign keys in databases, and they can be used for creating references between elements.

Figure 3 shows a part of the Observation element used to construct the Observation object from the complexType in Fig. 2. The attributes that are used as key/keyref pairs are indicated when the element is created. The keyref requires the information on the key it refers to and the corresponding attribute

```
<xsd:complexType name='Observation'>
 <xsd:sequence>
  <xsd:element name="name" type="xsd:string"  maxOccurs="1"
   minOccurs="1"/>
  <xsd:element name="obsPrecision" maxOccurs="2" minOccurs="2">
   <xsd:simpleType>
    <xsd:list itemType="xsd:string"/>
   </xsd:simpleType>
  </xsd:element>
 </xsd:sequence>
 <xsd:attribute name="oID" type="xsd:integer" use="required"/>
 <xsd:attribute name="cID" type="xsd:integer" use="required"/>
```

Fig. 2. Part of Observation object/complex type in XSD.

in the referring element where the key will be tied to (in this case cID under Observation-complexType). Figure 4 shows the corresponding key that is tied to the keyref in Fig. 3. As with the keyref, the key requires information on the attribute that is used for its value and where to find it (in this case cID under Citizen-complexType).

```
<xsd:element name="Observation" type="tns:Observation">
 <xsd:keyref name='citizenObservationRef'
  refer ='tns:citizenObservationKey'>
  <xsd:selector xpath ='./Observation' />
  <xsd:field xpath='@cID' />
 </xsd:keyref>
```

Fig. 3. Part of Observation element corresponding the Observation complexType.

```
<xsd:key name='citizenObservationKey'>
 <xsd:selector xpath='./Citizen' />
 <xsd:field xpath='@cID' />
</xsd:key>
```

Fig. 4. Key that connects to keyref in Observation element.

5   Web Interface with XSLT

In this section we introduce the method of creating a web interface using XSLT. XSLT is used to transform XML documents into new XML documents or to other formats. It is possible to insert XML compliant hypertext markup language (HTML) into XSLT documents to have the XML file transform into a webpage. XSLT is linked to the XML file by inserting it as a stylesheet in the beginning of the XML file and when that file is opened in a browser, the XSLT file is used to transform it into the specified format. It is possible to transform any XML document with the same XSLT document depending on the level of abstraction used and the detailed styling methods. The XML file used in this section is generated from the XSD created in the previous section.

5.1   Building the Interface

The interface is built using only basic HTML elements, such as table elements and rows. These basic elements are combined with basic style attributes using cascading style sheet (CSS) language to have some visual effects. CSS styles can be used in a separate file or implemented into the XSLT file if it is HTML compliant.

Figure 5 shows how XML elements can be displayed using HTML tags for table rows and elements.

```
<tr>
  <xsl:for-each select="root/Observation">
  <tr>
    <td><xsl:value-of select="./@oID"/></td>
    <td><xsl:value-of select="./name"/></td>
    <xsl:for-each select="key('obsToLocKey', @lID)">
      <td><xsl:value-of select="./lat"/></td>
      <td><xsl:value-of select="./lon"/></td>
    </xsl:for-each>
<xsl:key name="obsToLocKey" match="Location" use="@lID" />
```

Fig. 5. XSLT example of displaying observations and their locations with the key definition.

For XSLT to select specific values from an XML file, the value-of method needs to be used. When giving information to the method, keyword select is used and attributes are differentiated by using the @-character. Backlashes are used to access sub-tags if necessary. There are several basic functionalities that can be used in XSLT such as for-each loops and if-else clauses. While-loops need to be done using recursion.

In Fig. 5, each table row is made of specific values from the XML document under the Observation-tag. The observation information is linked to Location-tags with the key obsToLocKey. Keys can be defined inside the XSLT file and they work as key-value pairs. They are not the same as the key/keyref pairs in XSD. Keys in XSLT are used to simplify searching and value matching rather than enforcing bonds between values. obsToLocKey key fetches a Location-tag that has the same lID-attribute as the Observation. The definition of the key is shown at the end of Fig. 5 and it is made of three parts of information: name for the key, tags it searches and the value used to pairing.

Figures 6 and 7 are images of the web interface that is generated using XSLT. They are simplistic but they show that it is possible to create web interfaces using XSLT and XML compliant HTML. Creating a minimal web interface does not require much programming with XSLT. Only 30% of the XSLT file is for XSLT functions that selects the information from the XML file to display on the web interface and 70% of the file is used for the HTML tags that are necessary for displaying the information in a specific way. The same XSLT file can be used with different XML files as long as the XML file follows the same structures and tags used in the original. This means, that XML files can be split into multiple parts if the length of one file grows too long.

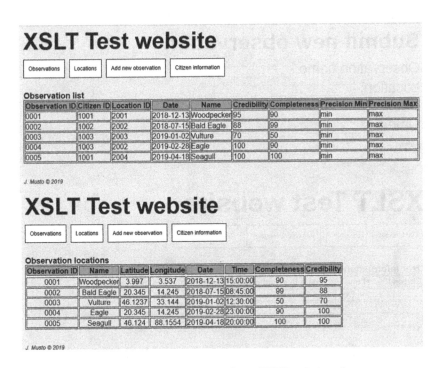

Fig. 6. Two interface images from XSLT web interface.

5.2 Discussions

Using XSLT has strengths and weaknesses when transforming XML files. XSLT is easy to use to create static webpages and the same XSLT file can be applied to any XML document. Depending on the level of abstraction in the XSLT, it is possible to apply the XSLT document to any XML following the same structure or to any XML following any structure. The latter example may be limited in the details of styling as abstraction can make it difficult to apply specific styles to specific information. Another strength with XSLT compared to regular HTML is that WYSIWYG editors support directly the displaying of XML. With HTML, editors may not show the XML file as it needs to be displayed using JavaScript and that can create difficulties when trying to display XML information. This also means, that XSLT requires less programming knowledge compared to HTML and JavaScript.

XSLT has several weaknesses as well. Figure 7 shows a form for submitting a new observation but there is no functionality for the actual submission and updating of the XML document. This is because the needed functionality is not possible to do with XSLT alone. Dynamic webpages are not possible to create

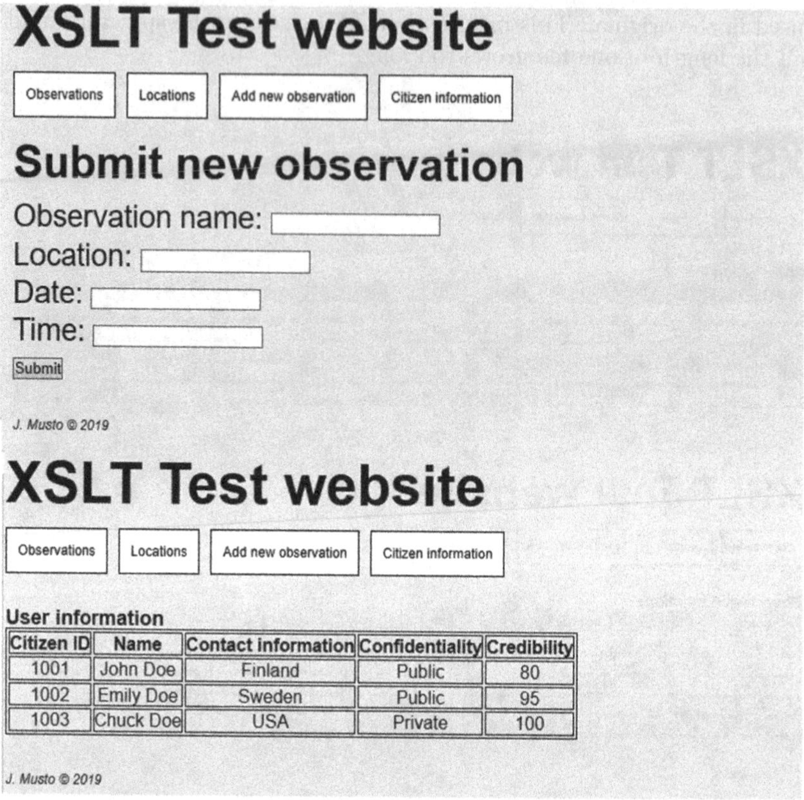

XSLT Test website

| Observations | Locations | Add new observation | Citizen information |

Submit new observation

Observation name: [_____]

Location: [_____]

Date: [_____]

Time: [_____]

[Submit]

J. Musto © 2019

XSLT Test website

| Observations | Locations | Add new observation | Citizen information |

User information

Citizen ID	Name	Contact information	Confidentiality	Credibility
1001	John Doe	Finland	Public	80
1002	Emily Doe	Sweden	Public	95
1003	Chuck Doe	USA	Private	100

J. Musto © 2019

Fig. 7. Two interface images from XSLT web interface.

with XSLT unless it is combined with another programming language, such as ASP.NET [10] or PHP [18]. Another weakness is that you can do same things with any other programming language albeit they may require more programming knowledge. For example, reading and displaying XML with JavaScript requires first getting the XML file and then parsing and displaying it with less intuitive methods. Some things that can be easily done in modern programming languages, such as adding new data to the XML, are more difficult to do using XSLT. This slows down the implementation process significantly as well. Final weakness relates to the handling of XML documents in general: Large XML files are difficult to process. This is true whenever processing XML files regardless of the chosen programming language or method.

Creating a good working web application from XML files require the application to be split into multiple parts. Following the MVC design pattern, XML or XSD files can be used as Model, XSLT as View, and PHP as Controller. This division requires more programming initially, but it provides more functionalities compared to only XSLT. MVC pattern allows changing the Model and View when necessary and the most programming heavy part, Controller, can be reused in different scenarios. Model is easy to change as long as the structure of the new model follows the same rules. If Controller and View have been made as generalized as possible, it is possible to include any type of Model regardless of the structure but there will be limitations to it. For example, creating a web application for observing wildlife is similar to observing botany and only the Model needs to be replaced. Observing wildlife and analyzing images are different types of projects and would require a new View with the new Model. A generalized View and Controller can work but their functionalities would be limited. This is probably the biggest reason why platforms, such as Zooniverse [24], that allow multiple citizen science projects only allow similar projects on their platforms.

Changing the View requires a new XSLT to be designed. There are commercial WYSIWYG editors for XSLT [15,16] that can be used if the creator does not have necessary knowledge to create one by hand. It is also possible to create pre-made templates for different situations with XSLT and just change which template is used. Templates can be used for whole pages or just specific parts of webpages, similar to building blocks. For example, there can be templates made for tables or lists and different templates have different styles. User would have to change only few words to switch to a new template.

Changing the controller requires the most programming knowledge in this setup. Best-case scenario would be to have a controller pre-made for specific scenarios that allows the changing of View and Model easily. This way, domain specialists can create applications for their own purposes and design the necessary View with minimal programming if they provide the Model they want to use.

Biggest issue using MVC pattern with a generalized controller is the usage of methods from the original OO model. XSLT 2.0 and PHP allow user defined functions but representing methods in XSD or XML is a problem. It can be done by giving the modeled functions the attribute or element type "function" but there is no standard for this and it needs to be designed specifically for the project and even then it may allow only simple methods.

While this research uses citizen science as an example, there is no reason why this could not be done in other applications outside citizen science. As long as there is an underlying XML file, web interfaces can be created with XSLT. Having a different type of application only means that the design for the interface has to be changed meaning the XSLT must be changed.

6   Conclusion and Future Work

This paper introduces a process for transforming an OO model into a working web interface. Using XML and XSD to create a text-based representation for an OO model is easy and straightforward. XML files can be used with many programming languages and they are even used as database files in some database management systems. Some of the steps can be automated to improve the speed of transformation. OO models, especially UMLs, can be automatically transformed into XSD or XML files with commercial tools. There are also commercial WYSIWYG editors to generate XSLT files.

Creating a web interface from XML using XSLT is possible but depending on the need, other options may be more viable. XSLT is a good option if your programming knowledge is limited and you want to create static web interfaces fast. The commands are fairly easy to comprehend and HTML tags are easy to use. XSLT is directly applied to XML so it only requires one file and if using WYSIWYG editors, you can see the XML data when designing rather than when trying to run it. However, if you have programming knowledge and want to create dynamic web interfaces, HTML and JavaScript are a better option. To get the most out of XSLT, it is highly recommended to combine XSLT with other programming languages to take advantage of the strengths while also countering the weaknesses.

The web interface created in this research is purely for example purposes to show that it is possible to generate a web interface from XML but to create an actual citizen science application would require more detailed design. The design of the model, XSD and finally XSLT can be improved to be more precise rather than a generalized version.

In a future work, a method of representing methods in XSD and XML to allow their transition from OO model to XSLT or to PHP will be implemented. XSLT will be combined with PHP to create a template that allows important functionalities such as adding or deleting observations from the underlying XML. The purpose for this template is to allow any XML file to be used to create a web interface with a simple plug-and-display installation. First step is to create the template for citizen science applications to test whether it works and how easy it is to set up. If the template proves to be successful, it would require only the change of an XML file. Final step to complete this work will be to combine tools to generate an XML from OO model and integrate it to the XSLT & PHP template to fully automate the process of generating a web interface from an OO model.

References

1. Altova: XML Editor: XMLSpy. https://www.altova.com/xmlspy-xml-editor
2. Cañadas, J., Palma, J., Túnez, S.: Model-driven rich userinterface generation from ontologies for data-intensive web applications. In: CEUR Workshop Proceedings, vol. 805 (2011)
3. Embley, D.W., Liddle, S.W., Pastor, Ó.: Handbook of Conceptual Modeling, pp. 3–16 (2011). https://doi.org/10.1007/978-3-642-15865-0
4. Tomassetti, G.: A Guide to Code Generation - Federico Tomassetti - Software Architect (2018). https://tomassetti.me/code-generation/
5. Genicore: Genicore. http://www.genicore.se/
6. ICT Group: Model Driven Engineering (MDE) software development methodology. https://ict.eu/model-driven-engineering/
7. Jonsson, T., Enquist, H.: CoreEAF - a model driven approach to information systems. In: CEUR Workshop Proceedings, vol. 1367, pp. 137–144 (2015)

8. Kolb, J., Hübner, P., Reichert, M.: Automatically generating and updating user interface components in process-aware information systems. In: Meersman, R., et al. (eds.) OTM 2012. LNCS, vol. 7565, pp. 444–454. Springer, Heidelberg (2012). https://doi.org/10.1007/978-3-642-33606-5_28
9. Machado, M., Couto, R., Campos, J.C.: MODUS: model-based user interfaces prototyping. In: Proceedings of the ACM SIGCHI Symposium on Engineering Interactive Computing Systems - EICS 2017, pp. 111–116 (2017). https://doi.org/10.1145/3102113.3102146
10. Microsoft: ASP.NET — Open-source web framework for .NET. https://dotnet.microsoft.com/apps/aspnet
11. Musto, J., Dahanayake, A.: Improving data quality, privacy and provenance in citizen science applications. In: Information Modelling and Knowledge Bases XXXI, Proceedings of the 29th International Conference on Information Modelling and Knowledge Bases, EJC 2019. Frontiers in Artificial Intelligence and Applications, Lappeenranta, Finland, 3–7 June 2019. IOS Press (2019, in press)
12. Obeo: UML to Java generator & reverse (2019). http://www.umldesigner.org/refdoc/umlgen.html
13. Pastor, O., Gómez, J., Insfrán, E., Pelechano, V.: The OO-method approach for information systems modeling: from object-oriented conceptual modeling to automated programming. Inf. Syst. **26**(7), 507–534 (2001). https://doi.org/10.1016/S0306-4379(01)00035-7
14. Sparx Systems: Enterprise Architect - XML Schema Generation (2019). https://sparxsystems.com.au/resources/xml_schema_generation.html
15. Stylus Studio: XSLT Editor (2019). http://www.stylusstudio.com/xslt-editor.html
16. SyncRO Soft SRL: XSLT Editor (2019). https://www.oxygenxml.com/xml_editor/xslt_editor.html
17. Thalheim, B.: Model-based engineering for database system development. In: Cabot, J., Gómez, C., Pastor, O., Sancho, M., Teniente, E. (eds.) Conceptual Modeling Perspectives, pp. 137–153. Springer, Cham (2017). https://doi.org/10.1007/978-3-319-67271-7_10
18. The PHP Group: PHP: Hypertext Preprocessor. https://php.net/
19. Visual Paradigm: Ideal Modeling & Diagramming Tool for Agile Team Collaboration. https://www.visual-paradigm.com/
20. W3C: W3C XML Schema Definition Language (XSD) 1.1 Part 1: Structures (2012). https://www.w3.org/TR/xmlschema11-1/
21. W3C: Extensible Markup Language (XML) (2016). https://www.w3.org/XML/
22. W3C: XSL Transformations (XSLT) Version 3.0 (2017). https://www.w3.org/TR/2017/REC-xslt-30-20170608/
23. Wordpress Project: Blog Tool, Publishing Platform, and CMS – WordPress. https://wordpress.org/
24. Zooniverse: Projects – Zooniverse. https://www.zooniverse.org/projects

Abstract Layers and Generic Elements as a Basis for Expressing Multidimensional Software Knowledge

Valentino Vranić[✉][iD] and Adam Neupauer

Institute of Informatics, Information Systems and Software Engineering,
Faculty of Informatics and Information Technologies,
Slovak University of Technology in Bratislava,
Ilkovičova 2, 84216 Bratislava 4, Slovakia
vranic@stuba.sk

Abstract. Enormous intellectual efforts are being invested into producing software in its executable form or, more precisely, a form from which this executable form can be automatically derived, commonly known as a source form (usually code, but may be a model, too). On the other hand, it is inherently complex to restore the ideas upon which software has been built. Moreover, it is usually not possible to produce software in its source form directly without producing a number of documents, diagrams, or schemes. All these artifacts, including the program code, represent software knowledge necessary for maintaining existing software and for building further systems in a given domain. But these software knowledge sources are disconnected from each other making it hard to navigate between them and to devise conclusions based on their relatedness, which is essential for their effective use. In this paper, a new approach to versatile graphical software modeling based on abstract layers and generic elements and its use in modeling multidimensional software knowledge and interrelating its pieces is proposed. The approach is supported by a prototype tool called InterSKnow, which targets mainly internal representation. This enabled to evaluate the approach from two perspectives: the efficiency of searching for software knowledge in software models and its comprehension. The results are generally plausible to the approach proposed here compared to Enterprise Architect as a representative of traditional, state-of-the-art software modeling tools.

Keywords: Software artifacts · Layers · UML ·
Domain specific modeling · Knowledge management

1 Introduction

Enormous intellectual efforts are being invested into producing software in its executable form [19,26] or, more precisely, a form from which this executable form can be automatically derived, commonly known as a source form (usually

© Springer Nature Switzerland AG 2019
T. Welzer et al. (Eds.): ADBIS 2019, CCIS 1064, pp. 232–242, 2019.
https://doi.org/10.1007/978-3-030-30278-8_26

code, but may be a model, too [31,33]). On the other hand, it is inherently complex to restore the ideas upon which software has been built. Moreover, it is usually not possible to produce software in its source form directly without producing a number of documents, diagrams, or schemes. All these artifacts, including the program code, represent software knowledge necessary for maintaining existing software and for building further systems in a given domain. But these software knowledge sources are disconnected from each other making it hard to navigate between them and to devise conclusions based on their relatedness [22,25], which is essential for their effective use.

To overcome this, yet another model is necessary: a model that will interrelate pieces of software knowledge residing in different dimensions. However, this *multidimensional software knowledge* is maintained in different tools used to manage code and all kinds of models, including text descriptions and spreadsheets. Software development processes, as they are implemented, typically rely on these tools. Therefore, in many cases it is not feasible to extract software knowledge from its original sources and enforce maintaining it further only within a provided tool no matter how sophisticated it might be. It is hard to expect that such a supertool would ever support everything that is supported by dedicated tools. Thus, the purpose of the interrelating model would be to interconnect software knowledge as it is captured in the corresponding tools and to enable enhancing it with further details. It should be noted that the interrelating model need not be populated only manually. Code [5] and runtime information analyzers [1] could be used to generate parts of it.

The area of knowledge management in software engineering is very broad with hundreds of approaches [6,21,23,29]. The approach to expressing multidimensional software knowledge by an interrelating model proposed here stems in software modeling without enforcing any particular format for recording software knowledge or prescribing its taxonomy (such as the one in an earlier attempt at establishing a software development knowledge base [10]).

The rest of the paper is structured as follows. Section 2 analyzes what it would take to support modeling multidimensional software knowledge. Section 3 explains the approach to versatile graphical software modeling based on abstract layers and generic elements. Section 4 reports on the tool prototype and approach evaluation. Section 5 discusses related work. Section 6 concludes the paper.

2 Modeling Multidimensional Software Knowledge

UML was envisaged as the ultimate modeling solution applicable not only to software. As such, it was expected to put an end to the proliferation of different notations and establish mutual comprehensibility thus becoming the lingua franca of (at least) software modeling. However, this did not happen [30]. Other notations are in use, too, and new notations are being invented. Often, an ad hoc notation is a handy way to capture and express ideas. This may be at the level of diagram sketching [32], but taken in a more organized way, such notations may develop into domain specific modeling languages, which are being increasingly

recognized as a more flexible basis for model driven software development than
(UML based) MDA is [9].

In a broader sense, a model is any artifact related to the software system
being developed, including text documents (in different formats) and spreadsheets. Usually, code is distinguished from models, although it actually may
be considered to be the final, executable model. Different models and code are
related in complicated ways. Leaving this to depend on human memory is not
feasible in the long run. It is not that only the new people joining the development process must discover the dependencies, but those that once knew the
dependencies are forced to rediscover them once they forget them.

As it has been already pointed out in the introduction, to deal with multidimensional software knowledge, yet another model is needed: an interrelating
model capable of expressing relationships between elements from the same or different models. Consider UML class diagram endless labyrinths. Understanding
these often requires relating distant classes, as displayed in Fig. 1.

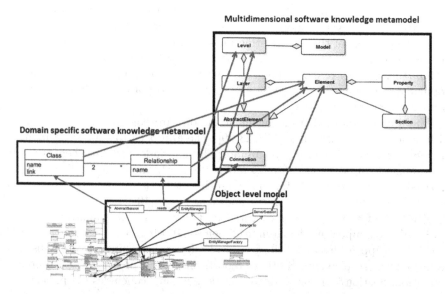

Fig. 1. Interrelating software knowledge: multidimensional software knowledge metamodel (the class diagram in the lower left part is a fragment from the EclipseLink
2.0 API UML class diagram (https://wiki.eclipse.org/EclipseLink/Development/
Architecture/EclipseLink/ClassDiagram)).

Relationships to external models and their elements may be expressed by
using projections of these elements in additional models as in model federa-
tions [14]. In general, a set of such interconnecting models may be used allowing
for multiple levels of interconnecting. Furthermore, the elements of these models
may be not just projections of the elements from the models being interrelated, but they may be enhancing the information contained in these models.

This might even happen within the elements that are projections by introducing additional properties within these elements.

Semantics of the interrelating model may vary significantly. While capturing structural or static dependencies is very important, interrelating models could be very useful at expressing time dependencies or, in general, a behavioral or dynamic view. Providing a particular set of element types and connectors would be of limited use. For the full flexibility, this interrelating modeling framework has necessarily to be tailored to the context and this has to happen dynamically while performing application modeling as in free modeling [15]. Multiple levels of metamodeling are possible and it seems these are most transparently treated via the instantiation relationship as in deep modeling [3].

Some of the scenarios in which interrelating software knowledge may help include:

Change impact estimation. An envisaged specification change could be tracked through all the levels of models, code, and tests to estimate its complexity and cost, which would be used as a backing of the decision whether the change is viable.

Exploratory testing. The order of the steps the tester performs in the exploration test uncovers model and code interrelations. This can help in exploratory testing automation [13].

Fulfilling regular development tasks. A developer can more easily understand a software system via the interrelating model.

Tracking the intent of a software system to code. While intent is understandable in high-level specification and analytical artifacts, it easily gets lost in models and code [36]. Interrelating these could help and might be seen as a noninvasive alternative to preserving use cases in code [8].

3   Abstract Layers and Generic Elements

Graphical software modeling serves a broad spectrum of purposes ranging from conceptualization of ideas to executable models. Apart from UML as a general purpose modeling notation, there is a growing tendency towards domain specific modeling languages. Freedom to chose different kinds of *visualization* some of which may be even three-dimensional, seems to be essential. What models actually mean and how they can be used further constitutes another varying perspective: *interpretation*. To accommodate this versatility in visualization and interpretation, a sufficiently general *internal representation* is necessary.

Pages in a book, sheets of paper in general, blackboards in a classroom, or even diagrams in contemporary software modeling tools, they all indicate layering is a way of coping with complexity natural to humans. Making layers and elements they consist of, along with their relationships, abstract and generic and allowing them to be visualized and interpreted in different ways might be a key to the internal representation we look for. From this point of view, it seems that the main problem of common approaches to graphical software modeling is interweaving visualization, interpretation, and internal representation, or giving priority to the former two over the internal representation.

Recall Fig. 1 from Sect. 2. Its upper right part depicts the multidimensional software knowledge metamodel, This is, actually, a simplified version of the metamodel that we implemented in out prototype tool called InterSKnow. The metamodel enables creating a connection between any two abstract elements, which means that it is possible to create connections between layers and elements, including layer–layer and element–layer ones. An element consists of any number of properties, which may represent anything relevant to it, possibly organized into sections.

What the metamodel from Fig. 1 does not show is that the elements can be of two kinds: native or proxy. While native elements can also be linked to other elements and, in some cases, represent them for the purposes of expressing further relationships between them, as depicted by the object level model in Fig. 1, proxy elements are meant to reflect the original elements. The reflected elements need not be just elements of other graph based software models: they can be any software artifacts, such as text documents or spreadsheets.

In general, proxy elements should reflect all content of the original artifact or, more often, just a part of it. Clicking or otherwise invoking such a proxy element should open the original artifact for editing in the corresponding tool. InterSKnow provides this behavior for Microsoft Word and Excel files, but it does not reflect the content in proxy elements.

4   Tool Support and Evaluation

In order to enable the evaluation of the concept of using abstract layers and generic elements as a basis for expressing multidimensional software knowledge, we developed a prototype modeling tool called InterSKnow. The implementation is based on the Java 8 platform with JavaFX 2.0.

The evaluation embraced two perspectives: the efficiency of searching for software knowledge in software models, presented in Sect. 4.1, and its comprehension, presented in Sect. 4.2. Threats to validity are discussed in Sect. 4.3.

4.1   Efficiency of Searching for Software Knowledge in Software Models

The efficiency of searching for software knowledge in software models was assessed in the context of solving given tasks. The following set of metrics was used:

– Number of context switches during the course of solving a task (context switches cause distraction, which decreases efficiency)
– Number of files opened during the course of solving a task (the necessity to open many files complicates work making it less efficient)
– Time necessary to solve a task

Analogically to Bystrický and Vranić's interpretation [8] related to source code, by a context switch in modeling we mean a necessity to look elsewhere than at the current diagram (within or outside the current model). This breaks a thread of thought, which negatively affects efficiency.

The experiment involved six participants, each of which had to solve three tasks, either with the InterSKnow tool or with Enterprise Architect as a representative of traditional, state-of-the-art software modeling tools. Each participant was provided with an equal computer configurations. The subject of the experiment was a model of a web based insurance system consisting of three separate applications called Insurance, Customer, and Notifier, with a common front end. The tasks performed by the participants were:

- Task 1: Finding and opening the corresponding class to add a new insurance policy type
- Task 2: Finding the unimplemented components that need to be removed
- Task 3: Finding the description and the class that corresponds to the Customer application interface

Table 1 summarizes the results. In a traditional setting, task 1 involved sequentially going through the models and their diagrams since the participants had to find a class whose name wasn't known to them. The results regarding the number of files opened only slightly favor InterSKnow because task 1 was focused on the interconnectedness of artifacts within one software modeling tool. Nevertheless, InterSKnow saves some effort by enabling to open directly the source code of a given class. The difference in context switches is significant. InterSKnow context switches count only for switches between layers, which are straightforward. Enterprise Architect required opening several models with the necessity to get back to previous ones since the participants tend to forget the relationships between these models. The time perspective also favors InterSKnow, which correlates with context switches.

Table 1. Efficiency of searching for software knowledge in software models.

	Context switches			Files opened			Time		
	Task 1	Task 2	Task 3	Task 1	Task 2	Task 3	Task 1	Task 2	Task 3
InterSKnow									
Participant 1	4	3	8	3	2	5	31	29	49
Participant 2	3	5	5	3	3	4	25	34	35
Participant 3	7	3	9	4	2	5	45	33	55
Enterprise Architect									
Participant 4	9	19	15	3	2	7	35	123	68
Participant 5	14	15	13	5	2	6	76	91	51
Participant 6	11	25	18	4	2	5	62	200	76

Task 2 results strongly favor InterSKnow in context switches and necessary time. This was expected, since Enterprise Architect does not express explicitly the relationships between components and packages, nor between packages and classes, which made the participants systematically open diagram by diagram.

Task 3 was focused on working with external software artifacts. In a traditional setting, the participants had to open all incriminated files one by one. With InterSKnow, the files were available directly from the tool via proxy elements. In many cases, it wasn't even necessary to click on the proxy element, since the content reflected by it was sufficient to determine whether the file is relevant or not.

4.2 Comprehension of Software Knowledge in Software Models

The comprehension of software knowledge in software models was assessed by verifying how much participants were able to learn from a given software model within a limited time using a questionnaire that included the following questions:

1. What version of the Spring boot is used by the Insurance application components?
2. Which application communicates via RESTS?
3. Which internal application communicates via the AMQP message system?
4. What types of messages are sent by the Notifier application?
5. What infrastructure technology is used in the insurance system (deployment/configuration)?
6. In what format does the Insurance application communicate via the external API?

The experiment involved eight participants. Each participant was shown the questionnaire for twenty seconds. Afterwards, they had three minutes to study the model. The same models as in the first experiment were used. Finally, the participants had to fill in the questionnaire they had been shown initially.

As can be observed in Table 2, the results speak in favor of InterSKnow. It only failed to outperform Enterprise Architect with respect to question 5. This can probably be attributed to the Enterprise Architect GUI as InterSKnow is only a prototype.

4.3 Threats to Validity

The fact that, unlike InterSKnow, Enterprise Architect was known to all participants represents a threat to internal validity. This was an advantage to Enterprise Architect. The results for an unknown tool would have probably been worse. To mitigate this, the participant had a brief demonstration of InterSKnow (on a different model than the one that was used in experiments, of course).

A low number of participants represents a threat to external validity. While we haven't had a possibility to increase the number of participants, we've taken care not to engage the same participants in experiments with both InterSKnow and Enterprise Architect.

The model in Enterprise Architect could have been created with some of the external software artifacts made available within the model itself using, for

Table 2. Comprehension of software knowledge in software models (correct answers indicated by the check mark).

	Question 1	Question 2	Question 3	Question 4	Question 5	Question 6
InterSKnow						
Participant 1	✓	✓			✓	
Participant 2		✓	✓	✓	✓	✓
Participant 3		✓	✓			
Participant 4		✓	✓	✓		✓
Enterprise Architect						
Participant 5		✓	✓			
Participant 6					✓	
Participant 7	✓				✓	
Participant 8			✓	✓	✓	

example, the UML note element. This can be viewed as another threat to external validity. However, an extensive use of UML notes to internalize external software artifacts is not a common practice.

5 Related Work

Melanee, Multilevel Modeling and Domain-Specific Language Workbench [2,3, 18], is a workbench for so-called multilevel or deep modeling. It supports creating both graphical and textual domain-specific modeling languages clearly separating their metamodeling levels on the instantiation basis. Differently than InterSKnow, Melanee provides no support of connecting to external software artifacts. Similarly to InterSKnow, Melanee decouples visualization, interpretation, and internal representation of the model, but it does so only partially. This involves employing the concept of layers (sometimes called levels, as in MetaCase [27]), which can also be observed in other approaches as well [11,24]. However, the layers are used there only as a means of distinguishing the level of abstraction. InterSKnow makes no limitations to the interpretation of layers.

Openflexo is an environment that makes possible forming a federation of software models contained in different tools by providing connectors to these tools [14,28]. In some cases, it is possible to maintain a live connection, i.e., upon changing an element or value in the external tool, its proxy element reflects the change automatically. The idea behind OpenFlexo itself is similar to our idea of interrelating software artifacts maintained in various external tools. InterSKnow goes beyond this by employing multiple levels of modeling in the interrelating model.

EMF Views is a tool that enables combining the information from different models in one view in the database (SQL) view fashion [4,7]. Some of the views are live, i.e., changes to proxy elements in a view are reflected in the actual elements. Again, as with Openflexo, this is similar to our idea of interrelating software artifacts maintained in various external tools. InterSKnow goes beyond this by making its interrelating model capable of introducing interconnections and further information not contained within the models being interrelated.

Among commercially available tools, IBM Rational Rhapsody Gateway enables elaborated connection to external software artifacts and their processing, mainly in the context of requirements engineering [20]. Different models and documents can be interconnected in this tool. Among supported ones are Microsoft Word, Microsoft Excel, and PDF formats. All these are transformed into a unified XML format within the import process that enables specifying what parts of these documents should be imported. Connected external software artifacts are available for opening directly from the graphical editor. However, all these documents are interpreted as sources of requirements, which makes it difficult to express other intentions. Moreover, Rhapsody Gateway itself does not support creation of software models, so even UML models have to be created elsewhere (usually in Rhapsody Modeler). InterSKnow, on the other side, aims at being a versatile software modeling tool.

6   Conclusions and Further Work

In this paper, a new approach to versatile graphical software modeling based on abstract layers and generic elements and its use in modeling multidimensional software knowledge and interrelating its pieces is proposed. The approach is supported by a prototype tool called InterSKnow, which targets mainly internal representation. This enabled to evaluate the approach from two perspectives: the efficiency of searching for software knowledge in software models and its comprehension. The results are generally plausible to the approach proposed here compared to Enterprise Architect as a representative of traditional, state-of-the-art software modeling tools.

The evaluation confirmed the importance of visualization which is largely underdeveloped in InterSKnow. A layered 3D visualization of software models [12,16,17] naturally fits the approach proposed in this paper. It could also be put into virtual reality [34,35]. Both possibilities would enhance opportunities for improving collaboration in distributed software development.

Acknowledgments. The work reported here was supported by the Scientific Grant Agency of Slovak Republic (VEGA) under the grant No. VG 1/0759/19 and by the Research & Development Operational Programme for the project Research of methods for acquisition, analysis and personalized conveying of information and knowledge, ITMS 26240220039, co-funded by the ERDF.

References

1. Asadi, F., Di Penta, M., Antoniol, G., Guéhéneuc, Y.G.: A heuristic-based approach to identify concepts in execution traces. In: Proceedings of 14th European Conference on Software Maintenance and Reengineering, CSMR 2010, Madrid, Spain. IEEE (2010)

2. Atkinson, C., Gerbig, R.: Flexible deep modeling with melanee. In: Proceedings of Modellierung 2016, Karlsruhe, Germany. LNI, GI (2019)
3. Atkinson, C., Gerbig, R., Kühne, T.: Comparing multi-level modelingapproaches. In: Proceedings of 1st Workshop on Multi-Level Modelling, co-located with 17th ACM/IEEE International Conference on Model-Driven Engineering Languages and Systems, MODELS 2014, Valencia, Spain, vol. 1286. CEUR (2014)
4. AtlanMod: EMF Views (2019). https://www.atlanmod.org/emfviews/
5. Bavota, G., et al.: The market for open source: an intelligent virtual open source marketplace. In: Proceedings of IEEE Conference on Software Maintenance, Reengineering and Reverse Engineering, CSMR-WCRE 2014, Antwerp, Belgium. IEEE (2014)
6. Bjørnson, F.O., Dingsøyr, T.: Knowledge management in software engineering: a systematic review of studied concepts, findings and research methods used. Inf. Softw. Technol. **50**(11), 1055–1068 (2008)
7. Bruneliere, H., Perez, J.G., Wimmer, M., Cabot, J.: EMF views: a view mechanism for integrating heterogeneous models. In: Johannesson, P., Lee, M.L., Liddle, S.W., Opdahl, A.L., López, Ó.P. (eds.) ER 2015. LNCS, vol. 9381, pp. 317–325. Springer, Cham (2015). https://doi.org/10.1007/978-3-319-25264-3_23
8. Bystrické, M., Vranić, V.: Preserving use case flows in source code: approach, context, and challenges. Comput. Sci. Inf. Syst. J. (ComSIS) **14**(2), 423–445 (2017)
9. Dalgarno, M., Fowler, M.: UML vs. domain-specific languages. Methods Tools **16**(2), 2–8 (2008)
10. Devanbu, P., Brachman, R., Selfridge, P.G., Ballard, B.W.: LaSSIE: a knowledge-based software information system. Commun. ACM **34**(5), 34–49 (1991)
11. Englebert, V., Heymans, P.: Towards more extensible MetaCASE tools. In: Krogstie, J., Opdahl, A., Sindre, G. (eds.) CAiSE 2007. LNCS, vol. 4495, pp. 454–468. Springer, Heidelberg (2007). https://doi.org/10.1007/978-3-540-72988-4_32
12. Ferenc, M., Polášek, I., Vincúr, J.: Collaborative modeling and visualisation of software systems using multidimensional UML. In: Proceedings of 5th IEEE Working Conference on Software Visualization, VISSOFT 2017, Shangai, China. IEEE (2017)
13. Frajták, K., Bureš, M., Jelínek, I.: Exploratory testing supported by automated reengineering of model of the system under test. Cluster Comput. **20**(1), 855–865 (2017)
14. Golra, F.R., Beugnard, A., Dagnat, F., Guerin, S., Guychard, C.: Addressing modularity for heterogeneous multi-model systems using model federation. In: MODULARITY Companion 2016, Companion Proceedings of 15th International Conference on Modularity, Málaga, Spain. ACM (2016)
15. Golra, F.R., Beugnard, A., Dagnat, F., Guerin, S., Guychard, C.: Using free modeling as an agile method for developing domain specific modeling languages. In: Proceedings of ACM/IEEE 19th International Conference on Model Driven Engineering Languages and Systems, MODELS 2016, Saint-Malo, France. ACM (2016)
16. Gregorovič, L., Polášek, I.: Analysis and design of object-oriented software using multidimensional UML. In: Proceedings of 15th International Conference on Knowledge Technologies and Data-Driven Business, Graz, Austria. ACM (2015)
17. Gregorovič, L., Polasek, I., Sobota, B.: Software model creation with multidimensional UML. In: Khalil, I., Neuhold, E., Tjoa, A.M., Da Xu, L., You, I. (eds.) CONFENIS/ICT-EurAsia 2015. LNCS, vol. 9357, pp. 343–352. Springer, Cham (2015). https://doi.org/10.1007/978-3-319-24315-3_35

18. Group, S.E.: Melanee project website. University of Mannheim (2014). http://www.melanee.org

19. Harman, M., Mansouri, S.A., Zhang, Y.: Search-based software engineering: trends, techniques and applications. ACM Comput. Surv. **45**(1), Article No. 11 (2012)

20. IBM: IBM rational rhapsody gateway (2014). https://www.ibm.com/support/knowledgecenter/SSB2MU_8.1.5/com.ibm.rhp.oem.pdf.doc/pdf/dassault/UserGuide.pdf

21. Indumini, U., Vasanthapriyan, S.: Knowledge management in agile software development - a literature review. In: Proceedings of 2018 National Information Technology Conference, NITC 2018, Colombo; Sri Lanka (2018)

22. Keivanloo, I., et al.: A linked data platform for mining software repositories. In: Proceedings of 9th IEEE Working Conference on Mining Software Repositories, MSR 2012, Zurich, Switzerland. IEEE (2012)

23. Khalil, C., Khalil, S.: Exploring knowledge management in agile software development organizations. Int. Entrep. Manag. J. 15 (2019)

24. de Lara, J., Guerra, E.: Deep meta-modelling with METADEPTH. In: Vitek, J. (ed.) TOOLS 2010. LNCS, vol. 6141, pp. 1–20. Springer, Heidelberg (2010). https://doi.org/10.1007/978-3-642-13953-6_1

25. Makedonski, P., Sudau, F., Grabowski, J.: Towards a model-based software mining infrastructure. ACM SIGSOFT Softw. Eng. Note **40**(1), 1–8 (2015)

26. Matharu, G.S., Mishra, A., Singh, H., Upadhyay, P.: Empirical study of agile software development methodologies: a comparative analysis. ACM SIGSOFT Softw. Eng. Note **40**(1), 1–6 (2015)

27. MetaCase: MetaCase (2019). https://www.metacase.com/

28. Openflexo: Openflexo project (2019). https://www.openflexo.org/

29. Ouriques, R., Wnuk, K., Gorschek, T., Berntsson Svensson, R.: Knowledge management strategies and processes in agile software development: a systematic literature review. Int. J. Softw. Eng. Knowl. Eng. **29**(3), 00153 (2018)

30. Petre, M.: UML in practice. In: Proceedings of 35th International Conference on Software Engineering, ICSE 2013, San Francisco, CA, USA. IEEE (2010)

31. da Silva, A.R.: Model-driven engineering: a survey supported by the unified conceptual model. Comput. Lang. Syst. Struct. **43**, 139–155 (2015)

32. Socha, D., Tenenberg, J.: Sketching software in the wild. In: Proceedings of 35th International Conference on Software Engineering, ICSE 2013, San Francisco, CA, USA. IEEE (2010)

33. Torchiano, M., Tomassetti, F., Ricca, F., Tiso, A., Reggio, G.: Relevance, benefits, and problems of software modelling and model driven techniques–a survey in the Italian industry. J. Syst. Softw. **86**(8), 2110–2126 (2013)

34. Vincúr, J., Návrat, P., Polášek, I.: VR City: software analysis in virtual reality environment. In: IEEE International Conference on Software Quality, Reliability and Security, QRS 2017, Prague, Czech Republic. IEEE (2017)

35. Vincúr, J., Polášek, I., Návrat, P.: Searching and exploring software repositories in virtual reality. In: Proceedings of ACM Symposium on Virtual Reality Software and Technology, VRST 2017, Gothenburg, Sweden. ACM (2017)

36. Vranić, V., et al.: Challenges in preserving intent comprehensibility in software. Acta Polytech. Hung. **12**(7), 57–75 (2017)

Model Suite and Model Set in Software Development

Igor Fiodorov[1](✉) ⓘ and Alexander Sotnikov[2] ⓘ

[1] Plekhanov Russian University of Economics, Moscow, Russia
Igor.Fiodorov@mail.ru
[2] Joint Supercomputer Center, Russian Academy of Sciences, Moscow, Russia
ASotnikov@jscc.ru

Abstract. Model-based systems engineering (MBSE) focuses on creating and exploiting multiple related models that help define, design, and document a system under development. The principal question about this approach concerns the level of association between these models. As our goal is a modeling of a system, we can assume that multiple models will form a tightly coupled suite, but in practice, partial models usually form a loosely bounded model kit. In this research we discuss two notions: model suite and model set, give them characterization based on mathematical theory.

Keywords: Model-driven development · Model suite · Model set · Algebraic system

1 Introduction

Modeling is one of the most important development stages; it results in a collection of models designed in a form suitable for further system implementation. The success of the entire development depends on how well these models are elaborated. A typical project involves numerous models in different notations. Often these models are arranged in sets. Models representing a view from a particular standpoint are usually called a perspective. It is obvious that the partial models that form the set should be well coordinated and well integrated with each other, otherwise, they can contradict. In most cases, interdependencies among singleton models in a set are not given in an explicit form. The term a model can mean either a singleton model as a part of a set or the entire set, making understanding ambiguous.

Thalheim contradistinguishes a loosely bounded model sets, used for distributed or collaborating IT and tightly coupled model suites with a consistent design and common data [1]. In his opinion, a model suite consists of a coherent collection of models representing different points of view and attention. He portrays a model suite as a kit of models with explicit associations among the models, with explicit controllers for maintenance of coherence of the models, with application schemata for their explicit

---

The work was done within the framework of the state assignment (research topic: 065-2019-0014 (reg. no. AAAA-A19-119011590097-1).

T. Welzer et al. (Eds.): ADBIS 2019, CCIS 1064, pp. 243–252, 2019.
https://doi.org/10.1007/978-3-030-30278-8_27

maintenance and evolution, and tracers for the establishment of their coherence. Changes within one model must be propagated to all dependent models. Each singleton model must have a well-defined semantics as well as a number of representations for a display of model content. The representation and the model must be tightly coupled. Thalheim and Tropmann-Frick defined model suite essential properties: well-formedness, adequacy, sufficiency. In their opinion a well-formed instrument is adequate for a collection of origins if (i) it is analogous to the origin to be represented according to some analogy criterion, (ii) it is more focused (e.g. simpler, truncated, more abstract or reduced) than the origin being modeled, and (iii) it is sufficient to satisfy its purpose [2]. In order to explain this definition we need to specify what is an analogy, how can a model be reduced, what is the purpose of modeling.

The notion of the model suite is essential for multi-model specifications that are widely used in model-based systems engineering (MBSE) [3]. Modern IT specification incorporates plenty of models in various notations. Unfortunately remains unclear how to distinguish a suite of tightly coupled models from a set of loosely connected ones. Until we give a comprehensive characterization of a model suite, we cannot expect progress in model-driven software development.

The object of this study would be multi-model specifications that unify numerous models in different modeling notations, while the subject would be a model suite constituted of tightly coupled submodels with a consistent design and common data. The objective of this study is to characterize the model suite and its essential properties. To achieve the goal we will use a mathematical models theory. We consider a model as a relational system—a set of abstract objects connected by relations. We investigate semiotic models used in technical science, i.e. in the computer industry. Compared to a natural language of human communication an artificial modeling notation sacrifices its beauty and imagery for the sake of accuracy and unambiguity. We limit observation to modeling of material objects, leaving ideal world out of concern.

2 Problem Statement

H. Stachowiak differentiates natural model, having the same physical nature as an original, and semiotic one made of signs [1]. Ch. Pierce proposed to distinguish between textual languages, the alphabet of which consists of letters, combined into meaningful words, and iconic languages where each sign represents a separate concept and provokes the emergence of a sensual image [4]. Harel and Rumpe attributed graphical modeling languages as iconic, their alphabet consists of a finite number of symbols, each carries its own semantic [5]. Schreider notes that universe of discourse (a set of specific things of the real world which are subject to modeling) is first simulated by the semiotic model, which is fashioned with another «mathematical» one [6]. The last does not explore the properties of natural things as such, but only relations between abstract «mathematical» objects. In mathematics, notions of a «structure» and a «model with a given structure» have a similar meaning. But when we investigate a universe of discourse, we have to distinguish between a model as a set of elements of a certain nature, connected to each other in a way that repeats the relations between thing which are subject to modeling, and a structure as an abstract category for which is no matter what the carrier set consists of and what is the real nature of relations. Following Schreider, we distinguish mathematical - $\mathfrak{W}_M$ and semiotic - $\mathfrak{W}_S$ models [7]

Figure 1 illustrates the proposed approach. A domain is formed by a set of material things. The right side illustrates the principle of how an analyst percept a reality. It is based on the Frege's triangle of reference, its sides can be interpreted as follows: the conceptualization mapping associates each thing with a certain concept; the semantic mapping relates a concept to a sign denoting it; the representation mapping render a sign to a thing, thus characterize a consistency between the model and the original. A concept abstract a notion related to a thing. The totality of all concepts constitutes ontology. A sign is a logical name assigned to the respective concept. A sum of all signs forms an alphabet of a language. Thus a sign of a modeling language denotes a thing if there is a concept associated with both [8]. The left side of this picture demonstrates the correspondence between a set of things and a set of abstract mathematical objects; we call this a formalization mapping. The correspondence between mathematical and semiotic models we call a reduction mapping.

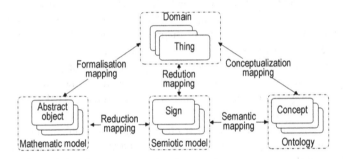

Fig. 1. Correspondence between mathematical and semiotic models

This article is structured as follows. First, we analyze the semiotic model and reduction properties. Later we investigate a correspondence between mathematical and semiotic models. Finally, we discuss the role of a signature for model comparison, give an interpretation of model suite essential properties, and argue the applicability of a proposed approach.

3   Model Reduction

In mathematics, a term a model (relational structure) means a set of abstract mathematical objects, together with a collection of relations defined on this set. Each relation is characterized by its arity to determine the number of objects participating in this relationship [9]. We will call a signature the collection of all relations (r) of the corresponding arity.

$$\mathfrak{W}_M = <\Theta; R>, \text{ where:} \tag{1}$$

Θ - a carrier set of abstract mathematical objects;
$R = \{o_r^{(J1)}, ..., r_n^{(Jn)}\}$ – a set of relations, of the corresponding arity.

To neatly judge the similarity of mathematical models we use a concept of isomorphism. Two models $\mathfrak{W}_{M1} = <\Theta1; R1>$ and $\mathfrak{W}_{M2} = <\Theta2; R2>$ are isomorphic if there is a one-to-one (bijective) mapping between the elements of their carrier sets $\Theta1$ and $\Theta2$ that preserve relations of both models. The isomorphism of models means the similarity of both structures. In other words, both models are formed by elements of different physical nature but are similarly structured. Two models are homomorphism if this mapping is injective. We agree to distinguish between the actual relations and the names of these relationships. The set of relationship names will be called a signature, it characterizes the model, determining from what symbols its expressions can consist and how they can be constructed.

According to Gurr [10] and Gastev [11] consistency between the model and the original can be characterized using the algebraic notion of homomorphism, understood by them as relations preserving mapping between two algebraic structures. If any property of the original can be uniquely mapped into the model and the model is in every respect equivalent to its prototype so that they are indistinguishable from each other, they are isomorphic. If the model is not the total equivalent of the original and discards some details not important for modeling objectives, the mapping is a homomorphism. Following Stachowiak we call it a model reduction [12].

We see two main mechanisms of reduction: a decrease of the model signature and a lessening of the model carrier set [9]. Let consider first mechanism, the carrier sets of both models coincide $\Theta1 = \Theta2$, and not empty signature of the first model is a subset of the signature of the second model $S1 \subseteq S2$. One can call the first model a projection of the second. For example, two maps: a geographical and a political can depict the same territory but a different relation between objects. Now we explore the second mechanism, the carrier set of the first model is a subset of the carrier set of the second one $\Theta1 \subseteq \Theta2$, while both signatures coincide and have the same arity $S1 = S2$. One can call the first model a submodel of the second one. For example, a detailed map depicts pathways while a small scale one leaves them out showing highways only. That means a set of things must be organized as an ordered list, to decline unnecessary details. Combining both mechanisms we can state that a finite projection of finite submodel is a local submodel [9]. Let us note that a reduction of the carrier set depends on the signature. One can reduce only an object that is insignificant in regard to a particular relation in this signature. Consider a pair of models with the same signatures, but having a different level of detalization, for example, two geographical maps differing in a scale and accuracy.

Figure 2 illustrates the reduction of a semiotic model. If a semiotic model reflects all things forming a domain and preserves all relations between them, then it is in all aspects equal to reality and isomorphic to it. However, it has a degree of complexity similar to the original which is not good for analyses. A semiotic model is usually a contraction of an original; it displays not all reality, but only a part of it, and does not convey all relations [12]. Thus the representation mapping connecting universe of discourse and a semiotic model is the homomorphism; at the same time the semiotic model is isomorphic to the local submodel, obtained by reduction of the full mathematical model.

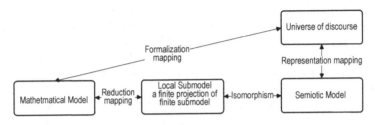

Fig. 2. Model reduction property

We can make following conclusions: two main mechanisms of model reduction are: (i) decrease of model signature and (ii) a lessening of the model carrier set, both mechanisms are interrelated and must be implemented in concert. A complex model can be represented as several simple projections that have a common carrier set.

4 Models Analyses by Means of a Signature

In this section we discuss multi-model specifications that are formed of several semiotic models in different modeling notations. A signature can be considered as an essential instrument for qualitative comparison and classification of semiotic models. Suppose different semiotic models are matched. Two models having identical signatures are considered to be of the same type, we can call them eponymous, they can be compared with each other even both they utilize different modeling notations. But if these signatures are different, then we conclude that models are incomparable. In case signatures coincide partially, these models can be compared only in terms of the relations of the same name. Now suppose comparing the modeling notation. One should first match the signatures of those models for the creation of which the studied notations are used. If the signatures coincide, it is possible additionally compare the expressive power of corresponding modeling languages [13, 14]. If the signatures of models do not match, the notations are not compatible. A similar problem can happen when trying to translate a semiotic model in one notation into a different language.

The following question should be taken into account: do different semiotic models, included in a single multi-model specification, use one, common to all of them, carrier set? If the carrier set is common, then individual semiotic models can be viewed as projections, each of the projections having a specific signature. In case individual projections do not contain common signatures, they can be considered mutually independent and can be developed separately from each other. Otherwise, if the signatures of the projections intersect, the projections turn out to be mutually interdependent, they must be designed consensually. Finally, it could happen that individual semiotic models forming one multi-model specification have dissimilar carrier sets. It reflects the fact that the universe of discourse is not a homogeneous collection of objects, but is partitioned in disjoint subsets of dissimilar type. From a mathematical point of view, it can be classified as a many-sorted model [15].

It is known that the model should be adequate for the purpose of modeling [16]. Let consider how to formulate modeling goal in terms of a signature. The SADT methodology states that the purpose of modeling is to obtain answers with a given degree of accuracy on a certain set of questions [17]. These questions are implied in the analysis and govern the creation of the model. If the model does not answer all questions or its answers are inaccurate, then modeling has not reached its goal. Thus, the purpose of modeling is determined by those questions this model must answer. We can see that the signature defines a set of relationships, each of them answer a particular question. For example, a model of an enterprise organizational structure displays all employees, their grouping and subordination. It answers four questions - who work in the organization, how the employees are grouped, to whom they subordinate, who is authorized to perform a specific unit of work? The signature of this model includes four relations. We make a conclusion, there is a one-to-one correspondence between the signature of a model and a list of questions it can answer, so we link the purpose of modeling and the signature of the corresponding model. If the signature includes the set of relationships required for the analysis, the corresponding model will be adequate to its goal; otherwise, if the model includes wrong relations it is considered inadequate to the purpose. The accuracy of the model can be associated with the reduction of its carrier set. If this type of reduction is absent, then the model has the maximum accuracy. As some elements of the carrier set are discarded, the accuracy of the model decreases. Thus, the degree of the base set elements reduction can characterize the model's accuracy.

However, it is necessary make the following comments on the applicability of the proposed approach. In this research we analyze the relational model under the assumption that operations on the carrier set of abstract objects are missing. But in reality, operations can exist. If we accept the existence of operations on the carrier set we should consider, whether the named set is closed under each operation? It can be argued that if the operations on the set do not lead to the emergence of new or the destruction of existing objects, the mentioned set is closed so that the above reasoning is correct. At the same time, there is a narrow class of models, in which, in result of a functional interaction, objects can emerge or destruct. In the last case, the reasoning should be clarified.

The motives why this class of models is not included in this research should be explained. We initially limited the subject matter to material things, excluding phenomena and events from the consideration. Phenomena imply a change in the things in the result of functional interaction with other things, and an event is associated with a change in the state of a thing. Thus, in order to be able to correctly analyze phenomena and events, it would be necessary to introduce two new concepts into consideration. First we will need a notion of a state; second, it will be necessary to describe a change of the state in the result of functional interaction. But, as we know, an abstract mathematical object neither possesses a state nor interacts with others. We believe that the named contradiction can be resolved as follows. First, it will require conducting an ontological study on the nature of the functional interaction between material things. Secondly, it will entail an investigation of the possibility to present functional interaction in the form of a mathematical relation.

5   A Model Suite and a Model Set

According to Dahanayake and Thalheim, information systems' modeling is based on a separation of concern such as separation into facets or viewpoints on the application domain from one side and separation of aspects from the other side. Facets and aspects are typically specified through different models that must be harmonized and made coherent [18]. We contribute to this definition by associating facets with a signature and aspects with multiple sorts of a carrier set.

We suggest differentiating a collection of models having one common carrier set, we can classify individual models as projections, and collection of models that have different disjoint carrier sets. We give an interpretation of a model suite as a multitude of partial models (projections) having a common carrier set. Each projection highlight specific relations between things constructing a carrier set. If different projections of the one model suite do not contain shared relations, they are considered autonomous and can be developed independently of each other. But if they contain common relations, these projections are mutually dependent and should be developed consistently. In the last case, there is a need for an additional controller to provide coherence between these projections. We also define a model set as a collection of models that have different carrier sets.

6   An Illustrative Example, Comparing Multi-model Specifications

Finally, we discuss some practical results of this study. Multi-model specifications are focused on creating and exploiting domain models as the primary means of information exchange between engineers, rather than on document-based information exchange [3]. For example, CIMOSA architecture suggests enterprises modeling using four perspectives [19]. The Zachman framework exploits six projections [20]. ARIS utilizes four perspectives [21]. Process modeling methodology includes four projections [22]. In all examples the perspectives are introduced empirically, are not substantiated theoretically, therefore it is difficult to compare these models. All specifications are inconcrete about providing consistency between projections.

This analyzes of MBSE approaches is provisional as we compare framework and method, also we consider ARIS as a technique but not as an instrumental tool that offers multiple modeling notations. We start with the well-known Zachman framework, which includes six perspectives. The author does not indicate the relations in each of them, but name every after a question it must answer. The "what" perspective is easy interpreted as the relationship between things that form the domain of material objects being processed. The "how" perspective describes the transformations that take place in things in the result of functional relations. The "who" perspective binds actors to work that induce the transformation. Note that the actors form a separate domain independent of the first one. The "where" perspective geographically locates actors executing a work. The "when" perspective links work to the timeline. Finally, the "why" perspective describes the goals of the work being performed. Note that the goal is usually formulated in financial terms - logical entities having a value in running a business, so we can talk about a third "business" domain. Thus, the Zachman framework uses three different disjoint domains: material things, actors and financial objectives. A similar situation is with CIMOSA and ARIS, which are based on two domains: material things and living persons. An interesting question is the level of integration between different domains. For example, ARIS method postulates that projections are integrated by means of control perspective, whereas the Zachman model does not consider the interrelationship of partial models. Thereby we categorize Zachman as a many-sorted algebraic structure [23]. Until the connections between the three domains are not well defined, it should be classified as a model set. However, if one will make some efforts to combine these three domains into one complex carrier set and will accurately describe the dependencies between domains, as well as analyze the dependencies between projections, to understand the degree of interdependence, the Zachman model will turn into a suite.

Nevertheless, we can see eponymous projections in all specifications above, for example, informational, organizational and functional perspectives are of the same name. Unfortunately, there is no common enumeration of relationships in each perspective. That is why analysts interpret these projections individually. Even within the same specification, the models of one projection implemented by different analysts can differ in signatures. If one would define a basic set of relations for each projection, it will eliminate the analyst's subjective understanding of the corresponding model, which, of course, will improve the modeling quality. The specifications above do not define languages used to build the corresponding perspectives; therefore we are not being able to analyze the grammars of the matching notations. However, we note that it is possible to set up a unified set of axioms for formal theories for all eponymous projections. This will make possible to define a general grammar for different languages used to describe the projections of the same name.

7 Conclusions

A singularity of this paper is in matching the mathematical and semiotic models. Its novelty is in applying the algebraic method to study an artificial modeling notation. Within frames of this discussion, we evaluate a model reduction; discuss the role of a signature in analyzing semantic models.

The results obtained in the paper are very important for model-based system engineering, which is based on the development of a system of interrelated models used to describe the requirements for the designed product. The main outcome is in the formulation of the model suite essential properties. If the multitude of partial models (projections) is designed in such a way that each reflects a certain relation between objects of one domain, common to all models, it can be called a suite. Otherwise, if partial models reflect relations between objects of different domains, these models form a set. If different projections of the one model suite do not contain shared relations, they are considered mutually independent and can be developed autonomously of each other. But if they contain common relations, these projections are mutually dependent and should be developed consistently. In the last case, there is a need for an additional controller to provide coherence between these projections. We show that a model set is a collection of models that have different carrier sets. Thus, well-known modeling methodologies, for example, Zachman and ARIS utilize several unrelated domains, therefore they form a model set and cannot be classified as a model suite. Therefore it would be important to continue this research by considering methods of connecting multi-sorted models that will allow transforming the collection of models into a model suite.

References

1. Thalheim, B.: The conceptual framework to multi-layered database modelling. In: Frontiers in Artificial Intelligence and Applications, EJC, Proc. Maribor, Slovenia, vol. 206, pp. 118–138 (2010)
2. Thalheim, B., Tropmann-Frick, M.: Models and their capability. In: Computational Models of Rationality, vol. 29, pp. 34–56 (2016)
3. Fisher, A., Friedenthal, S., Sampson, M., et al.: Model lifecycle management for MBSE. In: NCOSE International Symposium. Las Vegas, NV, vol. 24, pp. 207–229 (2014)
4. Atkin, A.: Peirce's Theory of Signs. The Stanford Encyclopedia of Philosophy (2013)
5. Harel, D., Rumpe, B.: Modeling Languages: Syntax, Semantics and All That Stuff, Part I: The Basic Stuff, Weizmann Science Press of Israel©, Jerusalem, Israel, Technical Report 2000, pp. 1–28 (2000)
6. Schreider, Y., Sharov, A.: Systems and Models. In Russian ed. 152 pp. Radio & Sviaz, Moscow (1982)
7. Schreider, Y.: Logika znakovyh system, p. 64. Znanie, Moscow (1974)
8. Chandler, D.: Semiotics for Beginners, p. 310. Routledge, Abingdon (2007)
9. Mal'cev, A.: Algebraic Systems, p. 315. Springer, Heidelberg (1973). https://doi.org/10.1007/978-3-642-65374-2
10. Gurr, C.: On the isomorphism, or lack of it, of representations. In: Marriott, K., Meyer, B. (eds.) Visual Language Theory, pp. 293–305. Springer, Heidelberg (1998). https://doi.org/10.1007/978-1-4612-1676-6_10
11. Gastev, Y.: Homorfizmy i modeli (in Russian), p. 152. Nauka, Moscow (1975)
12. Thalheim, B.: Towards a theory of conceptual modelling. J. Univ. Comput. Sci. **16**(20), 3102–3137 (2010)
13. Burton-Jones, A., Weber, R.: Building conceptual modeling on the foundation of ontology. In: Computing Handbook, Third edn. Computer Science and Software Engineering, pp. 1–15. Chapman and Hall (2014)
14. Rosemann, M., Green, P., Indulska, M., Recker, J.: Using ontology for the representational analysis of process modelling techniques. Int. J. Bus. Process Integr. Manag. **4**(4), 251–265 (2009)
15. Baader, F., Ghilardi, S.: Connecting many-sorted theories. J. Symb. Logic **72**(2), 535–583 (2007)

16. Thalheim, B.: Model adequacy. In: MOD-WS 2018 Workshops at Modellierung 2018. Braunschweig, Germany (2018)
17. Marca, D., McGowan, C.: SADT: Structured Analysis and Design Technique, p. 392. McGraw-Hill, New York (1988)
18. Dahanayake, A., Thalheim, B.: Co-evolution of (information) system models. In: Bider, I., et al. (eds.) BPMDS/EMMSAD -2010. LNBIP, vol. 50, pp. 314–326. Springer, Heidelberg (2010). https://doi.org/10.1007/978-3-642-13051-9_26
19. Vernadat, F.: Enterprise Integration: On Business Process and Enterprise Activity Modeling (1996)
20. Zachman, J.: The Zachman Framework: A Primer for Enterprise Engineering and Manufacturing. Zachman International (2003)
21. Software AG. Methods ARIS 7.0. Darmstadt (2011)
22. Curtis, B., Kellner, M., Over, J.: Process modeling. Commun. ACM **35**(9), 75–90 (1992)
23. Goguen, J., Burstall, R.: Institutions: abstract model theory for specification and programming . J. Assoc. Comput. Mach. **39**(1), 95–146 (1992)

Automatic Code Generator for Screen Based Systems

Katerina Korenblat and Elena V. Ravve[✉]

Ort Braude College, Karmiel, Israel
{katerina,cselenag}@braude.ac.il

Abstract. Definition and implementation of every system starts from its specification. After the developer received the verified specification, she can move to the code writing. Recently, modelling is used as the first step to the programming task. Unified Modeling Language (UML) is intensively exploited in order to standardize the code generation and to minimize the corresponding effort. Screen based interactive systems like smartphone applications or different booking services are of very special kind: GUI part of them is very big and mostly implements different screens and transitions from one screen to another. In our contribution, we provide general description and proof of concept of a graphical tool for such systems. First of all, the tool allows definition of the specification of the screen based systems in the most natural way - graphically. This avoid using of UML or similar formalisms as a less intuitive human dependent intermediate step. Then, given the specification of a screen based system, we show that a big part of the implementation (GUI code) may be generated automatically. We show how our general approach works for Android based implementation of the specification.

Keywords: Automatic code generator · Screen based applications · Model-Driven Architecture · Unified Modeling Language · Android OS

1 Introduction

The idea of automatic code generation is not really new, cf. [6]. One of the most recently developed branch in the general field is automatic code generation for Graphical User Interface (GUI). The corresponding contributions are mostly based on Model-Driven Architecture (MDA) as described in [4] (see also for more details, for example, [2]) and intensively exploit Unified Modeling Language (UML), cf. [5], or similar formalisms. As a rule, these approaches start from the assumption that the relevant UML diagram is given and it depicts the corresponding model, while the UML diagram should be manually composed, cf. [3,9–12]. Obviously, more natural way to describe **Graphical User Interface** is utilization of a **graphical** tool rather than UML diagrams, see a number of such tools in [7]. Some of these graphical tools suggest an option of code generation. On the other hand, writing a GUI code is more simple and effective using

© Springer Nature Switzerland AG 2019
T. Welzer et al. (Eds.): ADBIS 2019, CCIS 1064, pp. 253–265, 2019.
https://doi.org/10.1007/978-3-030-30278-8_28

some visual representation of GUI elements. For example, for Java based systems, JavaFX, cf. [1], provides a unified platform for the code developer, which allows representing and/or modifying the code in a visual way.

In this paper, we want to combine the whole process from GUI description to GUI code via checking of model fundamental properties such as GUI consistency. Our approach is more general, it is language independent, more visible, less confusing and may be directly used for both: specification verification and automatic code synthesis. The specification verification proofs its consistency and provides stub based verification environment, cf. [8]. Such an approach allows direct automatic implementation of a system from its visual specification, which significantly decreases the number of possible bugs in the system.

The proposed automatically generated code covers a big part of the implementation: the GUI code. In fact, any screen based system, like applications for smartphones or more general different booking systems are of a very specific kind. Actually, how do such applications work? The user pushes a GUI element, shown on a screen, and either the given screen is changed or a transfer to another screen is executed. In screen based systems, essential part of the specification describes screens, transitions between the screens and GUI elements, available on these screens. For such systems, we propose a new approach, where the specification may be formulated, using a graphical tool.

Now, we illustrate the immediate advantage of our approach. Take your smartphone. We are almost sure that you may activate or deactivate its Airplane Mode as well as switch its WiFi and Bluetooth ability from Off to On and vise versa. In fact there is at least one dependency between there choices: in Airplane Mode it should be impossible to switch WiFi or Bluetooth to On. We are almost sure that this requirement is ignored in your smartphone that may cause its confusing behaviour. Figure 1 shows how simple this feature (in fact, bug) may be prevented using our tool.

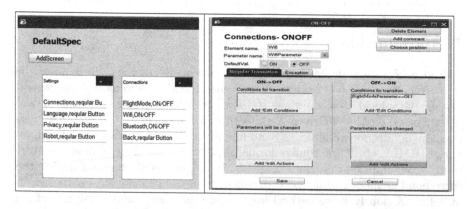

Fig. 1. Adding connection between different values

1.1   Defining Specification as a Set of Screens and Transitions

Now, we briefly describe the implemented graphical tool. The basic ability of its frontend is adding of a new screen to a specification of an application. In order to add a new screen, the user should press "add screen" button of the tool. Then, she sets the screen location inside an application window, defines the screen name and its description. The user can also add a GUI element to a screen, see Fig. 2.

In our implementation, we allow the following types of GUI elements of screens: Radio Button, Standard Button, Combobox, Text Field. Sure, the repertoire of the types may be extended, when needed. The internal terminology of the particular implementation of the tool names Radio Button as On/Off element, Combobox as List element and Text Field as Empty/NotEmpty element.

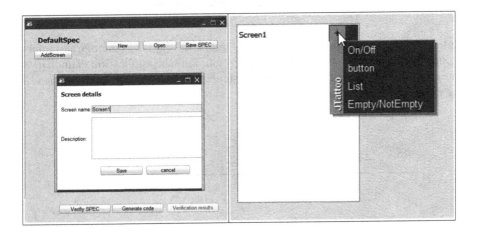

Fig. 2. Adding a new screen and a new element to the screen

For example, let us consider in more details specification of Standard Buttons, which are used in order to enable moving from one screen to another screen. They are specified by a field for Name, Conditions and the next screen to "Move to", see Fig. 3. We describe the graphical definition of each element of the specification in Sect. 2 in great details.

After the user created the specification of the system, the corresponding code may be generated automatically. In our particular implementation of the GUI code, we create a real Android Studio project (ASp), cf. [13] , which can pass compilation and run. Android Studio is a common development environment for Android developers, that is why we strived to create an Android Studio project. Bellow, we describe the process of the ASp creation.

In order to create an Android Studio project, we create the corresponding folders and files. Some of them are "constants" and never change. Others might be changed according to the project. The most important files in the project, which contain the real code of the application, are the .java and the .xml files.

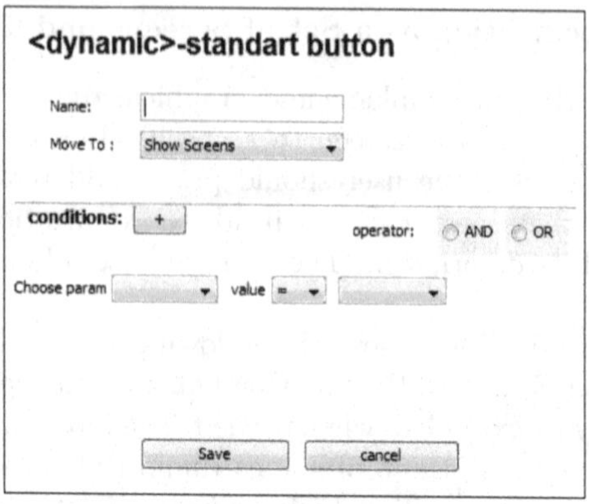

<dynamic>-standart button
Name:
Move To :
Show Screens
conditions:
+
operator:
AND
OR
Choose param
value
=
Save
cancel

Fig. 3. Adding a Standard Button

One more important file is the Android Manifest file. This file summarizes information about all the screens of the application, and the application cannot pass compilation and run without it. The file structure of the project and location of the described files is shown on Fig. 4.

ProjectName
.idea
.gradle
app
build
gradle
src
main
java
res
Android Manifest
.java files
.xml files

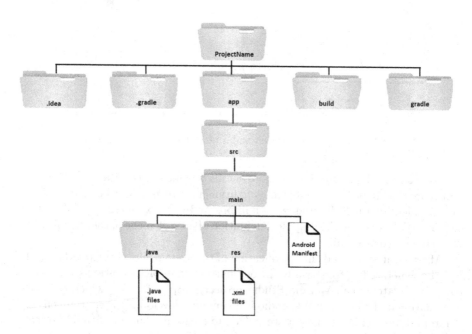

Fig. 4. Structure of an Android Studio project

The only thing, which we ask the user to create manually, is a new Android Studio project. Then, our tool alters and adds automatically the GUI code files, which are needed for the application to work. We describe the GUI code generation in grate details in Sect. 3. Section 4 summarises the paper.

In this paper, we propose a new graphical approach for definition of specifications of screen based systems, which extends and generalizes the know UML based approaches. As a proof of the concept, we show exactly what part of the GUI code can be generated automatically if the implementation assumes using of Android OS. We provide description of the corresponding GUI code generator, discuss its advantages and limitations.

2   Specification of Screen Based Systems

In this section, we give a detailed description of the graphical tool for definition of specifications of screen based systems. We show how to add different kinds of elements to the specification, and how to insert relevant information about them. As it can be seen on Fig. 2, in order to add a new screen, the user should press "add screen" button. Then, she sets the screen location, defines the name and the description and then press "save". The user should press "+" button in order to choose an element type from the menu bar. In this way, we translate the internal terms to the corresponding GUI elements. The menu bar includes the following options:

On-Off: this type allows activation or deactivation of some features. For this element type, we specified a field for the element name, a field for parameter name an associated action, and a default value.

List: if the user knows all the possible values of a parameter, then she can add them as a list. In this case, we specified a field for element name, a field for parameter name, a list of values, a default value, and an associated action.

Empty/NotEmpty: this type is defined for free text field. In this case, we specified a field for element name; field for parameter name and a default value.

Standard Button: this type is used in order to enable moving from screen to another screen. For this type, we specified a field for name, an availability condition and the next screen (see Fig. 3).

For each GUI element, we defined both: the element name, presented the text, which is associated in GUI to this element, as well as the parameter name for saving the corresponding data in the system, which may be different. For example, for the same List in one screen, we may use name "List of students", while, for another screen, the same data may be named as "Participants". For consistency of the specification, we have to show that, in fact, we are talking about the same list.

Finally, the composed specification is represented as a set of screens with the corresponding information and transitions between them. The specification is

saved for the further use for both verification and synthesis and may be loaded
and changed if needed

3   Code Generator for Screen Based Systems

For the given graphically composed specification, the corresponding GUI code
may be generated automatically. As a particular case, we explain how the translation from a specification to Android based code is done. Translation to other
formalisms may be done similarly. In Sect. 1.1, we explained how an Android
Studio project is built. Now, we show how the GUI elements are presented in
the project as well as what information, which we need, in order to generate the
corresponding code.

The constructed ASp contains executable GUI with a stub implementation
of the corresponding internal functions. Some features in the specification are
too common for automatic translation to the code (e.g., conditions on GUI
actions). We do not translate these features to real code. However, we add them
as comments in the relevant place in the code.

Bellow, we show how information from our graphically composed specification
is presented in the data structure of the "translator": its class diagram.

Generating Code of a Screen. Data structure, which is used for a screen
translation, is presented in class diagram of Fig. 5. The parameters, needed in
order to generate the corresponding Android code, are:

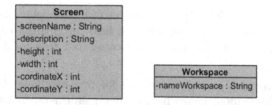

Fig. 5. Part of class "Screen" and class "Workplace" of the tool

A Screen Name. The screen name is saved in 'screenName' attribute of class
   Screen.
A* Name of xml file of the screen. If screen name (A) was 'HelloWorld', name
   of xml file (A*) will be 'activity_hello_world'.
B Project name, saved in attribute 'nameWorkspace' of class Workspace.

In order to generate a screen in Android code, we need to add 2 files. A java
class named A, and an xml file named A*, see Fig. 6. Note that every Android
application has a starting screen - a root screen, from which the application
starts. In our tool, the user must choose the root screen of the specification.

This root screen has a special name in Android code. The .java file of this screen will always be called MainActivity.java, and the .xml file will always be called 'activity_main.xml'.

Generating Code of a Standard Button. Data structure, which is used for a Standard Button translation, is presented in class diagram of Fig. 7. The parameters, needed in order to generate the corresponding Android code, are:

A Button text, saved in 'elementName' attribute of class Element.
B Button position.
C Name of the listener function (it also can be generated automatically).
D The next screen after pressing the button (screen name) to pass to, saved in attribute 'moveTo' of class StandardButton.

In order to generate the button in Android code, we need to add parts of code to both .java and .xml files of the relevant screen, see Fig. 8. This is the basic structure of the code that is needed for a button to work. Later, we will extend this code and show what can be added using our tool: actions and conditions.

In .java file, named A.java:
```
    package com.example.B;
    import android.support.v7.app.AppCompatActivity;
import android.os.Bundle;

public class A extends AppCompatActivity {
    protected void onCreate(Bundle savedInstanceState) {
        super.onCreate(savedInstanceState);
        setContentView(R.layout.A*);
    }
}
```

In .xml file, named A*.xml:
```
    <?xml version="1.0" encoding="utf-8"?>
<android.support.constraint.ConstraintLayout
xmlns:android="http://schemas.android.com/apk/res/android"
    xmlns:app="http://schemas.android.com/apk/res-auto"
    xmlns:tools="http://schemas.android.com/tools"
    android:layout_width="match_parent"
    android:layout_height="match_parent"
    tools:context="com.example.B.A">
</android.support.constraint.ConstraintLayout>
```

Fig. 6. Code for a screen

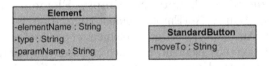

Fig. 7. Part of class "Element" and class "StandardButton" of the tool

In .java file of the relevant class

```
public void    C (View view) {
   Intent intent = new Intent(this,   D.class);
   startActivity(intent);
}
```

In .xml file of the relevant class

```
<Button
      android:id="@+id/A"
        android:layout_width="128dp"
        android:layout_height="58dp"
      android:layout_marginBottom="B"
      android:onClick="C"
      android:text="A"
      app:layout_constraintBottom_toBottomOf="parent"
      app:layout_constraintLeft_toLeftOf="parent"
      app:layout_constraintRight_toRightOf="parent" />
```

Fig. 8. Code for a Standard Button

Generating Code of an On/Off Element. Data structure, which is used for an On/Off element translation, is presented in class diagram of Fig. 9. The parameters, needed in order to generate the corresponding Android code, are:

Fig. 9. Part of class "Element" and class "OnOffElement" of the tool

A On/Off element text, saved in 'elementName' attribute of class Element.
B On/Off element position.
C Name of the parameter of the element, saved in 'paramName' attribute of class Element.

In order to generate a On/Off element in Android code, we need to add parts of code to both .java and .xml files of the relevant screen, see Fig. 10. Note that in order to 'listen' to the On/Off element, we need to set a specific listener and not a regular one. As it was shown above, in this listener, we update the value of a specific variable in the application. This variable is set by the user in our tool.

Generating Code of an Empty/NotEmpty Element. Data structure, which is used for an Empty/NotEmpty element translation, is presented in class diagram of Fig. 11. An Empty/NotEmpty element of our tool represents any text area in the application: any element, in which the user can insert text. We refer to the Empty/NotEmpty element as a simple text field (TextView) of Android.

The parameters, needed in order to generate the corresponding Android code, are:

A Empty/NotEmpty element text, saved in 'elementName' attribute of class Element.
B Button position.

In order to generate an Empty/NotEmpty element in Android code, we need to add parts of .xml code file of the relevant screen, see Fig. 12. In .java file of the relevant class, no special code is needed.

Generating Code of a List Element. Data structure, which is used for a List element translation, is presented in class diagram of Fig. 13. The parameters, needed in order to generate the corresponding Android code, are:

A List element name, saved in 'elementName' attribute of class Element.
B Button position.
C List element values, saved in 'values' attribute of class ListElement.

In order to generate a List element in Android code, we need to add parts of code to both .java and .xml files of the relevant screen, see Fig. 14.

In .xml file of the relevant class:

```
<Switch
    android:id="@+id/A"
    android:layout_width="121dp"
    android:layout_height="58dp"
    android:layout_marginEnd="16dp"
    android:layout_marginStart="16dp"
    android:layout_marginTop="B"
    android:text="A"
    app:layout_constraintEnd_toEndOf="parent"
    app:layout_constraintStart_toStartOf="parent"
    app:layout_constraintTop_toTopOf="parent" />
```

In .java file of the relevant class:

```
    Switch ASwitch = findViewById(R.id.A);
ASwitch.setChecked(MainActivity.C);
ASwitch.setOnCheckedChangeListener(new
CompoundButton.OnCheckedChangeListener() {
public void onCheckedChanged(CompoundButton compoundButton,
boolean b) {
    MainActivity.C = b;
    }
});
```

Fig. 10. Code for an On/Off element

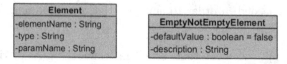

Fig. 11. Part of class "Element" and class "Empty/NotEmpty element"

In .xml file of the relevant class:

```
<EditText
        android:id="@+id/A"
        android:layout_width="113dp"
        android:layout_height="17dp"
        android:layout_marginBottom="B"
        android:layout_marginTop="8dp"
        android:text="A"
        app:layout_constraintBottom_toBottomOf="parent"
        app:layout_constraintEnd_toEndOf="parent"
        app:layout_constraintHorizontal_bias="0.501"
        app:layout_constraintStart_toStartOf="parent"
        app:layout_constraintTop_toTopOf="parent" />
```

Fig. 12. Code for a Empty/NonEmpty element

```
Element
-elementName : String
-type : String
-paramName : String
```

```
ListElement
-defaultValue : String = none
-values : String
```

Fig. 13. Part of class "Element" and class "ListElement" of the tool

Implementing Conditions and Actions. In our tool, for some elements, there is an option to add conditions and actions. As for conditions, we implemented the feature for Standard Button and On/Off elements. For the Standard Button, moving from screen to screen is conditioned. In fact, only if the corresponding conditions, which the user defined, are satisfied, then the pressing will change the screen. For On/Off element, switching from On to Off and vice versa is conditioned as well.

In order to implement conditions for a Standard Button or On/Off element, we altered the button listener. We added an 'if' statement before changing the screen. Inside the 'if' statement, the actions will be done, and then the screen will be switched. On Fig. 15, we show an example with the listener of a Standard Button. Only if the 'param' value is true, then the screen will be switched. On Fig. 16, we show another example with the listener of a On/Off element. Conditions are divided to tow parts: On-to-Off conditions, and Off-to-On conditions.

In .java file of the relevant class:

```
Spinner A = (Spinner) findViewById(R.id.A);
// Spinner click listener
A.setOnItemSelectedListener(new AdapterView.OnItemSelectedListener() {
    public void onItemSelected(AdapterView<?> parent, View view, int
position, long id) { }
    public void onNothingSelected(AdapterView<?> adapterView) {
    }
});
// Spinner values array
List<String> elements = new ArrayList<String>();

// Get spinner values from VerySPEC
for (int i=0;i<elements.length;i++) {
  elements.add(i,C.get(i));
}

ArrayAdapter<String> dataAdapter = new ArrayAdapter<String>(this,
android.R.layout.simple_spinner_item,elements);
dataAdapter.setDropDownViewResource(android.R.layout.simple_spinner_
dropdown_item);
A.setAdapter(dataAdapter);
```

In .xml file of the relevant class:

```
<Spinner
    android:id="@+id/A"
    android:layout_width="368dp"
    android:layout_height="wrap_content"
    android:layout_marginBottom="B"
    app:layout_constraintBottom_toBottomOf="parent"
    app:layout_constraintLeft_toLeftOf="parent"
    app:layout_constraintRight_toRightOf="parent" />
```

Fig. 14. Code for a List element

```
public void   button_Listener (View view) {
    Intent intent = new Intent(this,  Screen2.class);
    if ( MainActivity.param == true ) {
        startActivity(intent);
    }
}
```

Fig. 15. Adding condition for a Standard Button

In this example, there was only an On-to-Off condition, which is shown in the second 'if' after the On-to-Off comment. Only if 'param' is true, the On/Off state will be changed to Off. In a similar way, conditions and actions may be added and translated for other GUI elements, if it is needed.

Android Manifest File. This file describes the structure of the project, screen-wise. An example of an Android Manifest file for an application with four screens, is shown on Fig. 17.

```
    switch.setChecked(MainActivity.b);
    switch.setOnCheckedChangeListener(new
CompoundButton.OnCheckedChangeListener() {
        public void onCheckedChanged(CompoundButton
compoundButton, boolean b) {
            // off to on conditions:
            if (b == true)
                MainActivity.b = b;
            // On To Off conditions:
            if (b == false) {
                if ( MainActivity.param == true ) {
                MainActivity.b = b;
                bbSwitch.setChecked(b);
            }
            else {
                MainActivity.b = !b;
                bbSwitch.setChecked(!b);
                }
            }
        }
    });
```

Fig. 16. Adding condition for an On/Off element

```
<?xml version="1.0" encoding="utf-8"?>
<manifest
xmlns:android=http://schemas.android.com/apk/res/android
        package="com.example.defaultspec">
<application
                    android:allowBackup="true"
                    android:icon="@mipmap/ic_launcher"
                    android:label="@string/app_name"
                    android:roundIcon="@mipmap/ic_launcher_round"
                    android:supportsRtl="true"
                    android:theme="@style/AppTheme">
                    <activity android:name=".MainActivity">
                            <intent-filter>
                                <action
android:name="android.intent.action.MAIN" />
                                <category
android:name="android.intent.category.LAUNCHER" />
                            </intent-filter>
                    </activity>
                    <activity android:name=".Screen2" />
                    <activity android:name=".Screen3" />
                    <activity android:name=".Screen4" />
        </application>
</manifest>
```

Fig. 17. Android Manifest file for an application with four screens

4   Conclusion

In this paper, we generalize approach for GUI code generation of cf. [3, 9–12]. Our approach is graphical based, language independent and platform independent. We do not limit ourselves by Android based applications but rather use them as a particular implementation of a more general approach. To do so, we introduce the notion of screen based systems. Smartphone application is only a special case of them. We present a general definition as well as a partial implementation of the graphical tool, which is aimed to replace commonly used UML approach or other similar formalisms. The tool allows graphical definition of the specification as transfers from one well-defined screen to another one. Then, the tool allows verification of the specification as well as generation of as big as possible part of the corresponding GUI code.

Acknowledgements. We would like to thank our students S. Namih, A. Mnasra, Y. Dubinsky and A. Zobedat for implementation of the tool.

References

1. JavaFX: https://openjfx.io/. Accessed 05 May 2019
2. Atkinson, C., Kühne, T.: Model-driven development: a metamodeling foundation. IEEE Softw. **20**(5), 36–41 (2003)
3. Benouda, H., Essbai, R., Azizi, M., Moussaoui, M.: Modeling and code generation of Android applications using Acceleo. Int. J. Softw. Eng. Appl. **10**, 83–94 (2016)
4. Object Managment Group: MDA Guide Version 1.0.1. https://www.omg.org/news/meetings/workshops/UML_2003_Manual/00-2_MDA_Guide_v1.0.1.pdf. Accessed June 2003
5. Object Managment Group: OMG Unified Modeling Language. https://www.omg.org/spec/UML/2.2/Superstructure/PDF. Accessed 2009
6. Herrington, J.: Code Generation in Action. Manning Publications Co., Greenwich (2003)
7. Keshtcher, Y.: Top 22 prototyping tools for UI and UX designers 2019. https://blog.prototypr.io/top-20-prototyping-tools-for-ui-and-ux-designers-2017-46d59be0b3a9. Accessed 23 June 2019
8. Korenblat, K., Ravve, E.: Automatic verification and (partial) implementation of specifications of smartphone applications. (in preparation)
9. Kraemer, F.A.: Engineering Android applications based on UML activities. In: Whittle, J., Clark, T., Kühne, T. (eds.) MODELS 2011. LNCS, vol. 6981, pp. 183–197. Springer, Heidelberg (2011). https://doi.org/10.1007/978-3-642-24485-8_14
10. Lachgar, M., Abdali, A.: Modeling and generating the user interface of mobile devices and web development with DSL (2015)
11. Monte-Mor, J., Ferreira, E., Campos, H., da Cunha, A., Dias, L.: Applying MDA approach to create graphical user interfaces. In: 2011 Eighth International Conference on Information Technology: New Generations, pp. 766–771 (2011)
12. Sabraoui, A., Koutbi, M.E., Khriss, I.: GUI code generation for Android applications using a MDA approach. In: 2012 IEEE International Conference on Complex Systems (ICCS), pp. 1–6 (2012)
13. Android Studio Site: User Guide. https://developer.android.com/studio/intro/index.html. Accessed 05 May 2019

Formalizing Requirement Specifications for Problem Solving in a Research Domain

Nikolay A. Skvortsov[(⊠)] and Sergey A. Stupnikov

Institute of Informatics Problems, Federal Research Center
"Computer Science and Control" of the Russian Academy of Sciences,
Moscow, Russia
nskv@mail.ru, sstupnikov@ipiran.ru

Abstract. The paper presents the research of a methodology of conceptual scheme development to solve problems in subject domains. A semantic approach to domain model specifications building is principal for it. The development process involves formulating a model of requirements to the domain from verbal specifications of domain requirements, developing a domain ontology, transforming it into a conceptual scheme, and reusing domain knowledge specifications in the domain. Relevant data sources are mapped to conceptual schemes of domains in data infrastructures. Requirement specifications are implemented over conceptual schemes for entity resolution and problem-solving in domains using accessible data sources.

Keywords: Conceptual modeling of problem domains ·
Domain specifications · Conceptual scheme · Requirement model

1 Introduction

Research of methods and means of problem-solving over heterogeneous information resources in data infrastructures are conducted today. They lead to conclusions of necessity to form communities of researchers working in certain subject areas and solving certain classes of problems. Communities develop domain specifications, acquire heterogeneous information resources, integrate them to make materialized data repositories or virtual executive environments. This approach allows to describe problems in terms of the subject domain within data infrastructures and to solve them using integrated information resources.

A reasonable formation of domain specifications themselves for solving problems in data infrastructures is a subject of investigations. Very often development of domain models is deemed being too expensive and unnecessary. Nevertheless, domain specifications may become a basis of problem-solving, interoperation and collaboration of researchers within communities. The aim of this work is to propose a semantic approach to constructing conceptual schemes of domains and using them in data infrastructures. We begin with specifications of the requirements of domain problems formulated in a natural language, involve descriptions of the domain knowledge and hold the direction to formalizing specifications to solve domain problems.

© Springer Nature Switzerland AG 2019
T. Welzer et al. (Eds.): ADBIS 2019, CCIS 1064, pp. 266–279, 2019.
https://doi.org/10.1007/978-3-030-30278-8_29

This theme touches investigation having a long history of development [17, 24, 25]. Conceptual schemes of databases are logical representations of information structures, understandable to human and independent of implementation in information systems. As data models develop, the notion of conceptual schemes changed for including behavior specifications too and scoping not only in databases but in artificial intelligence, domain modeling, data structures in programming and other areas [7]. Object-oriented analysis and design technologies use these results for the development of object specifications in information systems. Research on ontologies was inspired by the problems of semantic interoperability of systems and operates with concepts understandable to both human and machine. Unfortunately, this direction was significantly mingled with data representation development, rare research returned to differences of ontologies and data representation schemes [11, 20].

Generally, conceptual analysis of domains takes place implicitly by the developers and results in programs over some data representation. Formal domain specifications are usually developed separately for some specific needs and these processes are poorly associated with one another. We combine formalization of domain models with heterogeneous data integration, semantic-based entity resolution and building solutions of domain problems in terms of unified conceptual data representation.

The paper largely follows and enhances ideas presented in [26] in Russian. That investigation represented domain specifications in a single expressive specification language SYNTHESIS [14] having defined formal semantics and making possible reasoning of specification refinement [4]. Here we use other, standard languages and means for different kinds of specifications. Ontological specifications are constrained by expressive features of OWL [2] language. For specification of conceptual schemes, a relational model is used, or OWL can be used too. To express object constraints and behavior specifications in conceptual schemes, RIF [5] language recommended by W3C is applied. It allows unifying the representation of rules over conceptual structures. It also is used for mapping data sources to domain conceptual schemes.

Further presentation is organized as follows. Section 2 presents a brief retrospective analysis of the studies in mentioned technologies of information system analysis and design. Section 3 describes the semantic-based process of developing conceptual specifications as a whole. Sections from 4, 5 and 6 describe details of domain and problem analysis stages including development of a model of requirements for solving problems, development of a domain ontology and transformation of ontology into a conceptual scheme for data representation and object behavior constraining in the domain and problem solving over multiple sources including identification of domain objects from different sources.

2   Related Research

Among the studies related to designing conceptual schemes of databases and information systems, well-known specification notation UML [3] should be mentioned. To develop an information system scheme in UML, firstly the agents and their activities are considered. Then sequences of their interaction are described by means of function signatures and an object scheme specifying class structures and methods according to agent activities is defined. Function specifications are specified in a constraint language (for example, OCL). Thus, development in UML is primarily focused on function signatures defining components of a problem-solving process.

Such an approach considering activities of agents as black boxes has been criticized by Milopoulos et al. [13], and as a solution, a requirement-driven methodology of process design has been proposed [16]. An information system specification process includes constructing and refining a tree of goals formulated in natural language and then formalizing these specifications as operations in a process language and information objects produced by them. This approach provides reasonable development of behavioral specifications and structures for problem-solving based on requirement specifications.

Semantic-based and requirement-driven approaches to conceptual scheme design should draw upon ontological modeling [10, 19, 22]. General principles of ontology development include the analysis of the domain for selection of essential concepts, discovering their relationships, and definition of concept constraints with satisfiability control. Reverse engineering is used too if the domain ontology should be restored from data schemes of existing information systems.

Various projects related to conceptual scheme design based on knowledge derived from ontologies use different perceptions of both ones. For example, in [15, 23] principles of development of ER models and UML diagrams on the basis of semantic relations of certain types or concept hierarchies are is presented. In [8] a method of concept, relation, and restriction selection from a wide domain ontology and construction of a conceptual scheme for the specific problem solution is described.

3　An Approach to Development of Domain Specifications

An approach to domain conceptual scheme design and solving problems over them proposed here joins requirement modeling, formalization of ontologies and construction of conceptual schemes. The design process includes the following steps:

- development of models of requirements for tasks and problems in the domain;
- building a domain glossary for terms mentioned in requirement specifications and commonly used in the domain;
- building domain ontology based on the glossary;
- making a conceptual scheme from the ontology and the requirement model;
- integration of related data sources using domain specifications;
- generation of identification criteria for resolution of objects from different sources;
- formulation of solutions for problems over the conceptual scheme of the domain.

This approach allows to consistently refine and formalize the semantics of the domain and some classes of problems in it. At the same time, it provides creation of a formal domain model which may be reused continuously in data infrastructures.

4   Requirement Model Creation

In discussing problems expecting to be solved with experts of a domain community and problem statement analysis, requirements are recorded, which are conditions or abilities which the system should satisfy in order to solve the problem or to achieve a goal [1]. Requirements are refined by decomposition to more specific requirements in the form of a tree. Decomposition is performed as long as the requirement can be formulated without specifying implementation issues and usage of data resources.

Functional requirements are decomposed by splitting it into a set of jointly fulfilled specific requirements (AND-decomposition), or into a set of alternative requirements (OR-decomposition), some of them can be selected as alternative branches in the tree [18]. So the problem can be solved by fulfilling all leaves of the tree without rejected alternative branches. The requirements included in the execution plan are specified with pre-and post-conditions for their interactions.

As an example, we present a fragment of the tree of requirements (Fig. 1) for an astronomical problem of searching for stellar systems of high multiplicity. Components (single stellar objects) and pairs within a multiple stellar system can be observed not only visually using photometry, but also by spectroscopic, interferometric and other methods. A stellar system may consist of binary stars (pairs) of different observational types. Generally, each type of observation is a special knowledge domain with its own observed parameters and features of the studied objects. Data on pairs of different observational types are stored in different data sources with their own structure.

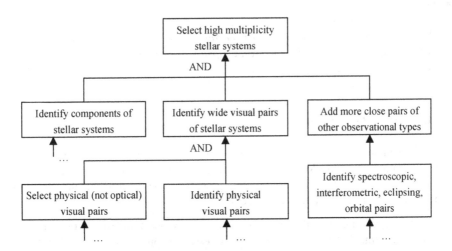

Fig. 1. Requirement tree fragment

To find multiple stellar systems consisting of components and pairs of different observation types, it is necessary to identify all components and pairs of them. So we begin from matching and identifying components of the systems as single objects from different data sources. Then for data on visual binary stars, it is verified whether they are physical visual pairs: we try to remove from the consideration pairs located close optically but at different distances from the observer. Physical visual pairs are matched and identified to deduplicate data from different sources. Finally, more close pairs are added which are not observed photometrically as binary objects, but show duality in other methods of observation. Such pairs should be identified in different data sources as well. Thus, multiple systems of high multiplicity are composed of pairs of different observational types.

Functional requirements for identification of components, selection of physical visual pairs, identification of visual pairs, and identification of close pairs are further decomposed in the requirement tree to formulate identification criteria for each type of object and detailed process of stellar system detection. Different branches of the requirement tree use different knowledge domains, such as astrometry, photometry, spectroscopy, orbital movement. Leave requirements of the tree and some parent goals are operationalized. Their inputs and outputs, pre- and post-conditions are specified like the following:

- Requirement: *"Select physical (not optical) visual pairs"*;
- Operation name: *selectPhysicalPairs*;
- Input: *"a set of pairs"*;
- Output: *"a set of pairs"*;
- Precondition: *"pairs are visual ones"*;
- Post-condition: *"pairs having known parameters of orbital movement, or observational type of a close pair (spectral, eclipsing or orbital), or similar parallaxes of components are physical visual ones"*.

5   Ontology Development

For a domain specification, firstly, its conceptualization should be held. For this purpose, building a domain glossary that contains domain terms and their verbal definitions can be very useful. A glossary building process can be performed at the same time with a requirement tree refinement process:

- essential terms are identified in problem statements and in the tree of requirements;
- for each term, text fragments defining its essential properties or constraining its interpretations are selected, the terms are provided in the glossary with definitions;

- new essential terms derived from other term definitions in the glossary are added to the glossary with their definitions until the glossary is saturated with terms essential for the domain or for given classes of problems.

In the example above (Fig. 1), the requirement tree contains the following essential terms for the domain glossary: *Astronomical Object, Star, Stellar System, Component, Pair, Observational Type, Visual Pair, Physical Visual Pair*. These terms should be verbally defined with essential property, relation, and constraint descriptions. For example, visual pairs and physical pairs are described as follows:

"Visual pair is a pair of stellar objects observed photometrically as two visually distinct components. Observational type of such pairs is visual. Observation of relative positions of visual binary stars components includes angular separations and position angles. Among visual pairs, optical and physical ones are distinguished. A physical pair can be recognized by common parallaxes of components or orbital movement of its components, or if the pair is also observed as one of more close type of pairs, for instance spectroscopic, eclipsing one."

Definitions and constraints of some terms can be selected from this description. Then the glossary is completed with derived terms in subdomains, for example:

- *Coordinate, Coordinate System, Equatorial Coordinate System, Right Ascension, Declination, Precession, Parallax, Proper Motion*;
- *Photometric System, Passband, Magnitude*;
- *Orbital Movement, Period, Position Angle, Angular Separation, Excentricity*.

The glossary is completed with thesaurus semantic relationships: synonymy, hypernymy/hyponymy, part/whole, dependence, associative relationships revealed from verbal definitions or from external thesauri. Kinds of terms are determined: entities, roles of entities, characteristics of entities, measurable characteristics, groups, processes, etc. Further analysis of verbal definitions serves to transform the glossary into an ontology, which consists of concepts, their relations, and constraints. Ontological concepts are produced in accordance with terms of particular kinds. Hypernymy/hyponymy relations determine concept hierarchy. Binary and more complex relations of concepts are determined. Terms from the requirement tree branches can be useful for ontology modularization. The problem of verbal definition analysis for ontology formalization is out of the paper issues. It has been investigated for the Russian language as mentioned in [26]. The investigation shows that it is possible to make a set of rules for glossary definition analysis, most of which are independent on a certain domain. Some approaches for the English language can be seen in [9, 12]. Formalization of constraints related to concepts depends on the ontology language expressive power.

In the following example represented in the abstract syntax of OWL language [2], ontological concepts *Pair* and *VisualPair* are defined in accordance with verbal definitions of the glossary terms.

```
Class(Pair partial CompountObject
  restriction(isPairOf
    allValuesFrom(StellarSystem))
  restriction(hasPrimaryComponent
    allValuesFrom(Component) maxCardinality (1))
  restriction(hasSecondaryComponent
    allValuesFrom(Component) maxCardinality (1))
  restriction(hasObservationType
    allValuesFrom(ObservationalType))
  restriction(hasOrbit
    allValuesFrom(Orbit) maxCardinality (1))
  restriction(hasComponent
    allValuesFrom(Component) maxCardinality(2)))
Class(VisualPair partial Pair
  restriction(hasRelationalPosition
    allValuesFrom(RelationalPosition))
  restriction(hasObservationalType
    hasValue(Visual)))
```

The concept *Pair* is defined as having relations (OWL properties): *hasObservationalType, hasPrimaryComponent, hasSecondaryComponent, hasOrbit*. Its subconcept *VisualPair* is defined with the relational position of the components. Some of the relations are generated from verbal definitions of glossary terms, others are implied from the verbal definitions of related terms. For example, the *isPairOf* property is inverse to the *hasPairs* property in the *StellarSystem* concept. There are some relations inherited from the superconcept in the *Pair* concept: *hasIdentifier, hasCoordinate, isComponentOf, hasEvolutionaryStatus, hasParameter,* and others.

6   Conceptual Scheme Generation

Conceptual schemes define information structures and behavior of information objects in domain communities. So, specifications of conceptual schemes include definitions of abstract data types with their structure, their extensions as sets of objects, integrity constraints, methods; and processes that determine the behavior of objects in the domain and problem-solving processes.

Using knowledge from a domain ontology, it is possible to construct a conceptual scheme for solving domain problems. This process can be semi-automated and performed by an expert, who can be offered of preferable transformations of the scheme. The principles of scheme generation are described in terms of abstract data types [14] since this notion includes abstract structure and behavior specifications of objects. But in this paper, we use a relational data model for structures and rule specifications for constraint and behavior specifications. Another natural approach is to use OWL for both ontologies (as domain concept specifications) and data structure specifications. General principles of conceptual scheme generation from an ontology are as follows:

- for ontological concepts of entities and roles, accompanying abstract data types of conceptual schemes are created; subconcept relations are copied into the conceptual scheme; named relations of concepts become attributes in types (later value types of these attributes will be defined: scalar, enumeration, association type); experts can choose which concepts should be accompanied by abstract data types;
- extensions (or classes) are created for sets of independent domain entities, multiple classes can be created for the same type;
- if a type has subtypes without additional relations defined in them, subtypes can be replaced by an attribute with an enumeration of values corresponding to the subtypes or of Boolean type if there is a single subtype;
- supertypes that are not used directly, can be removed: their structures and constraint are replaced into each of their subtype;
- types that are not directly used, linked by named relations with other types, can be removed; their structural descriptions are replaced into each associated type; relation multiplicities should be taken into account;
- concepts of measurable parameters can be accompanied with types with a scalar type attributes; a type having this attribute can be removed, and its attribute replaces every named relation to a removed type in all linked types;
- functional requirements are operationalized; also functions can be defined for the concepts of processes, activities, and dependent values;
- if a function receives a single instance of a data type as a parameter, it can become a method of that data type;
- type integrity (invariants), pre- and post-conditions of functions and methods correspond to the constraints defined in the concepts,
- constraints are formalized for text requirements which have not been formalized before;
- experts can rename any elements.

Using these principles, the ontology of the domain of multiple stellar systems could be used to generate a conceptual scheme including relational structures for object types *StellarSystem*, *Component*, and *Pair* with sets of attributes for all their observed measurable parameters and foreign keys for associations (attributes with values of object types) between them Additional relational structures *Coordinate*, *RelPosition* and *Orbit* are created but not used without the object type structures above, they contain coordinate, relational position of components, and pair orbit parameter attributes respectively.

The *Pair* structure contains the following attributes:

- *pairOf* (foregn key (FK) referencing a *StellarSystem* tuple);
- *primaryComponent* (FK to *Component*);
- *secondaryComponent* (FK to *Component*);
- *coordinate* (FK to *Coordinate*)
- *orbit* (FK to *Orbit*);
- *relPosition* (FK to *RelPosition*, it is used for visual physical pairs);

- *obsType* (FK to *ObsType* structure generated in the relational model for multiple values of observational types of pairs);
- *parallax* (measured parameter);
- *properMotion* (measured parameter);
- *epoch* (time of observation);
- *isPhysical* (boolean, generated instead of the subtype).

Alternatively, using the same principles of ontology transformation into a conceptual scheme, another structure of the conceptual scheme could be reasonable for this domain. It could have a single type *StellarObject* for either systems, pairs, components and single stars with an enumeration attribute with values *StellarSystem*, *Component*, and *Pair*. A single type *Parameter* generated for all parameter types could contain *parameterType* attribute with enumeration vocabulary of all parameters, the *stellarObject* attribute and the *value* attribute.

The scheme should be completed with functions and object methods. They are created for operationalized functional requirements, process/activity concepts, and dependences of measured parameters. Pre- and post-conditions of functional requirements and type constraints are used to generate pre- and post-conditions of functions and methods. Parameters of functions are generated from inputs and outputs of requirements and relations of concepts. If conditions are described in details, the conceptual scheme can contain full specifications of functions that can be implemented without manual programming.

If conceptual scheme specifications are created using ontological knowledge, it is possible and necessary to store the ontological provenance of specification elements as semantic annotations of the scheme. Types, attributes, functions, and parameters should be annotated with ontological concepts. During scheme transformation, these annotations should be tracked. So, the semantics of elements collected in the ontological level are not lost. Semantic annotation rules virtually define subconcepts of the domain ontology even for a single instance, which is an annotated element. It allows not only declaring instances of ontological concepts but using expressions in terms of the ontology. For example, the parallax of a pair in the scheme is annotated by a rule defining a concept *PairParallax* using a parallax of its primary component in terms of the ontology (here and lower RIF-FLD [6] framework features are used):

```
Forall ?x (
  ?x#PairParallax
  :-
  Exists ?y ?z (
    And (?y#Pair ?z#Component
      ?y[hasPrimaryComponent->?z]
      ?z[parallax->?x])))
```

Used data source scheme elements are annotated too. Based on formal inference over semantic annotations, relevant elements of the source schemes and the conceptual scheme of the domain can be found. At this stage, conflict resolution rules can be developed and used to map data source schemes into the conceptual scheme [21].

An execution environment over conceptual scheme specifications using the mappings of data sources and a query language or a workflow system over the conceptual scheme support implementation of the scheme. It may be a materialized repositories or virtual mediators. Some fragments of the conceptual scheme can be implemented as a new data source or as a program or a workflow over the scheme.

Requirements are specified formally as functions with pre- and post-conditions. For example, requirements for physical visual pair selection are:

- orbital movement is recognizable, and/or
- a visual pair is spectroscopic, eclipsing or orbital, and/or
- parallaxes of components in the pair are similar.

The last requirement is specified formally in terms of the conceptual scheme below. The rule defines that *isPhysical* is true if parallaxes of components are similar:

```
Forall ?p ?c1 ?c2 ?x1 ?x2 (
  And ( ?p#Pair ?c1#Component ?c2#Component
    ?p[observationalType->Visual
        primaryComponent->?c1
        secondaryComponent->?c2]
    ?c1[parallax->?x1] ?c2[parallax->?x2]
    componentParallaxesAreSimilar(?x1 ?x2))
    :-
    ?p[isPhysical->true])
```

The *componentParallaxesAreSimilar* predicate is not detailed to shorten the specification, it should take into account errors of parallax measurements. Is the rule, ":-" is implication, terms with "?" are variables, "#" is an instancing relation, "[...]" defines a frame as a set of pairing attributes and values. "Forall", "And" are special terms.

Another kind of requirements is identification criteria for entity resolution and matching. They use two variables of objects of the same type. Criteria for physical visual pair identification are developed from the detailed model of requirements:

- limiting difference of position angles changed in time;
- similarity of angular separations, for closer pairs angular separation similarity is ignored since orbital motion can be very fast;
- similarity of proper motion of pairs
- similarity of parallaxes of pairs.

The last criterion is specified by *pairParallaxesAreSimilar* predicate definition in terms of the conceptual scheme. The criterion is false if parallaxes ($?x1$, $?x2$) of both pairs exist and differ too much. The rule uses reverse implications to define equality condition. The ε symbol is used instead of taking into account measurement errors of parallaxes to shorten the specification.

```
Forall ?p1 ?p2 (
  And (?p1#Pair ?p2#Pair
    Exists ?x1 ?x2 (
      And (?p1[parallax->x1] ?p1[parallax->x1]
        pred:numeric-greater-than(fn:abs(?x1 - ?x2) ε)
        :-
        pairParallaxesAreSimilar(x1 x2) = fn:false)
    Exists ?x1 ?x2 (
      And (?p1[parallax->x1] ?p1[parallax->x1]
        pairParallaxesAreSimilar(x1 x2) = fn:false
        :-
        pred:numeric-greater-than(fn:abs(?x1 - ?x2) ε))))
```

The following rule specifies *parallaxCriterion* function that reduces a set of candidates *?l1* to identification with the pair *?p* by the parallax similarity criterion.

```
Forall ?p ?l1 ?l2 ?i (
  And (?p#Pair
  ?i#Pair pred:list-contains(?l1 ?i)
  func:intersect(?l1 ?l2) = ?l2
  Or (And (pred:list-contains(?l2 ?i)
      pairParallaxesAreSimilar(?p ?i))
     (And Neg(pred:list-contains(?l2 ?i)
      pairParallaxesAreSimilar(?p ?i) = fn:false))
  :-
  parallaxCriterion(?p ?l1) = ?l2))
```

The rules above are part of conceptual scheme specification. Their implementations can be used for data integration and domain problem solving process specification.

7 Discussion

There is an implementation of the same problem of multiple stellar system identification in a programming language without using semantic technologies. This solves the problem too, but comparing approaches can be useful. The development of software to solve this problem manually required the following actions:

- development of internal data structures for representation of data on stellar systems, components, and pairs in the project;
- implementation of access to data sources, parsing data on binary stars of different observation types, and transformation of data into the internal representations;
- development of an algorithm to go through the data and identify all pairs of stars from different sources;

- implementation of matching pairs by similarity of coordinates and relative positions components;
- development of the result structure and creation of a new catalog with stellar system identification results.

However, to do these works, all the same processes that have been presented in this paper had to be implicitly and mentally done by a developer during the problem analysis and the solution implementation. In textual specifications of the problem, it was necessary to identify necessary objects, to understand their essential properties and relations in order to develop data structures for internal representation of data in the project. Then they had to understand structures of data source catalogs and how elements of catalogs semantically correspond to the internal representation of data in order to correctly transform data. Analysis of the problem was necessary to develop identification criteria for different types of objects and an approach to correct application of those criteria. To simplify the implementation of criteria, significant knowledge of domain objects were not taken into account.

After solving the problem by means of programming there are no explicitly expressed shared formal specifications for reasoning certain decisions. Therefore, it is impossible to control correctness of data transformation and entity identification. Changing a set of data sources requires changes in programs. It is difficult to solve similar problems in the domain with the same means, The implementation is not designed for reuse of the developed tools and for integration to problems solved in a research community of the domain. After all, the lack of formal specifications eliminates possibility of automated operations with developed tools.

8 Conclusion

We described the design process of domain specifications necessary for community collaboration in research data infrastructures for domain problem solving with reuse of heterogeneous information resources. The proposed approach includes problem requirement model building, domain ontology formalization, conceptual scheme generation, operationalization of requirements in terms of conceptual scheme, mapping data sources into the conceptual scheme and entity resolution for data from multiple sources. The approach is distinguished by a focus on identification and preservation of semantics of the domain in the developed conceptual scheme, and on domain modeling for interoperability and problemsolving.

Acknowledgments. The work has been supported by the Russian Foundation of Fundamental Research (grants 18-07-01434, 18-29-22096, 19-07-01198).

References

1. IEEE Standard Glossary of Software Engineering Terminology. IEEE Std 610.12-1990
2. OWL 2 Web Ontology Language Document Overview, Second edn. W3C (2012). http://www.w3.org/TR/owl-overview/

3. About the Unified Modeling Language Specification Version 2.0. OMG (2011). https://www.omg.org/spec/UML/2.0/
4. Abrial, J.-R.: The B-Book. Cambridge University Press, Cambridge (1996)
5. Boley, H., Kifer, M. (eds.): RIF Framework for Logic Dialects. W3C Recommendation, 2nd edn. W3C (2013)
6. Boley, H., Kifer, M. (eds.): RIF Basic Logic Dialect. W3C Recommendation, 2nd edn. W3C (2013)
7. Brodie, M.L., Mylopoulos J., Schmidt, J.W. (eds.): On Conceptual Modelling: Perspectives from Artificial Intelligence, Databases, and Programming Language. Springer, New York (1984). https://doi.org/10.1007/978-1-4612-5196-5. ISBN 978-1-4612-9732-1
8. Conesa, J., Olivé, A.: Pruning ontologies in the development of conceptual schemas of information systems. In: Atzeni, P., Chu, W., Lu, H., Zhou, S., Ling, T.W. (eds.) ER 2004. LNCS, vol. 3288, pp. 122–135. Springer, Heidelberg (2004). https://doi.org/10.1007/978-3-540-30464-7_11
9. De Nicola, A., Missikoff, M.: A lightweight methodology for rapid ontology engineering. Commun. ACM **59**(3), 79–86 (2016). https://doi.org/10.1145/2818359
10. Fernández-López, M., Gómez-Pérez, A.: Overview and analysis of methodologies for building ontologies. J. Knowl. Eng. Rev. **17**(2), 129–156 (2002)
11. Fonseca, F., Martin, J.: Learning the differences between ontologies and conceptual schemas through ontology-driven information systems. J. Assoc. Inf. Syst. – Spec. Issue Ontol. Context IS **8**(2), Article 3 (2007). http://www.personal.psu.edu/faculty/f/u/fuf1/publications/Fonseca_Martin_Ontologies_and_Schemas_Abstract.pdf
12. Helbig, H., Gnörlich, C.: Multilayered extended semantic networks as a language for meaning representation in NLP systems. In: Gelbukh, A. (ed.) CICLing 2002. LNCS, vol. 2276, pp. 69–85. Springer, Heidelberg (2002). https://doi.org/10.1007/3-540-45715-1_6
13. Jiang, L., Topaloglou T., Borgida, A., Mylopoulos, J.: Goal-oriented conceptual database design. In: Conference on Requirements Engineering (RE 2007), Delhi (2007)
14. Kalinichenko, L.A., Stupnikov, S.A., Martynov, D.O.: SYNTHESIS: a Language for Canonical Information Modeling and Mediator Definition for Problem Solving in Heterogeneous Information Resource Environments, 171 p. IPI RAN, Moscow (2007)
15. La-Ongsri, S., Roddick, J.F.: Incorporating ontology-based semantics into conceptual modelling. Inf. Syst. **52**, 1–20 (2015). https://doi.org/10.1016/j.is.2015.02.003
16. Lapouchnian, A., Yu, Y., Mylopoulos, J.: Requirements-driven design and configuration management of business processes. In: Alonso, G., Dadam, P., Rosemann, M. (eds.) BPM 2007. LNCS, vol. 4714, pp. 246–261. Springer, Heidelberg (2007). https://doi.org/10.1007/978-3-540-75183-0_18
17. McLeod, D., Smith, J.M. (eds.): Abstraction in databases. In: Proceedings of the Workshop on Data Abstraction, Databases and Conceptual Modeling, SIGMOD Record, vol. 11, no. 2 (1981)
18. Sebastiani, R., Giorgini, P., Mylopoulos, J.: Simple and minimum-cost satisfiability for goal models. In: Persson, A., Stirna, J. (eds.) CAiSE 2004. LNCS, vol. 3084, pp. 20–35. Springer, Heidelberg (2004). https://doi.org/10.1007/978-3-540-25975-6_4
19. Serna, E.M., Serna, A.A., Bachiller, O.S.: A framework for knowledge management in requirements engineering. Int. J. Knowl. Manag. Stud. **9**(1), 31–50 (2018). https://doi.org/10.1504/IJKMS.2018.089694
20. Spyns, P., Meersman, R., Jarrar, M.: Data modelling versus ontology engineering. SIGMOD Rec. **31**(4), 12–17 (2002)

21. Stupnikov, S.: Rule-based specification and implementation of multimodel data integration. In: Kalinichenko, L., Manolopoulos, Y., Malkov, O., Skvortsov, N., Stupnikov, S., Sukhomlin, V. (eds.) DAMDID/RCDL 2017. CCIS, vol. 822, pp. 198–212. Springer, Cham (2018). https://doi.org/10.1007/978-3-319-96553-6_15

22. Suárez-Figueroa, M.C., Gómez-Pérez, A., Fernandez-Lopez, M.: The NeOn methodology framework: a scenario-based methodology for ontology development. Appl. Ontol. 10(2), 107–145 (2015). https://doi.org/10.3233/AO-150145

23. Sugumaran, V., Storey, V.: The role of domain ontologies in database design – an ontology management and conceptual modeling environment. ACM Trans. Database Syst. 31(3), 1064–1094 (2006)

24. Smith, J.M., Smith, D.C.P.: Database abstraction: aggregation and generalization. ACM TODS 2(2), 105–133 (1977)

25. Sundgren, B.: An infological approach to data bases. Urval, N7. National Central Bureau of Statistics, Stockholm, Sweden (1973)

26. Vovchenko, A.E., et al.: From specifications of requirements to conceptual schema. In: Proceedings of the 12th Russian Conference on Digital Libraries RCDL 2010, pp. 375–381. Kazan Federal University, Kazan (2010). (in Russian)

ADBIS 2019 Workshop: Modern Approaches in Data Engineering and Information System Design – MADEISD

Customer Value Prediction in Direct Marketing Using Hybrid Support Vector Machine Rule Extraction Method

Suncica Rogic$^{(\boxtimes)}$ and Ljiljana Kascelan

University of Montenegro, 81000 Podgorica, Montenegro
suncica@ucg.ac.me

Abstract. Data mining techniques can aid companies in evaluation of customers that generate highest amount of revenue in a direct marketing campaign. Most commonly, customer value is evaluated by a uniform segmentation of customers (20% for each segment) based on buying behavior using recency, frequency and monetary (RFM) attributes, whereby for direct campaigns the segments with the highest score of these attributes are subjectively selected. In this paper, the method of k-means clustering, according to RFM attributes is proposed, based on which the customer value can be more objectively determined. The most valuable customers, as a rule, are the smallest group compared to other clusters, so the problem of class imbalance occurs. In order to overcome this problem, a hybrid Support Vector Machine Rule Extraction (SVM-RE) method is proposed for predicting which customer belongs to a cluster, based on data on consumer characteristics and offered products. The SVM classifier is known as a good predictor in case of class imbalance, but does not generate an interpretable model. Therefore, the Decision Tree (DT) method generates rules, based on the prediction result of the SVM classifier. The results of the empirical case study showed, that using this hybrid method with good classification performance, customer value level can be predicted, i.e. targeting existing and new buyers for direct marketing campaigns can be efficiently done, regardless of the class imbalance problem. It's also shown that using the hybrid SVM-RE method, it is possible to obtain significantly better prediction accuracy than using the DT method.

Keywords: Direct marketing · Customer classification · Class imbalance · SVM rule extraction

1 Introduction

Direct marketing is a process of promoting and selling products, where promotional materials and incentives are sent to individual customers via social networks, or directly, by e-mail, post, phone call and the like. The main goal of direct marketing is identification, i.e., targeting customers from an existing base (but also new ones) that will most likely respond to a particular marketing campaign, which increases revenue and reduces the cost of the campaign. Methods for customer targeting can be divided into two groups: segmentation and scoring methods [1, 2]. Segmentation methods

© Springer Nature Switzerland AG 2019
T. Welzer et al. (Eds.): ADBIS 2019, CCIS 1064, pp. 283–294, 2019.
https://doi.org/10.1007/978-3-030-30278-8_30

divide customers into groups (segments), using appropriate explanatory variables, so that segment members are as homogeneous as possible, regarding the expected response to a direct marketing campaign [3, 4]. The offer is sent to customers from the segments that have the highest likelihood of response. In the case of customer response models [5, 6], each customer receives an appropriate score based on the predicted probability of the response to the offer. However, the high likelihood of response does not necessarily mean a high profit. The most important scoring methods include methods for predicting customer profitability [7–10]. Many authors, as the most important criterion for targeting, emphasize the customer's recency [4, 11]. In this sense, it would be best to predict customer value by including all three dimensions – recency, frequency of the response, and profitability. In this paper, a classification method for customer segmentation is proposed. Recency, frequency, and profitability are used to segment the customers in order to determine the degree of their value for the company. Then, with predictive classification, based on the characteristics of customers and products that are being purchased, the customer value level (CV-level) is predicted.

Segmentation methods, which are most commonly applied, split a customer data set using RFM attributes. They are based on various techniques, ranging from the simplest cross-tabulation technique, to more complex weighted techniques [4, 12]. These techniques generally require a subjective assessment for the necessary parameters. For this reason, data mining methods, such as K-means or Artificial Neural Network (ANN) clustering, can give more objective results for RFM customer segmentation [13–15].

Since the most valuable customer cluster is usually the smallest, there is a problem of class imbalance. This problem in most predictive classification methods leads to bias toward small classes and most often to their misclassification [16, 17]. The previous literature confirms that in case of class imbalance, the SVM method has the best predictive classification performance and can be used as a pre-processor that balances classes for other classifiers [18]. However, the SVM classifier is a "black-box", i.e. does not generate a model that can be interpreted. This deficiency can be solved by a hybrid approach, where the SVM is combined with rules extracting techniques [19]. In order to solve the subjectivity problems of RFM segmentation and class imbalance, in this study k-means clustering and the SVM-RE method were used for customer value prediction. The ultimate goal is to generate predictive classification rules for different CV-levels, which can result in more efficient customer selection and lower costs of direct marketing campaigns.

The paper is organized as follows: The second section gives an overview of related papers. Section three shows the proposed methodology, and the fourth section presents the empirical test of the case study and discusses the results, which is followed by concluding observations.

2   Related Papers

This section provides an overview of previous research related to the proposed SVM-RE method, as well as the previous papers dealing with customer value analysis in direct marketing on which our research is based.

2.1   SVM Rule Extraction Method

For linearly inseparable classes, Vapnik [20] proposed a SVM method that maps data (viewed as n-dimensional vectors) from the original space into a space larger by one dimension (n + 1-dimensional space), where the classes can be separated by means of a hyperplane. Finding such a hyperplane is realized by minimizing the distance between its end position (so that the gap between the classes i.e. the margin is greater) and the closest points (support vectors). Instead of an explicit mapping function in a larger dimension space, a kernel function is used, which allows calculating the scalar product of the vectors (i.e. the distance of the support vector from the hyperplane) in the original space (kernel trick). Various kernel functions can be used, but Radial Basis Function (RBF) is applied most often [21]:

$$K(x_i, x_j) = exp(-\gamma \parallel x_i - x_j \parallel^2) \tag{1}$$

The training of the SVM classifier comes to the selection of the optimized values of the gamma parameter for the RBF kernel, and parameter C, which represents the boundary for the margin, i.e. empty space between classes. Selecting lower values for parameter C reduces over-fitting and increases the generality of the SVM model, i.e. its predictive performance.

In addition to solving the problem of linear inseparability of classes, the advantage of this method is that, in the case of class imbalance, it exhibits better predictive performance than the standard methods, such as logistic regression [18]. The literature confirms that the SVM can successfully remove the noise, i.e., class overlapping from data. Namely, the parameter C can be set so that a number of examples of a larger class, which are close to the example of a lower class (which means that they are similar), are declared as examples of the lower class. For this reason, the SVM can be used as a pre-processor that balances and purifies data, thus providing higher classification accuracy [18, 22]. However, the SVM does not generate an interpretable model, which is usually very important in application. This problem has been solved in the literature by means of rule extraction techniques that enable generating the rules from the SVM results [19, 23]. According to Barakat and Bradley [19], SVM-RE techniques are grouped into two categories: those based on the components of the SVM model, and those that do not use the internal structure of the SVM model, but draw the rules from the SVM output. When the SVM model is interpreted, or SVM is used as a data pre-processor, authors recommend techniques from the second group because they provide more under-standable rules. In line with this recommendation, our research uses rule extraction from SVM output. Namely, customer targeting rules are derived from SVM output using a classification DT method [24].

The DT method divides the data set by attributes values, so that subgroups contain as many examples of one class. The criterion by which division is made (measure of quality of division) can be information gain [25], gain index [26], gini index [27], or accuracy of the whole tree. The attributes that provide the best division according to the given criterion are chosen. During the inductive division, a tree-shaped model is formed. The path from the root to the leaves defines if-then classification rules in the terms of the predictive attributes (tree nodes). The complexity and accuracy of the generated model depends on the depth of the tree, the minimum size of the node by which the division can be made (the number of examples in its subgroup), the leaf size, and the defined minimum gain achieved by the node division. The smaller depth, the larger the minimum size for the split, the larger the leaf size and the higher minimum gain, lead to less complex tree, but also a tree with smaller accuracy.

SVM rule extraction is not a new method in literature and it was applied in some previous economic studies [22, 24, 28], but for the topic of direct marketing, i.e., to solve the problem of the minor class of the most valuable customers in CV-value prediction, it is applied for the first time in our research.

2.2   Customer Value Analysis in Direct Marketing

One of the most commonly applied customer value analyzes in direct marketing is the RFM analysis, defined by Hughes [29] in 1994. The RFM model is based on the behavior of consumers, recorded in the database. Recency is the length of the time period since the last purchase; frequency indicates the number of purchases made in the specified period, while monetary defines the total value of customer's transactions during that period [30]. RFM analysis begins by sorting data on customer transactions according to recency - the database is divided into five equal parts (of 20% of customers), taking into account the Pareto principle ("80% of sales come from 20% of clients" [31]). Then, 20% of customers who have bought the product most recently get score 5, the next 20% get score 4, etc. The next step involves sorting consumers within all quintiles according to frequency, and as in the first step, a score of 5 to 1 is assigned. After the second step, the database is divided into 25 groups, so in the last step, each of them is divided into five parts according to the monetary indicator, which will ultimately result in a database divided into 125 groups according to RFM values [4]. Finally, the best consumer segment will have a score of 555, while the worst will have a score of 111. Which segments will be targeted in the direct campaign will be determined subjectively. Based on RFM results, consumers are grouped into segments that can be further analyzed by customer characteristics, product specific variables or by profitability, most commonly using data mining techniques [32–34].

In their research, Cheng and Chen [34] used k-means clustering [35] for segmenting customers by scaled RFM attributes, using a uniform scaling approach (by dividing data into segments of 20% each). Since k-means clustering operates with numerical attributes, our study suggests that only the attribute R (i.e. dates) should be scaled in this way because the F and M attributes are already numerical. This avoids the loss of important information and subjectivity in assessing whether the highest ratings for F and M should be assigned to the first 20% or more/less customers. Further, they tested the approach by creating 3, 5, and 7 clusters, while our study suggests the estimate of the centroid cluster model based on the Davies-Bouldin Index (DB) [36], which guarantees maximum homogeneity within the cluster and maximum heterogeneity between the clusters. The DB index measures the Euclidean distance from the centroid inside and between the clusters. At the level of each cluster, a maximum coefficient of scattering within the cluster is taken and a measure of separation from other clusters. Then the average for all clusters is calculated. The lower absolute values of the DB index mean better clustering quality.

Cheng and Chen [34] use a rough set and LEM2 rule extraction method to generate a set of explicit rules that can be used to target customers using their characteristics (the region and credit debt, in this case). In order to achieve a high accuracy rate, the predictive attributes also use RFM attributes that have the greatest impact on the classification, because clusters are formed on this basis. Therefore, the rules may not show some customer characteristics that are very important in targeting (they can be absorbed by the effect of RFM attributes). In our study, the characteristics of customers and product data are used as predictive attributes, which can provide predictive rules with more useful information for customer targeting [33].

To assess classification performance of rough set LEM2 methods, Cheng and Chen [34] used the accuracy rate exclusively (the percentage of precisely predicted examples within all examples). Clusters do not contain the same number of customers, as there is usually the smallest number of customers with the highest CV-level. Thus, with the predictive classification, there is a problem of class imbalance, which can lead to a low class precision (the percentage of precisely predicted examples within a predicted class) and/or class recall (the percentage of accurately classified examples within the actual class) for the smallest class, which is the most important for this issue. As it was stated in the previous section, the SVM method successfully solves the problem of class imbalance and is suggested in our study as suitable for such prediction. Since the SVM does not generate rules necessary for group customer targeting, a hybrid SVM-DT rule extraction method is proposed that addresses the non-interpretability problem.

3   Methodology

3.1   Conceptual Framework for Customer Value Prediction

In this study, a model for customer value prediction in direct marketing is proposed. First step is collection of data on purchasing transactions from direct campaigns, which can include customer data such as (gender, age, region, wealth, etc.), product data such as (type, category, purpose, etc.), and purchasing behavior data - RFM attributes. RFM attributes are defined as follows: R-date of the last order, F-total number of orders in the considered period and M- monetary amount spent in the considered period. The R attribute is encoded so that for 20% of the most recent dates, score 5 is assigned, the next 20% less recent dates are given score 4 and so on until score 1. Attributes F and M are retained in their original form. In the end, all three attributes were normalized with 0–1 range transformation.

Using normalized RFM attributes, by means of k-means clustering, customers are divided into clusters (cluster members have similar purchasing behavior), and then the appropriate CV-level is assigned to the customers of individual clusters.

The CV-level of the customer (i.e., belonging to the appropriate cluster) is then predicted using the classification method. Therefore, it is possible to predict customer's CV-level, if the information about the customer and products offered in the campaign is available. In the case of customer value classification, there is a problem of the smallest and most important class (consists of most valuable customers, which is, as a rule, the smallest). For direct marketing purposes, it is useful to discover rules that describe clusters of customers with a higher CV-level in terms of customer data and preferred products. These rules can be used for group customer targeting as well as targeting new potential customers. It is therefore of great importance that the predictive classification method is interpretable.

In accordance with set conceptual model, this study will test the following hypothesis:

1. Using the hybrid SVM-RE method, CV-level can be predicted (i.e. whether the customer will be targeted or not), with high accuracy rate, class precision and class recall, regardless of class imbalance.
2. The SVM method pre-processing of data on purchase transactions (by removing noise i.e. overlapping classes) improves the classification performance of the DT method
3. Pre-processed DT validly interprets the SVM model i.e. generates rules from the SVM prediction results with a high fidelity.

3.2   Predictive Classification Procedure

For assessment of predictive classification performance, accuracy rate, class precision and class recall are used (these indicators are explained at the end of Sect. 2.2), obtained by k-fold cross-validation with stratified sampling. In addition to generating predictive rules with high accuracy, for customer targeting it is important to classify existing consumers more accurately, so class recall is an important indicator of model performance.

The k-fold cross-validation procedure with stratified sampling implies that the starting data set is split into subgroups, taking care that percentage of class representation in subgroups corresponds to percentages of class representation in the entire set of data. Then k-1 subsets are used for training the model (training set), while one of the subsets is used for validation, i.e., testing how this model works on an unknown set of data (test set). The procedure is repeated k times, so that each of the subsets is a test set. At each iteration, the parameters for classification (accuracy rate, class precision, and class recall), are calculated and finally their average value is found.

The procedure of predictive classification of customers by CV-level consists of the following steps:

1. Data preparation (calculation of RFM attributes).
2. Segmentation of customers, i.e. transactions, by normalized RFM attributes using k-means clustering with the evaluation of the centroid cluster model based on the DB index.
3. Cluster description via RFM attributes and assignment of CV-level to consumers, based on belonging to the appropriate cluster (the CV-level class label is added to the starting set of data). For example, customers belonging to the cluster with the highest values of the RFM attributes are assigned CV-level = 1, those belonging to the cluster with lower values of these attributes are assigned CV-level = 2, etc.
4. Generating a DT model for predicting CV-level based on purchase transactions attributes. This step involves finding optimal parameters for the DT model (the criterion for split evaluation, the minimum size for split, the minimum leaf size, the

minimum gain, the maximum depth) in the k-fold cross-validation procedure, so that the highest accuracy of the prediction is obtained.

5. Generating a SVM model for predicting CV-level based on purchase transactions attributes. The training of SVM requires that an optimal combination of a gamma parameter from the RBF kernel function (1) and margin C to be determined, which will give the highest average classification accuracy during k-fold cross-validation. To select the optimal combination of parameters, Grid-Search approach is used.

6. Generating a class label, based on the SVM prediction (the SVM class label is added to the data set). At this step, customers are assigned a CV-level predicted by the SVM classifier.

7. Generating a DT model for predicting CV-level based on purchase transactions attributes and SVM class label. The DT generated at this step is an SVM model interpreter i.e. it performs rule extraction from the SVM model. Similar to the fourth step, optimal DT parameters for maximum accuracy are selected based on the k-fold cross-validation. Compared with the DT model from point 4, this DT model should have a significantly higher predictive accuracy.

8. Generating if-then rules from the DT model obtained in the previous step. The consistency of the derived rules depends on the fidelity (a percentage of the examples for which the class label predicted by the DT matches the SVM class label). These rules explicitly indicate the purchasing transactions attributes that predict the appropriate CV-level, and, therefore, can help to make the new customers more easily targeted.

A defined predictive procedure was implemented using Rapid Miner.

4 Empirical Testing and Discussion of Results

For the empirical testing of the proposed procedure, a data set of on-line purchasing transactions from direct campaigns of Sport Vision Montenegro (part of the Sport Vision system - leading sport retailer in the Balkans) was used, for the period from the beginning of September 2018 to the end of January 2019. The data was prepared by calculating and normalizing the RFM attributes according to the procedure proposed in the previous section (step 1).

By clustering the starting data set using k-means method and normalized RFM attributes, following results are obtained - shown in Table 1 (step 2). It can be seen that the best DB index (minimum absolute value) is achieved for a 3-cluster model. This cluster model is shown in Table 2.

Table 1. Selection of number of clusters (parameter k) for k-means clustering

K	2	3	4	5	6	7	8	9	10
DB	−1.025	**−0.811**	−0.983	−0.958	−0.909	−1.02	−0.98	−0.976	−0.96

Table 2. Centroid cluster model for RFM segmentation of customers

	R	F	M	Items
cluster_0	0.766807	0.565126	0.677733	238
cluster_1	0.735348	0.088065	0.179499	819
cluster_2	0.137318	0.101734	0.211345	548

Note: RFM attributes are normalized (0–1 range transformation)

From Table 2, it can be seen that cluster_0 consists of the most recent, most frequent and most profitable customers (CV-level = 1), cluster_1 consists of recent, but less frequent and less profitable customers (CV-level = 2), while cluster_2 is made of non-recent customers, that are less frequent and less profitable (CV-level = 3). The most valuable customer cluster contains significantly less items than the other two clusters (238 versus 819 and 548), so the problem of class imbalance is evident. Hence, step 3 has been implemented.

Using the optimization of parameters gained through Grid-Search method and 10-fold cross-validation, for predicting CV-level, the DT model (optimal parameters: criterion = gini_index, min_size_for_split = 5, min_leaf_size = 6, max_depth = 11, min_gain = 0.1) was generated (step 4). Then, for the same purpose, the SVM model is trained using the Grid-Search method and 10-fold cross-validation (optimal parameters: gamma = 400.0, C = 400.0) (step 5). Predicted values for CV-level are added to the starting data set as a new class label (SVM CV-level) (step 6). Then, a DT model with such class label was generated (pre-processed DT), using the Grid-Search parameter selection and a 10-fold cross-validation (optimal parameters: criterion = accuracy, min_size_for_split = 7, min_leaf_size = 2, max_depth = 15, min_gain = 0.3) (step 7). The classification performance of these models are shown in Table 3.

Table 3. Results of testing of the predictive classification procedure

DT	SVM	Pre-processed DT (SVM-RE)
Accuracy: **61.27% +/- 2.58% (mikro: 61.28%)**	Accuracy: **61.31% +/- 3.28% (mikro: 61.31%)**	Accuracy: **85.73% +/- 2.52% (mikro: 85.75%)**
Class recall: **4.27%**, 80.00%, 57.77%	Class recall: **19.75%**, 81.44%, 49.27%	Class recall: **62.70%**, 93.07%, 77.53%
Class precision: **33.33%**, 62.69%, 60.08%	Class precision: **51.65%**, 61.42%, 63.08%	Class precision: **77.45%**,86.27%, 86.46%

Note: The parameters we have selected to evaluate predictive performance of the models in Sect. 3.2 are displayed

From the Table 3, it can be noticed that the SVM-RE method has significantly better classification performance than the DT method. The DT method correctly targeted only 4% of the most valuable customers, while SVM-RE successfully targeted 63% of them. This practically means that, among existing customers only 9 out of 238 most valuable customers will be identified, which are highly likely to respond to the campaign. SVM-RE will target 150 of the 238 most valuable customers for the campaign and increase the potential revenue from the campaign by 16.7 times, compared to DT targeting.

Also, all considered classification performances are better with the SVM-RE model than with DT. The class precision of the most valuable customers for DT is only 33%, which means that the company will have unnecessary campaign costs for 67% of wrongly classified customers. Precision of SVM-RE model for the class is 77%, which means that only 23% of the offers sent are likely to be unanswered. Therefore, SVM-RE will, in relation to DT, reduce the cost of the campaign. It can be concluded that, with the high accuracy of CV-level prediction (86%), the proposed SVM-RE method managed to solve the problem of class imbalance, thus confirming the first hypothesis.

The results show that SVM as the pre-processor of data on purchase transactions eliminated noise, so that more precise classification is possible (the DT classification accuracy is increased by 25% - the accuracy for the initial DT is 61% and for pre-processed DT 86%; mean class recall for DT is 47%, and after SVM data preprocessing it is 78%; mean class precision has increased from 52% to 77% after SVM preprocessing), which confirmed the second hypothesis.

Given the high fidelity (cross-validation accuracy extraction rules from the SVM prediction amount to 86%), the rules performed by pre-processed DT validly interpret the SVM model, thus confirming the third hypothesis. Table 4 shows some of the 39 derived primary rules (the rules that cover a large number of examples are shown).

On the basis of derived rules, it can be noticed that customers with CV-level = 1 are male buyers who mainly buy: lifestyle clothes for adult men, from licensed brands (brands for which Sport Vision has licensed production and distribution, such as: Champion, Umbro, Lonsdale, Ellesse, Slazenger, Sergio Tacchini, etc.) and with a discount of 25% to 45%; clothing and trainers for basketball, for teenage boys, from licensed brands; as well as: sports equipment for women, from licensed brands.

Customers of CV-level = 2 mainly buy clothes and trainers from A brands (brands for which the company is a distributor, such as: Adidas, Nike, Under Armor, Reebok, Converse etc.) for adult men with a discount of less than 35%, or lifestyle clothing and trainers for licensed brands, for adult men with a discount of 35%–45%.

The least valuable customers (CV-level = 3), mostly buy trainers for adults from A brands, with a high discount (>45%). In a direct campaign, customers from the appropriate cluster can be offered products that are identified by the rules. Also, based on the characteristics of the buyer and the products offered, CV-level can be predicted for that customer, with a probability of about 86% and, therefore, it would be known if they are more or less likely to respond to the campaign.

Table 4. Some of the predictive rules derived from pre-processed DT

Rule	Prod_gend	Prod_cat	Prod_age	Prod_brand	Prod_type	Gender	Discount	CV-level	Confid
R2	For boys	Lifestyle	For teens (8–14)	Licence	Apparel	M	**	1 (9/0/1)*	90%
R3	For boys	Basket	For teens (8–14)	Licence	Footwear			1 (6/0/0)	100%
R4	For men		For adults	A brands	Apparel	M	35%–45%	1 (7/2/0)	77.78%
R5	For men	Lifestyle	For adults	Licence	Apparel	M	25%–45%	1 (32/0/0)	100%
R6	For women	Lifestyle	For adults	Licence	Equipment	M		1 (6/0/0)	100%
R9	For men		For adults	A brands	Apparel	M	25%–35%	2 (3/80/1)	96.39%
R12	For men		For adults	A brands	Footwear	M	<35%	2 (0/66/3)	95.65%
R14	For men	Lifestyle	For adults	Licence	Apparel		35%–45%	2 (0/63/0)	100%
R15	For men	Lifestyle	For adults	Licence	Footwear		25%–45%	2 (0/51/0)	100%
R31	For men	Lifestyle	For adults	A brands	Footwear	M	>45%	3 (0/0/43)	100%
R36	For women	Lifestyle	For adults	A brands	Footwear		>45%	3 (0/0/32)	100%

* The R2 rule accurately classifies 9 most valuable customers, none of the most valuable customers is misclassified in medium-valuable customers, while 1 of the most valuable customers is misclassified in the least valuable customers, so the confidence of this rule is 90%.

** An empty field in the table means that this attribute does not appear in a rule (DT induction selects the attributes that give the purest division in relation to the target classes, which means that not all attributes must appear in the generated tree, i.e., rules).

5 Conclusions

In this paper, an efficient method for customer value prediction in direct marketing is proposed. The predictive procedure involves the classification of customer clustering. Namely, customers are clustered (using the k-means algorithm) based on their purchasing behavior (more precisely, using RFM attributes). Customers belonging to different clusters have a higher or lower customer value level and, hence, a greater or lesser likelihood of responding to a direct campaign. Then, using the SVM-RE method, based on the characteristics of the customer and the data on the products they purchased, consumer's affiliation with one of the clusters, i.e., consumer's appropriate CV-level is predicted. On this basis, it can be decided whether this customer will be targeted for the campaign or not. In addition, the SVM-RE method extracts the classification rules that can target new customers in the campaign and offer the appropriate products.

Empirical testing has shown that the proposed method successfully solved the problem of class imbalance that often leads to misclassification of the smallest class (most valuable customers). By increasing the class recall and class precision for the most valuable customer class, the SVM-RE method can significantly increase revenue and reduce the cost of a direct campaign. It has also been shown that the SVM method can be used as a data preprocessor that successfully solves the problem of class overlapping and improves the classification performance.

In future research, this method can be tested on other data sets to verify or improve its efficiency (by including more customer attributes, clearer rules for targeting new customers can be obtained).

References

1. Jonker, J., Piersma, N., Van den Poel, D.: Joint optimization of customer segmentation and marketing policy to maximize long-term profitability. Expert Syst. Appl. **27**, 159–168 (2004)
2. Kaymak, U.: Fuzzy target selection using RFM variables. In: Proceedings Joint 9th IFSA World Congress and 20th NAFIPS International Conference (Cat. No. 01TH8569)
3. Hughes, A.: Strategic Database Marketing. McGraw-Hill, New York (2005)
4. McCarty, J., Hastak, M.: Segmentation approaches in data-mining: a comparison of RFM, CHAID, and logistic regression. J. Bus. Res. **60**, 656–662 (2007)
5. Olson, D., Cao, Q., Gu, C., Lee, D.: Comparison of customer response models. Serv. Bus. **3**, 117–130 (2009)
6. Olson, D., Chae, B.: Direct marketing decision support through predictive customer response modeling. Decis. Support Syst. **54**, 443–451 (2012)
7. Cui, G., Wong, M., Wan, X.: Targeting high value customers while under resource constraint: partial order constrained optimization with genetic algorithm. J. Interact. Market. **29**, 27–37 (2015)
8. Kim, D., Lee, H., Cho, S.: Response modeling with support vector regression. Expert Syst. Appl. **34**, 1102–1108 (2008)
9. Otter, P.W., Scheer, H.V.D., Wansbeek, T.: Optimal selection of households for direct marketing by joint modeling of the probability and quantity of response. s.n. University of Groningen, CCSO Centre for Economic Research, Working Papers (2006)
10. Malthouse, E.: Ridge regression and direct marketing scoring models. J. Interact. Market. **13**, 10–23 (1999)
11. Wu, J., Lin, Z.: Research on customer segmentation model by clustering. In: Proceedings of the 7th International Conference on Electronic Commerce - ICEC 2005 (2005)
12. Drozdenki, R., Drake, P.: Optimal Database Marketing. Sage Publications, Thousand Oaks (2002)
13. Hosseini, S., Maleki, A., Gholamian, M.: Cluster analysis using data mining approach to develop CRM methodology to assess the customer loyalty. Expert Syst. Appl. **37**, 5259–5264 (2010)
14. Sarvari, P., Ustundag, A., Takci, H.: Performance evaluation of different customer segmentation approaches based on RFM and demographics analysis. Kybernetes **45**, 1129–1157 (2016)
15. Khalili-Damghani, K., Abdi, F., Abolmakarem, S.: Hybrid soft computing approach based on clustering, rule mining, and decision tree analysis for customer segmentation problem: real case of customer-centric industries. Appl. Soft Comput. **73**, 816–828 (2018)
16. Kim, G., Chae, B., Olson, D.: A support vector machine (SVM) approach to imbalanced datasets of customer responses: comparison with other customer response models. Serv. Bus. **7**, 167–182 (2012)
17. Miguéis, V.L., Camanho, A.S., Borges, J.: Predicting direct marketing response in banking: comparison of class imbalance methods. Serv. Bus. **11**, 831–849 (2017)
18. Farquad, M., Bose, I.: Preprocessing unbalanced data using support vector machine. Decis. Support Syst. **53**, 226–233 (2012)
19. Barakat, N., Bradley, A.P.: Rule extraction from support vector machines: a review. Neurocomputing. **74**, 178–190 (2010)
20. Vapnik, V.N.: The Nature of Statistical Learning Theory. Springer, New York (2010)
21. Sanderson, M., Manning, C.D., Raghavan, P., Schütze, H.: Introduction to Information Retrieval. Cambridge University Press, Cambridge (2008). Nat. Lang. Eng. **16**(1), 100–103 (2010)

22. Martens, D., Huysmans, J., Setiono, R., Vanthienen, J., Baesens, B.: Rule extraction from support vector machines: an overview of issues and application in credit scoring. In: Diederich, J. (ed.) Rule Extraction from Support Vector Machines. Studies in Computational Intelligence, vol. 80, pp. 33–63. Springer, Heidelberg (2008). https://doi.org/10.1007/978-3-540-75390-2_2

23. Diederich, J.: Rule extraction from support vector machines: an introduction. In: Diederich, J. (ed.) Rule Extraction from Support Vector Machines. Studies in Computational Intelligence, vol. 80, pp. 3–31. Springer, Heidelberg (2008). https://doi.org/10.1007/978-3-540-75390-2_1

24. Martens, D., Baesens, B., Gestel, T.V., Vanthienen, J.: Comprehensible credit scoring models using rule extraction from support vector machines. Eur. J. Oper. Res. **183**, 1466–1476 (2007)

25. Quinlan, J.R.: Induction of decision trees. Mach. Learn. **1**, 81–106 (1986)

26. Quinlan, J.R.: C4.5 - Programs for Machine Learning. Kaufmann, San Mateo (1992)

27. Breiman, L.: Classification and Regression Trees. Wadsworth International Group, Belmont (1984)

28. Kašćelan, L., Kašćelan, V., Jovanović, M.: Hybrid support vector machine rule extraction method for discovering the preferences of stock market investors: evidence from Montenegro. Intell. Autom. Soft Comput. **21**, 503–522 (2014)

29. Hughes, A.M.: Strategic Database Marketing: The Masterplan for Starting and Managing a Profitable, Customer-Based Marketing Program. Irwin, Chicago (1994)

30. Wang, C.-H.: Apply robust segmentation to the service industry using kernel induced fuzzy clustering techniques. Expert Syst. Appl. **37**, 8395–8400 (2010)

31. Marshall, P.: The 80/20 Rule of Sales: How to Find Your Best Customers. https://www.entrepreneur.com/article/229294

32. Hsieh, N.-C.: An integrated data mining and behavioral scoring model for analyzing bank customers. Expert Syst. Appl. **27**, 623–633 (2004)

33. Tsai, C.-Y., Chiu, C.-C.: A purchase-based market segmentation methodology. Expert Syst. Appl. **27**, 265–276 (2004)

34. Cheng, C.-H., Chen, Y.-S.: Classifying the segmentation of customer value via RFM model and RS theory. Expert Syst. Appl. **36**, 4176–4184 (2009)

35. MacQueen, J.: Some methods for classification and analysis of multivariate observations. In: Proceedings of the Fifth Conference on Mathematical Statistics and Probability, vol. 1, pp. 281–297 (1967)

36. Davies, D.L., Bouldin, D.W.: A cluster separation measure. IEEE Trans. Pattern Anal. Mach. Intell. PAMI **1**, 224–227 (1979)

Crowd Counting á la Bourdieu
Automated Estimation of the Number of People

Karolina Przybylek[1] and Illia Shkroba[2(✉)]

[1] University of Warsaw, Warsaw, Poland
karolina.m.przybylek@gmail.com
[2] Polish-Japanese Academy of Computer Technology, Warsaw, Poland
is@pjwstk.edu.pl

Abstract. In recent years, sociologists have taught us how important and emergent the problem of crowd counting is. They have recognized a variety of reasons for this fact, including: public safety (e.g. crushing between people, trampling underfoot, risk of spreading infectious disease, aggression), politics (e.g. police and government tend to underestimate the number of people, whilst protest organisers tend to overestimate it) and journalism (e.g. accuracy of the estimation of the ground truth supporting an article). The aim of this paper is to investigate models for crowd counting that are inspired by the observations of famous sociologist Pierre Bourdieu. We show that despite the simplicity of the models, we can achieve competitive result. This makes them suitable for low computational power and energy efficient architectures.

Keywords: Crowd counting · Deep learning · Mall dataset

1 Introduction

Nowadays, due to the increasing degree of urbanization, crowd management and control is a key issue for human life and security. For example, stampedes at the Kumbh Melas used to kill hundreds of people each event, until appropriate crowd management policy has been applied. In fact, many such problems related to crowds can be prevented or completely solved if aspects of crowd management and control are well organized. Crowd estimation is the first and foremost task in every crowd management process. Other crucial applications of crowd counting include politics and journalism: the number of people that take part in a given event indicates the strength and the impact of the event. Moreover, in our democratic world the number of people is the chief argument in every discussion. For this reason, crowd counts in presidential inauguration ceremonies (e.g. Obama in 2009 vs. Trump in 2017), demonstrations (e.g. "Black march" in 2018 in Poland) or festivals (e.g. LGBT Film Festival in 2011) are highly disputed. Automated crowd counting methods provide objective and indisputable estimation of the size of crowd.

The names of the authors are arranged in alphabetical order.

© Springer Nature Switzerland AG 2019
T. Welzer et al. (Eds.): ADBIS 2019, CCIS 1064, pp. 295–305, 2019.
https://doi.org/10.1007/978-3-030-30278-8_31

The aim of crowd counting methods is to provide an accurate estimation of the number of people (the crowd) presented on a given picture. This task is extremely challenging for a number of reasons: scaling factor and perspective (people that are nearer to the camera appear much bigger), resolution of fragments containing a single person (i.e. few pixels per person), occlusion and clutter, illumination issues, distinct ambient environments, to name a few. Over last decades we witnessed a significant progress in automated crowd counting.

Two most successful classes of methods for crowd counting are: regression based and detection based methods. Regression based methods aim at estimating the density of the crowd by computing (either directly or indirectly) the heat-map of a picture. For example:

- A state-of-the-art regression model is described in [24].
- A very prominent approach to crowd counting by regression is described in [10]. The authors propose a multi-output regression model that learns how to balance local with global features (i.e. so called "spatially localised crowd counting").
- The first method capable of reliably counting thousands of people was proposed in [13]. The authors combine three sources of information to construct a regression model: Fourier transformation, interest points identification and head detection.
- A deep convolutional model to estimate the density map of an image is described in [4]. Actually, the main idea of the paper is to combine both deep and shallow convolutional neural networks to achieve good scaling effect.

On the other hand, detection based methods try to recognise some identifying features of each single person on a picture and sum up the number of recognized in this way people. For example:

- State-of-the-art pedestrian detection based on multiple features detection is described in [16]. The authors focuses on the problem of overlapping people as one of the main challenge to successful object-based detection. The method is optimised to count low density crowd and whose individuals have enough detail.
- The authors in [23] describe an end-to-end method of identification of pedestrians on images by using a novel architecture based on LSTM recurrent neural network. The results are competitive on TUDCrossing and Brainwash datasets. One may hope to further improve the method by using dynamic RNNs and Luong attention model as indicated in [6].
- It is well-known that naive transfer-learning does not work well in crowd counting. This issue is investigated in [25]. The authors proposes some methods to overcome such problem.

Our approach is a detection based method with explicit segmentation. We split an image into small segments, build a neural network to estimate the probability that a given segment contains a human, apply a cut-off point value and sum up all of the predictions. This approach was implicitly suggested by Pierre

Bourdieu in [5]. The sociologist argued that habitual complexity of individuals is present at different levels of scale, and, therefore, one may identify a person (or even the social class of a person) by looking only at some of her tiny details in isolation. We show that the models that we build, despite their simplicity, are competitive on Mall Dataset [1] (also see: [8,9,17] and [10]). The simplicity of our models makes them suitable for mobile and embedded architectures, where computational power and space are limited. We note that other methods inspired by social observations were successfully used in the past, for example: [3] and its extensions [18,19].

The source code of our methods is available at a public repository:
https://github.com/s14028/engineer.

The paper is organised as follows. In the next section we describe the experimental environment. The dataset is described in Subsect. 2.1 and our transformations of datasets together with necessary preprocessing (i.e. data segmentation and data augmentation) in the following two subsections. Subsection 2.4 describes our approach to image augmentation. Subsection 2.5 describes the hardware that we used to train our models. Subsection 2.6 describes the error metrics that we use to assess the quality of the models. In Sect. 3 we describe two models of increasing complexity: the first one is described in Subsect. 3.1 and operates on raw images, whereas the second of the models, described in Subsect. 3.2, utilises additional information from the perspective map. We summarise the results and conclude the paper in Sect. 4.

2 Experimental Environment

We evaluate our models and compare against the state-of-the-art model on Mall Dataset [1].

2.1 Dataset

The data from Mall Dataset consists of:

- 2000 images taken from the same camera with the same perspective and with resolution 640 × 480 pixels each;
- the coordinates of people's faces on images;
- the perspective map of the images (see Subsect. 2.3).

The distribution of the number of people in the database is shown on Fig. 1(a). The distribution clearly resembles the Gaussian distribution with mean 31.16 and standard deviation 6.94. Basic statistics for the whole dataset are as follows:

- maximum number of people on an image: 53;
- minimum number of people on an image: 13;
- average number of people on an image: 31.16;

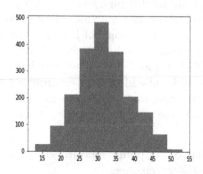

(a) Distribution of crowd sizes across samples

(b) Sample image

Fig. 1. Mall dataset

- median number of people on an image: 31;
- standard deviation of the number of people on an image: 6.94.

A sample image from the dataset is shown on Fig. 1(b).
The dataset was randomly split into:

- the training set consisting of 1500 images;
- the validation set consisting of 300 images;
- the test set consisting of 200 images.

2.2 Data Segmentation

Since we were interested in low memory cost of crowd counting, we could not work with full images. We decided to split each image into smaller segments of the same shapes and with each segment associate a binary value indicating whether or not a person is present on the segment. We accomplished this by creating a map that to the coordinate of the centre of a person's face assigns the segment that contains it. In this process we lost some information, but the accuracy of the final classifiers shows that the loss was not that big.

We examined several splitting methods: we split images into 100, 400 and 1600 non-overlapping rectangular segments of the same shape[1]. Figure 2 shows segments of an image after splitting it on a various number of segments. It is worth noticing, that a person's sweater is barely visible when we split the image on 1600 segments. We have found that splitting into 400 segments performs best for our models. One may argue that this splitting method has a reasonable ratio of segments with a person to segments without a person at around 0.679, and at the same time it has a decently low loss rate. Table 1 summarises the statistics for other splitting methods.

---

[1] Notice, that we did not use content-based image segmentation.

Table 1. Number of segments

Segments	Ratio	Mean loss	Std loss
100	0.2105	−10.1035	3.7835
400	0.0679	−3.9995	2.1153
1600	0.0188	−1.0040	1.0144

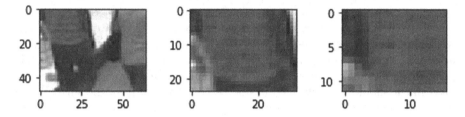

Fig. 2. Segments after splitting an original image on 100, 400 and 1600 fragments.

2.3 Perspective Map

In Mall Dataset the camera that took the pictures was not positioned perpendicularly to the scene, therefore objects on the pictures may have significant distortion. Mall Dataset provides this information in the form of perspective map. The perspective map indicates the real size of each pixel of the images. Figure 3 shows a sample image from the dataset whose fragments are adjusted according to the perspective map.

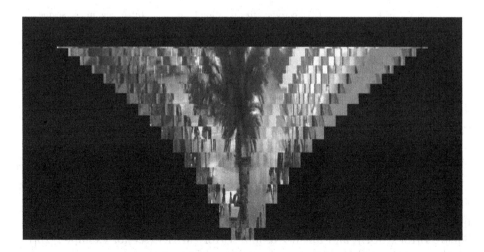

Fig. 3. An image with adjusted perspective.

2.4   Data Augmentation

Due to the fact that the number of segments with a person is much smaller than the number of segments without a person (the ratio is equal to 0.0679), the two classes are unbalanced and a classifier may favor the larger class. Therefore, some augmentation techniques had to be used to produce more balanced classes.

The segments were transformed using random rotations of images by an angle smaller or equal to $\frac{\pi}{12}$. The problem with the standard method of data augmentation is, that after a rotation the segment is cropped to the shape of (a non-rotated) rectangle, whereas new fragments which occur after the rotation are filled out with meaningless values. To prevent this and to create completely augmented segments, a larger surrounding fragment was rotated. Thanks to this approach we did not lose any data after rotation. Figure 4 shows how a rotation of a segment should be performed to not lose information from the surrounding image.

There are also segments for augmentation that lie on the edges of an image—they were augmented in the same way, except that areas of the square which were out of the image were filled with zeros.

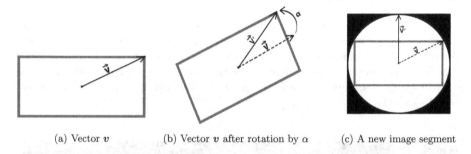

(a) Vector v (b) Vector v after rotation by α (c) A new image segment

Fig. 4. A segment within the surrounding image.

2.5   Hardware

We trained our models on an Intel Xeon workstation with 256 GB RAM and equipped with four NVIDIA Tesla K80 units (a single K80 unit consists of two interconnected K40 units).

2.6   Error Metrics

In order to compare the models, we use two fundamental error metrics—Mean Square Error (MSE), and Mean Absolute Error (MAE) (see for example: [26]), which are calculated according to the following formulas:

$$MSE = \frac{1}{n} \sum_{i=1}^{n} (y_i - x_i)^2$$

$$MAE = \frac{1}{n} \sum_{i=1}^{n} |y_i - x_i|$$

Where $\langle x_1, x_2, \ldots, x_n \rangle$ are the predicted values, whereas $\langle y_1, y_2, \ldots, y_n \rangle$ are the reference (i.e. true) values. In addition, we calculate the accuracy of a prediction x_i as follows:

$$acc(x_i, y_i) = \begin{cases} 1 - \frac{|x_i - y_i|}{y_i} & \text{if } x \in \langle 0, 2y_i \rangle \\ 0 & \text{if } x \in (2y_i, +\infty) \end{cases}$$

and the accuracy of a model as the average over all its predictions.

3   Models

As mentioned earlier we focused on building possibly simple models that can be run on low computational power/memory devices, without sacrificing much on the accuracy. Two major limiting factors for our architectures were: (a) we could not afford working with full images, therefore the images had to be split into smaller segments, (b) in deep neural network architectures, we could not afford any variant of skipping connections (because they force the device to cache the data along the forward-pass of the network). We experimented with variety of architectures of neural networks to predict if a single segment of an image contains a person. We trained 17 different models in total. Two best models are presented in the following subsections (see Fig. 5). The first one operates on raw (segments) of images, whereas the second makes use of perspective map. Other models mostly differ in the values of regularisation coefficients and the number of segments each image was splitted[2].

To get the final prediction of the size of crowd, we use a cut-off point of each prediction, and then sum over all values of segments that constitute the image.

3.1   Convolutional Model v1

The architecture of Convolutional model v1 is presented on Fig. 5(a). The building blocks of the model are as follows: 2-dimensional convolutional layers [15,21] with receptive field size of 3×3, stride of 1, $(2,2)$-max pooling layers [11,20], Batch normalization layers [14], Dropout layers [2,22] with parameter 0.5 and dense (i.e. fully connected) layers. Next to the name of the layers, the numbers specify the size of the input and the size of the output respectively (width, height and depth). We used segmentation technique as described in Subsect. 2.2 (each image was split into 400 segments) and data augmentation as described in Subsect. 2.4 to balance the classes of positive and negative appearances.

Because we used a convolutional layer as the first layer in the network, the system had to store segments as full tensors rather than mere vectors. In the

---

[2] We also tested a split into 1600 segments, which gave much worse results.

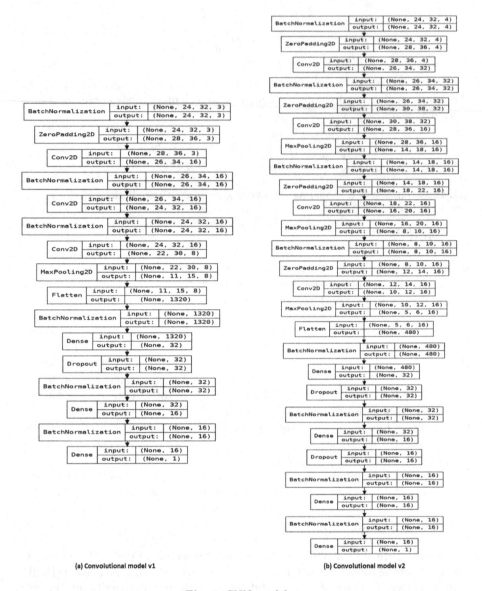

(a) Convolutional model v1   (b) Convolutional model v2

Fig. 5. CNN models

result the underlying filters had to be represented as full tensors as well and used in convolution process. Such situation resulted in high memory consumption, and we were forced switch in the training process from batches to mini-batches. The neurons from the hidden layers used Rectified Linear Unit as the activation function, whereas the output layer used the sigmoid function. We used a batch normalization behind each layer which allowed us to set higher values of learning rate at the beginning of neural network learning process, and also prevented

Internal Covariate Shift process (the issue is discussed in [14]). The training process utilises two regularisation techniques: dropout connections as described in the above, and l2 regularisation. We tested several regularisation coefficients for the learning process, but found that the value of 10^{-7} performs best.

We trained the model with Adam optimiser initially setting the learning rate to 10^{-1} to speed up training process. High value of learning rate was acceptable due to Batch Normalization usage. We trained the model for about 75 epochs till accuracy for validation set started getting worse. Then we applied a learning rate value 10^{-3} and continued training for next 30 epochs. At that point we exceeded previous results. We decided to settle much lower learning rate - 10^{-5} and continue the process for next 20 epochs. At that point we reached the best weights. Further training caused overfitting.

The cut-off point has been chosen to 0.3. The overall accuracy of the model is presented in Table 2. We achieved Mean Absolute Error (MAE) of 3.44 and Mean Squared Error (MSE) of 18.87.

3.2 Convolutional Model v2

Convolutional model v2 (see Fig. 5(b)) is an enhanced version of the model from the previous subsection. Together with a 3-channel image it additionally utilises the fourth channel with the perspective map of the image. Because the size of the input has significantly grown, a deeper model than the previous one performed better. All of the parameters of the architecture remained unchanged. Data augmentation had to be modified to incorporate the perspective map.

The training process of convolutional model v2 was similar to the training process of convolutional model v1. At the beginning we settled learning rate at value 10^{-3} and started training for 30 epochs. After that, we lowered learning rate to value 10^{-4} to move slowly to local minimum and continued for the next 20 epochs. Then we lowered learning rate to the value 10^{-5} and continued for the next 80 epochs. At that point we reached local minimum and further training had not give any better results.

We experimented with several l2 regularisation parameters for this model as well. Value 10^{-6} performed best for this model.

The cut-off point has been chosen to 0.6. The overall accuracy of the model is presented in Table 2. We achieved Mean Absolute Error (MAE) of 3.35 and Mean Squared Error (MSE) of 18.33, which is slightly better than with the previous model.

Table 2. Accuracy and error of the models

Model	Accuracy	MSE	MAE	Epochs
Convolutional model v1	88.73	18.87	3.44	125
Convolutional model v2	88.90	18.33	3.35	130
State-of-the-art	NA	NA	2.01	NA

4   Conclusions and Future Work

The accuracy and the errors of each of our models are summarised in Table 2. The last line in the table shows MAE of state-of-the-art solution described in [24] (the authors do not provide the accuracy nor MSE of their model).

As it turns out, convolutional model v1 and convolutional model v2 show competitive results, despite the simplicity and low-computational-cost of their architectures. Although our MAE of 3.35 is significantly worse than MAE of 2.01 (state-of-the art solution), this is almost negligible from the perspective of crowd-sizes of about 50 persons. Therefore, they may be successfully run on devices that have relatively small resources available. Moreover, we did not use any ensemble methods to tune up our models. By the utilisation of the fourth channel with the perspective map, convolutional model v2 behaves slightly better than its convolutional model v1 predecessor (at the cost of increased complexity).

More optimization techniques may be applied to our models to obtain even lighter architectures (e.g. reduce the size of the color channels, MobileNet architectures [7,12]). We leave this for the future work. From the perspective of crowd management another interesting direction of research is to identify more details about crowd beside the mere number of the participants. The theory developed by Pierre Bourdieu in [5] tells us that the aspects like gender, race or social class of the individuals may play crucial role in crowd management. We also leave this for future work.

References

1. Mall dataset (2014). http://personal.ie.cuhk.edu.hk/~cloy/downloads_mall_dataset.html
2. Ba, J., Frey, B.: Adaptive dropout for training deep neural networks. In: Advances in Neural Information Processing Systems, pp. 3084–3092 (2013)
3. Bonyadi, M.R., Michalewicz, Z., Przybylek, M.R., Wierzbicki, A.: Socially inspired algorithms for the travelling thief problem. In: Proceedings of the 2014 Annual Conference on Genetic and Evolutionary Computation, pp. 421–428. ACM (2014)
4. Boominathan, L., Kruthiventi, S.S., Babu, R.V.: Crowdnet: a deep convolutional network for dense crowd counting. In: Proceedings of the 24th ACM International Conference on Multimedia, pp. 640–644. ACM (2016)
5. Bourdieu, P.: Distinction: A Social Critique of the Judgement of Taste. Routledge, New York (2013)
6. Brzeski, A., Grinholc, K., Nowodworski, K., Przybylek, A.: Evaluating performance and accuracy improvements for attention - OCR. In: 18th International Conference on Computer Information Systems and Industrial Management Applications (2019)
7. Brzeski, A., Grinholc, K., Nowodworski, K., Przybylek, A.: Residual mobilenets. In: Workshop on Modern Approaches in Data Engineering and Information System Design at ADBIS (2019)
8. Change Loy, C., Gong, S., Xiang, T.: From semi-supervised to transfer counting of crowds. In: Proceedings of the IEEE International Conference on Computer Vision, pp. 2256–2263 (2013)

9. Chen, K., Gong, S., Xiang, T., Change Loy, C.: Cumulative attribute space for age and crowd density estimation. In: Proceedings of the IEEE Conference on Computer Vision and Pattern Recognition, pp. 2467–2474 (2013)
10. Chen, K., Loy, C.C., Gong, S., Xiang, T.: Feature mining for localised crowd counting. In: BMVC, vol. 1, p. 3 (2012)
11. Graham, B.: Fractional max-pooling (2014). arXiv preprint arXiv:1412.6071
12. Howard, A.G., et al.: Mobilenets: efficient convolutional neural networks for mobile vision applications (2017). arXiv preprint arXiv:1704.04861
13. Idrees, H., Saleemi, I., Seibert, C., Shah, M.: Multi-source multi-scale counting in extremely dense crowd images. In: Proceedings of the IEEE Conference on Computer Vision and Pattern Recognition, pp. 2547–2554 (2013)
14. Ioffe, S., Szegedy, C.: Batch normalization: accelerating deep network training by reducing internal covariate shift (2015). arXiv preprint arXiv:1502.03167
15. Krizhevsky, A., Sutskever, I., Hinton, G.E.: ImageNet classification with deep convolutional neural networks. In: Advances in Neural Information Processing Systems, pp. 1097–1105 (2012)
16. Leibe, B., Seemann, E., Schiele, B.: Pedestrian detection in crowded scenes. In: 2005 IEEE Computer Society Conference on Computer Vision and Pattern Recognition (CVPR 2005), vol. 1, pp. 878–885. IEEE (2005)
17. Loy, C.C., Chen, K., Gong, S., Xiang, T.: Crowd counting and profiling: methodology and evaluation. In: Ali, S., Nishino, K., Manocha, D., Shah, M. (eds.) Modeling, Simulation and Visual Analysis of Crowds. TISVC, vol. 11, pp. 347–382. Springer, New York (2013). https://doi.org/10.1007/978-1-4614-8483-7_14
18. Przybylek, M.R., Wierzbicki, A., Michalewicz, Z.: Decomposition algorithms for a multi-hard problem. Evolut. Comput. **26**(3), 507–533 (2018)
19. Przybylek, M.R., Wierzbicki, A., Michalewicz, Z.: Multi-hard problems in uncertain environment. In: Proceedings of the Genetic and Evolutionary Computation Conference 2016, pp. 381–388. ACM (2016)
20. Scherer, D., Müller, A., Behnke, S.: Evaluation of pooling operations in convolutional architectures for object recognition. In: Diamantaras, K., Duch, W., Iliadis, L.S. (eds.) ICANN 2010. LNCS, vol. 6354, pp. 92–101. Springer, Heidelberg (2010). https://doi.org/10.1007/978-3-642-15825-4_10
21. Simonyan, K., Zisserman, A.: Very deep convolutional networks for large-scale image recognition (2014). arXiv preprint arXiv:1409.1556
22. Srivastava, N., Hinton, G., Krizhevsky, A., Sutskever, I., Salakhutdinov, R.: Dropout: a simple way to prevent neural networks from overfitting. J. Mach. Learn. Res. **15**(1), 1929–1958 (2014)
23. Stewart, R., Andriluka, M., Ng, A.Y.: End-to-end people detection in crowded scenes. In: Proceedings of the IEEE Conference on Computer Vision and Pattern Recognition, pp. 2325–2333 (2016)
24. Walach, E., Wolf, L.: Learning to count with CNN boosting. In: Leibe, B., Matas, J., Sebe, N., Welling, M. (eds.) ECCV 2016. LNCS, vol. 9906, pp. 660–676. Springer, Cham (2016). https://doi.org/10.1007/978-3-319-46475-6_41
25. Wang, M., Li, W., Wang, X.: Transferring a generic pedestrian detector towards specific scenes. In: 2012 IEEE Conference on Computer Vision and Pattern Recognition, pp. 3274–3281. IEEE (2012)
26. Willmott, C.J., et al.: Statistics for the evaluation and comparison of models. J. Geophys. Res. Oceans **90**(C5), 8995–9005 (1985)

A Holistic Decision Making Framework for a Vehicle Sharing System

Selin Ataç[iD], Nikola Obrenović[(✉)][iD], and Michel Bierlaire[iD]

École Polytechnique Fédérale de Lausanne, Lausanne, Switzerland
{selin.atac,nikola.obrenovic,michel.bierlaire}@epfl.ch

Abstract. The vehicle sharing systems (VSSs) are becoming more and more popular due to both financial and environmental effects. On the other hand, they face many challenges, such as inventory management of the vehicles and parking spots, imbalance of the vehicles, pricing strategies, and demand forecasting. If these are not addressed properly, the system experiences a significant loss of customers and therefore revenue. Although efficient methods to solve these problems are well-studied in the literature, there does not exist any work in the literature which considers a VSS as a whole, and identifies and analyzes all of its components and their relations, to the best of our knowledge. Therefore, this work provides a new framework for a VSS management from a wider perspective by addressing the components and their relations with the inclusion of a time dimension. The proposed framework is aimed to apply for any kind of VSS. After addressing as many problems as possible related to a VSS, we will focus on the application of the framework to the light electric vehicle (LEV) sharing system.

Keywords: Vehicle sharing systems · Decision making framework · Strategic planning · Tactical planning · Operational planning

1 Introduction

The idea of vehicle sharing systems (VSSs) bases back 1940s [9,18]. However, due to the lack of identification of the customers, the constructed systems were not as practical as nowadays. With technological improvements, the VSSs are now able to identify the customers through a mobile phone application, a magnetic card, etc. Therefore, the notion of vehicle-sharing has become more and more popular during the last 20 years. The car-sharing systems (CSSs) are available in over 600, where the bicycle-sharing systems (BSSs) in more than 700 cities in several countries [16,17]. For CSSs, for instance, as of February 2018 car2go, which is the largest CSS company in the world, announced that their system serves to 3 million registered members, of which more than a half being in Europe, on its own. They also claim that they experience 30% growth in car2go membership year-over-year [3].

© Springer Nature Switzerland AG 2019
T. Welzer et al. (Eds.): ADBIS 2019, CCIS 1064, pp. 306–314, 2019.
https://doi.org/10.1007/978-3-030-30278-8_32

The increasing usage of VSSs brought many challenging questions. The VSS companies try to maximize the profits by analyzing their costs and revenues. The number of vehicles to be used in the system, the rebalancing structure to be deployed, the demand estimation for a vehicle or for a parking spot and pricing schema of the trips can be counted among the most common problems investigated in the literature. However, to the best of our knowledge, the literature lacks a holistic analysis of the system. In other words, there is no work providing the components of this system in a nutshell and identifying inter- and intrarelations of these components. Therefore, this work aims to fill this gap in the literature and provide a general framework for the VSSs.

While identifying the components of the system, it is also important to think about the time horizon and the corresponding decisions such as strategic, tactical, and operational. The two literature surveys from Laporte et al. [10,11] talk about these decision levels and provide a summary of existing works and place them under these levels, but do not discuss the relations within or between these decision levels or the problems discussed. Our work does not only construct the framework itself but also defines the relations within decision levels and the problems.

Moreover, the future work aims to apply this proposed framework for the newly introduced light electric vehicles (LEVs). As the existing studies are focused on BSSs and CSSs, the existing methodologies became inapplicable for the LEVs. The rebalancing methods, for instance, are not convenient for the LEVs since a LEV is not as small as a bicycle, making it unsuitable for rebalancing with a truck, and has only one seat, preventing the transport of staff, which is common in car rebalancing operations. Also, demand forecasting becomes a challenging task since LEVs are allowed to be parked on any designated spot in the city. LEVs also serve for a higher portion of the population since they do not require a driving license. Moreover, as the vehicle is electric there should exist a fleet of workers who replace the batteries. It is also good to note that although the future work consists of an application on LEVs, the framework is aimed to apply not only to conventional vehicles but also other vehicle types which might be introduced in the future.

The paper is organized as follows: The second chapter presents a brief literature review on VSSs and their components. In the third chapter the proposed framework is presented. The last chapter includes the conclusions and future work.

2   Literature Review

There exists numerous studies in the literature about VSSs. In this section, we talk about the studies that are related to our context. The reader may find other literature surveys in [10,11].

One of the most studied problems in VSSs is the imbalance of the vehicles observed in the system. People using the system may not find a spot to park their vehicles in the destination, or they may not find a vehicle in the origin.

There have been a considerable set of recent studies on bike rebalancing. In BSSs the rebalancing is usually performed using trucks or similar vehicles [7,12,14], which relocate the bikes from station with high availability and low demand to the stations with high demand and low availability. Therefore, the bike rebalancing problem consists of two major parts: estimation of the required inventory level of stations or city zones, and the routing of relocating vehicles. The relocating vehicles routing is most often formulated as an optimization problem based on either capacitated traveling salesman problem (TSP) [14] or vehicle routing problem (VRP) [7,12].

Resolving this issue in CSSs involves staff members to redistribute the vehicles between stations. This, however, yields the subsequent problem of relocating the staff itself between two stations and two car balancing operations. In most of the reviewed literature, these two problems are tackled jointly, by defining optimization problems whose solutions determine simultaneously the rebalancing of both vehicles and staff [5,13]. The strength of [13] and [5] is that they are both evaluated on real case studies in Toronto, Canada, and Nice, France, respectively. However, neither of the approaches [5,13] include the forecasting of the demand and relocation of vehicles according to it where in [8] it is emphasized that the relation between these is important. In [5], authors account for the demand uncertainty, but in the case of high demand, the vehicle requests are denied, which implies loss of demand. As the loss of demand comes with many drawbacks such as bringing the company into disrepute, the constructed framework should be able to predict the demand and rebalance vehicles in advance in order to reduce the demand loss as much as it is allowed by the available data. With respect to the used methodology, in [13] the problem definition is based on the multi TSP, while in [5] the authors have tailored a specific Mixed Integer Linear Programming (MILP) for this purpose.

The demand estimation problem can be addressed by machine learning algorithms used for forecasting [12], by simulating the demand with a Poisson process [7], or even by calculating the worst-case demand, as the solution of optimization problem, and optimizing the rebalancing strategy according to it [7]. Combining these two components of the system, demand forecasting and rebalancing, the problem can be formulated as a two player game [7] or only sequentially forecasting the demand and rebalancing the vehicles afterwards [12]. In the case of two player game, one player is creating a high demand, while the other is rebalancing the vehicles to reduce the demand loss as much as possible.

On the other hand, the demand and rebalancing problems can also be manipulated by different pricing schemas. For instance, decreasing the price of the trip from a low demand area to a high one triggers users to utilize that option. By this way, the system does not only encourage customers to use the system but also rebalances itself. However, in some cases it may end up with demand loss because of the high pricing for the trips from high demand areas to low ones. Therefore, this trade-off should be analyzed in detail. In practice, the companies tend to use a fixed value for a starting price and a variable amount which increases with the time and/or the distance covered. In theory, there exist different approaches for dynamic pricing. The authors in [8] tackled the vehicle imbalance problem by defining a pricing schema which motivates the users to do trips which lower the imbalance and brings the system closer to the equilibrium state. Their work showed that using only pricing strategy, i.e. without any relocation, can improve the balance of the system, but will serve less demand. The authors in [6] assign dynamic prices independently of their origin whereas in [20] the price is set as soon as the itinerary of the customer is revealed and fixed till the end of the trip. The approach in [20] is further extended with another approach using a fluid approximation [19]. In [15], the authors applied their methodology on a case study on a BSS in London, and it is shown that the level of service was improved with the introduction of dynamic pricing schema for the weekends. However, during the weekdays, because of the rush hours, they could not come up with a pricing schema that will improve the performance of the system. Therefore, the literature still lacks research in terms of pricing in VSSs.

One of the most recent surveys conducted by Laporte et al. [11] puts emphasis on different decision levels of the VSSs as well as the problems faced. They come up with a two dimensional classification where one is the type of the problem and the other is the decision level. Their results show that there still exists lack of research in some specific areas such as pricing incentives and routing problems at strategical level or locating stations in tactical and operational levels. For instance, they claim that determining the optimal inventory level at each station within a theoretical framework has not received much attention although it is closely related to the rebalancing problem.

Putting all these together, there do not exist any studies which take all the components of the system into account to the best of our knowledge. Moreover, they do not consider the time dimension within the system. We think that a proper optimization framework for VSSs should provide a decision support in each of the levels, i.e. strategic, tactical, and operational. Furthermore, according to [11], the existing studies on the VSSs do not consider vehicles other than cars or bicycles. However, with the recent introduction of new type of vehicles [1], the proposed methodologies became inapplicable. Therefore, it is important to construct a framework for any kind of VSS which is another aim of this work. The next section provides the details on the proposed framework.

3   Proposed Framework

Before going into the details of the framework, we would like to give the idea behind. As in every decision model, we first gather data, then we construct models to be able to represent the data, and finally, take actions according to the outputs of these models for the future. These notions form each decision problem and represent one dimension of the proposed framework. Since we want to solve decision problem on both supply and demand side, these form the second framework dimension. Finally, in order to introduce the time dimension to our framework, we analyze supply and demand decision problems on all planning levels, i.e. strategic, tactical, and operational. The proposed framework has been illustrated in Fig. 1. By this way, we are able to place each problem component in one of the 18 boxes. We represent the interaction between the components with white, the dependence with blue, and intra-level interactions with a dashed line. Next, we review the main problems observed in a VSS and place those in boxes of Fig. 1.

The inventory management of the vehicles and parking spots include works on (1) optimizing the fleet size of the vehicles that will serve to the customers [4], (2) deciding the optimal location and the size of the parking facilities in order to prevent both overstocking and understocking of the vehicles [4], (3) optimizing the routing of the fleet of workers who are responsible for the maintenance of the vehicles and their fuel/battery level depending on the type of the vehicle [5]. The first corresponds to tactical level decisions and can be changed in mid-term. The second, on the other hand, is generally decided at the beginning of the system installment in the case of station-based systems, which places under strategic level. Daily or hourly decisions are made to overcome the third problem, which makes it to be placed under operational level decisions. For instance, for electric vehicles, the battery is also an issue for the developer. There are several approaches for keeping the battery level sufficient for each user. Some examples are: (1) the users are required to charge the batteries in certain locations if it is under a certain threshold, (2) the company replaces the batteries of the vehicles by monitoring their levels and (3) the company hires staff to drive the vehicles to the charging stations. The first one is not user-friendly as it needs time and makes the user responsible from an act where the second and the third put responsibility on the company. These decisions are made on a daily basis.

Because of the dynamics in the city, these systems also experience imbalance during the day. The vehicle rebalancing can be dynamic, where the relocation is performed during the system operation, or static, where the relocation is done when the system is closed (e.g. over night) [11]. To minimize the cost of such an implementation, this problem is layered into several subproblems in the literature: (1) optimizing the staff allocation and relocation who are responsible for the rebalancing, (2) routing of vehicles performing rebalancing in the case of BSSs, (3) routing of the relocated vehicles in the case of CSSs [5,13]. The decisions regarding the type of rebalancing strategy can be made in both strategic and tactical level: the former level decisions correspond to the type of vehicles used and the latter the time of the operation. After, the daily decisions regarding that strategy should be addressed under operational level decisions. Also, the decision maker should decide the level of service to be provided in the strategic level.

From pricing point of view, there exist many applications in the industry. Some companies work with a fixed price to reserve the vehicle and it increases with the distance and/or time the customer travels with the vehicle. Some others also try to encourage people so that they return the vehicles to the place where they actually picked up to serve balancing issues [2]. On the other hand, there also exist studies in the literature where they assume that dynamic pricing is possible [9]. With such an approach the company aims to manipulate the market so that the system will need less rebalancing while the revenue is not sacrificed. The pricing component can be placed under both tactical and operational level decisions. The tactical level decisions can be thought as the pricing strategy, i.e. dynamic or fixed, to be applied and the offers that will be presented to the customers, and the short-term decisions can be made through deciding the actual price. Furthermore, the determination of budget for advertisement and market placement can be listed in the strategic level decisions that will relate pricing in lower decision levels.

Last but not least, one of the main problems faced in VSSs is forecasting the demand. First of all, with different type of vehicles the people are expected to behave differently. For instance, LEVs are available for a higher portion of the population since they do not require a driving license as in a car or no effort to ride it as in a bicycle. Second of all, the type of the stations, which can be fixed and free-floating [11], also affects the forecasting procedure. The type of the stations should be decided at strategic level. After, the historical demand data helps the decision maker to design the network accordingly, i.e. placing the stations and deciding their capacities for the fixed station case, allocating parking spots throughout the city for the free-floating case. Note that the capacities are determined with the help of the forecasting done on mode- and destination-choice. In the second level, the mid-term demand forecasting model should be constructed to be able to come up with the pricing strategy. The short-term decisions regarding this problem correspond to forecasting the demand for the vehicles and the parking spots -which can be considered as forecasting the supply- per station or zone.

The type of the data used for the models also varies with the decision level. At supply side the geographical characteristics of the city cannot be changed and therefore is an input at strategical level. However, the seasonality or the important events taking place in the city may affect mid-term decisions. Within the operational level, we deploy current state of the system as the input at supply side. The demand side of the information relies on the historical demand and the aggregation is mostly related to the level of decision.

Figure 1 provides the overall picture discussed above and the relationship between the components of the aforementioned problems with time dimension. The vertical relations are represented with one-way white arrows since the data is an input to the constructed models and the models with the input data help the decision maker to decide on the action. The horizontal interaction at the *Models* level is two-way and represented with a white arrow because these models interact with each other, where in *Data* level it is a two-way blue relation since these information depend each other. The interactions between the decision levels are from the *Actions* component of the upper level to the *Models* of the lower level. These interactions represent the fact that the chosen actions on the upper planning level determine to a great extend the used models and their outputs in the lower planning level. For instance, the pricing strategy to be applied affects the approach taken in both tactical and operational levels. Therefore, this relation is represented with dashed one-way arrows.

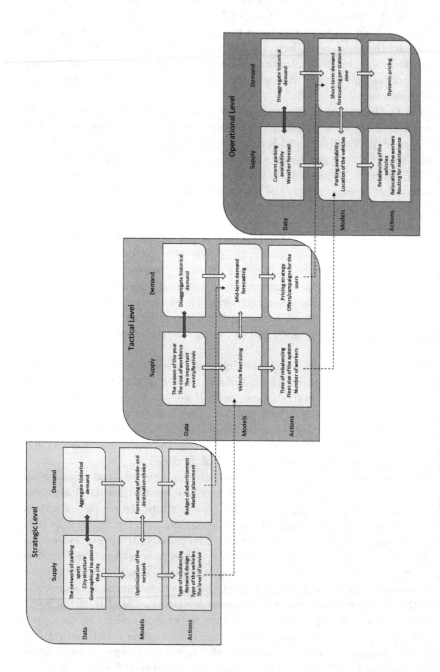

Fig. 1. Vehicle sharing system framework (Color figure online)

4 Conclusion and Future Work

Through the review of available literature, we have identified a lack of a unified approach of modeling all VSS aspects, with respect to different planning horizons, and a holistic solution approach to the related problems. Consequently, the goal of our work is to create a framework for VSS management that will encompass all decision-making tasks of the system and provide the best possible solution to the problems related to them. In order to achieve this, we have to simultaneously take into account all aspects of the system, i.e., to consider the impact the solutions of different problems have on each other. The contribution of this paper is to provide a wider perspective on the design and operations of shared mobility systems. While the literature has focused mainly on specific problems such as the routing aspect, many other methodological challenges are associated with such systems. The proposed framework provides a methodological map that may not be comprehensive, but attempts to cover the most important challenges of these mobility systems.

Our further goal is the apply the framework to a system of shared LEVs, and design framework components tailored to the unique characteristics of such vehicles. To the best of our knowledge, the specific problems arising in the LEV sharing systems have not yet been addressed in the literature, and this paper represents the first consideration of such system. We have also seen that the literature lacks of disaggregate demand forecasting in the operational level. Therefore, we are going to focus firstly on the demand modeling and forecasting. We will try to avoid unrealistic assumptions to represent the real-world system better. Moreover, the dynamic pricing module is also of interest of future work.

References

1. How it works. https://www.enuu.ch. Accessed 25 Feb 2019
2. Mobility explained in brief - how it works. https://www.mobility.ch/en/how-it-works/mobility-in-brief/. Accessed 25 Feb 2019
3. We are 3 million! (2019). https://blog.car2go.com/2018/02/07/we-are-3-million/. Accessed 27 Feb 2019
4. Boyacı, B., Zografos, K.G., Geroliminis, N.: An optimization framework for the development of efficient one-way car-sharing systems. Eur. J. Oper. Res. **240**(3), 718–733 (2015). https://doi.org/10.1016/j.ejor.2014.07.020. http://www.sciencedirect.com/science/article/pii/S0377221714005864
5. Boyacı, B., Zografos, K.G., Geroliminis, N.: An integrated optimization-simulation framework for vehicle and personnel relocations of electric carsharing systems with reservations. Transp. Res. Part B Methodol. **95**, 214–237 (2017). https://doi.org/10.1016/j.trb.2016.10.007. http://www.sciencedirect.com/science/article/pii/S0191261515301119
6. Chemla, D., Meunier, F., Pradeau, T., Calvo, R.W., Yahiaoui, H.: Self-service bike sharing systems: simulation, repositioning, pricing (2013)
7. Ghosh, S., Trick, M., Varakantham, P.: Robust repositioning to counter unpredictable demand in bike sharing systems (2016)

8. Jorge, D., Correia, G.: Carsharing systems demand estimation and defined operations: a literature review. Eur. J. Transp. Infrastruct. Res. **13**(3) (2013). https://doi.org/10.18757/ejtir.2013.13.3.2999. https://journals.library.tudelft.nl/index.php/ejtir/article/view/2999

9. Jorge, D., Molnar, G., de Almeida Correia, G.H.: Trip pricing of one-waystation-based carsharing networks with zone and time of day price variations. Transp. Res. Part B Methodol. **81**, 461–482 (2015). https://doi.org/10.1016/j.trb.2015.06. 003. http://www.sciencedirect.com/science/article/pii/S0191261515001265. Optimization of Urban Transportation Service Networks

10. Laporte, G., Meunier, F., Wolfler Calvo, R.: Shared mobility systems. 4OR **13**(4), 341–360 (2015). https://doi.org/10.1007/s10288-015-0301-z

11. Laporte, G., Meunier, F., Wolfler Calvo, R.: Shared mobility systems: an updated survey. Ann. Oper. Res. **271**(1), 105–126 (2018). https://doi.org/10.1007/s10479-018-3076-8

12. Liu, J., Sun, L., Chen, W., Xiong, H.: Rebalancing bike sharing systems: a multi-source data smart optimization. In: Proceedings of the 22nd ACM SIGKDD International Conference on Knowledge Discovery and Data Mining, KDD 2016, pp. 1005–1014. ACM, New York (2016). https://doi.org/10.1145/2939672.2939776

13. Nourinejad, M., Zhu, S., Bahrami, S., Roorda, M.J.: Vehicle relocation and staff rebalancing in one-way carsharing systems. Transp. Res. Part E Logist. Transp. Rev. **81**, 98–113 (2015). https://doi.org/10.1016/j.tre.2015.06.012. http://www.sciencedirect.com/science/article/pii/S1366554515001349

14. Pal, A., Zhang, Y.: Free-floating bike sharing: solving real-life large-scalestatic rebalancing problems. Transp. Res. Part C Emerg. Technol. **80**, 92–116 (2017). https://doi.org/10.1016/j.trc.2017.03.016. http://www.sciencedirect.com/science/article/pii/S0968090X17300992

15. Pfrommer, J., Warrington, J., Schildbach, G., Morari, M.: Dynamic vehicle redistribution and online price incentives in shared mobility systems. IEEE Trans. Intell. Transp. Syst. **15**(4), 1567–1578 (2014)

16. Shaheen, S., Cohen, A.: Worldwide carsharing growth: an international comparison. Institute of Transportation Studies, UC Davis, Institute of Transportation Studies, Working Paper Series 1992, January 2008. https://doi.org/10.3141/1992-10

17. Shaheen, S.A., Martin, E.W., Chan, N.D., Cohen, A.P., Pogodzinski, M.: Public bikesharing in North America during a period of rapid expansion: understanding business models, industry trends and user impacts (2014)

18. Shaheen, S.A., Sperling, D., Wagner, C.: Carsharing in Europe and North America: past, present and future (1998)

19. Waserhole, A., Jost, V.: Vehicle sharing system pricing regulation: a fluid approximation (2013). http://hal.archives-ouvertes.fr/hal-00727041

20. Waserhole, A.: Vehicle sharing systems pricing optimization. Ph.D. thesis, Université de Grenoble (2013)

Residual MobileNets

Adam Brzeski[1,2] ⑩, Kamil Grinholc[1,3], Kamil Nowodworski[1,4],
and Adam Przybylek[1(✉)] ⑩

[1] Faculty of Electronics, Telecommunications and Informatics,
Gdansk University of Technology, Narutowicza 11/12, 80-233 Gdansk, Poland
adam.brzeski@pg.edu.pl, kamil.grinholc@gmail.com,
kamilnowodworski@gmail.com, adam.przybylek@gmail.com
[2] CTA.ai, Gdansk, Poland
[3] Spartez, Gdansk, Poland
[4] IHS Markit, Gdansk, Poland

Abstract. As modern convolutional neural networks become increasingly deeper, they also become slower and require high computational resources beyond the capabilities of many mobile and embedded platforms. To address this challenge, much of the recent research has focused on reducing the model size and computational complexity. In this paper, we propose a novel residual depth-separable convolution block, which is an improvement of the basic building block of MobileNets. We modified the original block by adding an identity shortcut connection (with zero-padding for increasing dimensions) from the input to the output. We demonstrated that the modified architecture with the width multiplier (α) set to 0.92 slightly outperforms the accuracy and inference time of the baseline MobileNet ($\alpha = 1$) on the challenging Places365 dataset while reducing the number of parameters by 14%.

Keywords: MobileNet · Shortcut connections · Skip connections · CNN

1 Introduction

Since AlexNet [13] won the ImageNet challenge in 2012, the general trend of convolutional neural network (CNN) design has been to find deeper models to get higher accuracy. The most accurate CNNs have hundreds of layers and achieve remarkable performance in a wide variety of computer vision tasks including image classification [3,4,13,15], object detection [4,17], semantic segmentation [19] and photo OCR [2,12]. Nevertheless, the large memory footprint and slow inference time of these networks prevent them from being deployed on mobile or embedded devices. It also seems unlikely that new mobile hardware will meet computational requirements on a limited power budget in near future. Therefore, much recent research has focused on building more efficient, lightweight architectures [14,16].

This work has been partially supported by Statutory Funds of Electronics, Telecommunications and Informatics Faculty, Gdansk University of Technology.

© Springer Nature Switzerland AG 2019
T. Welzer et al. (Eds.): ADBIS 2019, CCIS 1064, pp. 315–324, 2019.
https://doi.org/10.1007/978-3-030-30278-8_33

The state-of-the-art compact CNN is MobileNet, which exploits depthwise separable convolutions [11] as the key building block to reduce computation and the model size. A standard convolution filters and combines inputs in one step to produce a new feature [9]. Depthwise separable convolution which is a form of factorized convolution, decomposes a standard convolution into a depthwise convolution followed by a pointwise 1×1 convolution. The depthwise convolution applies a single filter per each input channel. The pointwise convolution then applies a 1×1 convolution to combine the outputs of the depthwise convolution [9]. Effectively depthwise separable convolution reduces computation compared to standard convolution by almost a factor of k^2 [18], where k is the size of kernel (MobileNet uses $k = 3$). Indeed, the experiments [9] show that MobileNet is able to be 4% better than AlexNet on ImageNet while being 45× smaller and 9.4× less compute than AlexNet.

In this paper, we present an improved version of MobileNet, called Residual MobileNet. Inspired by ResNet [7], we incorporated shortcut connections into MobileNet to improve information flow throughout the network. We found out an architecture that outperforms the baseline MobileNet regarding accuracy, inference time, and the number of parameters.

2   Related Work

One of the first lightweight architectures was SqueezeNet, which exceeds AlexNet accuracy with 50× fewer parameters [10]. The basic building block of SqueezeNet consists of a squeeze layer (which has only 1×1 filters) followed by an expand layer that has a mix of 1×1 and 3×3 convolution filters (in the primary configuration the number of 1×1 and 3×3 filters is the same). The squeeze layer is responsible for decreasing the number of input channels to the expand layer. Every squeeze layer has 8× fewer output channels than the accompanying expand layer. When the input and output of the block are of the same dimensions, they are connected by a shortcut. In practice, it means that shortcuts are added around 4 out of 9 blocks.

The work on SqueezeNet was continued by Gholami et al. [5], who (I) used a more aggressive channel reduction by incorporating a two-stage squeeze layer; (II) decomposed the 3×3 convolutions into two separable convolutions of size 1×3 and 3×1 to further reduce the model size, and remove the additional 1×1 branch in the expand layer; and (III) exploited a bottleneck design proposed by [7]. The new architecture, named SqueezeNext, is able to deliver about 2.5%-point improvement on ImageNet accuracy compared to SqueezeNet without increasing model size.

Zhang et al. [20] proposed a new architecture called ShuffleNet, which utilizes a bottleneck design, group convolution and channel shuffle to reduce computation cost and parameters. However, their architecture is about two times slower than MobileNet due to the costly channel shuffle operation, which is required to enable information flow between different groups of channels.

Concurrently with our work, Sandler et al. [18] introduced MobileNetV2, which is based on the MobileNetV1 architecture, but adopts and adapts a bottleneck design. Their novel building block named inverted residual with linear bottleneck consists of 3 layers: a 1×1 pointwise convolution responsible for increasing dimensions, a 3×3 depth-wise convolution, and a 1×1 pointwise convolution responsible for reducing dimensions. This block takes as an input a low-dimensional compressed representation which is first expanded to high dimension, then filtered with a lightweight depthwise convolution, and in the end features are projected back to a low-dimensional representation [18]. Shortcuts are used directly between the bottlenecks. The first and second convolution in the block is followed by ReLU. The overall architecture contains the initial standard convolutional layer, 17 residual bottleneck blocks, 1 pointwise convolutional layer, and 1 fully-connected layer. Although MobileNetV2 is a bit more accurate than its predecessor it is also slower [1].

3 Experimental Setup

3.1 Datasets

We evaluate our modifications to MobileNet on the Places365-Standard dataset [21]. The train set of Places365-Standard contains 1.8 million images from 365 scene categories, where there are at most 5000 images per category. To speed up the training of exploratory experiments, we reduce the number of images in each category to 1200. We refer to this reduced dataset as Places365-Reduced. However, our final experiments (Sect. 5.3) are conducted on the whole Places365-Standard dataset. The validation set has 100 images per category and we do not change it.

3.2 Training Settings

We use a single Tesla V100 GPU on NVIDIA DGX Station for training. We train our models with a batch size of 32 for 15 epochs using the Nadam optimizer with 10% dropout in a fully connection layer. The initial learning rate is set to 0.001 and multiplies with a factor 0.95 in each epoch. We tuned the learning rate and decay by grid search using $lr = \{0.001, 0.0001\}$ and the decay strategy as specified above or no-decay. We believe that further hyper-parameter tuning could increase accuracy, but would not be able to change the relative performance of the analyzed models. We also do not perform any data augmentation.

In turn, to evaluate the inference speed of the models, we built a Docker container with 1 CPU, 2 GB RAM, and no GPU. Every model is run 13 times with a batch size of 1 for 365 consecutive forward passes (1 image from each category). The images, as well as the model, are preloaded before running the benchmark.

4   MobileNet Architecture

MobileNet is composed of 1 standard convolutional layer, 13 depthwise separable convolutional layers, and 1 fully-connected layer. All convolutions are followed by a batch normalization and a ReLU activation with the exception of the final fully connected layer which has no nonlinearity and feeds into a softmax layer for classification [9]. Down sampling is handled with strided convolution in the depthwise convolutions as well as in the first layer. A final average pooling reduces the spatial resolution to 1 before the fully connected layer [9].

Although MobileNet is already compact, it introduces two hyper-parameters: width multiplier (α) and resolution multiplier (ρ) which can trade-off between the computational cost and accuracy. The width multiplier α slims a model uniformly at each layer. For a given layer and width multiplier α, the number of input channels M becomes αM and the number of output channels N becomes αN, where $\alpha \in (0; 1]$. The resolution multiplier ρ is applied to the input image and the internal representation of every layer is subsequently reduced by ρ [9].

5   Approach

5.1   Exploration of the Building Block Design Space

Batch normalization and ReLU are essential for proper gradient backpropagation. However, the performance can vary depending on the locations and the number of these layers within a convolutional block [6,8]. As for compact CNNs, it was found that MobileNet2 performs better without ReLU in the projection layer [18], while ShuffleNet performs better without ReLU after depthwise convolution [20].

In this section we experiment with various configurations of a depthwise separable convolutional block as shown in Fig. 1. For each configuration we built a MobileNet network, train it three times on the Places365-Reduced dataset and then the instance that achieved the best Top-1 validation accuracy was

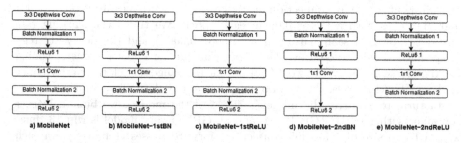

Fig. 1. The structure of a depthwise separable convolutional block. (a) original MobileNet block; (b) MobileNet without the first BN; (c) MobileNet without the first ReLU; (d) MobileNet without the second BN; (e) MobileNet without the second ReLU.

Table 1. Performance of MobileNet with different configurations of the building block.

	MobileNet	1stBN	1stReLU	2ndBN	2ndReLU
Top-1 [%]	42.72	39.08	42.74	39.8	40.05
Inference time [s]	81.1	56.7	81.4	47.6	81.2
Images per second	4.5	6.4	4.5	7.7	4.5

run 13 times with a batch size of 1 for 365 consecutive forward passes. Table 1 reports the best Top-1 validation accuracy, the median inference time for each configuration and the corresponding number of images processed per second.

According to our expectations, removing either the first or the second Batch Normalization layer adversely affects accuracy. At the same time, this significantly reduces the inference time of the network. As for ReLU layers, remaining only the second one is sufficient to ensure nonlinearity. However, somehow surprisingly, when a ReLU layer is removed, the inference time is slightly longer. This is probably a consequence of cheaper multiplication in the subsequent convolution due to a higher number of zeros in the feature maps. To summarize, if inference time is a critical factor, one can consider removing the second Batch Normalization layer. However, in the subsequent experiments we build on the original MobileNet block.

5.2 Shortcut Connections

Recently, the adoption of various types of shortcut connections into CNNs has received increasing attention [6–8,10]. Shortcut connections were proposed by He et al. [7] in their residual network (ResNet) as a remedy to the degradation of training accuracy caused by the increasing depth of networks. Shortcut connections are those skipping one or more layers and, in the straightforward case, are implemented by performing the element-wise addition on the input and output feature maps, channel by channel. Since shortcut connections strengthen information propagation across the network, they make it possible to avoid the degradation problem when training very deep architectures (with more than 1000 layers) [7].

Although we do not struggle with the degradation problem, we use shortcut connections to improve the prediction accuracy of MobileNet. We explore six different variants of MobileNet, each with a different approach to introduce shortcut connections. We refer to these variants as Mod1-6 (see Table 2) and illustrate them in Fig. 2.

To examine the influences of the modifications on the performance of MobileNet, we conducted experiments on Places365-Reduced. The results are presented in Table 3. As for Top-1 accuracy, we report the maximum achieved validation accuracy over 3 runs for each architecture.

Note, that projection shortcuts introduce extra parameters to the model, while identity shortcuts do not. Thereby, projection shortcuts bring in a

Table 2. Modifications to the MobileNet architecture, which incorporate shortcut connections.

Mod1	We insert identity shortcuts if the input and output of the block are the same size. As a result, 8 out of 13 blocks have shortcut connections
Mod2	Identity shortcuts are used in the same way as in Mod1 and projection shortcuts are additionally used for increasing number of channels, but only if the input and output feature maps have the same size. Projection shortcuts are implemented as a 1×1 convolution with the number of filters set equal to the number of output channels, followed by batch normalization. In practice, projection shortcuts are added around block 1
Mod3	Identity shortcuts are used in the same way as in Mod1 and projection shortcuts are additionally used to match the dimensions. If projection shortcuts go across feature maps of two sizes, they are performed with a stride of 2. Accordingly, all blocks have shortcut connections
Mod4	We use identity shortcuts to connect all blocks' outputs that have the same dimensions
Mod5	We insert identity shortcuts if the input and output feature maps have the same size. Zero-padding is used for increasing number of channels. As a result, one more skip path is added as compared to Mod1
Mod6	We use the same shortcuts as in Mod2, but additionally we added a ReLU layer after each addition with the skip path

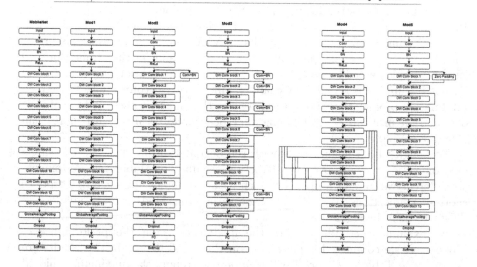

Fig. 2. MobileNet and its variants with shortcut connections

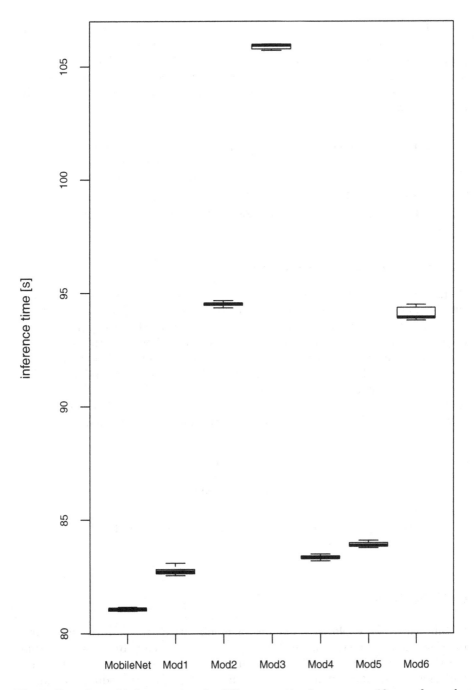

Fig. 3. Box plots of inference time for 365 consecutive images over 13 runs for each model

possibility of overfitting. Indeed, Mod3, which extensively utilizes projection shortcuts, is the only one that does not improves the validation accuracy, but achieves the best training accuracy. All other Mods yield a similar improvement over the original MobileNet. Figure 3 reveals that models with projection short-cuts are much slower than the baseline, while identity shortcuts introduce only a minor slowdown.

Table 3. Performance of different shortcut connection implementations.

	MobileNet	Mod1	Mod2	Mod3	Mod4	Mod5	Mod6
Top-1 [%]	42.72	42.77	43.45	41.65	43.12	43.26	43.37
Params [K]	3581	3581	3583	4283	3581	3581	3583

5.3 Adjusting the Width Multiplier

By introducing shortcut connections, we improved the accuracy, but at the same time slowed down the inference time. In this section, we experimentally try to adjust the width multiplier (α) to find out a model, which outperforms the original MobileNet in regard to accuracy, inference time, and the number of parameters. Since the most promising architectures seem to be Mod2 and Mod5 we train them with a reduced α on the whole Places365-Standard dataset. The results are presented in Table 4. It turns out that Mod5 with $\alpha = 0.92$ dominates over the baseline MobileNet ($\alpha = 1$). However, the main advantage is reduction in model size, which is crucial to avoid overfitting on tasks with a smaller dataset.

Table 4. Performance of the best models.

	MobileNet $\alpha = 1.0$	Mod2 $\alpha = 0.83$	Mod5 $\alpha = 0.92$
Top-1 [%]	49.78	49.92	49.99
Params [K]	3581	2523	3062
Inference time [s]	81.1	81.3	80.3

6 Conclusions

In this paper, we explored a number of modifications to improve MobileNet. First, we investigated the impact of Batch Normalization and ReLU layers within the MobileNet building block on the accuracy and inference time. We found that removing the second Batch Normalization layer significantly reduces the infer-ence time, but at the same time decreases the accuracy by almost 3% points. Then, we investigated six different approaches to introduce shortcut connections into MobileNet. We found that the approaches that use projection shortcuts sig-nificantly slow down the network and may lead to the overfitting problem. On the other hand, the approaches that are based on identity shortcuts introduce only a minor slowdown, while can improve the accuracy. Finally, we demonstrated that it is possible to adjust the width multiplier (α) so that the proposed architecture (Mod5) outperforms the baseline MobileNet regarding accuracy, inference time, and the number of parameters. Although the improvement in accuracy and inference time is practically negligible, the reduction in the number of parameters by 14% is meaningful, especially if there is a need to train the model on a smaller dataset.

For those who want to replicate our experiments on different datasets we made our architectures publicly available at https://github.com/ElpistoleroPL/MobileNetExperiments.

References

1. Bianco, S., Cadene, R., Celona, L., Napoletano, P.: Benchmark analysis of representative deep neural network architectures. IEEE Access **6** (2018). https://doi.org/10.1109/ACCESS.2018.2877890

2. Brzeski, A., Grinholc, K., Nowodworski, K., Przybylek, A.: Evaluating performance and accuracy improvements for attention-OCR. In: 18th International Conference on Computer Information Systems and Industrial Management Applications (CISIM 2019), Belgrade, Serbia (2019)

3. Byra, M., et al.: Impact of ultrasound image reconstruction method on breast lesion classification with neural transfer learning (2018). arXiv:1804.02119

4. Cychnerski, J., Brzeski, A., Boguszewski, A., Marmolowski, M., Trojanowicz, M.: Clothes detection and classification using convolutional neural networks. In: 22nd IEEE International Conference on Emerging Technologies and Factory Automation (ETFA), Limassol, Cyprus (2017)

5. Gholami, A., et al.: SqueezeNext: hardware-aware neural network design. In: ECV Workshop at CVPR 2018, Utah, USA (2018)

6. Han, D., Kim, J., Kim, J.: Deep pyramidal residual networks. In: 30th IEEE Conference on Computer Vision and Pattern Recognition (CVPR), Honolulu, HI (2017)

7. He, K., Zhang, X., Ren, S., Sun, J.: Deep residual learning for image recognition. In: 29th IEEE Conference on Computer Vision and Pattern Recognition (CVPR), Las Vegas, NV (2016)

8. He, K., Zhang, X., Ren, S., Sun, J.: Identity mappings in deep residual networks. In: Leibe, B., Matas, J., Sebe, N., Welling, M. (eds.) ECCV 2016. LNCS, vol. 9908, pp. 630–645. Springer, Cham (2016). https://doi.org/10.1007/978-3-319-46493-0_38

9. Howard, A.G., et al.: MobileNets: efficient convolutional neural networks for mobile vision applications (2017). arXiv:1704.04861

10. Iandola, F.N., Han, S., Moskewicz, M.W., Ashraf, K., Dally, W.J., Keutzer, K.: SqueezeNet: AlexNet-level accuracy with 50x fewer parameters and <0.5 MB model size (2016). arXiv:1602.07360

11. Jaderberg, M., Vedaldi, A., Zisserman, A.: Speeding up convolutional neural networks with low rank expansions. In: 2014 British Machine Vision Conference, Nottingham, UK (2014)

12. Janczyk, K., Czuszynski, K., Ruminski, J.: Digits recognition with quadrant photodiode and convolutional neural network. In: 11th International Conference on Human System Interaction (HSI 2018), Gdansk, Poland (2018)

13. Krizhevsky, A., Sutskever, I., Hinton, G.E.: ImageNet classification with deep convolutional neural networks. In: Advances in Neural Information Processing Systems, pp. 1097–1105 (2012)
14. Mehta, S., Rastegari, M., Shapiro, L., Hajishirzi, H.: ESPNetv2: a lightweight, power efficient, and general purpose convolutional neural network (2018). arXiv:1811.11431
15. Podlodowski, L., Roziewski, S., Nurzynski, M.: An ensemble of deep convolutional neural networks for marking hair follicles on microscopic images. In: 2018 Federated Conference on Computer Science and Information Systems (FedCSIS 2018), Poznan, Poland (2018). https://doi.org/10.15439/2018F389
16. Przybylek, K., Shkroba, I.: Crowd counting á la Bourdieu. In: Workshop on Modern Approaches in Data Engineering and Information System Design at ADBIS 2019, Bled, Slovenia (2019)
17. Redmon, J., Divvala, S., Girshick, R., Farhadi, A.: You only look once: unified, real-time object detection. In: 29th IEEE Conference on Computer Vision and Pattern Recognition (CVPR), Las Vegas, NV (2016)
18. Sandler, S., Howard, A., Zhu, M., Zhmoginov, A., Chen, L.: MobileNetV2: inverted residuals and linear bottlenecks (2018). arXiv:1801.04381
19. Siam, M., Gamal, M., AbdelRazek, M., Yogomain, S., Jagersand, M., Zhang, H.: A comparative study of real-time semantic segmentation for autonomous driving. In: 2018 IEEE/CVF Conference on Computer Vision and Pattern Recognition Workshops, Utah, USA (2018)
20. Zhang, X., Zhou, X., Lin, M., Sun, J.: ShuffleNet: an extremely efficient convolutional neural network for mobile devices (2017). arXiv:1707.01083
21. Zhou, B., Lapedriza, A., Khosla, A., Oliva, A., Torralba, A.: Places: a 10 million image database for scene recognition. IEEE Trans. Pattern Anal. Mach. Intell. **40**(6), 1452–1464 (2018)

A Blockchain-Based Decentralized Self-balancing Architecture for the Web of Things

Aleksandar Tošić[1,2]([mail]) [iD], Jernej Vičič[2] [iD], and Michael Mrissa[1,2] [iD]

[1] InnoRenew CoE, Livade 6, 6310 Izola, Slovenia
{aleksandar.tosic,michael.mrissa}@innorenew.eu
[2] Faculty of Mathematics, Natural Sciences and Information Technology,
University of Primorska, Glagoljaška ulica 8, 6000 Koper, Slovenia
{aleksandar.tosic,jernej.vicic,michael.mrissa}@famnit.upr.si

Abstract. Edge computing is a distributed computing paradigm that relies on the computational resources of end devices in a network to bring benefits such as low bandwidth utilization, responsiveness, scalability and privacy preservation. Applications range from large scale sensor networks to IoT, and concern multiple domains (agriculture, supply chain, medicine, etc.). However, resource usage optimization is a challenge due to the limited capacity of edge devices and is typically handled in a centralized way, which remains an important limitation. In this paper, we propose a decentralized approach that relies on a combination of blockchain and a consensus algorithm to monitor network resources and, if necessary, migrate applications at run-time. We integrate our solution into an application container platform, thus providing an edge architecture capable of general purpose computation. We validate and evaluate our solution with a proof-of-concept implementation in a national cultural heritage building.

Keywords: Edge computing · Internet of Things ·
Decentralized applications · Blockchain

1 Introduction

In the last few years, edge computing has received a lot of attention as an alternative to cloud computing, due to the multiple advantages it offers, such as low bandwidth usage, responsiveness, scalability [10], and privacy preservation [17]. Edge computing has becomes possible due to the evolution of devices that offer more computational power than ever. Combined with application container platforms such as Docker [3] that mask heterogeneity problems, it becomes possible for connected devices to form a homogeneous distributed run-time environment. Additionally, orchestration engines (i.e., Kubernetes[1]) have been developed that

---

[1] https://kubernetes.io/.

© Springer Nature Switzerland AG 2019
T. Welzer et al. (Eds.): ADBIS 2019, CCIS 1064, pp. 325–336, 2019.
https://doi.org/10.1007/978-3-030-30278-8_34

manage and optimize usage of network, memory, storage, or processing power for edge devices and improve the global efficiency, scalability and energy management of edge platforms. However, such solutions are centralized, which means that they represent a single point of failure (SPOF), which entails several drawbacks, such as lack of reliability and security. The problem is so critical that developments for high availability have been explored, for instance with Kubernetes[2].

This paper proposes a solution that uses a decentralized algorithm that monitors network resources to drive application execution to address this problem. Our solution relies on an original combination of blockchain, a consensus algorithm, and a containerized monitoring application to enable run-time migration of applications, when relevant, according to the network state. It provides several advantages, such as verifiable optimal usage of all devices on the network, better resilience to disconnection, independence from cloud connection, improved privacy and security.

The remainder of this paper is organized in 7 sections. Section 2 introduces our motivating scenario related to a cultural heritage building and shows the need for a decentralized approach. Section 3 overviews relevant related work and highlights the originality of our approach. Section 4 details our proposed architecture and shows how it drives run-time migration of applications on the edge. Section 4.2 presents our network monitoring application and shows how monitoring takes place. In Sect. 5, we propose a technical implementation, and we validate and evaluate our solution with a proof-of-concept prototype related to our cultural heritage scenario. Section 6 discusses the results obtained and gives insights for possible future work.

2   Motivating Scenario

In this section, we illustrate the relevance of our approach with a scenario related to a Slovenian cultural heritage building located in Bled, Slovenia. This building has been equipped with multiple sensors to monitor its dynamic environment that affects the building and its contents. The collected data includes temperature, CO_2, relative humidity, Volatile Organic Compounds (VOC), ambient light and atmospheric pressure. In this scenario, the following constraints motivate the need for a fully decentralized edge computing approach:

- Privacy: collected data about the state of the technological solution being deployed is classified as sensitive information. Although data about the building could be sent to the cloud, data about the state of resources needs to remain local and only accessible for administration purpose and for the deployed solution to self-manage.
- Reliability: centralized orchestration is not appropriate as data collection needs to be resilient to failure of any device. The network of devices needs to adjust to device disconnection at any time and keep operating in an optimal way.

---

[2] https://kubernetes.io/docs/setup/independent/setup-ha-etcd-with-kubeadm.

- Cost: reducing the overall cost by avoiding investing in a cloud infrastructure that involves monthly payments and permanent connection to maintain.
- Scalability: as the number of devices will evolve over time, it is necessary for the solution to be able to adjust to changes and homogeneously spread the computation over the network.
- Performance: reactivity to external events is improved if processing is performed on-site.
- Cost effectiveness: using existing devices that control sensors to perform necessary processing reduces the resource requirements of cloud based solutions, which reduces cost.

In this context, it is relevant to equip devices with the capacity to run applications locally and to self-manage the global network load and distribute it over connected devices, according to the state of the network. In the next section, we present related work and show the need for a decentralized self-managed platform on the edge. We also overview existing solutions to abstract from platform heterogeneity and justify the technological choice of a container platform to support our solution.

3   Background Knowledge and Related Work

A recent study by Taherizadeh et al. [19] shows that no widely-used cloud monitoring tools yet provide an integrated monitoring solution within edge computing frameworks, as some monitoring requirements have not been thoroughly met by any of them. Diallo et al. [6] present AutoMigrate, which incorporates a selection algorithm for deciding what services to migrate that maximizes the availability of migration. The system addresses most of the problems that are discussed in our paper. However, it relies on a single agent to manage services introducing a Single Point Of Failure (SPOF). The most notable difference in our implementation is a decentralized architecture that eliminates the SPOF.

3.1   Choreography Solutions for Edge Computing

Strictly observing the definition of orchestration, it always represents control from one party's perspective. This differs from choreography, which is more collaborative and allows each involved party to describe its part in the interaction [16]. However, to the authors' knowledge, there are no choreography solutions that tackle the problems defined in the previous section. Existing orchestration solutions typically rely on a master/slave model where a node is put in charge of the network and decides to allocate applications to nodes according to an optimization algorithm.

Containers as used in the purpose of this paper are run as a group of namespaced processes within an operating system, avoiding the overhead of starting and maintaining virtual machines (at the same time providing most of the functionalities).

The selected platform for our research was Docker [3] as it is the most widely used platform and one of the few that can migrate apps at runtime and enables easy communication. The migration is done by pausing the container, dumping the context of the paused container, transferring the context on a different host that can resume the execution given the context.

3.2   Decentralized Self-managing IoT Architectures

Kubernetes [8] is the most widely used orchestration tool, it is the go-to tool for orchestration in the Google cloud, and is the most used in the Microsoft Azure platform and similar products. It is also the most feature-filled orchestration tool available [12]. It has strong community support across many different cloud platforms (in addition to Google cloud, OpenStack, AWS, Azure).

AWS Elastic Container Service (AWS ECS) [1], Amazon's native container orchestration tool, is the best option for orchestration of AWS services as it is fully integrated into the Amazon ecosystem. It thus integrates easily with other AWS tools. The biggest limitation is that it is limited to Amazon services.

Docker Swarm[3] ships directly with Docker (integrates with Docker-compose) and is supposed to have the simplest configuration. However, it lacks some advanced monitoring options as compared to other products like Kubernetes.

Apache Mesos' based DC/OS[4] is a "distributed operation system" running on private and public cloud infrastructure that abstracts the resources of a cluster of machines and provides common services.

All presented architectures still have a common flaw: single point of failure and a lack of integration with edge computing.

There have been some proposed solutions that enable fully decentralized self-managing architectures for the IoT. For example, [11] focuses on a decentralized solution for energy management in IoT architectures connected to smart power grids. In [7], the authors propose a distributed IoT approach for electrical power demand management problems based on "distributed intelligence" rather than "traditional centralized control," with the system improving on many levels. Suzdalenko et al. [18] further develop the former approach by creating a decentralized distributed model of an IoT; where consumers can freely join and leave the system automatically at any time. Niyato et al. [13] present a system that uses machine-to-machine (M2M) communication to reduce the cost of a home energy management system. A distributed and decentralized microscopic simulation that eliminates the central entity and thus avoids the bottleneck in synchronization is presented in dSUMO [4]. In [2], the authors demonstrate the effectiveness of utilizing a publish/subscribe messaging model as connection means for indoor localization utilizing Wireless Sensor Networks (WSNs) through a middle-ware, the results showed that RSS reaches an acceptable level of accuracy for multiple types of applications.

---

[3] https://github.com/docker/swarm.
[4] https://dcos.io/.

However, all the aforementioned contributions are different from the solution we propose in this paper, at two levels. First, they mostly focus on a single specific aspect and find an optimal solution for it, without considering the fact that an IoT architecture involves multiple criteria that require optimization. In our work, we already consider multiple criteria to optimize application migration, while envisioning that this number of criteria can increase in the future. Second, as far as we know, there is no approach that combines a blockchain data structure with a consensus algorithm in a single framework with the objective to drive application migration at run-time on the edge, which is the main contribution of this paper.

4 A Decentralized Self-managing Architecture

In the following, we describe the general architecture that support our edge computing platform. Devices on the edge are nodes running node software and containerization software. A node can join the network by following a network protocol for exchanging known nodes and participate by executing the consensus algorithm. Nodes keep discovering the network by asking connected nodes for peers. For the sake of simplicity, in this paper we consider that the number of nodes remains reasonably limited, so that large scale discovery issues remain out of the scope of this paper.

4.1 General Architecture

Our devices are equipped to allow a specific containerized application (called node app) to introspect the state of the node and handle the diffusion of this information over the network. It also is responsible for maintaining the information about the other nodes up to date, for participating in the consensus algorithm, and for listening to messages coming from the exposed node API.

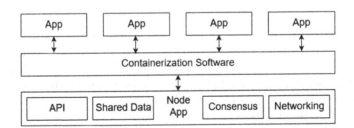

Fig. 1. Architecture of an edge device software platform: a Node App that deals with the consensus algorithm, accesses shared data and exposes the querying API is deployed into the container (in our case Docker).

Figure 1 shows the key components of Nodes in the system. The node software is deployed into the container, in our case Docker. The container mounts a direct socket to the containerization service for querying the state of the system and managing local containers. Docker is useful here to alleviate from the typical heterogeneity problems encountered in the IoT world (different processors and OSes).

4.2   Node Application

Every 500 ms, each device collects information about the state of its neighbours. Typically, a state is a vector of scores that describes the device state and the applications being executed by the node. In this work we define a state to be a matrix of vectors

$$S \ (APP, \ CPU, \ RAM, \ DISK, \ NETWORK, \ TIMESTAMP)$$

where each vector represents an application being executed by the node and the corresponding resource consumption. Resources are reported as a fraction of the total available. In order to have comparable values between nodes, reporting on CPU usage and network utilization the CPU is normalized with the number of cores whereas network bandwidth (download/upload) is measured when transferring containers between nodes.

Monitoring resources within the P2P network is done by having nodes maintain a list of scores of all other nodes (neighbours or not). All nodes periodically send digitally signed messages containing their score to all neighbour nodes. All nodes follow simple P2P broadcasting rules that guarantee finality and efficiency in message propagation.

- If elapsed time greater then ΔST, send signed a message containing own score to all neighbour nodes.
- When receiving a new score message, check if the message was received before (compare digital signatures).
- If the message was not seen before, send it to all connected nodes with the exception of the originating node.

Where ΔST is the time interval in which the container statistics are collected and it is configurable and should depend on the time interval of the consensus algorithm. The score pool hence contains scores of all nodes participating in the network. Each score has a corresponding time-stamp which is later used by elected nodes to create a migration strategy.

Messages containing blocks can become relatively large when the number of applications in the system increases. For improved efficiency, every score message broadcast is prefaced with a "Do you need this" (DYNT) message coupled with the digital signature of the message only. Messages are sent to nodes that reply to the DYNT message to minimize bandwidth use.

Consensus Algorithm. The network requires a consensus algorithm to avoid race conditions when migrating applications. The choice of a consensus algorithm depends on the requirements of the implementation and domain of application. In general, any consensus based on leader election can be plugged in. Examples of such consensus algorithms are Paxos [9], Raft [15], PoET [14], etc. However, in our implementation PBFT [5] was used as it is relatively simple to implement and all its properties satisfy our demands. The only real drawback of the algorithm is that the number of messages increases exponentially with the number of nodes, so it is not applicable to large networks. It was a viable alternative for our proof-of-case implementation with a limited number of nodes. The elected leader is responsible for creating a migration plan and including the resource consumption estimates in a block. The block gets digitally signed so other nodes can verify it originates from the elected leader. Nodes receiving a new block must verify the migration plan by computing it locally and comparing the results. If the migration plan is equal, they act on it, otherwise discard the block and wait for a new one. With these simple protocol rules in place the network is Byzantine fault tolerant [5]. The block verification step is necessary to minimize accidental network forks. A migration strategy is analogous to blocks in block-chain based systems. Blocks contain all the data shared among nodes in the network and include a digital signature of the previous block thus creating a block chain. In order to create a digital signature of block $n + 1$ a node needs to have the digital signature of node n. A well formed block can be verified by other nodes that also have block n. In case of a malformed block, verification will fail, and nodes will reject the block, thus forcing the nodes to agree on the shared data. The block serves as an instruction set mapping applications to nodes. Consider a case with 4 nodes in set N denoted by A, B, C, and D respectively. All nodes share their score and keep a local copy of reported scores of other nodes. Each node stores a vector of applications $v \in V$ that need to be executed. Each node has a canonical list of block B of size k where k is the current block height. Table 1 shows an example of a block k which assigns every $v \in V$ to a node $n \in N$ To create block $k + 1$ a node elected as leader computes an assignment such that the use of resources is optimized (improved). The input to the Algorithm 1 is limited to block data to ensure determinism that can enforce consensus. The Algorithm 1 depends on the application domain and exploring available possibilities will be subject to future work. In this paper, we use the simple described in Algorithm 1, which is deterministic and can only take the block data as input for computation. Once a block is created, currently reported scores are included that will be used to compute block $k + 2$. Additionally, blocks are equipped with meta-data like block hash, previous block hash, etc. to facilitate their utilization.

5   Implementation and Evaluation

5.1   Technical Implementation

As described in Sect. 2, we have implemented and evaluated our solution with a set of sensors deployed in the cultural heritage building Mrakova Domačija in Bled, Slovenia. Each sensor is connected to a Raspberry Pi device that hosts a Linux Alpine OS in a Docker container. The container has access to the docker

Data: BlockData
Result: Migration plan
$Max \leftarrow FindMaxLoadedNode(BlockData)$;
$Min \leftarrow FindMinLoadedNode(BlockData)$;
if *!AppQueue.isEmpty()* **then**
    while *!AppQueue.isEmpty()* **do**
        $Min \leftarrow FindMinLoadedNode(BlockData)$;
        $Min.addApp(AppQueue.dequeue())$;
    end
else
    $AppToMigrate \leftarrow Max.MaxLoadApp$;
    $CurrentDeltaScore \leftarrow (Max.score - Min.score)$;
    $FutureDeltaScore \leftarrow (Max.score - AppToMigrate.score) - (Min.score + AppToMigrate.score)$;
    if $Math.abs(CurrentDeltaScore > FutureDeltaScore)$ **then**
        Migrate $AppToMigrate$ to Min;
    end
end

Algorithm 1. Deterministic migration plan generation algorithm

Table 1. Block data

V	Node	RAM	DISK	CPU	Average latency
v_0	A	50%	23%	90%	23 ms
v_1	B	47%	87%	23%	33 ms
v_2	C	12%	25%	15%	51 ms
v_3	A	35%	14%	56%	101 ms
v_4	D	25%	74%	16%	9 ms

daemon via unix socket. We developed our node application inside a container, it relies on the Docker introspection capacity (`docker stats` command called from our Java program) to collect information about each device. The devices simply collect temperature and relative humidity measurements and calculate their averages. It also hosts a HTTP server[5] that exposes a RESTful API providing access to the system. In such a decentralized system, interaction can be done by any node in the network as follows:

– HTTP GET gives a representation of the target node, which includes information about the state of the device as well as all the necessary information about the node (i.e., last connection time, average connection time, etc.). HTTP GET enables users to view the shared pool of resource stats nodes maintain. Most importantly, it gives a list of all applications in the system.

---

[5] Please note that CoAP could be used for energy saving purposes.

- HTTP PUT/POST enables users to queue an application to be run by the system.
- HTTP DELETE is utilized when an application must be deleted from the queue.

In order to deploy our prototype, we use 5 Raspberry Pi 3 Model B+ connected to Arduino Nano via USB (Universal Serial Bus), the Arduino is connected to the sensors via UART (Universal Asynchronous Receiver Transmitter) ports. We have connected DHT22 sensors to the Arduino boards to capture temperature and humidity.

5.2 Validation and Evaluation

To validate the feasibility of our approach and test its scalability we ran performance simulation test cases. In each test case, a fixed number of nodes formed a P2P network. Nodes were assigned applications to execute. Each application had a random execution time and preset resource consumption expressed in fractions between 5%–40%. For the sake of simplicity, only one resource was used (CPU). The simulation ran for 100 blocks with a block time of 1 s. Applications were queued until the average load of the entire system rose above 90%. The migration strategy was implemented based on the algorithm described in Sect. 4.2. Applications arrived in the queue with certain probability, which was gradually increased with the number of nodes in the system. From the reported resource loads of nodes (reported in %), we compute the standard deviation as a measure of how balanced resource consumption is.

In Fig. 2, we observe that the standard deviation remains low even when the number of applications in the system grows. The lower load cases where we can observe higher swings in standard deviations are expected due to the low number of applications in the system. The crossover happens when the number of applications exceeds the number of nodes and migrations can be beneficial. Below the threshold, there are bound to be nodes that do not run any applications. We can observe from Fig. 2a that as the number of nodes is low, resource balancing between nodes is effective earlier, which explains why the measures are less marked than with the other figures, that correspond to test cases where it takes the simulation a longer time to reach the point of crossover where a higher number of applications is distributed over a lower number of nodes.

Figures 2b–d show that the architecture can scale with the growing number of nodes in the network. Additionally, the naive algorithm for creating a migration strategy performed well in distributing load across the system.

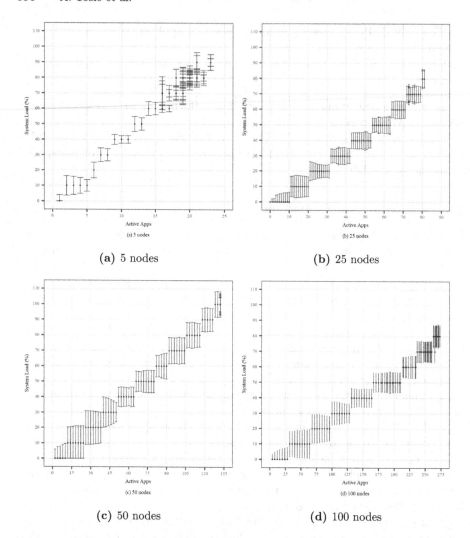

(a) 5 nodes

(b) 25 nodes

(c) 50 nodes

(d) 100 nodes

Fig. 2. Simulation results, error bars are standard deviation of the system load

6   Discussion and Conclusion

In this paper, we propose a decentralized solution to the resource usage optimization problem, a typical issue in edge computing. Our solution avoids the single point of failure that centralized architectures suffer from and improves network resilience as it does not depend on a master node. To design our solution, we have combined a blockchain shared data structure and a consensus algorithm with a monitoring application that runs on top of the Docker platform. Such combination allows edge devices to check at run-time if there is a need for migrating an application, and to reach consensus on a decision to do so. With our contribution, edge devices become a completely decentralized and distributed run-time platform. We have implemented and evaluated our solution with a set of sensors deployed in a cultural heritage building in Bled, Slovenia.

Results show that our approach is able to adjust and normalize the application load over a set of nodes. It also provides, thanks to the fact that the algorithm we use is deterministic and that all the data is stored in a distributed structure, the possibility to verify all the decisions that have been taken to optimize the usage of edge devices. The consensus algorithm that we use also allows adjustments to the global network behaviour for entering or leaving nodes.

Several limitations have been identified that give insights for future work. First, it is important to observe how adding and removing devices affects network behaviour and to explore how scalable our approach is over a large number of devices. Second, it seems appropriate to find out what specific aspects of use cases can help determine which consensus algorithm is most suitable for deploying our solution, in order to best match the use case requirements. Third, it includes semantically describing applications and the services that edge devices offer, to support application migration, and combine in the same architecture the need for efficiently managing network resources together with the needs of applications in terms of functionality and quality of service.

Acknowledgment. The authors gratefully acknowledge the European Commission for funding the InnoRenew CoE project (Grant Agreement #739574) under the Horizon2020 Widespread-Teaming program and the Republic of Slovenia (Investment funding of the Republic of Slovenia and the European Union of the European regional Development Fund). The first author also acknowledges the support of the ARRS grant N1-0093.

References

1. Acuña, P.: Amazon EC2 container service. In: Acuña, P. (ed.) Deploying Rails with Docker, Kubernetes and ECS, pp. 69–98. Apress, Berkeley (2016). https://doi.org/10.1007/978-1-4842-2415-1_4
2. Al-Madani, B.M., Shahra, E.Q.: An energy aware plateform for IoT indoor tracking based on RTPS. Procedia Comput. Sci. **130**(C), 188–195 (2018)
3. Anderson, C.: Docker [software engineering]. IEEE Softw. **32**(3), 102-c3 (2015)
4. Bragard, Q., Ventresque, A., Murphy, L.: Self-balancing decentralized distributed platform for urban traffic simulation. IEEE Trans. Intell. Transp. Syst. **18**(5), 1190–1197 (2017)
5. Castro, M., Liskov, B., et al.: Practical byzantine fault tolerance. In: OSDI, vol. 99, pp. 173–186 (1999)
6. Diallo, M.H., August, M., Hallman, R., Kline, M., Slayback, S.M., Graves, C.: Automigrate: a framework for developing intelligent, self-managing cloud services with maximum availability. Cluster Comput. **20**(3), 1995–2012 (2017)
7. Higgins, N., Vyatkin, V., Nair, N.K.C., Schwarz, K.: Distributed power system automation with IEC 61850, IEC 61499, and intelligent control. IEEE Trans. Syst. Man Cybern. Part C (Appl. Rev.) **41**(1), 81–92 (2011)
8. Hightower, K., Burns, B., Beda, J.: Kubernetes: Up and Running: Dive Into the Future of Infrastructure. O'Reilly Media, Inc., Sebastopol (2017)

9. Lamport, L., et al.: Paxos made simple. ACM SIGACT News **32**(4), 18–25 (2001)
10. Mach, P., Becvar, Z.: Mobile edge computing: a survey on architecture and computation offloading. IEEE Commun. Surv. Tutor. **19**(3), 1628–1656 (2017)
11. Maior, H.A., Rao, S.: A self-governing, decentralized, extensible internet of things to share electrical power efficiently. In: 2014 IEEE International Conference on Automation Science and Engineering (CASE), pp. 37–43. IEEE (2014)
12. Medel, V., Rana, O., Bañares, J.Á., Arronategui, U.: Modelling performance & resource management in Kubernetes. In: 2016 IEEE/ACM 9th International Conference on Utility and Cloud Computing (UCC), pp. 257–262. IEEE (2016)
13. Niyato, D., Xiao, L., Wang, P.: Machine-to-machine communications for home energy management system in smart grid. IEEE Commun. Mag. **49**(4), 53–59 (2011). https://doi.org/10.1109/MCOM.2011.5741146
14. Olson, K., Bowman, M., Mitchell, J., Amundson, S., Middleton, D., Montgomery, C.: Sawtooth: An Introduction. The Linux Foundation, January 2018
15. Ongaro, D., Ousterhout, J.: In search of an understandable consensus algorithm. In: 2014 USENIX Annual Technical Conference (USENIX ATC 2014), pp. 305–319 (2014)
16. Peltz, C.: Web services orchestration and choreography. Computer **36**(10), 46–52 (2003)
17. Satyanarayanan, M.: The emergence of edge computing. Computer **50**(1), 30–39 (2017)
18. Suzdalenko, A., Galkin, I.: Instantaneous, short-term and predictive long-term power balancing techniques in intelligent distribution grids. In: Camarinha-Matos, L.M., Tomic, S., Graça, P. (eds.) DoCEIS 2013. IAICT, vol. 394, pp. 343–350. Springer, Heidelberg (2013). https://doi.org/10.1007/978-3-642-37291-9_37
19. Taherizadeh, S., Jones, A.C., Taylor, I., Zhao, Z., Stankovski, V.: Monitoring self-adaptive applications within edge computing frameworks: a state-of-the-art review. J. Syst. Softw. **136**, 19–38 (2018)

A Business-Context-Based Approach for Message Standards Use - A Validation Study

Elena Jelisic[1(✉)] ⓘ, Nenad Ivezic[2] ⓘ, Boonserm Kulvatunyou[2] ⓘ,
Nenad Anicic[1] ⓘ, and Zoran Marjanovic[1] ⓘ

[1] Faculty of Organizational Sciences, University of Belgrade, Belgrade, Serbia
{elena.jelisic,nenad.anicic,zoran.marjanovic}@fon.bg.ac.rs
[2] National Institute of Standards and Technology, Gaithersburg, MD, USA
{nivezic,serm}@nist.gov

Abstract. While necessary for a successful integration of enterprise services and business-to-business (B2B) applications, message standards can be difficult to use. This paper proposes an innovative, business-context approach and a software tool that we believe can overcome those difficulties. To accomplish both, the paper shows how the business-context approach and the new tool address difficult issues common to traditional approaches of using message standards. The paper also identifies research questions that need to be addressed to increase scalability of the new approach and tool.

Keywords: Enterprise applications · B2B · Services · Integration · Business context · Software

1 Introduction

Message standards are key to the integration of many applications and services, especially in this age of numerous, emerging, ecosystems of services [1,2]. Yet, message standards usage is complex because the standards are very large superset specifications. Moreover, those specifications were developed to address integration requirements from several business processes and multiple industry domains. Consequently, to use the standards, it is necessary to (1) know the implementation-specific language for the standard and (2) have capability to create a *profile* (relevant subset of) the standard for the specific integration situation. Recently, a new software tool [3], based on a UN/CEFACT international standard, has been developed to address both necessities. The tool creates an implementation-neutral representation based on the business context in which the messages are used. The neutral specification facilitates integration; the business context facilitates profiling. In other industries where message standards are used, however, there is currently lack of knowledge of the tool, profiling method, and the impact on different integration scenarios.

© Springer Nature Switzerland AG 2019
T. Welzer et al. (Eds.): ADBIS 2019, CCIS 1064, pp. 337–349, 2019.
https://doi.org/10.1007/978-3-030-30278-8_35

This paper documents an initial validation of the tool and method based on a small but realistic integration use case from the Electric Power industry. The validation is based on the ISO 15000-5 (Core Component Technical Specification) [4] and the Score tool [3]. The validation compares the traditional method of using and integrating message standards with our proposed method. The comparisons focus on the functions associated with profiling including (1) removing the need to know implementation language for the integration; (2) enhancing the ability to capture intent for use of the message standard; (3) decreasing the likelihood of generating new, superfluous, and redundant variations of message standards and their components; and (4) increasing the coherence of the standard by reducing the need for ad-hoc re-use of the its components.

The rest of the paper is organized as follows. Section 2 includes background information about concepts used in the paper. Section 3 describes both a traditional and the newly proposed approach and provides a use case-based analysis of the two on the task of message profiling. Section 4 provides an analysis of approaches on the task of message profiling. Section 5 provides discussion of the analysis results and proposes next research steps. Section 6 completes the paper with conclusions.

2   Background

Score [3] is a novel tool supporting message-standards development and use. It was created in cooperation between Open Applications Group Inc. (OAGi) [5] and National Institute of Standards and Technology (NIST) [6]. The tool, which is based on the Core Component Technical Specification (CCTS) [4], has been used in the development of the latest version of the Open Applications Group Integration Specification (OAGIS) [7]. CCTS is an implementation-neutral, standardization approach that offers two types of data modeling components: Core Components (CCs) and Business Information Entities (BIEs). Together, these components capture both the structure and contents of information-exchange models [8]. CCs are context-independent, conceptual, data-model components. BIEs are logical, data-model components that restrict the underlying CCs to specific business context. Business context (BC) is used to capture intent of a created BIE. UN/CEFACT defines a BC by a set of the context values associated with their corresponding context categories [8]. UN/CEFACT proposed eight BC categories. In this paper we will use four of those categories: Business process role, Geo-political, Activity (Business process), and Industry.

OAGIS uses XML schemas to normatively define the structure of its message specifications in what are called Business Object Documents (BODs). The specifications leave the usage intent of these BODs largely open. BODs follow a standard architecture, developed by OAGi, that contains two main areas: an application area and a data area. The application area conveys integration context information, while data area carries the business and engineering contents. The data area is described using a verb and a noun. A verb is used to define the type of operation that should be conducted using the exchanged business content. A noun is defined using a set of components, which can either be a simple field or a compound field that envelops a set of fields [9].

The Score tool implements an innovative approach that brings new efficiencies to the size of OAGIS. This novel approach increases the reusability of OAGIS's library of standard CCs both at the modeling and XML-Schema levels [10]. This is accomplished using the business context that describes the intent of a profiled message standard. Such an enhanced version of OAGIS is referred to as "business-context-aware OAGIS".

3   Traditional vs. New Approach to Enterprise Applications Integration: A Use Case-Based Comparison

Guided by our intent to validate the use of the new approach, we position the validation activities within accepted practices in business-to-business integration (B2B). This section starts with a simple use case utilized for comparison of two approaches. The first one is a traditional approach, where OAGIS is used as a message standard for defining interoperable structures of an interchanged business documents (i.e. messages). The second one is the proposed approach, where business-context-aware OAGIS will be used for the same purpose. Then, we describe steps that are common to both the traditional and the new approach, which occur prior to the profiling step. Next, we discuss the profiling step in the traditional approach and identify issues in this approach. Finally, we provide details of the profiling step in the proposed approach.

3.1   A Use Case: Complaints Processing

An existing enterprise in Serbia, denoted as Company A, designs new functionality to handle complaints. This enterprise operates in the Electric power generation, transmission, and distribution industry. Besides its power-servicing activities, this enterprise also manages public procurements. The subject of a public procurement can be anything, but this paper will focus on integration with enterprises that supply products to the company. This supplier is denoted as Company B. For ease of understanding, we focus only on the Complaint business document that is created because of the processing of complaints upon receiving a shipment. Figure 1 shows a complete, public-procurement process using a System Context Diagram (SCD). The rectangles in the figure are the actors who share information - the edges in the figure - that is needed to create the public-procurement, complaint document. That document is represented by the thick edge, **Complaint**, from the pubic-procurement function and the business partner, **Supplier** actor. The message, the one that contains the compliant documents, will be the focus of our validation.

Fig. 1. Public procurement System Context Diagram (SCD).

3.2   Traditional or New Approach: Pre-profiling Steps

Within the validation study, there are seven pre-profiling steps that are common to both the traditional and the new approaches.

1. *Identify business partner(s).*
2. *Identify messages (documents) to exchange with business partner(s).*
3. *Identify the part of a business system that is subject of integration process.*
4. *Identify public processes.*
5. *Agree on business document structure.*
6. *Agree on a message standard usage.*
7. *Create a source serialization format – an XML Schema.*

For realization of the first five steps, the FonLabis [11] information-system-design methodology (a best-practice system-identification approach) was used. For identifying business partners and messages, the FonLabis approach was used to create the above SCD. Steps three and four are identified using Data Flow Diagrams (DFDs). In the fifth step, the FonLabis approach was used to access the data models and data dictionaries needed to recognize the structure of the business document that should be interchanged. In the sixth step, review of existing message (or document) standards against both business and technical requirements was conducted. In the seventh step, required database tables, and specific columns inside those tables, are chosen and transformed into XML Schema. Following are some details of the use case-specific realization.

Step 1. Company A manages procurements and collaborates with partners. Some of these partners are internal while others, such as suppliers, are external. Since B2B is the focus of this paper, only external partners will be considered. The supplier in our example is Company B from USA.

Step 2. Since the scope of business partners is narrowed, the only messages considered are those exchanged with Company B. This paper focuses on interchange of the Complaint business document.

Step 3. We have used DFDs to decompose the public procurement process into three sub-processes: Contracting, Realization, and Processing of Complaints.

Since we focused on suppliers and the Complaint business document, the only process of interest is Processing of Complaints.

Step 4. Processing of Complaints is further decomposed into three sub-processes: Receipt Processing, Complaint Sending, and Response Processing. Only the former two processes are communicating with a supplier by exchanging messages. These processes are called public [11]. At this point we define integration domain as *Sending a complaint to Supplier.*

Step 5. Data Dictionary is used to describe the business document structure. The complaint business document structure is presented in Fig. 2. Each field is described by its domain and constraints.

```
Complaint:<Complaint_Number, Customer, Supplier, Receipt_Number, Contract_Number,
Invoice_Number, Date, Complaint_Explanation, Employee>

Customer: <Customer_Name, Customer_Tax_ID, Customer_Address>
Supplier:<Supplier_Name, Supplier_Tax_ID, Supplier_Address>
Customer_Address: <Country, ZIP_Code, Town, Street_Name, Number, [Appartment_Number]>
SupplierAddress: < Number, Street_Name, [Appartment_Number], Town, State, ZIP_Code>
Employee :<Employee_Name, Employee_Surname>
```

Fig. 2. Complaint message data dictionary.

Step 6. OAGIS is a widely used message standard for B2B integration and it is adopted for creating an interoperable Complaint business message structure.

Step 7. By analyzing Company A's database design, we have identified tables and columns inside those tables that are necessary in this use case realization. Thereafter, an SQL view is created over all identified tables and columns. Finally, XML Schema is generated using a user-defined function that takes an existing view and creates a corresponding XML Schema document.

3.3   Traditional Approach: Profiling Step

These activities take place within the Profiling step of the traditional approach:

1. Understand the selected message standard use.
2. Define the message standard intent.
3. Select an adequate message standard schema.
4. Profile the message standard schema.

The first activity needs to answer questions such as (a) What is the structure of a business message? (b) How the message can be customized? (c) What common XML Schemas are used? (d) What integration scenarios exist? (e) What messages are used in the scenarios? Next activity chooses an integration scenario in which the message will be used. Based on the scenario, the user selects an appropriate, standard, message schema. Finally, that selected message schema needs to be profiled meaning that for each element in the source XML Schema, a counterpart must be selected in the standard message schema. Following are details of the use case-specific realization along with issues discovered in the traditional approach.

Step 1. As discussed in Sect. 2, OAGIS BODs cover many functional areas of enterprise. Consequently, OAGIS messages have a very modular and elaborate structure that requires reuse of existing components or fields, as shown in Fig. 3. The graph shows a part of the ***ItemNonconformance*** noun. Nodes colored black are defined in the ***ItemNonconformance*** schema. White and light gray nodes respectively reflect reuses of ***Components*** and ***Fields*** defined in the corresponding modular schemas. The dark gray node ***IdentificationType*** is a reusable node defined in the ***Meta*** schema[1].

Issue 1. In Fig. 3 there is a referenced element ***Party***, defined in ***Components*** schema. This element is described through a set of complex types in an inheritance hierarchy starting from the ***PartyIdentification***, ***PartyBaseType***, and then ***PartyType***. This pattern also repeats in its sub-elements such as ***Location***. As we can see, a long list of complex types and their extensions have to be navigated to get to the whole structure of only one element, ***Party***. Even in this very simple example, it is evident that the structure of messages (1) can be far from simple AND (2) hard to trace through either automatically or manually.

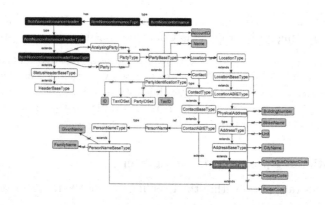

Fig. 3. An illustration of the ItemNonconformance noun and the Party component structures.

Step 2. OAGIS offers several, generic, integration scenarios as starting points for designing use case-specific integration solutions [7]. Each such scenario proposes a set of generic messages to as the basis for integration. By analyzing the OAGIS-provided list of proposed integration scenarios, we have concluded that, for our use case, the most fitting one is Scenario #64 Item Nonconformance.

---

[1] The XML-Schema structure shown in Fig. 3 is for this paper only. It is but one possible structure for ***ItemNonconformance***. It does not conform to any known standard notation.

This scenario hints at several possible integration flows for reporting item or product non-conformance [7]. These flows represent the two possible reasons for a complaint. Two nouns are selected from the scenario including *ItemNonconformance* and *EngineeringChangeOrder*. These nouns are meant to be used with three, selected, use case verbs: *Process*, *Notify* and *Show*.

Issue 2. Scenarios that exist in the OAGIS standard, as stated, are used only as a guide providing starting points for defining one's own specific integration flow. This means that messages identified in the scenario can be used in several, different, integration use cases. Hence, there is a problem of finding and selecting the use case, and its existing messages, that is most relevant to our use case.

Step 3. All identified messages from the chosen scenario (Scenario #64) were inspected and after analyzing their structures, we have selected *NotifyItemNonconformance* BOD for our example.

Issue 3. This type of analysis needs to be done for every new integration use case. Because of the ad-hoc nature of the step, existing message profiles might not be found; even if there were a similar use case and intent from such an existing profile.

Step 4. For illustration, we focus only on profiling the *Party* component from *ItemNonconformance* schema. This element is referenced, and its first-level type extends *PartyBaseType*. For our mapping we needed only *TaxID*, *Name*, *Location* and *Contact* elements. For *Contact* we needed *GivenName* and *FamilyName* elements, and for *Location* all elements that are presented in Fig. 3 are needed.

Issue 4. Figure 3 shows that complex types can be directly restricted or extended (through XML Schema extension/restriction); but, such restrictions or extensions do not apply to complex types of the referenced elements. These types must be restricted (or extended) through new elements with new complex types. So, even though all the needed elements already exist in the original OAGIS schema, we still must introduce a new element *Party* with a newly defined complex type. Similarly, a new element *Contact* with new complex type would also need to be created to restrict *PersonNameType* to *FamilyName* and *GivenName* elements. The same procedure needs to be applied for all reused elements and underlying types that need restrictions or extensions. Consequently, the connections between the newly added elements and their original versions, which are used to give them their intended semantics, are lost.

3.4 New Approach: Profiling Step

Within the *Profiling* step of the new approach, the following activities take place:

1. Define Business Context.
2. Profile message standard component or noun for the defined business context.
3. Export profiled message or noun using XML expression.
4. Reference profiled components from the profiled BOD.

The first activity needs to (a) define business categories, (b) define lists of values for each category, and (c) define requested BCs by choosing values for BC categories. In the second activity, one needs to create new profiles by using the existing components and messages. In addition, we reference the BC in which the components and messages are intended to be used. Next, all these profiled objects are exported using XML expression. When we finally get to XML Schemas, all, and only, the needed profiled components are referenced from the target BOD profile in the final step. Following are details of the use case-specific realization.

Step 1. In this use case, four relevant context categories have been identified, including Business Process Role, Geo-political, Activity, and Industry. Relevant roles in the use case that become context values are Procurement director from Company A, and Sales manager from Company B. Since our enterprises are from Serbia and USA, these two values were defined in the list of values for the Geo-political BC category. Other category values could be countries from all around the world. Using International Standard Industrial Classification of All Economic Activities (ISIC) [12] we have defined the list of values for Industry BC category. Lastly, the list of values for the Activity BC category are obtained from the business processes modeled in the SCD and DFDs. Once all the BC category values are known, the needed BCs are created as shown in Table 1.

Table 1. Business contexts A and B.

BC category	Business contexts	
	BC_A	BC_B
Business process role	Procurement director	Sales manager
Geo-political	Serbia	USA
Activity	Sending a complaint	Receiving a complaint
Industry	Electric power generation, transmission and distribution	Manufacturing

Step 2. A logical data model of BIEs is created using UML class diagram and is presented in Fig. 4. The prefix to a BIE name indicates the BC in which the BIE is intended to be used. All BIEs are obtained by restriction of the underlying CCs. The CC library contains all terms used in the OAGIS messages (whether the type of term is component, field, BOD, or data type). Since the Score tool normalizes the inheritance hierarchy of each CC when a corresponding BIE is created, it was easy to restrict all needed CCs through the corresponding BIEs. There are no new standard components introduced.

Step 3. All created profiles are exported as separate XML Schemas. The Score tool uses a specialization of XML Schema, Naming and Design Rules Technical Specification for transforming BIEs into XML schema elements [10, 13].

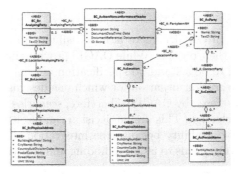

Fig. 4. Logical data model of the ItemNonconformance BIE.

Step 4. In the exported ***ItemNonconformance*** profiled schema, we have imported the other two created schemas for profiled components. Finally, ***ItemNonconformance*** schema elements ***Party*** and ***AnalysingParty*** are changed into referenced elements, thus referencing appropriate elements from the imported schemas.

4   An Analysis of Traditional and New Approaches

This section presents a detailed analysis of the two presented approaches applied to the profiling task. Each use case-specific issue identified in the profiling step of the traditional approach is generalized. In this section we comment on these generalized issues and then show how the proposed approach resolves the issue.

1. Understand the selected, message-standard use.

Generalized Issue: The OAGIS message standard is designed to maximize reusability and extensibility of its components. It is an attractive solution for applications and services integration across functional domains within or between enterprises as well as within and across various industry verticals. This causes the OAGIS structure to be highly compositional. The complex structure demands significant investment to learn to extend and correctly use OAGIS when developing and profiling messages. Also, like virtually all message standards today, OAGIS is an implementation-specific standard and requires extensive XML or JSON knowledge.

New Approach Resolution: The new approach adopts a modular, syntax-independent, CCTS-based representation to manage the OAGIS standard in a model-based manner. Consequently, the OAGIS structure is simplified and made more comprehensive by using the CCTS conceptual and logical components instead of the XML-specific concepts. These components can be presented using UML class diagrams. A UML profile is available to enrich diagrams with CCTS stereotypes.

2. Define the message standard intent.

Generalized Issue: In OAGIS, there is no possibility to record the intent of message standards or their profiles. While OAGIS provides integration scenarios, they are not intended to be a standard specification but only a starting point to describe a similar integration situation for which one designs one's own solution based on the content offered by OAGIS. Therefore, scenarios only give guidelines for OAGIS message selection. This design decision can negatively affect the message standard profile reusability because the integration situation is described imprecisely. This, in turn, leads to issues in finding and selecting existing relevant profiles.

New Approach Resolution: The new approach addresses these issues by enabling a BC definition. BC is an innovative way to define a message standard intent, based on the international CCTS standard. The standard suggests a combination of eight BC categories for representation of a BC. BC can be defined both at the message and the component level. Consequently, BC enables precise description of the integration intent and the component and message profile reusability in related integration scenarios.

3. Select an adequate standard message schema.

Generalized Issue: OAGIS provides a list of integration scenarios that can help users find an appropriate message schema. Using XML tools, every message identified in the scenarios should be analyzed to find the most suitable one. The selected message schema must have a counterpart for each element in the relevant business document. These steps are repeated in every integration use case. This means that redundant and superfluous standard messages are likely to be generated.

New Approach Resolution: The new approach bypasses this issue by using the CCTS representation of OAGIS standard. An appropriate message is chosen by analyzing conceptual, data-model components. Needed components from the conceptual model will be used for creating a logical data model for a specified BC. When the corresponding logical data model is created, it can be transformed into XML Schema or JSON Schema. This logical model can be reused with similar integration scenarios.

4. Profile the standard message schema.

Generalized Issue: Because of the complex structure of the OAGIS canonical message standards in XML Schema, profiling them in that syntax is a demanding undertaking. OAGIS messages are created using inheritance and reference methods. Inherited types can either be restricted or extended, but referenced elements need to be treated separately. The user may use one of the following methods to customize referenced elements (a) Introduce new schema elements with new defined types, or (b) Reference needed elements directly and not through components that contain them. These two resolutions lead to bloated message standards that contain new, superfluous and redundant components. Also, those new standards can cause loss of semantics by ad-hoc combination of existing components.

New Approach Resolution: The user is not working at the XML Schema level; instead he is working at the syntax-independent, conceptual, and logical levels (see Fig. 4). Since a BIE normalizes the inheritance hierarchy of each CC, it is as if each CC exists independently. Moreover, it is easy to restrict all needed CCs through corresponding BIEs, without introducing new ones. The BIEs' names are prefixed with the BC in which they are intended to be used. This increases reusability of existing components. When a logical model is created, it can be transformed into an implementation-specific model. At this time, the tool supports XML Schema and JSON schema serializations. In other words, all the XML Schema-specific requirements, where users had to introduce new elements and types in order to customize original OAGIS Schema, are avoided. This is accomplished by building an implementation-neutral, context-aware model of OAGIS.

5   Discussion and Next Steps

5.1   Profiling Task - Discussion of Analysis Results

We have seen how the traditional approach to message standards use has several issues. Message standards are implementation-specific and require knowledge of an implementation-specific language. In addition, their structure can be very complex, causing the task of message standards profiling to be an exceptionally demanding undertaking. Above all, the most significant shortcoming is that message standards alone do not provide a mechanism to track message standards intent; and there is no existing tool to do so either. As a result, message standards become bloated, grow very large, and become difficult to maintain and use. The newly proposed approach, which is implemented in the Score tool, resolves these issues by providing a simplified and implementation-neutral model for the OAGIS message structure. This model is realized using the CCTS components to build both conceptual and logical data models. Such a representation of the OAGIS messages removes all XML-specific issues and makes message standards more accessible to the integration developer, architect, or the enterprise end user. The new approach provides benefits over a pure, type-based approach. Most significantly, it increases reusability of the Core Components standard's library both at the modeling and schema levels [10].

However, we have also identified research questions that come with the newly proposed approach. These questions need to be addressed for the scalable use of the new tool. Although BC is a valuable concept, it brings new concerns. The lists of categories that describe a BC must be chosen carefully since they should enable the unique identification of the BC. Special attention needs to be paid to given values for each category. Combinations of these values should enable reuse of messages and their components, but only in the right situations. Using many abstract BCs could result in higher chances of inappropriate BIE reuses. Creating rigorous a BC requires in depth analysis of enterprise operating environment and could be time consuming. In addition, BC, at least in parts, should not be defined manually or ad-hoc, but should be defined automatically in some way.

The mapping problem in the traditional approach, however, exists in the proposed approach as well. The ambiguous situations where some elements from the actual business document can be mapped to multiple OAGIS Schema elements still remain and need to be addressed through future research.

5.2   Next Steps

Considering the issues identified in the proposed approach, future research may examine data mining techniques to accomplish BC definition. Namely, such techniques would help identify combinations of context categories expected to adequately define a specific BC. For example, one could envision probabilistic models that take context categories and check the possibility that a given specific BC is plausible. This could allow a user to classify message instances automatically (i.e., assign BC and serve as an outlier detection). For example, one could be alerted if (1) the probability of occurrence of a BC is very low for the given context categories, or (2) that there are additional context categories which are commonly used but were not specified.

There have been some attempts to automate BC definition using business process models. Notably, the collaboration between NIST and OAGi, has led to the development of the Business Process Cataloging and Classification System (BPCCS). The BPCCS tool was developed to create and manage Context Model and to provide a user interface to the Business Process Analyst. The resulting context model is specified along with additional semantic constraints on the process model [14].

The mapping problem could be attacked by introducing the CCTS in the database-design process. This would be achievable only for new, information-system designs where conceptual data models would be created using Core Components. Also, a set of logical data models for different BCs would be provided. This would enable automatic translation of the underlying data structures into OAGIS messages without any risk that incorrect mappings could occur.

6   Conclusion

This paper validates a newly proposed context-based approach for message standards use. A simple, yet realistic integration use case is used as a foundation of the validation. In the traditional approach, the OAGIS message standard is used for defining the structure of exchanged business documents. A number of issues arise in the traditional approach. The paper shows that the new approach can address all the issues by using the novel Score tool to bring context-awareness to the OAGIS standard. Also, although BC is a valuable concept, the paper points at new research challenges that come with the proposed approach. Definition and management of BCs need to be done carefully, not manually or ad-hoc, hence the processes should be automated in some way. The proposed approach still needs to address the mapping problems. The important conclusion is that existence of standards is simply not enough to ensure a cost-efficient, successful application integration. The new approach shows to be a promising avenue to meet these goals.

Disclaimer
Any mention of commercial products is for information only; it does not imply recommendation or endorsement by NIST.

References

1. Barros, A.P., Dumas, M.: The rise of web service ecosystems. IT Prof. **5**(8), 170–177 (2006). https://doi.org/10.1109/MITP.2006.123
2. Cardoso, J., Voigt, K., Winkler, M.: Service engineering for the internet of services. In: Filipe, J., Cordeiro, J. (eds.) ICEIS 2008. LNBIP, vol. 19, pp. 15–27. Springer, Heidelberg (2009). https://doi.org/10.1007/978-3-642-00670-8_2
3. Context-Driven Message Profiling. https://oagi.org/. Accessed 10 June 2019
4. UN/CEFACT Core Components Technical Specification CCTS, version 3.0. https://www.unece.org. Accessed 10 May 2019
5. The Open Applications Group Inc. https://oagi.org
6. The National Institute of Standards and Technology (NIST). https://www.nist.gov
7. OAGIS 10.5 Enterprise Edition documentation. https://oagi.org/
8. Novakovic, D.: Business Context Aware Core Components Modeling. Publikationsdatenbank der Technischen Universität Wien (2014). https://publik.tuwien.ac.at/
9. Kulvatunyou, B., Ivezic, N., Srinivasan, V.: On architecting and composing engineering information services to enable smart manufacturing. J. Comput. Inf. Sci. Eng. **16**(3), 15–27 (2016)
10. UN/CEFACT XML Naming and Design Rules Technical Specification. https://www.unece.org/. Accessed 12 May 2019
11. Jankovic, M.: Specifikacija aspekata interoperabilnosti u metodološkim pristupima razvoju IS. Univerzitet u Beogradu (2016). http://nardus.mpn.gov.rs
12. International Standard Industrial Classification of All Economic Activities (ISIC), Rev. 4. https://oagi.org/, https://doi.org/10.18356/8722852c-en
13. OAGi Specification for Serializing BIE to XML Schema. https://oagi.org
14. Ivezic, N., Ljubicic, M., Jankovic, M., Kulvatunyou, B., Nieman, S., Minakawa, G.: Business process context for message standards. In: CEUR Workshop Proceedings, pp. 100–111 (2017)

A Two-Tiered Database Design Based on Core Components Methodology

Elena Jelisic[1]([✉]) [iD], Nenad Ivezic[2] [iD], Boonserm Kulvatunyou[2] [iD],
Marija Jankovic[3] [iD], and Zoran Marjanovic[1] [iD]

[1] Faculty of Organizational Sciences, University of Belgrade, Belgrade, Serbia
{elena.jelisic,zoran.marjanovic}@fon.bg.ac.rs
[2] National Institute of Standards and Technology, Gaithersburg, MD, USA
{nivezic,serm}@nist.gov
[3] Centre for Research and Technology Hellas, Thessaloniki, Greece
jankovicm@iti.gr

Abstract. The number of Industry 4.0, Internet of Things, and cloud service implementations are growing rapidly. In the resulting, emerging, cross-industry, cooperative environments, a common understanding of message standards will be necessary to enable better semantic interoperability among both traditional enterprise applications and business-to-business applications. In this paper, we first discuss issues with the current state of interoperability, which is based on industry sector-based message standards. We then propose a new database design, which incorporates the Core Components Technical Specification (CCTS) methodology, to resolve those issues. Finally, we analyze benefits that come with the new database design and identify new challenges that should be considered through future research.

Keywords: Database design · Business context ·
Enterprise application · Integration · Global standard

1 Introduction

Evolving fabrication and information technologies are enabling greater cross-industry collaborations. For example, an enterprise specializing in additive manufacturing can provide parts to both aerospace and medical-device sectors. In another example, a vendor of data-analytics can provide those same services to many different business and industry sectors. Today, and for the foreseeable future, such collaborations are, and will continue to be, fueled by the implementations of three concepts: Industry 4.0, Internet of Things, and cloud services. To a large extent, the success of such cross-industry collaboration will depend on solving the semantic interoperability problems associated with those implementations.

Semantic interoperability is not a new problem. It has existed, at least within each vertical, industry sector, for decades. Many existing message standards[1]

---

[1] Message standards are also referred to as document standards and content standards.

© Springer Nature Switzerland AG 2019
T. Welzer et al. (Eds.): ADBIS 2019, CCIS 1064, pp. 350–361, 2019.
https://doi.org/10.1007/978-3-030-30278-8_36

support semantic interoperability within each industry sector. Those sector-based standards, however, are incompatible. They cannot provide the common reference model needed to achieve cross-industry semantic interoperability. Neither individually nor collectively. Recent attempts to develop a common reference model are based on transforming existing message standards to a global message standard. This transformation has been supported by the Core Component Technical Specification (CCTS) methodology [1]. CCTS is an implementation-neutral standardization method that offers two types of data modeling components – Core Components (CCs) and Business Information Entities (BIEs). Together, these two components can capture both the structure and the contents of information exchange models [2].

Inspired by these transformation attempts, we are intrigued by the opportunity to introduce the CCTS **itself** as the basis for a new database. A database that is designed specifically to facilitate cross-industry, semantic interoperability. The core idea is that if the CCs are stored in a common registry, it would be possible to create a **universal**, conceptual, data model. A data model that could be contextualized easily, giving rise to BIEs, to create logical and physical data models. This idea will be elaborated throughout this paper, which documents a preliminary analysis of both the concept and the benefits of creating such a CCTS-based database. The analysis has been performed from two perspectives: using the database to facilitate cross-industry, semantic interoperability and comparing the database to existing approaches that adopt the existing message standards. In this analysis, we particularly focus on the same complex mapping processes that cause semantic interoperability failures between partners within the same industry sector. In performing that analysis, we made four assumptions (1) all considered business partners are using message standards that have adopted the CCTS methodology; (2) semantic interoperability between these standards is achieved by their transformation into a global standard; (3) all Core Components are stored in a common, open-standards based registry that is accessible to all [3]; and (4) a new information system is considered with CCTS-based database design.

The rest of the paper is organized as follows. Section 2 gives a background information about concepts used in the paper. Section 3 describes the considered integration solutions for new information systems and shows a use case that will be used for the analysis of proposed solutions. Section 4 provides discussion of the presented approach and proposes next research steps. Section 5 gives conclusions of the paper.

2   Background

2.1   Industry Sector Efforts

There have been some attempts to create common, reference, data models that could assure semantic interoperability between business partners within a vertical industry sector. In this paper we will mention only two reference models in the healthcare industry. One of them is the Clinical Information System (CIS), which provides a basis for Electronic Health Record (EHR) systems. CIS is a computer-based system that is designed for collecting, storing, manipulating and communicating available clinical information important to the healthcare delivery process [2]. Another is the Health informatics – HL7 v3 Reference Information Model (RIM) [4]. These standards, even though they are from the same industry sector, are far from being completely harmonized and adopted.

Other attempts to create a common reference model have tried using industry-independent, data-modeling languages. In [5], for example, the authors stated that UML is an industry-independent standard that has been used by Standards Development Organizations (SDOs) as a language to formally represent the semantics. Unfortunately, UML, just like all other standardized datamodeling languages, has not been consistently implemented in any industry where it has been adopted.

We draw two conclusions from these examples. First, semanticinteroperability in a given industry sector is still an unsolved problem. Second, neither industry-specific nor industry-independent data modeling languages cannot be the basis for the common reference model needed for cross-industry semantic interoperability.

2.2   Cross-Industry Efforts

In the past few years, cross-industry efforts have focused on transforming sectorspecific message standards into global message standards. There are multiple ways to achieve such a transformation. In this paper, we use the Core Component Technical Specification (CCTS) methodology [1] developed by the UN Centre for Trade Facilitation and Electronic Business (UN/CEFACT). CCTS is an implementation-neutral standardization approach that offers two types of data modeling components – Core Components (CCs) and Business Information Entities (BIEs), which capture the structure and contents of information exchange models [7]. CCs are used for creating conceptual data models, and they are context-free. BIEs are context-specific and they are used for creating logical data models. BIEs restrict the underlying CCs for a specified business context.

UN/CEFACT provides a list of available CCs that can be used for description of exchanged content. This list is called the Core Components Library (CCL). It contains more than 7,000 business entities that can be reused in many scenarios [8]. Example CCs from that list include Document, Contact, Contract, and Person. In our view, the CCL can be the foundation for that missing, common reference model.

Business context (BC) is used to capture the intent of a message. UN/CEFACT defines a business context to be a set of the context values associated with their corresponding context categories [1]. UN/CEFACT provides eight business context categories that can be used for business context description. BIEs are obtained by specifying values or constraints on the values for each of the selected eight business context categories.

SDOs, such as Chemical Industry Data eXchange (CIDX), the Open Applications Group Incorporated (OAGi), Automotive Industry Action Group (AIAG),

Universal Business Language (UBL), and RosettaNet, have already incorporated CCTS methodology or are in the process of adopting the CCTS methodology and standards stack [3]. This transition means that, for example, all components of the Open Application Group Integration Specification (OAGIS) standard are mapped to CCs [6].

From these examples we came to the following conclusions. First, global standard can provide cross-industry semantic interoperability, but this holds only for those message standards that have adopted a common data model, like CCTS, in its core. Second, global standard cannot solve all problems in the traditional approach, like mapping problems, which we discuss below.

3   A Foundation for CCTS-Based Database Design

Inspired by transformations of message standards to global standards, our paper presents a new approach in database design that would incorporate multiple promising techniques towards achieving overall semantic interoperability. First, this section describes a simple use case that will be used as a basis for analysis of a new database design. Then, by analyzing integration requirements, we distinguish two alternatives for achieving semantic interoperability (1) a specific message standard selection, and (2) adoption of CCTS methodology in database design. We consider both alternatives, and then focus our discussion on the latter one. Accordingly, we propose a new foundation for the considered alternative realization.

In performing that analysis, we will use three of the eight categories for BC definition: Geo-political, Activity (Business process) and Industry. When BC is applied on some of the provided CCs, it gives it a necessary semantics in a specific integration scenario. For example, if BC is defined as *Invoicing business process in chemical industry in Serbia*, and is applied on the *Document CC*, we can say that this abstract, implementation-neutral concept *Document* in this specific BC represents an Invoice business document for that business context of the *invoice processing in the chemical industry in Serbia*.

3.1   A Use Case

Company A, a logistics enterprise, wants to provide logistics services to two different kinds of enterprises that currently operate in different European countries, called B and C. Enterprise B is a chemical company that has adopted the CiDX messages standards for their business processes. Enterprise C is an automotive manufacturing company that has adopted the OAGIS message standards for theirs. A System Context Diagram (SCD) that captures the business processes that define the relationships between Company A and Companies B and C is presented in Fig. 1. The rectangles represent the two business partners, B and C. Each business partner is described by its name and the industry sector to which it belongs (in brackets). The oval represents Company A that has its own business process, which differs from both B's and C's business processes.

The directed arcs in the diagram present messages exchanged between Company A and its business partners. Each message is labeled; there are two output *Invoice* messages from Company A; and two, input *PurchaseOrder* messages to Company A. This means that Company A must have two sub-processes: *Receiving a Purchase Order and Sending an Invoice.* Businesses B and C will have similar sub-processes for *Sending a Purchase Order* and *Receiving an Invoice.* Our case study focuses on the *Sending an Invoice* sub-process in Company A.

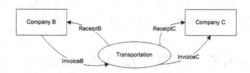

Fig. 1. Company A - Transportation business process SCD.

The problem with implementing this simple diagram is that each company represents *Invoices* (and *PurchaseOrders*) in completely different modeling languages. Company A creates an *Invoice* business document, after any transportation activity is successfully completed, to charge for its services. The structure of the message containing that *Invoice* should be compliant with, and easily converted into, the two semantic interoperability standards used by the business partners: CiDX and OAGIS. We investigated two alternatives for achieving the required cross-industry, semantic interoperability.

1. Company A should adopt one of the two vertical standards either CiDX or OAGIS.
2. Company A should adopt neither vertical standard. Instead, it should incorporate our proposed CCTS-based database.

3.2   Alternative 1: Select One Vertical Standard

This selection means that the existing database design and data modeling language used by Company A to create the structure of an *Invoice* message remain the same. The structure and content of this message do not have to necessarily be compatible with the structure and content of either CiDX or OAGIS. Nevertheless, both CiDX and OAGIS have already incorporated CCTS into their core. This means that, theoretically at least, these industry-specific standards have the basis for a transformation into a single, global standard. If successful, this global standard would further imply that any message created in one of the two standards should be **correctly interpretable** in the other one. Defining the necessary real-world mappings needed for "correctly interpretable" still requires additional, and sometimes difficult, work. So, generally speaking, if Company A were to select one of these vertical standards to represent its *Invoice* message, semantic interoperability should be achievable at a much-reduced cost and risk.

The constraint of this approach, of course, is the difficulty in defining those mappings. This undertaking brings in many issues that can have a negative influence on interoperability. In other words, for each local concept in the local data model, an appropriate counterpart should be found in the chosen message standard. Otherwise, failure to find this match could cause semantic interoperability problems between partners. Also, local data might be lost if there is no adequate counterpart in the message standard structure. These issues can be addressed by adopting a new, database design based on CCTS, as described next.

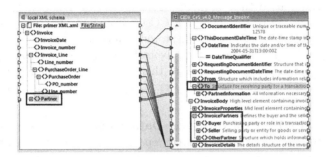

Fig. 2. Partner details mapping.

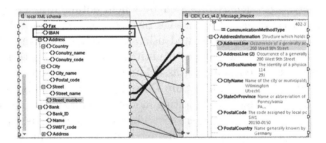

Fig. 3. Address and IBAN mapping.

Figure 2 presents the first constraint of this approach - finding an appropriate counterpart. XML Schema that presents the structure of *Invoice* business document from the local database is on the left side, and on the right side is *CIDX Invoice message schema*. This figure shows that Partner details can be mapped to multiple elements in *CIDX Invoice message schema - To, Buyer, Seller,* and *OtherPartner*. Which one is going to be used depends on user preferences. In addition, this mapping can be implemented differently each time. In other words, there is no consistency in mapping. Figure 3 presents two elements from a local XML schema (Street and IBAN) that do not have adequate counterpart. Actually, *Street name* and *Street number* have the only one fitting element in *CiDX Invoice message schema*, and that is the *AddressLine*. By default, there is only one *AddressLine* element, but in order to map both street details from local schema we had to duplicate *AddressLine* element, otherwise some data from local schema would be lost. The other element that does not have an appropriate counterpart is IBAN. This figure depicts the second issue when local data might be lost if there is no adequate counterpart in the message standard structure.

3.3   CCTS-Based Database Design

The second alternative that Company A considers is a new database design that would make use of the emerging CC environment. This environment (1) provides a collection of CCs and BIEs stored in a common registry accessible to all and (2) enables the creation of a universal, conceptual, data model [3]. This paper proposes a foundation, based on this environment, for designing such a database. That foundation has two tiers: a conceptual, data-model tier and a business-context (BC) tier.

The First Tier - Integral Invoicing Data Model
To develop the integral data model that represents the concept *Invoice*, we propose a three-step process.

1. Examine the CiDX data model of the concept *Invoice*
2. Examine the OAGIS data model of the concept *Invoice*
3. Create a new data model using CCs

Step 1. By analyzing schemas provided by Chem eStandards 4.0 [9] we have decided to use CIDX_CeS_v4.0_Message_Invoice schema. Figure 4 shows the structure of Invoice message presented using UML class diagrams. Those diagrams contain a CiDX-specific, CC list of entities. Those entities describe the various properties and their relationships associated with the concept *Invoice*. Those properties include Invoice Number, Ship Date, Language Code, Issue Date, and Invoice Date, to name a few. The resulting model gives an *Invoice* message structure that is defined inside CiDX Chem eStandards. In other words, this is Invoice structure that is specific for chemical industry. As we stated in Sect. 2, each of the entities presented in Fig. 4 has some predefined CC as its basis.

Step 2. By analyzing schemas provided by OAGIS 10.5 Enterprise Edition [10] we have decided to use the GetInvoice schema. Figure 5 presents the structure of this schema using UML class diagram. This diagram presents OAGIS - specific Invoice message structure. An Invoice is described using Invoice Line and Invoice Header. Invoice Header holds general data, like Tax, Total Amount, details about Supplier, Customer, Billing Party and so on. Invoice Line presents Item-specific data - Quantity, Unit Price, Amount Discount to name a few. As before, each of the entities presented in Fig. 5 has some predefined CC in its basis.

Step 3. The previously described data models, their concepts, and their underlying CCs[2], provide the basis for creating a CCTS-based, conceptual, data model.

---

[2] The list of all CCs is available in [8].

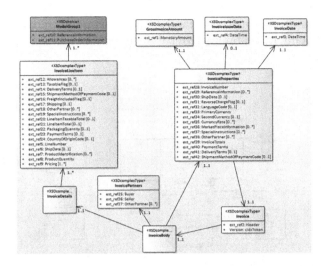

Fig. 4. CiDX Chem eStandards Invoice message UML class diagram.

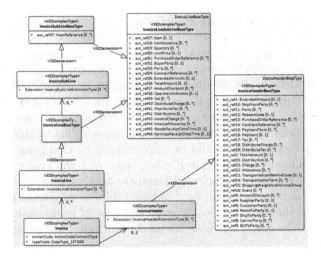

Fig. 5. OAGIS Invoice message UML class diagram.

By manually analyzing the entire list of CCs, we have concluded that the two
individual *Invoice* structures can be interpreted using (1) *Document* Aggregate
Core Components (ACC) that has a list of Basic Core Components (BCCs)
and (2) Association Core Components (ASCCs) used for association with other
ACCs. For easier reading, we will call the BCCs as "fields", the ACCs as "components", and ASCCs will be referred to as "associations". Since the resulting
CCTS-based document is supposed to support the structures of all, Invoice-

related, business document types, its data model is quite complex. Thus, this paper presents only its representative parts without any "fields" specifications.

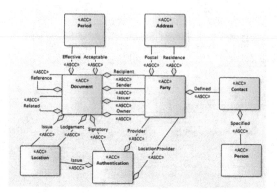

Fig. 6. Document ACC - Conceptual data model.

The resulting conceptual data model is presented using UML class diagrams in Fig. 6. In this model we can see that *Document* has associations with several components including *Party, Period, Location* and *Authentication*. With each of these components, multiple associations can be created. For example, for *Document* we can identify details about its sender party, its recipient, its owner and its issuer. Further, *Party* has details about its *Address* and *Contact Person*. So, using the conceptual data model presented in Fig. 6, complete CiDX or OAGIS Invoice message structures can be interpreted. Finally, this tier presents the database structure that would be implemented by Company A. Using the conceptual data model presented in Fig. 6, Company A will create its own physical data model, specific for a selected Database Management System (DBMS). The main idea is that there will be database tables as presented in conceptual data model and they will be used to store any type of business document used by Company A. In our case, those documents include the *PurchaseOrder* and the *Invoice*.

The Second Tier - Business Context (BC) Definitions

The second tier involves two functions. First, it provides information about all BCs in which the target enterprise operates. Second, it holds definitions for the structure of each business document type. These definitions will be stored in a BC repository that can be implemented inside the individual database schemas.

In Fig. 7, a data model that supports business context definitions is presented as a UML class diagram. In this model, we can see that each BC is described by a set of BC categories. Each BC category has its list of available values. There is also an association class named List of Values that further has an association with BC Category Value. This association is used to denote which BC Category's specific value is applied in a specific BC. Through List of Values association class values for each BC Category are defined to describe some BC.

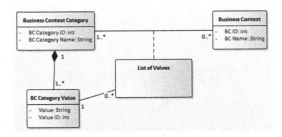

Fig. 7. Data model for Business context tier.

As we stated, the second tier will also hold definitions for each business document type. This function will be implemented through a list of SQL views for each identified business-document type. For example, an SQL view for Invoice would select a list of fields and components (from conceptual data model in Fig. 4) that are needed to describe this type of message. In addition, the same SQL view will reference the BC in which it is supposed to be used. This resulting SQL view will use concepts introduced in the conceptual data model (*Document, Party, Location* etc.) and their names will not be message standard-specific. The assumption is that a business partner will be able to interpret any message defined in this way since it has adopted CCTS-based message standard.

For our use case, we have defined BC_B and BC_C, that denote business environment in which Company A cooperates with its business partners (Companies B and C respectively). These BCs are presented in Table 1. Since our business partners are from Europe, the relevant BC value is defined in the list of values for the Geo-political BC category. Other category values could be countries from all around the world. Using International Standard Industrial Classification of All Economic Activities (ISIC) [11] we have defined the list of values for the Industry BC category. For these BC categories we have assigned values of chemical and automotive, since these are the ones in which Company A's business partners operate. The list of values for the Activity BC category is defined by the business processes of Company A. In our case, this category has two values: *Receiving a Purchase Order* and *Sending an Invoice*.

In Fig. 8 an overall architecture for proposed approach in database design is presented.

Table 1. Business contexts B and C.

BC category	Business contexts	
	BC_B	BC_C
Geo-political	Europe	Europe
Activity	Receiving a Purchase Order	Sending an Invoice
Industry	Chemical industry	Automotive industry

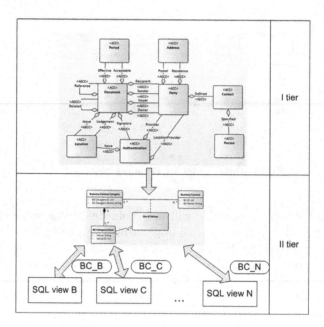

Fig. 8. Database design - an overall architecture.

4   Discussion and Next Steps

In this paper we have seen possible alternatives for achieving cross-industry semantic interoperability. One presented alternative is to adopt a message standard that some of business partners have already incorporated in their business. In this case, local data models need to be mapped to a message structure of the chosen message standard. The paper named some issues that arise in this mapping process. These issues have negative effects on achieving cross-industry, semantic interoperability.

We have also presented another alternative that does not require choosing any specific message standard. In this approach, semantic interoperability is achieved through an integral, conceptual, data model that is based on CCTS. The result is that collaboration can be achieved with any message standard that has adopted CCTS at its core. In addition, a CCTS-based design bypasses mapping process, thus eliminating the identified issues.

This paper also opens three questions that need to be addressed through future research. First, the presented solution is applicable only to information systems being designed from scratch. The future work will consider applying this approach to existing information systems. Second, currently, the unifying conceptual data model can only be created manually and, thus, is error prone. Further research needs to focus on the possibility of automated conceptual data model creation. Third, we can see that the new approach does not provide information about the message standard that is adopted by a business partner. Thus, future research may consider including one additional BC category that would denote message standard that the business partner is using.

5   Conclusion

In this paper, we analyzed alternative integration solutions that are available for new information systems. Each solution is considered in turn and potential problems were identified for each. In particular, this paper presented a foundation for a CCTS-based environment that could support newly proposed approach to information systems database designs. This foundation could contribute to significant improvements in cross-industry semantic interoperability. The paper identifies also a new collection of research questions that need to be addressed to realize CCTS-based database designs.

Disclaimer

Any mention of commercial products is for information only; it does not imply recommendation or endorsement by NIST.

References

1. UN/CEFACT Core Components Technical Specification CCTS, version 3.0. https://www.unece.org. Accessed 10 May 2019
2. Clinical Information System. Biohealthmatics. http://www.biohealthmatics.com/. Accessed 12 June 2019
3. CiDX Core Component White Peper. Open Applications Group Inc. https://oagi.org/. Accessed 14 May 2019
4. HL7 Reference Information Model. HL7 International. http://www.hl7.org/. Accessed 12 June 2019
5. Lopez, D.M., Blobel, B.G.M.E.: Enhanced semantic interoperability by profiling health informatics standards. Methods Inf. Med. **2**(48), 170–177 (2009)
6. Context-Driven Message Profiling. https://oagi.org/. Accessed 10 June 2019
7. Novakovic, D.: Business Context Aware Core Components Modeling, 2nd edn. Publikationsdatenbank der Technischen Universität Wien (2014). https://publik.tuwien.ac.at/
8. UN CCL. The United Nations Economic Commission for Europe. http://tfig.unece.org/. Accessed 10 May 2019
9. Chem eStandards 4.0. Open Applications Group Inc. https://oagi.org/. Accessed 15 May 2019
10. OAGIS 10.5 Enterprise Edition documentation. https://oagi.org/. Accessed 15 May 2019
11. International Standard Industrial Classification of All Economic Activities (ISIC), Rev. 4. https://oagi.org/10.18356/8722852c-en

ADBIS 2019 Workshop: Semantics in Big Data Management – SemBDM and Data-Driven Process Discovery and Analysis – SIMPDA

Using Rule and Goal Based Agents
to Create Metadata Profiles

Hiba Khalid[1,2]([✉]) and Esteban Zimányi[1]

[1] Université Libre De Bruxelles, Brussels, Belgium
{Hiba.Khalid,Esteban.Zimanyi}@ulb.ac.be
[2] Poznan University of Technology, Poznań, Poland

Abstract. Good quality metadata can be a contributing factor when it comes to large scale data integration. In order to minimize data fetch and access request in data lakes, metadata can present adequate solutions that require minimal data access provided metadata exists. Metadata discovery can help us understand how data semantics operate, intrinsic and extrinsic data relationships as well as features that guide query processing, data management, and data integration. Metadata is mostly generated using manual annotation or is discovered through data profiling. What we are looking to explore as a part of our research is to understand available metadata and create profiles that can serve as 'menu card' for the other datasets in the data lake. In this paper, we present a technique for generating metadata profiles using goal based and rule-based agents. To this end, we apply simple rules and guide agents with actionable goals to attain an automatic categorization of a metadata file. Our technique was evaluated experimentally, the results show that applied techniques allow comparing multiple metadata profiles in order to compute similarity and difference measures.

Keywords: Metadata management · Metadata representation · Rule based agents · Metadata categorization

1 Introduction

Data integration is a process that lies at the core of all data management and data analysis. The ever increasing number of large datasets imposes unique challenges such as multi-threaded processing, coherent application access, data integration, data provenance management and data analysis to name a few. The entire process of data integration is error-prone and difficult in terms of handling complexity and managing data. With the numerous big data sets the data integration process and available architectures have become more complex, primarily because of the data heterogeneity and complexity. The industry acceptable and acclaimed architecture for managing big data is a data lake. However, managing and querying data lakes is difficult due to (1) data heterogeneity, (2)

Supported by Erasmus Mundus IT4BI-DC.

data duplication, (3) data volume and (4) data quality. Almost all data engines have a query engine that is designed to understand the content of a data lake (DL) [7], convert data types and formats so they are usable, the engine should also understand the data models and applicable formats and to visualize query results. To fictionalize aforementioned processes, high quality content must be available for the query engine. Such content is commonly regarded as metadata. Generally, there are three main categories of metadata namely (1) Descriptive metadata, (2) Administrative metadata and (3) structural metadata [4]. *Descriptive* metadata is a collection of general information about the data source such as domain descriptions, titles, data summaries, keywords. *Structured* metadata define resource organization and element alignment inside a resource. *Administrative* metadata fulfills the resource management requirements such as system components, schema representation, users, access rights, data characteristics and performance metrics amongst many others. In short, metadata are crucial components in data processing life cycle (data ingestion, pre-processing, delivery) [3,5,9]. Metadata also needed for data consistency and provenance [11,12].

For data integration, the system should be able to access the available and analyze similarities and differences between data sources We propose rule and goal based agents to create metadata profiles, which can be further utilized to execute comparisons with other metadata profiles inside a data lake. Our goal is to answer the following (1) can rule based agents accurately categorize available flat files into categories?, (2) can goal based agents work towards achieving metadata categories for advanced uses such as finding keywords, data range values etc.?, (3) can created metadata profiles be compared and evaluated to find similarities and differences between data sources?. We evaluated our approach on road safety datasets acquired from UK Gov and Kaggle. The experiments demonstrate that proposed method logy functions adequately for finding relevance between different metadata profiles in Sect. 4.

2   Related Work

To understand the importance of metadata we first have to understand the complexity and necessity of data integration [10]. Data integration is inevitable, but it has become costly due to rapid big data production and dynamic changes in large scale systems [8]. Different solutions have been devised to address large scale data integration challenges, but there are still challenges that need to be dealt with using intuitive techniques. Amidst all of techniques, metadata has always been the biggest priority for data integration tasks. However, the challenge with metadata is its availability and quality. Most datasets do not have good quality metadata, which means it needs to be discover so it can be utilized for data integration. For metadata discovery, data profiling [1] is by far the most efficient and accurate technique. It is also noteworthy that metadata assisted management and integration has been successfully accomplished (with due limitations) by researchers in different domains such as in hyper-graphs [3]. Once the issue of metadata availability [9] (manual, profiled) is resolved, its management

[5] is the next big task. Along with metadata management, meta-models [6] can facilitate the process of data integration. The industry giants such as Google have dealt with metadata management and integration cases. Google GOODS [2] is an example of such implementation that takes into account the usage of metadata and its role in managing and organizing Google's datasets effectively. To further the research, automatic and semi-automatic techniques are needed that can learn and function on their own. To best of our knowledge there are no existing techniques that use creation of metadata profiles using AI agents that we propose in our paper. After analyzing the state of the art, we conclude, that an automatic methodology is required for establishing metadata profiles by using available or discoverable metadata.

3 Metadata Profiles

Metadata (MD) comes in many forms and types. It is crucial to differentiate between metadata categories and evaluate which type of metadata contributes the most to data integration. There are numerous challenges (e.g. identifying MD types, identifying data structures, modifying data, metadata preparation, data profiling etc.) when it comes to understanding data integration in regards to MD. The biggest challenge is creating MD, substantially most MD is created manually to date or discovered through data profiling. The second biggest challenge is to use the power of existing MD for facilitating data (analysis, storage, retrieval, integration).

We respond to this challenge by using rule and goal based AI agents to create MD profiles. Figure 2 overviews our approach. In the first step, we evaluate the availability of MD. In context of this research paper, we only focus on the case where we have metadata available. If the datasets under consideration contain MD files, the lower flow is executed i.e. (1) MD pre-processing, this evaluates MD validation such as headers and labels, it also ensures to look for misspelled data, missing values etc. (2) in the second step, MD annotation is carried out to provide stronger meaning to headers, features, textual descriptions or different sections in a MD file. If the MD is not available then the upper flow is executed that ensures discovery of MD using various techniques such as data analysis and data profiling, however the details and implementation of metadata discover (i.e. upper flow) is beyond the scope of this research paper. In the third step (3) we add all accessible MD files to an agent manager or collaborator, whose task is to assign tasks to two types of agents. The first group of agents operates on set of rules to categorize MD files called rule based agents or (RBA) and the second group of agents called goal based agents or (GBA) operate on goals assigned to them to distinguish between different types of MD and add sections to MD profiles based on defined goals. To this end, the following tasks are executed (1) metadata collection, that collects different metadata files whether manually created or profiled (2) metadata pre-processing that ensures no invalidation persists within the MD files before they are handed over to agents (3) agent collaborators, that contains a group of agents programmed to function on a set of rules and goals to categorize metadata and generate profiles that summarizes them. Both flows result in creation of metadata profiles that can further be utilized for matching and analysis in data repositories and data lakes (Fig. 1).

Fig. 1. The metadata profile creation process overview

We propose the concept of creating metadata profiles that can act as summary indexes for preliminary metadata matching. Metadata profiles are similar to data profiles or social media profiles. They provide insight into the content of metadata as well as what can be expected from the corresponding data file. The profiles serve the purpose of access free data matching i.e. when a new dataset is ingested in the data lake it does not have to be evaluated completely. There are two cases (a) if the ingested dataset say X, has an existing metadata source, then the metadata file can be matched with profiles to find compatibility or references and case (b) is designed to extract metadata using EDA and data profiling and then generate metadata profiles, which is beyond the scope of this research paper. In this context, we assume that all metadata for considered datasets is available. It is also evident that not all datasets have metadata profiles by default. These profiles can be extracted from metadata files by deploying RBA and GBA. Once a metadata profile has been created, it is made available for viewing. The viewing concept is a property of profile similar to broadcasting. The broadcasted metadata profiles can be viewed by incoming data sources, and can be evaluated based on keyword matching (KW) represented as 'KW' in Fig. 2, percentage similarity as 'S' or by manual annotations presented as 'M' in Fig. 2. Px is a new profile that has been ingested in the data lake, it can either randomly select from available profiles. It can also send a 'topic sentence' i.e. based on domain and data titles or keywords to different profiles for finding similarity between them as depicted in Fig. 2.

In our experimental setup we focused on datasets with metadata (acquired or profiled). **UK Gov** (https://data.gov.uk/) data files that were downloaded did not contain high quality metadata. However, the dataset was associated with data description and small textual summary. We extracted the summary and descriptions along with data publishing dates etc. The dataset comprises of three main file categories (accidents, casualties, vehicles). The dataset contained label description file (that described different labels, variables in dataset). **Kaggle** (https://www.kaggle.com/datasets) the data files acquired came with some

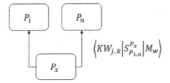

Fig. 2. The metadata profile broadcasting overview

metadata files that contained information such as column count, data types, total number of rows, domain of the data, data publishing data, the origin of dataset, label definitions and additional metadata descriptions. For example, the Kaggle accidents file had data previews, column descriptions, the creation date, data labels, data variables etc. The metadata extraction and management process involves header identification. Thus, words like 'content', 'context', 'manifest' and 'inspiration' are managed as metadata headers. The data source is provided as a collection from 2005–2015. This provided us with a unique future opportunity to access the data overlapping challenge for identifying duplicated using metadata and metadata profile.

3.1   Rule and Goal Based Agents

Our work focuses on minimizing manual effort required to complete metadata descriptions. Rule based systems present a very unique opportunity in this case. These systems are designed with rules that create a knowledge representation upon which actions are conducted. Typically, any rule based system can be utilized for purposes of storage, categorization and retrieval. These rules can be generated in two ways (a) rules created by humans that serve as policy or law book for AI agents to follow and execute commands based on these rules and, (b) rules that are created by a self-taught AI learner. In our research, we have developed rules in accordance with method (a). However, we were creating MD profiles thus, we crafted general rules to provide autonomy to agents and create a more generic aspect for agents. Meaning, the set of rules crafted apply to multiple domains and are effective for most metadata categories. The very first step is to create a rule base that is dynamically manageable. This is mostly defined as 'knowledge base' in AI and expert systems terminology. The next step is to develop a semantic interpreter to understand the rules enforced and actionable response of rules. In terms of rule based systems, this is called as an 'Inference engine'. The inference engine is responsible for executing a cycle to accomplish the task. Its task is to execute and maintain the order in which production rules are to be executed. The cycle comprises of three main phases: (1) matching phase, focuses on executing the production rules or more precisely the IF ELSE THEN statements (2) conflict resolution, it is invoked if several rules match in the production. This phase is responsible for highlighting any contradictory events that might have occurred or might occur in execution. This is also the logical resolution since its job is to look for conflicts or disparities amongst execution strategies laid out for the engine and (3) action: this phase focuses on executing the recorded or stored actions as a result of positive assessment. The negative assessment also has a set of actions such as warning message, throwback calls etc. For example, in our selected datasets, the available metadata comprised of column names, data types, data domain, textual description, missing values, row count, column count, date of creation, and data versions etc. We created a rule system that would handle the most common types of metadata. Access the metadata file under consideration and create a structured profile that can be used for mapping. From the Kaggle dataset, we were able to retrieve aforementioned metadata. The retrieved metadata was collected and rule agents were set to create distinctions between different categories i.e. whether this is descriptive metadata, administrative or structural. Furthermore, all rule sets focused on identifying most common types of metadata. Some examples of the rules are provided below in Table 1. As descried in Table 1 the agents operates on these rules to create profiles. The Table 1 only provides some of the rules to give an overview on how rules are set down. For example, the metadata identifier rule contains multiple rules as its subset. The agent is prioritized to execute identifier rules first. The order of execution can e changed and decided on spot. We wrote a python utility (version 3.7, Numpy, Pandas and scikit-learn libraries) script to create an execution order for agents. This first option in the script functioned on priority based rule execution i.e. when the most critical aspects are handled first such as identifying if there is a metadata file, what is the file name etc. The second option was to let all agents randomly execute all rules and the final execution option was to allow 2 agents at a time to deal with a certain task i.e. if the task was to find attribute lists and features, then the two agents were dedicated to identifying features or attributes within the metadata file (MDF).

Table 1. Sample rule definitions designed for creating metadata profiles

Rule indicator	Rule identifier	Rule description
RBA_101	Data type identifier	Identifies data types (int, text, real etc.)
RBA_102	Column calculator	Identifies attribute names including hyphens or abbrv. terms
RBA_103	Attribute counter	Identifies total number of attributes in MDF
RBA_104	Attribute lists	Generates list of distinct attribute names
RBA_105	Row counter	Identifies total rows from MDF description
RBA_106	Missing data values	Identifies missing values using textual indicators
RBA_107	Creation_Date	Identifies creation data and trace information in MDF
RBA_108	Version management	Identifies data versions, version creation dates
RBA_109	Domain identifier	Identifies keywords describing data domain
RBA_110	Metadata identifier	Identifies MD files, or hierarchy of MDFś
RBA_111	Metadata types	Classifies MD types (structural, administrative etc)

The next phase of our research was to design goal based agents (GBA). These agents have more autonomy as compared to the rule based agents (RBA). The intuition behind designing GBA was to observe how agents can alter their course of action to achieve the task at hand. The agents were designed to create meta-data profiles. Thus, the goals assigned to these agents were of specific nature. We defined three broad goals for the agents. The first goal was to attain a keyword list for each available metadata file. The agents working on this task would both search and select various options to accomplish this task. For example, to analyze a MD file an agent can decide to splice the document in different sections and look for keywords in parallel with collaborating agents. Another course of action would be to scan the MD file and generate keywords. The third course of action would be to look for a header that contains the word keyword or keyword list etc. This can be accomplished by using a word pair dictionary and synonym list. Once the goal has been accomplished, the agent submits a completion report and keeps a log and carries onto the next task. The intention behind log management was to ensure that agents do not loop back to repetitive tasks. It was also designed to incorporate work memory for agents, this working memory ensures that the AI agents can learn from their successful and failed tasks. As a part of coordinated research we aim to develop self-learning AI agents that can exhibit experience and learning without explicit programming, to understand in more detail please see Sect. 5.

Another important goal for our work was to design goals for agents that can find existing connections or representations withing the metadata files. Meaning, if a metadata file contained hyperlinks or references to other relevant metadata files, then our agents should be able to highlight these connections and submit in the report with appropriate headers. We accomplished this by establishing goals such as 'look for: relevant metadata', 'look for: related data', 'look for: relevant data sources' etc. The look for is a defined function that instructs the agent to search and exploit. On the other hand, we supplied synonym dictionary so agents can formulate pairs of words that might occur together and might represent the same context (e.g, relevant, related both words point to possibility of closely related data sources or metadata). The third type of goal we provided to our agents was to look for ranges of data. This goal is associated with data values and data types that are initially determined by rule based agents (RBA). The output from RBA serves as a feed input for these agents. For example, if the available metadata file contains headers title as 'data ranges' etc. then it is easier for the agent to extract this information. However, we wanted to deal with cases where data ranges and value ranges are missing. Thus, the last goal we set for our agents was to determine data range for specific columns such as primary key attributes, foreign key attributes etc. The agent can actually prompt the user to initiate a data search and calculate data range values and populate the metadata file. This indeed queries the data and in cases of large data files this becomes expensive. However, for population of metadata file and for generating metadata profiles, agent can request for a data query. If the user does not want to query the data the it populates the value 'Not available' against data range in metadata profile.

4   Experimental Results

The approach presented in this paper was evaluated by experiments. The experiments were designed to understand the working of rule based agents and goal based agents. The secondary purpose of experiments was to understand how accurately the agents classified the available metadata into profiles. In particular the goals of the experiment were (1) to create metadata profiles (2) to analyze metadata files and categorize them under metadata profile features using rule based and goal based agents (3) to evaluate the correctness of values categorized by designed agents. All experiments were designed in Python including file preparation and metadata pre-processing. The experiments were run on a PC: Intel(R) Core i7, 16 GB of RAM, and Windows 10. The UK Gov and Kaggle datasets were used, as described in Section 3.

4.1 Analyzing Rule and Goal Based Agents

The first step is to gather all the metadata files and dump them in a common place so agents can work on the files and associate different categories (vehicles, casualties, accidents) with file titles, details etc. The next step was to allow rule based agents to access different set of rules and apply them to start extracting information such as (titles, file names, attribute name, categories, data types etc.). For each file that is analyzed by a rule based agent some basic information is recorded and maintained in the log. This information serves as history and record keeping. The information that we collect in this initial phase includes (1) agent number and type, (2) rule set applied, (3) file name on which the rules are applied and (4) date and time of execution.

Table 2. Selected agent performance overview

Agent ID	Recall	Precision	Accuracy	F-1 score
RBA1	0.94	1	0.95	0.96
RBA3	1	1	1	1
RBA6	0.92	0.86	0.85	0.88
GBA4	0.66	1	0.8	0.79
GBA8	0.8	0.8	0.6	0.8
GBA10	1	1	1	1

In our experiment, we gather metadata from the file as a part of the process. Thus, total number of columns or attribute names are extracted using the rules from rule set and similarly the year of publishing is also retrieved amongst many other metadata characteristics. Different agents can take up different rules. Every time an agent is invoked to apply a rule and categorize a metadata file, we record the initiation as well as its correct categorization and incorrect categorization. To figure out whether agents accurately categorized metadata based on rule sets, we manually created a metadata profile for each file the (casualties, vehicles, accidents). Thus, once agents completed the action of applying rules. We would take their generated metadata profile and cross-reference it with a manually created metadata profile by an expert. Figure 3 demonstrates the performance of 9 rule based agents over 20 initiations. Initiations is the total number of time an agent was invoked to process. To limit results for comprehension we selected 20 initiations data for each agent. The Fig. 3 has a total of 9 graphs. Each graph represents an application of a rule set indicated at the top by titles such as RBA_101, RBA_102 etc. The agents correct rule application is recorded as '1' and incorrect categorization is recorded as '0'. For example, the first graph in Fig. 3 is for application of RBA_101 by a rule based agent, this agent was invoked 20 times i.e. the agent applied the rule RBA_101 for 20 times (there were more initiations, but for simplicity we have limited the result set to 20 initiations). Out of these 20 initiations, the agent was able to apply the RBA_101 correctly for 16 times. The incorrect categorization was very low, only 4/20 times the agent was incorrect in categorizing or did not provide any categorization at all. Similarly, for rule application RBA_109 that the agent applied the rule with a all correct categorization.

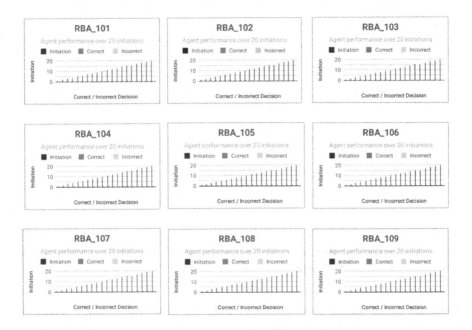

Fig. 3. The overview of agent initiation, correct categorization and incorrect categorization.

The Fig. 4 represents the results of 10 goal based agents. Each agent was assigned five goals. These goals correspond to methodology for finding keywords, finding data ranges etc. see Sect. 3.1. Each agent was evaluated for attaining a goal and this was recorded as its True state i.e. '1'. If the agent failed in attaining a goal then it was recorded as False i.e. '0'. Every time an agent made the wrong decision or failed in completing its goal the agent log was updates. This is regarded as memory for agent to record its failure and update its info regarding the decision and process. In Fig. 4 you can see that each agent is given a title like GBA_101, GBA_102 etc. For instance GBA_108 was able to accomplish all goals assigned, but we experimentally validated if all accomplished goals were accurate or not see Table 3.

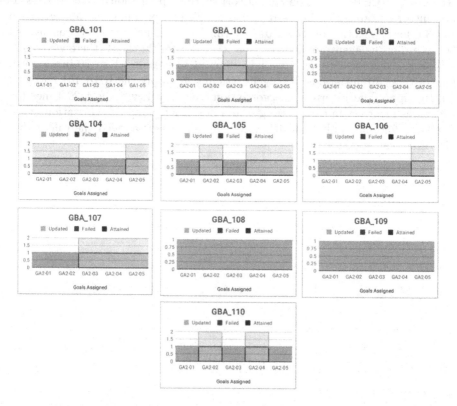

Fig. 4. The overview of goal based agents performance. Each agent was assigned 5 set of goals to test their abilities and correct categorizations.

Analyzing Metadata Profile Matching. We also evaluated the metadata profiles that were generated by the rule and goal based agents. A collection of metadata profiles were generated for evaluation, for simplicity and comprehension we limit the details to already explained examples in Figs. 3 and 4. The Table 2 presents an overview of selected agents performances. For each agent its recall, precision, accuracy and F-1 score were evaluated and recorded. Our Cross-matching or analyzing methodology was to evaluate the generated profiles with manual testing. Thus, each categorization was evaluated by an expert to provide a decision. In this way, we were able to evaluate our agents performance and how accurately they categorized. To further evaluate our results we designed an manual profiles (MP) for all datasets considered. these profiles were expert generated thus were accurate. We also created an anomaly profile (AP) as shown in Table 3 to match against agent generated profiles. the anomaly profiles contains some correct information similar to datasets and some completely wrong metadata information that did not belong to road safety datasets. These profiles were matched against agent generated files to see how dissimilar these profiles were and whether the agents were able to comprehend the metadata accurately or not. The Table 3 presents a detailed overview of profiles matched with each other and how similar they were in content. MDP (MDP1, MDP2, MDP3, MDP4, MDP5) represents metadata profiles generated by rule and goal based agents, AP1 represents the intentionally incorrect metadata profile designed to test agent generated profiles. Finally, MP (MP1, MP2) represent the expert generated metadata profiles to be compared with agent generated profiles, again to test the correctness of agents in generating metadata profiles.

Table 3. Metadata profile matching.

Metadata profiles	Comparison profile	Percentage similarity
MDP1	MDP2	45%
MDP2	MDP3	10%
MDP3	MDP5	37.50%
MDP4	MDP3	22.60%
MDP5	MDP4	42.89%
AP1	MDP1	17%
AP1	MDP2	34%
AP1	MDP3	22%
AP1	MDP4	9%
AP1	MDP5	24%
MP1	MDP1	62%
MP1	MDP2	67%
MP1	MDP3	45%
MP1	MDP4	35%
MP1	MDP5	46%
MP2	MDP1	44%
MP2	MDP2	47%
MP2	MDP3	34%
MP2	MDP4	52%
MP2	MDP5	62%

5   Conclusion and Future Work

In this paper, we proposed a solution that can semi-automate the process of preliminary analysis amongst various datasets in a data lake provided the metadata is available. We contributed by developing two different types of agents and allowed these agents to create metadata profiles that serve as a preamble or information page for other metadata profiles and datasets to interact with. To the best of our knowledge, there are no artificial intelligence agent based systems that propose or resolve the matter of data interactions using metadata profiles. With experiments we demonstrated the effectiveness of using goal and rule based agents instead of manual annotators and human labelling experts. We also demonstrated the accuracy rate of both types of agents which is dominant and provides positive outcome. Finally, we presented results of the comparison between multiple metadata profiles. We conclude, that adequate metadata profiles tend to provide ease and take away expensive data to data matching. Our approach as simple in its methodology, still proved to be accurate and applicable for large scale systems. The future research will be focused on developing agents that can learn from their experiences and actions to minimize failure rate and maximize learning rate.

Acknowledgment. This research has been funded by the European Commission through the Erasmus Mundus Joint Doctorate Information Technologies for Business Intelligence-Doctoral College (IT4BI-DC).

References

1. Abedjan, Z., Golab, L., Naumann, F.: Data profiling. In: IEEE International Conference on Data Engineering (ICDE), pp. 1432–1435 (2016)
2. Halevy, A.Y., et al.: Goods: organizing Google's datasets. In: ACM SIGMOD International Conference on Management of Data, pp. 795–806 (2016)
3. Hewasinghage, M., Varga, J., Abelló, A., Zimányi, E.: Managing polyglot systems metadata with hypergraphs. In: Trujillo, J.C., et al. (eds.) ER 2018. LNCS, vol. 11157, pp. 463–478. Springer, Cham (2018). https://doi.org/10.1007/978-3-030-00847-5_33
4. IEEE Standards Association: IEEE Big Data Governance and Metadata Management (BDGMM). https://standards.ieee.org/industry-connections/BDGMM-index.html
5. Kolaitis, P.G.: Reflections on schema mappings, data exchange, and metadata management. In: ACM SIGMOD-SIGACT-SIGART Symposium on Principles of Database Systems (PODS), pp. 107–109 (2018)
6. Poole, J., Chang, D., Tolbert, D., Mellor, D.: Common Warehouse Metamodel. Developer's Guide. Wiley, Hoboken (2003)
7. Russom, P.: Data lakes: purposes, practices, patterns, and platforms. TDWI White Paper (2017)
8. Suriarachchi, I., Plale, B.: Provenance as essential infrastructure for data lakes. In: Mattoso, M., Glavic, B. (eds.) IPAW 2016. LNCS, vol. 9672, pp. 178–182. Springer, Cham (2016). https://doi.org/10.1007/978-3-319-40593-3_16

9. Varga, J., Romero, O., Pedersen, T.B., Thomsen, C.: Analytical metadata modeling for next generation BI systems. J. Syst. Softw. **144**, 240–254 (2018)
10. Wiederhold, G.: Mediators in the architecture of future information systems. IEEE Comput. **25**(3), 38–49 (1992)
11. Wu, D., Sakr, S., Zhu, L.: HDM: optimized big data processing with data provenance. In: International Conference on Extending Database Technology (EDBT), pp. 530–533 (2017)
12. Wylot, M., Cudré-Mauroux, P., Hauswirth, M., Groth, P.T.: Storing, tracking, and querying provenance in linked data. IEEE Trans. Knowl. Data Eng. (TKDE) **29**(8), 1751–1764 (2017)

On Metadata Support for Integrating Evolving Heterogeneous Data Sources

Darja Solodovnikova[(⊠)], Laila Niedrite, and Aivars Niedritis

Faculty of Computing, University of Latvia, Riga 1050, Latvia
{darja.solodovnikova, laila.niedrite,
aivars.niedritis}@lu.lv

Abstract. With the emergence of big data technologies, the problem of structure evolution of integrated heterogeneous data sources has become extremely topical due to dynamic and diverse nature of big data. To solve the big data evolution problem, we propose an architecture that allows to store and process structured and unstructured data at different levels of detail, analyze them using OLAP capabilities and semi-automatically manage changes in requirements and data expansion. In this paper, we concentrate on the metadata essential for the operation of the proposed architecture. We propose a metadata model to describe schemata and supplementary properties of data sets extracted from sources and transformed to obtain integrated data for the analysis in a flexible way. Furthermore, the unique feature of the proposed model is that it allows to keep track of all changes that occur in the system.

Keywords: Big data · Data warehouse · Evolution · Metadata

1 Introduction

In recent years, the concept of big data has been increasingly attracting attention and interest from researchers and businesses around the world. There are different approaches to the processing and analysis of big data and one of them is to use a data warehouse and OLAP techniques. Various solutions based on Hadoop and similar frameworks allow to implement a data warehouse also to support analysis of big data.

In the context of relational data warehouses, evolution problems have been known for a long time. Evolution can be caused by changes in data sources of the data warehouse, and changes in requirements when additional information becomes necessary for decision making. In the big data world, evolution problems have become even more topical [1, 2] as big data are more dynamic, diverse and can be generated at higher speeds, but the solution to the big data evolution problems is a more challenging task for several reasons. First, there is currently no standard data warehouse architecture that should be used to support big data analysis. Second, in the context of big data, data sources are very often unstructured or semi-structured, and tracking and processing changes in such data sources is a complex task. Finally, in big data systems, the data could be generated in real time, which means that the changes could occur in real time, so they should be processed in real time too.

© Springer Nature Switzerland AG 2019
T. Welzer et al. (Eds.): ADBIS 2019, CCIS 1064, pp. 378–390, 2019.
https://doi.org/10.1007/978-3-030-30278-8_38

In this paper, we propose a data warehouse architecture over big data that can adapt to evolving user requirements and changes in data sources. We present the model that allows to store the metadata describing schemata of involved data sets and their changes that are essential for the operation of the proposed architecture.

The paper is structured as follows. In Sect. 2 the related work is discussed. Section 3 introduces a data warehouse architecture over big data. The main contribution of this paper is presented in Sect. 4, where the metadata model is described. Section 5 discusses evolution support in our proposal. Section 6 is devoted to the description of the case study system. We conclude with directions for future work in Sect. 7.

2 Related Work

The problem of representing metadata of heterogeneous data sources has been studied extensively in the context of data lakes, where structure and other characteristics of data are not defined a priori and are discovered during the data analysis or processing. The paper [3] describes challenges that developers of data lakes are often facing. One of such challenges is obtaining metadata about data extracted from heterogeneous sources. Along with other challenges, the authors mention data evolution when data representation changes over time or the same data are provided in different formats.

The authors of the paper [4] propose a metadata classification to support a data mining process in big data. Although the evolution is not mentioned in the classification and some types of metadata described are specific to data mining tasks, the proposed classification may be used complementary to our metadata model.

The paper [5] proposes a formal metadata model for data lakes. The model is represented as a graph based on the metadata classification proposed by [6]. The model is applied in two algorithms used to obtain a structure from unstructured data lake sources and to integrate sources of different formats.

A metadata model that represents 3 types of metadata describing data lake sources: structure metadata, metadata properties, and semantic metadata is proposed in the paper [7]. The structure metadata describe the schemata of data sources. The authors distinguish between 2 types of structures: matrices and trees. Metadata properties are descriptive information about the contents of data source files, such as file name, author, date modified, including information from the file itself, such as date, information about the experiment for which the source file contains data. Semantic metadata contain annotations of any source elements that could be expressed as URIs that point to a concept in an ontology. The authors use data unit templates to obtain metadata properties that may have different structures. We adapted the metadata model proposed in [7] for our big data warehouse architecture and extended it with metadata necessary for evolution support.

3 Big Data Warehouse Architecture

To solve the topical problem of big data evolution in the data warehousing context, we propose to design a big data management system according to the big data warehouse architecture shown in Fig. 1. The detailed description of the architecture is given in [8].

The architecture consists of a data processing pipeline (*Data Highway* in the figure). The idea and the concept of the data highway was first presented in [9]. Data in the system are gathered from various heterogeneous data sources and loaded in their original format at the first level of the data highway which may be considered as a data lake. Each next level data are obtained from the previous level data by ELT processes by means of applying transformation operations, adding structure to unstructured and semi-structured data, integrating and aggregating data. The number and contents of the levels of the data highway depend on the system requirements. The final level of the data highway is a data warehouse which stores structured aggregated information ready for OLAP operations. Since the volume of data stored in the data warehouse may be too large to provide a reasonable performance of data analysis queries, the architecture may be augmented by the cube engine component, which precomputes various dimensional combinations and aggregated measures for them.

To support big data evolution, we expanded the big data warehouse architecture with an adaptation component that is responsible for handling changes in data sources and levels of the data highway. The main idea of this component is to generate several potential adaptation options to schema of the data highway levels affected by each change in a data source of other level of the highway and to allow a developer to choose the most appropriate option that must be implemented.

One of the central components of the architecture is the metastore that incorporates six types of interconnected metadata necessary for the operation of various components of the architecture. In this paper, we concentrate on the metadata support to maintain information about big data evolution. Three types of such metadata are highlighted with a dark font color in the figure. Schematic metadata describe schemata of data sets stored in the different levels of the highway. Mapping metadata define the logic of ELT processes. Information about changes in data sources and data highway levels is accumulated in the evolution metadata. Such information may be obtained from wrappers or during the execution of ELT processes.

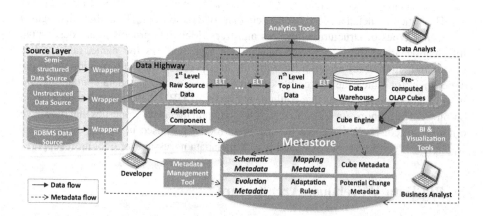

Fig. 1. Big data warehouse architecture

There are also other three types of metadata used for other purposes. Cube metadata describe schemata of precomputed cubes and are leveraged not only during the cube computation process but also for execution of queries. Adaptation rules specify adaptation options that must be implemented for different types of changes. Finally, potential change metadata accumulate proposed changes in the data warehouse schema.

4 Metadata Model

To accumulate metadata about the structure of data sources as well as data sets included at various levels of the data highway and to maintain information about changes that occur in them, we propose the use of metadata that is designed according to the conceptual model presented in Fig. 2.

4.1 Schematic Metadata and Mappings

In this section, we concentrate on the elements of the model that are used to describe schemata of data sources and data highway levels. A class *Data Set* is used to represent a collection of *Data Items* that are individual pieces of data. The *Data Set* class is split into three subclasses according to the type and format. *Structured Data Set* represents a relational database table where data items correspond to table columns. *Semi-structured Data Set* reflects files where data items are organized into a schema that is not pre-defined. The attribute *Type* of a data item incorporated into a such data source indicates the position of it in the schema. For example, XML documents are composed of *Elements* and their *Attributes*, JSON documents may contain *Objects* and *Arrays*. Semi-structured proprietary file formats of software tools or hardware devices may be also described by the *Semi-structured Data Set* class. *Unstructured Data Sets* include data that do not have any organization or acknowledged schema, such as text files, images, other multimedia content, etc. Usually, an unstructured data set is represented by a single data item. However, pieces of supplementary information like keywords or tags that are used to get an idea of the data set content may be available. They are represented as additional data items associated with the corresponding data set in the model.

A data set may be obtained from a *Data Source* or it may be a part of a *Data Highway Level*. We also maintain the information about the rate at which data in the data set are collected or updated by assigning one of the velocity types and frequency attribute of the *Data Set* class. If a data set is a part of a multidimensional model of a data warehouse, its role (dimension or fact table) is assigned to the attribute *Role* and data items contained in such a data set get roles of either dimension attribute or fact table measure.

There usually exist relationships between data items in the same data set or across different data sets determined by the format of these data sets. For example, elements in XML files are composed of other subelements and attributes, objects in JSON files may contain other objects as property values, foreign keys relate columns of relational tables, predicates relate subjects and objects in RDF. There may exist a link between an unstructured data set and structured/semi-structured data. We modelled such relationships by an association class *Relationship* that connects child and parent data items and assigns the corresponding relationship type. The relationship type *Equality* is assigned if two items of different data sets contain the same data. Equality relationships help integrate separate data sets.

Our proposed architecture of the big data system implies that data sets contained in various data highway levels are obtained either from data sources in their original format or from data sets at other data highway levels by performing transformation, aggregation and integrating related data items. To maintain metadata about provenance of data sets within the data highway and make it possible to follow their lineage, an association class *Mapping* was introduced in the model. A mapping defines a way how a target data item is derived from origin data items by a transformation indicated in the attribute *Operation* of the class *Mapping*. Not only our model allows to maintain provenance metadata of data items in the highway calculated directly from source data, but it also supports such cases when previously processed data sets of the highway are further transformed to obtain new data sets at subsequent data highway levels.

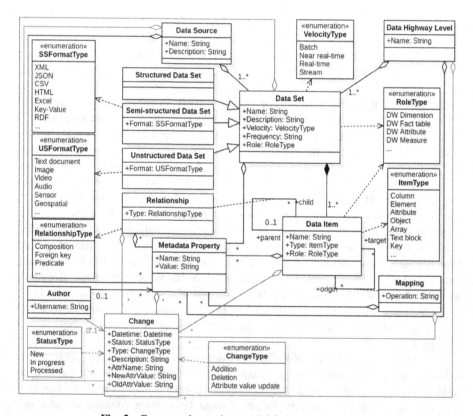

Fig. 2. Conceptual metadata model for evolution support

4.2   Metadata Properties

The classes and associations of the model described above reflect mainly the schematic metadata of the system whereas other characteristics of the data should also be represented in the model. For this purpose, we included a class *Metadata Property* that is used to store various characteristics of different elements of the model. Each metadata property is represented by a name:value pair to allow for some flexibility as metadata properties of different classes may vary considerably. Examples of metadata properties include file or table name, dates created and updated, location, file size, character set, version, check constraints for data sets; type, length, precision, scale, nullable, range for data items; mechanism used to retrieve data from a data source (for instance, API request).

Certain metadata properties may be obtained in a machine-processible form automatically, but others may be only entered manually. Furthermore, an experienced user may discover some new metadata properties of data sets obtained from a data source during data analysis or transformation and of processed data sets that might be valuable for other users of the system. In such a case, the user may augment the metadata by the discovered properties. Such conversational [3] metadata are associated with a user who recorded the property represented by the class *Author* in the model.

4.3   Evolution Metadata

In this section, we discuss classes and associations introduced in the model to store information about the evolution of both the schemata of data sources and data highway data sets and their supplementary characteristics represented as metadata properties. Examples of changes that are supported by the model are given in more detail in the next section.

Evolution is reflected in the model by a class *Change* that is connected by associations with other classes. These associations determine the element of the model that was affected by each change. In the metadata, we store the date and time when the change took place, *Type* of the change, *Status* that determines whether that change is new, processed (propagated in the system) or in progress (being currently processed). In case if evolution was caused by a change in a value of any attribute of a model element including metadata property, we record the name of the affected attribute as an attribute *AttrName* of the class *Change* and both the value before the change (attribute *OldAttrValue*) and after it (attribute *NewAttrValue*). If a change was performed manually by a known user, the corresponding *Author* is associated with such change.

5   Evolution Support

In the proposed big data architecture, data from heterogeneous data sources are integrated and transformed gradually to obtain a data warehouse level of the data highway. Various kinds of changes to the data employed in each step of this process must be recorded in the metadata model. The list of atomic changes classified according to the part of the metadata model they affect is given in Table 1. For each change, the table describes classes that are connected with the instance of the class *Change* in the model by an association, the key attributes of the class *Change* with their value and additional metadata that must be recorded in the model.

Table 1. Supported atomic changes

Change	Associated classes	Key attribute values	Additional metadata
Schematic changes			
Addition of a data source	Data Source	Type: Addition	For each data set included in the new data source, the corresponding schematic metadata and metadata properties must be added
Unavailable data source	Data Source	Type: Deletion	
Addition of a data highway level	Data Highway Level	Type: Addition	For each data set included in the new data highway level, the corresponding schematic metadata and metadata properties must be added along with mappings that define the origin of data
Deletion of a data highway level	Data Highway Level	Type: Deletion	
Addition of a data set	Data Set	Type: Addition	For each data item included in the new data set, the corresponding schematic metadata and metadata properties must be added
Unavailable data set	Data Set	Type: Addition	
Change of data set format	Data Set Data Item Relationship	Type: Deletion Type: Addition	Such change must be recorded as two changes: deletion and addition of a data set. After that, new relationships with the type *Equality* between the corresponding data items of the removed data set and the new data set must be added to the metadata
Renamed data set	Data Set	Type: Attribute value update AttrName: Name	The attributes *OldAttrValue* and *NewAttrValue* of the class *Change* are filled with the previous and updated name of the data set
Addition of a data item	Data Item	Type: Addition	A change is associated with the new instance of the class *Data Item*

(continued)

Table 1. (*continued*)

Change	Associated classes	Key attribute values	Additional metadata
Renamed data item	Data Item	Type: Attribute value update AttrName: Name	The attributes *OldAttrValue* and *NewAttrValue* of the class *Change* are filled with the previous and updated name of the data item
Change of a data item type	Data Item	Type: Attribute value update AttrName: Type	The attributes *OldAttrValue* and *NewAttrValue* of the class *Change* are filled with the previous and updated type of the data item
Deletion of a data item from a data set	Data Item	Type: Deletion	
Addition of a relationship	Relationship	Type: Addition	A new instance of the association class *Relationship* connects two related data items
Deletion of a relationship	Relationship	Type: Deletion	
Addition of a new mapping	Mapping	Type: Addition	A new instance of the class *Mapping* that is being added within the change and associated with target and origin data items defines the way the target data item is obtained in the attribute *Operation*
Deletion of a mapping	Mapping	Type: Deletion	
Changes in metadata			
Addition of a metadata property	Metadata Property	Type: Addition	
Deletion of a metadata property	Metadata Property	Type: Deletion	
Update of an attribute value	A class containing an updated attribute	Type: Attribute value update	The attribute values before and after the update are recorded in the attributes *OldAttrValue* and *NewAttrValue* of the class *Change*, which is associated with the model class instance affected by the change

When elements of the model are deleted, they remain in the model, but information about deletion is maintained as instances of the class *Change* with the type *Deletion*. In case of a deletion of a relationship, such change may affect integration of heterogeneous data sets. This aspect must be considered during change propagation. If a data item or a data set is deleted from the system individually or by deletion of the data source or data highway level containing it, such change affects data sets that were populated with data from a deleted model element (determined by mappings) before the change. If there are any alternative ways how affected data sets may be obtained from other data sets described in metadata, new mappings should be defined by a change "Addition of a new mapping". When a mapping is deleted, it should be replaced by a new mapping to provide successful execution of ELT processes and maintain data lineage. If a replacement is not possible, a change of deletion of a mapping must be registered.

In case of addition of a new element to the system, a new instance of the corresponding class described in the column Associated Class of Table 1 is created in the metadata. If an element containing child elements (such as a data set, data source or data highway level) is added to the system, only one instance of the class *Change* with the type *Addition* is created and associated with the new class containing other child elements.

When a new change is discovered automatically by a source wrapper or during the execution of ELT processes or observed or generated by a user, the corresponding metadata about it must be entered in the metadata repository as an instance of the class *Change* with the status *New*. Thereafter, the adaptation component of the big data warehouse architecture must inject the change and generate possible adaptation scenarios. At this stage the status of the change must be updated to *In progress*. Finally, when a developer or administrator of the system accepts any of the adaptation scenarios proposed by the adaptation component and change propagation is complete, the status of the change must be adjusted to *Processed*.

6 Case Study

As a proof of concept, we have applied our proposed approach to the publications big data system with the purpose to validate the metadata model. The goal of the system is to integrate data about publications authored by employees and students of the University of Latvia from multiple heterogeneous sources and to provide these data for analysis in a data warehouse. The architecture of the developed system that includes data sources and data highway levels is shown in Fig. 3.

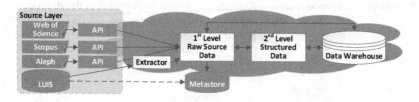

Fig. 3. Architecture of the publication data warehouse

6.1 Data Sources

We integrated data from four structured and semi-structured data sources. The sources contain complementary information about publications, therefore integration of all sources is necessary to obtain the unified view of all publications authored by employees and students of the university.

LUIS is the university data management system implemented in a relational database Oracle. Data about employees and students of the university as well as data about publications entered by LUIS users in the database are gathered and loaded into the first level of the data highway. Aleph is the library data management system. Bibliographical data about publications are obtained from it using API in XML format. All data sets gathered from Aleph have the same structure. Scopus is an indexation system and data from this source are obtained using API in XML format. We used four data set types from this source: publication bibliographical data, author data, affiliation data, and data about publication citation metrics. Web of Science (WOS) is another indexation system. We use API to gather data from WOS in XML format. One type of a data set is available to the university and it contains information about publications, which also includes limited author data (names, surnames and ResearcherID field).

6.2 Data Highway

There are three levels in the data highway of the case study system. Data from the sources are ingested and loaded into the first raw data level. We use Scoop to extract and transfer data from the relational database LUIS into Hive tables. Data from other sources are first pulled from the API and saved in Linux file system and then data are transferred into HDFS using a script with HDFS commands.

At the 2nd level of the data highway, XML files are transformed into structured Hive tables. Data at this level are not yet fully integrated. It is not necessary to transform structured data from the relational database, therefore such data are not included at the 2nd level of the data highway.

Finally, the 3rd level of the highway is a data warehouse implemented in Hive. Data from external data sources that are partially transformed at the 2nd level are integrated with LUIS data directly from the raw source data level.

For the sake of simplicity, we will discuss the fragment of the data warehouse schema represented as a dimensional fact model in Fig. 4. Aggregated data from all four sources are used to populate this star schema. The measures in the fact table contain summarized values classified across four dimensions. Category dimension with the same attribute is used to classify publications by types, such as journal article, book, article in a conference proceedings, etc. Time is a traditional dimension in data warehouses with the quarter of a year granularity in our case study. Faculty dimension contains a hierarchy from a department within a faculty to represent the organizational structure used at the university. Journal metrics dimension stores attributes that characterize a journal or conference proceedings where publications were published.

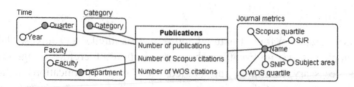

Fig. 4. Dimensional fact model of the publications data warehouse

6.3 Metadata

We implemented the metastore as a relational database Oracle in accordance with the proposed metadata model. Since we have access directly to LUIS data source, we embedded a procedure directly into the data source system that collects metadata about the structure and other metadata properties of tables used to populate the publications system. After source data in XML format are loaded from other data sources, we run a procedure that collects metadata about the structure of XML documents and other metadata properties. Both procedures are able to compare discovered metadata with the information in the metastore and register in the evolution metadata any detected differences.

We defined mappings between data items of the 3 data highway levels. Examples of mappings are given in Table 2. For the sake of clarity, we prefixed names of data items with the corresponding data sources or data highway levels and names of data sets. Question marks denote placeholders for origin data items in the order as they appear in the column *Origin Data Items*.

Table 2. Example mappings of the publication system

Target data item	Origin data items	Mapping operation	Description
Level3.Faculty. Department	Level1.LUIS_person. Affiliation	?	The department of an author of the publication is determined from the column Affiliation of the table LUIS_person
Level3. Publications. Number of Scopus citations	Level2.Scopus_publ. citedby_count	SUM(?)	The number of Scopus citations is calculated as a sum of citations of individual publications extracted from Scopus data source
Level2.Aleph. doc_number	Level1.Aleph. doc_number	?	The item doc_number is obtained from the same item from the 1st level of the data highway
Level3. Publications. Number of publications	Level1.LUIS_publ. Publ_ID Level2.Aleph. doc_number Level2.Scopus_publ. identifier Level2.WOS.uid	COUNT (?) + COUNT (?) + COUNT (?) + COUNT(?)	The number of publications is calculated as a sum of counts of identifiers from all sources. In this case relationships are defined to avoid counting the same publication from multiple sources several times

6.4   Evolution

During operation of the publication system, several changes were discovered during the comparison of the metadata present in the metastore and structure and properties of the data incoming from the data sources. The list of changes with the corresponding metadata additions is presented in Table 3.

Table 3. Changes in the publication system

Change	Change processing
Addition of a data item *citeScoreYearInfoList* to the data set *Scopus_metrics*	A new data item (XML element) was composed of several subelements that were also absent in the previously gathered data sets, however, only one change was created in the metadata for this case and assigned to the uppermost ancestor of the subelements
Deletion of a data item *IPP* from a data set *Scopus_metrics*	This changed affected the data loading process. Since it was not possible to substitute the deleted data item by another data item present in any of the data sources, a deletion of a mapping was recorded in the metadata
Addition of a data source *DSpace*	According to new analysis requirements, the system must have been supplemented by data that contain pre-prints or published full texts of papers. A new data source contained unstructured data (full text files) and metadata associated with them as tags. Files and tags were added as individual data items. The change (addition of a new data source) was associated only with the new data source
Update of a metadata property *API request* value of a data set *Scopus_metrics*	This change was discovered during the execution of the script that extracts data from the API. It had to be processed manually since the new API request could not be discovered automatically. The change was processed as a change in the value of the attribute *Value* of the class *Metadata Property*

7   Conclusions and Future Work

In this paper, we presented a system architecture to address the topical problem of evolution in heterogeneous big data sources. The architecture consists of data highway levels that contain data extracted from sources in their original format and gradually transformed and integrated to obtain a structured data warehouse data. The main contribution of this paper is the flexible metadata model that describes the structure and other characteristics of such data sources and data highway levels as well as changes in the structure not only of data sources, but also of data highway levels for the purpose to support data expansion and evolution of user requirements for data. We defined a set of atomic changes and their representations in the model and applied the model in the case study system.

The directions for future work include full implementation of the proposed architecture. We will develop algorithms for automatic and semi-automatic change treatment. We also plan technology improvements to our case study system by incorporating Presto or Spark SQL for data processing to improve performance.

Acknowledgments. This work has been supported by the European Regional Development Fund (ERDF) project No. 1.1.1.2./VIAA/1/16/057.

References

1. Ceravolo, P., et al.: Big data semantics. J. Data Semant. **7**(2), 65–85 (2018)
2. Kaisler, S., Armour, F., Espinosa, J.A., Money, W.: Big data: issues and challenges moving forward. In: Proceedings of 46th Hawaii International Conference on System Sciences, pp. 995–1004 (2013)
3. Terrizzano, I.G., Schwarz, P.M., Roth, M., Colino, J.E.: Data wrangling: the challenging Yourney from the wild to the lake. In: Proceedings of 7th Biennial Conference on Innovative Data Systems Research (CIDR 2015), Asilomar, CA, USA (2015)
4. Bilalli, B., Abelló, A., Aluja, T., Wrembel, R.: Towards intelligent data analysis: the metadata challenge. In: Proceedings of the International Conference on Internet of Things and Big Data - Volume 1, IoTBD, pp. 331–338, Rome, Italy (2016)
5. Diamantini, C., Lo Giudice, P., Musarella, L., Potena, D., Storti, E., Ursino, D.: A new metadata model to uniformly handle heterogeneous data lake sources. In: New Trends in Databases and Information Systems, ADBIS 2018 Short Papers and Workshops, Budapest, Hungary, pp. 165–177 (2018)
6. Oram, A.: Managing the Data Lake. O'Reilly, Sebastopol (2015)
7. Quix, C., Hai, R., Vatov, I.: Metadata extraction and management in data lakes with GEMMS. Complex Syst. Inform. Model. Q. **9**, 67–83 (2016)
8. Solodovnikova, D., Niedrite, L.: Towards a data warehouse architecture for managing big data evolution. In: Proceedings of the 7th International Conference on Data Science, Technology and Applications (DATA 2018), Porto, Portugal, pp. 63–70 (2018)
9. Kimball, R., Ross, M.: The Data Warehouse Toolkit: The Definitive Guide to Dimensional Modeling, 3rd edn. Wiley, Hoboken (2013)

The Impact of Event Log Subset Selection on the Performance of Process Discovery Algorithms

Mohammadreza Fani Sani[1(✉)], Sebastiaan J. van Zelst[1,2], and Wil M. P. van der Aalst[1,2]

[1] Process and Data Science Chair, RWTH Aachen University, Aachen, Germany
{fanisani,s.j.v.zelst,wvdaalst}@pads.rwth-aachen.de
[2] Fraunhofer FIT, Birlinghoven Castle, Sankt Augustin, Germany

Abstract. Process discovery algorithms automatically discover process models on the basis of event data, captured during the execution of business processes. These algorithms tend to use all of the event data to discover a process model. When dealing with large event logs, it is no longer feasible using standard hardware in limited time. A straightforward approach to overcome this problem is to down-size the event data by means of sampling. However, little research has been conducted on selecting the right sample, given the available time and characteristics of event data. This paper evaluates various subset selection methods and evaluates their performance on real event data. The proposed methods have been implemented in both the ProM and the RapidProM platforms. Our experiments show that it is possible to speed up discovery considerably using ranking-based strategies. Furthermore, results show that biased selection of the process instances compared to random selection of them will result in process models with higher quality.

Keywords: Process mining · Process discovery · Subset selection · Event log preprocessing · Performance enhancement

1 Introduction

Process discovery, one of the main branches of *process mining*, aims to discover a process model that accurately describes the underlying process captured within the event data [1]. Currently, the main research focus in process discovery is on quality issues of the discovered process models; however, at the same time, the ever-increasing size of the data handled in process mining leads to performance issues when applying the existing process discovery algorithms [2]. Some process discovery algorithms are impractical in big data settings. Moreover, some process mining tools impose constraints on the size of event data, e.g., the number of events. Also, in many cases, we do not require the whole event log, and an approximation of the process can already be discovered by only using a small fraction of the event data.

© Springer Nature Switzerland AG 2019
T. Welzer et al. (Eds.): ADBIS 2019, CCIS 1064, pp. 391–404, 2019.
https://doi.org/10.1007/978-3-030-30278-8_39

In real life, process discovery is often of an exploratory nature, that means sometimes we need to apply different process discovery algorithms with several parameters to generate different process models and select the most suitable process model. When the discovery algorithms are used repeatedly, such an exploratory approach makes sense only if performance is reasonable. Thus, even a small improvement in performance may accumulate to a significant performance increase when applied several times. Furthermore, many process discovery algorithms are designed to also generalize the behavior that is observed in the event data. In other words, these algorithms are able to reproduce process behavior extends beyond the example behavior used as input. Therefore, it may still be possible to discover the underlying process using a subset of event data.

This research studies the effectiveness of applying biased sampling on event data prior to invoking process discovery algorithms, instead of using all the available event data. In this regard, we present and investigate different biased sampling strategies and analyze their ability to improve process discovery algorithm scalability. Furthermore, the techniques presented allow us to select a user-specified fraction of inclusion of the total available event data. Using the ProM-based [3] extension of RapidMiner [4], i.e., RapidProM, we study the usefulness of these sampling approaches, using real event logs. The experimental results show that applying biased sampling techniques reduces the required discovery time for all discovery algorithms.

The remainder of this paper is structured as follows. In Sect. 2, we discuss related work. Section 3 defines preliminary notation. We present different biased sampling strategies in Sect. 4. The evaluation and corresponding results are given in Sect. 5. Finally, Sect. 6 concludes the paper and presents some directions for future work.

2   Related Work

Many discovery algorithms such as the Alpha Miner [5], the ILP Miner [6,7], and the Inductive Miner [8] first create an abstraction of the event data, e.g., the directly follows graph, and in a second step discover a process model based on it. The performance of all these algorithms depends on different factors such as the number of process instances and the unique number of activities.

Recently, preprocessing of event data has gained attention. In [9,10], techniques are proposed to increase the quality of discovered process models by cleansing the event data. Also, in [11,22] and [12] we have shown that by removing/modifying outlier behavior in event logs, process discovery algorithms are able to discover process models with higher quality. Moreover, [13] uses data attributes to filter out noisy behavior. Filtering techniques effectively reduce the size of the event data used by process discovery algorithms. However, sometimes the required time for applying these filtering algorithms is longer than the process discovery time. Also, these techniques have no accurate control on the size of the sampled event log.

Filtering techniques focus on removing infrequent behavior from event data; however, sampling methods aim to reduce the number of process instances and increase the performance of other algorithms. Some sampling approaches have been proposed in the field of process mining. In [14], the authors recommend a random trace-based sampling method to decrease the discovery time and memory footprint. This method assumes that process instances have different behavior if they have different sets of directly follows relations. However, using a unique set of directly follows relations may different types of process behavior. Furthermore, [15] recommends a trace-based sampling method specifically for the Heuristic miner [16]. In both of these sampling methods, there is no control on the size of the final sampled event data. Also, they depend on the defined behavioral abstraction that may lead to the selection of almost all the process instances. In this paper, we analyze random and biased subset selection methods with which we are able adjust the size of the sampled event data.

3 Preliminaries

In this section, we briefly introduce basic process mining terminology and notations that ease the readability of this paper. Given a set X, a multiset M over X is a function $M : X \to \mathbb{N}_{\geq 0}$, i.e., it allows certain elements of X to appear multiple times. $\overline{M} = \{e \in X \mid M(e) > 0\}$ is the set of elements present in the multiset. The set of all possible multisets over a set X is written as $\mathcal{M}(X)$.

Let X^* denote the set of all possible sequences over a set X. A finite sequence σ of length n over X is a function $\sigma : \{1, 2, ..., n\} \to X$, alternatively written as $\sigma = \langle x_1, x_2, ..., x_n \rangle$ where $x_i = \sigma(i)$ for $1 \leq i \leq n$. The empty sequence is written as ϵ. The concatenation of sequences σ and σ' is written as $\sigma \cdot \sigma'$. We define the frequency of occurrence of σ' in σ by $freq : X^* \times X^* \to \mathbb{N}_{\geq 0}$ where $freq(\sigma', \sigma) = |\{1 \leq i \leq |\sigma| - |\sigma'| \mid \sigma'_1 = \sigma_i, ..., \sigma'_{|\sigma'|} = \sigma_{i+|\sigma'|}\}|$. For example, $freq(\langle b \rangle, \langle a, b, b, c, d, e, f, h \rangle) = 2$ and $freq(\langle b, d \rangle, \langle a, b, d, c, e, g \rangle) = 1$.

Event logs describe sequences of executed business process activities, typically in the context of some cases (or process instances), e.g., a customer or an order-id. The execution of an activity in the context of a case is referred to an *event*. A sequence of events for a specific case is referred as a *trace*. Thus, it is possible that multiple traces describe the same sequence of activities, yet, since events are unique, each trace itself contains different events. Let $\mathcal{A}$ be a set of activities. An event log is a multiset of sequences over $\mathcal{A}$, i.e., $L \in \mathcal{M}(\mathcal{A}^*)$. Moreover, we let each $\sigma \in \overline{L}$ describe a *trace-variant* whereas $L(\sigma)$ denote how many traces of the form σ are presented within the event log. S_L is a subset of event log L, if for any $\sigma \in S_L$, $S_L(\sigma) \leq L(\sigma)$. We call S_L as a sampled event log of L.

Different types of behavior abstractions in an event log could be defined. One abstraction is the directly follows relation between activities that can be defined as follows.

Definition 1 (Directly Follows Relation). *Let a and $b \in \mathcal{A}$ be two activities and $\sigma = \langle x_1, .., x_n \rangle$ a trace in the event log. A directly follows (DF) relation from a to b exists in σ, if there is $i \in \{1, .., n - 1\}$ such that $x_i = a$ and $x_{i+1} = b$ and is denoted by $a >_\sigma b$.*

We can map an event log to a directed graph whose vertices are activities and edges are directly follows relations and we call it directly follows graph (DFG). So, if there is a $a >_\sigma b$ in the event log, there is also a directed edge from a to b in the corresponding DFG of this event log.

In [12], it shows that the occurrence of a low probable sub-pattern, i.e., a sequence of activities, between pairs of frequent surrounding behavior, which we refer to it as behavioral contexts has negative effects on the results of process discovery algorithms.

Definition 2 (Behavioral Context). *We define the set of behavioral contexts present in event log L according to subsequence σ', i.e., $\beta_L \in \mathcal{P}(\mathcal{A}^* \times \mathcal{A}^*)$, as follows:*

$$\beta_L(\sigma') = \{(\sigma_l, \sigma_r) \in \mathcal{A}^* \times \mathcal{A}^* \mid \exists \sigma \in L, \sigma' \in \mathcal{A}^* (\sigma_l \cdot \sigma' \cdot \sigma_r \in \sigma)\}. \tag{1}$$

For example, in trace $\sigma = \langle a, b, c, d, e, f, h \rangle$, $\langle a, b \rangle$ and $\langle e \rangle$ are two subsequences that surround $\langle c, d \rangle$; hence, the pair $(\langle a, b \rangle, \langle e \rangle)$ is a behavioral context. We inspect the probability of contextual sub-patterns, i.e., behavior that is surrounded by the frequent behavioral contexts and denoted by σ' in Eq. 1. Thus, we simply compute the empirical conditional probability of a behavioral sequence, surrounded by a certain context.

Definition 3 (Conditional Contextual Probability). *We define the conditional contextual probability of σ_s, w.r.t., σ_l and σ_r in event log L, i.e., representing the sample based estimate of the conditional probability of σ_s being surrounded by σ_l and σ_r in L. Function $\gamma_L \colon \mathcal{A}^* \times \mathcal{A}^* \times \mathcal{A}^* \to [0,1]$, is defined as:*

$$\gamma_L(\sigma_s, \sigma_l, \sigma_r) = \frac{\sum\limits_{\sigma \in L} \left(|\sigma_{\sigma_l \cdot \sigma_s \cdot \sigma_r}| \right)}{\sum\limits_{\sigma \in L} \left(\sum\limits_{\sigma' \in \mathcal{A}^*} |\sigma'_{\sigma_l \cdot \sigma' \cdot \sigma_r}| \right)} \tag{2}$$

On the basis of these probabilities, we are able to detect unstructured behavior in traces.

4   Subset Selection

In this section, we present different subset selection strategies to improve the discovery procedure's performance. Different behavioral elements of an event log, e.g., events, directly follow relations, traces, and variants can be used for sampling. However, not all of them are useful for the purpose of process discovery. By selecting events, it is possible to consider events from different parts of a process instance that results in imperfect traces that are harmful for all process discovery algorithms. Selecting DF relations is useful for some process discovery algorithms like the Alpha Miner. But, they are insufficient for other process discovery algorithms. Thus, here we make subset of event logs only based on traces and variants. Consequently, these subset selection methods take an event log as an input and return a sampled event log. The schematic of subset selection methods is illustrated in Fig. 1.

Note that in *XES* standard [3], variants are not stored in event logs separately. However, there are other structures that we can keep variants and their frequencies as metadata [17] that are more efficient for process discovery algorithm. Here, we consider *XES* standard; however, in sampled event logs, we keep only one trace for each selected variant. Consequently, the frequency of each trace in the sampled event log equals to 1.

In many process discovery algorithms such as the ILP Miner, the family of Alpha miners and the basic Inductive Miner, the frequencies of traces variants (i.e., $L(\sigma)$) have no important effects on discovered process models. Therefore, here we mainly focus on selecting variants; but, all these methods can easily be extended to trace-based subset selection methods. We also used just control-flow related information that is available in all event logs and this is consistent with the way.

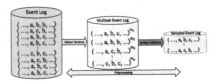

Fig. 1. Schematic overview of event log subset selection

One of the most important characteristics of a sampled event log is the number of its traces, i.e., $|S_L|$. When it is the same as the original event log, there is no reduction in the size. We can set the size of the sampled event log (i.e., the sampling threshold) as $c = \frac{|S_L|}{|L|}$ that $0 < c \leq 1$[1]. Moreover, the less required subset selection time is more desirable as it is considered as a preprocessing phase.

We can select traces in an event log randomly or based on some strategies. In the following, we will explain both of these methods.

4.1 Random Sampling

In this method, we randomly select $c \times |L|$ traces in the event log without replacement and return these traces or just unique the trace-variants among them. This method is fast because we do not need to traverse the original event log. However, it is possible that many of sampled traces have similar behavior and we keep just a few unique variants in the sampled event log. The statistical sampling method [14] works based on this approach.

As an alternative method, we can first find all the unique variants in an event log; afterward, randomly select $c \times |\overline{L}|$ variants from them. This approach is a bit slower; however, it is able to return much behavior compared to the previous approach.

4.2 Biased Sampling Strategies

In general, traversing a big event log is not too time-consuming compared to the time of process discovery. Therefore, as it shown in Fig. 1, instead of randomly

---

[1] Here, we select only one trace for each variant.

selecting the variants, we are able to first find all variants in an event log and use more advanced strategies (biases) to select them. In this type of approaches, we first rank all variants of an event log based on different strategies. Afterward, we return the top $c \times |\overline{L}|$ variants with the highest rank in the sampled event log. We are able to use different ranking strategies that will be discussed as follows. These ranking strategies have different preprocessing time and result in different sampled event logs.

As we select variants, the frequency of behavior in the sampled event log will be unusable. To consider these frequencies, we can benefit from other event log standards that are able to keep frequencies of variants like [17] or instead of returning $c \times |\overline{L}|$ variants, we should return $c \times |L|$ traces that correspond to these high ranked variants.

Frequency-Based Selection: The first ranking strategy is selecting variants based on their frequencies in the original event log. This ranking strategy gives higher priority to a variant that has a higher occurrence frequency in the event log. So, we sort the variants based on their frequencies or $L(\sigma)$ and return the top $c \times |\overline{L}|$ of variants as a sampled event log. The trace-based version of this strategy is already presented in many process mining tools that helps users to keep the top most frequent behavior of the event log. However, in some event logs, the majority of process instances have a unique trace-variant which makes this subset selection method unusable.

Length-Based Selection: We are able to rank variants based their length (i.e., $|\sigma|$). So, in this strategy, we sort variants based on their length and choose the longest or the shortest ones first. By using the *longer* strategy, we keep much behavior in our sampled event log and at the same time leave out many of incomplete traces, that may improve the quality of resulted process models. However, if there are self-loops and other loops in the event log, there is a high probability to choose many infrequent variants with the same behavior for process discovery algorithms. On the other hand, by applying *shorter* strategy, there will be less behavior in the sampled event log; but, it is possible to keep many incomplete traces that leads to an unsuitable process model.

Similarity-Based Sampling: If we are interested in retaining the main-stream behaviors of the event log, we need to rank variants based on the similarity of them to each other. In this approach, we first find common behavior of the event log. For this purpose, we can use different types of behavior; however, the simplest and the most acceptable type of behavior for process discovery is the DF relation. Thus, we compute the occurrence probability of each directly follows relation (a, b) (that $a, b \in \mathcal{A}$) according to the following equation:

$$Prob(a, b) = \frac{|\sigma \in \overline{L}|a >_\sigma b|}{|\overline{L}|}. \tag{3}$$

So, we compute the occurrence probability of all of the DF relations in a variant. If $Prob(a, b)$ is high enough (i.e., be higher than a defined threshold T_P), we expect that sampled variants should contain it. So, any variant that contains such a high probable behavior, will give a $+1$ to its rank. Otherwise, if a variant does not contain a probable behavior, we decrease its rank by -1. Contrariwise, if a variant contains a low probable behavior (i.e., $Prob(a, b) \leq 1 - T_P$), we decrease its rank by -1. To normalize the ranking values, we divide them by the variant length. By using this ranking strategy, we are looking for variants with much of high probable behavior and less of low probable ones. Note that it is possible that some DF relations be neither high probable nor low probable that we do not consider them in the ranking procedure. Finally, we sort the variants based on their ranks and return the $c \times |\overline{L}|$ ones with the highest rank.

The main advantage of this method is that it helps process discovery algorithms to depict the main-stream behavior of the original event log in the process model. However, it needs more time to compute the similarity score of all variants. Especially, if we use more advanced behavioral data structures such as eventually follows instead of DF relations, this computation can be a limitation for this ranking strategy.

Structure-Based Selection: In this subset selection method, we consider the presence of unstructured behavior (i.e., based on Definition 3) in each variant. In this regard, we first compute the occurrence probability of each sub-patten σ' among its surrounding contextual contexts $\beta_L(\sigma')$ (i.e., $\gamma_L(\sigma_s, \sigma_l, \sigma_r)$). If this probability is below the given threshold, i.e., T_S, we consider it as unstructured behavior. We expect that unstructured subsequences have problematic effects on process discovery algorithms and make discovered process models inaccurate and complex [12]. Thus, for each unstructured behavior in a variant, we give a penalty to it and decrease its rank by -1. Consequently, a variant with unstructured behavior receives more penalties and it is not appealing to be placed in the sampled event log.

The main reason to use this ranking strategy is that it results in the main skeleton of the process models. It is designed to reduce improbable parallel behavior and having simpler process models. However, this subset selection strategy requires longer time to rank variants in event logs.

5   Evaluation

In this section, we aim to find out the effects of subset selection methods on the performance of process discovery algorithms. Moreover, we will analyze the quality of process models that are discovered via sampled event logs.

To apply the proposed subset selection methods, we implemented the *Sample Variant* plug-in in ProM framework[2]. In this implementation, we used static

---

[2] Sample Variant plug-in in: https://svn.win.tue.nl/repos/prom/Packages/LogFiltering.

thresholds for both similarity and structure-based ranking strategies. Using this plug-in, the end user is able to specify her desired percentage of sampling variants/traces and the ranking strategy. It takes an event log as an input and returns the top fraction of its variants/traces. In addition, to apply our proposed method on various event logs and use different process discovery algorithms with their different parameters, we ported the *Sample Variant* plug-in to RapidProM which extends RapidMiner with process analysis capabilities. In our experiment, we also used the statistical sampling method [14]; however, as we consider only work-flow information, its relaxation parameter is ignored.

Information about real event logs that are used in the evaluation is given in Table 1[3]. To differentiate between the activities that are occurred at the starting and ending parts of traces, we added artificial *Start* and *End* activities to all of the traces. For process discovery, we used the Inductive Miner [18],

Table 1. Details of real event logs

Event log	Activities#	Traces#	Variants#	DF#
BPIC-2012	26	13087	4336	138
BPIC-2017	28	31509	1593	178
BPIC-2019	44	251734	11973	538
Hospital	20	100000	1020	143
Road	13	150370	231	70
Sepsis	18	1050	846	115

the ILP Miner [7], and the Split Miner [19]. On the sampled event logs, we applied process discovery algorithms just without their built-in filtering mechanisms.

We investigate the probable improvement of subset selection methods on the performance of process discovery algorithms. To measure this improvement, we apply the following equation:

$$DiscoveryTimeImprovement = \frac{DiscoveryTime(WholeLog)}{DiscoveryTime(SampledLog) + SamplingTime(WholeLog)}. \tag{4}$$

Figure 2 shows improvements in the performance of process discovery algorithms when we select subset of event logs. Here, for ranking-based strategies, we used sampling threshold (i.e., c) equals to 0.1. Each experiment was repeated four times (because the discovery and sampling times are not deterministic) and the average values are shown. For some event logs, the improvement is more than 100 by sampling event logs. It is shown that the improvement is significantly higher when the statistical sampling method is used because it does not need to traverse the input event log. However, for *Sepsis* that has few traces, ranking based subset selection methods are faster. Note that the structure-based strategy sometimes has no improvement because it requires higher sampling time. The sampling time of different methods are depicted in Fig. 4. As we expected, the statistical sampling method is much faster than other subset selection methods and the structure-based is the slowest one.

---

[3] https://data.4tu.nl/repository/collection:event_logs_real.

The improvement in performance of process discovery algorithms may be driven by reducing (1) the number of activities, (2) the number of traces, or (3) the amount of unique behavior (e.g., DF relations or variants). By event log subset selection, it is possible that some of infrequent activities are not placed in the sampled event log. Moreover, by subset selection we reduce the size of event logs (i.e., $|S_L| \leq |L|$) and also possible behavior in the event log. Figure 3 shows the process discovery time of sampled event logs with different sampling thresholds when we used the ILP miner. When we used the sampling threshold equals to 0.99, we will have almost all behavior of the original event log in the sampled event log; however, the number of traces in the sampled event log is significantly lower than the original event log. Results show that for many event logs, the main reason of the improvement in performance of the process discovery is gained by reducing the number of variants. However, for *Road* and *Hospital* event logs that there are high frequent variants, reducing the number of traces has higher impact on the performance of the process discovery algorithms.

As explained, the amount of behavior in the event log has an important role on the performance of process discovery algorithms. The remained percentage of DF relations in the sampled event logs for different subset selection methods are given in Fig. 5. We can see that for most of the event logs the similar and structure based methods keep fewer DF relations. However, according to their ranking policy, they keep the most common DF relations among the original event log.

In the previous experiment, the sampling threshold for ranking-based subset selection methods equals to 0.1. Note that there is no such control on the statistical sampling method and it aims to keep as much as DF relations in sampled event logs. However, for most of process discovery algorithms variants are more important compared to only DF relations. Even the basic Inductive miner that uses DF relations may result in different process models for event logs that have identical sets of DF relations. For example, $L_1 = [\langle a, b, c \rangle, \langle a, c, b \rangle]$ and $L_2 = [\langle a, c, b, c \rangle, \langle a, b \rangle]$ have the same sets of DF relations; however, their process models are different. Figure 6 indicates the average percentage of remaining variants in sampled event logs using the statistical sampling method. It shows that, this method is able to just keep few percentage of variants of an event log. For ranking-based strategies, the number of remaining variants can be adjusted by c.

In Fig. 7, we compared the average of preprocessing time for trace filtering methods and the similarity-based subset selection method. For this experiment, we filter event logs with six different filtering settings and iterate the experiments for four times. Also, we used the similarity-based strategy with the threshold in $[0.01, 0.05, 0.1]$. It is clear that the subset selection method preprocessed the event logs faster. Also, with the trace filtering methods, we do not have control over the size of the filtered event logs [22].

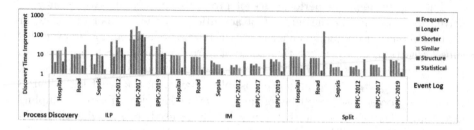

Fig. 2. Discovery time improvement for discovering process using subset selection.

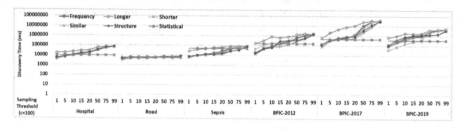

Fig. 3. The average of discovery time of the ILP miner for different subset selection methods and different sampling thresholds.

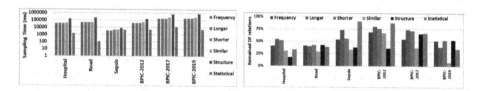

Fig. 4. Sampling time of different subset selection methods.

Fig. 5. Remained percentage of DF relations in the sampled event logs.

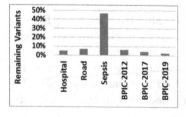

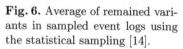

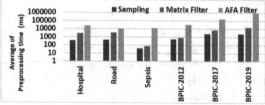

Fig. 6. Average of remained variants in sampled event logs using the statistical sampling [14].

Fig. 7. The average of preprocessing time when we used similarity-based strategy and two state of the arts trace filtering methods [11,20].

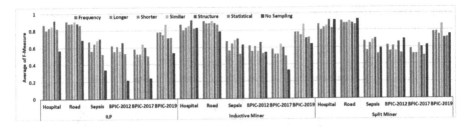

Fig. 8. Comparing the F-Measure of discovered process models with different subset selection methods. For ranking-based strategies, we used the sampling threshold in [0.01, 0.05, 0.1] and the average value of F-Measure is presented.

To analyze the quality of process models that are discovered from sampled event logs we can use *fitness* and *precision*. Fitness measures how much behavior in the event log is also described by the process model. Thus, a fitness value equal to 1, indicates that all behavior of the event log is described by the process model. Precision measures how much of behavior, that is described by the process model, is also presented in the event log. A low precision value means that the process model allows for much behavior compared to the event log. There is a trade-off between these measures [21], sometimes, putting aside a small amount of behavior causes a slight decrease in the fitness value, whereas the precision value increases dramatically. Therefore, we use the F-Measure metric that combines both of them according to the following formula:

$$\text{F-Measure} = \frac{2 \times \text{Precision} \times \text{Fitness}}{\text{Precision} + \text{Fitness}}. \tag{5}$$

Figure 8 compares the quality of best process models that are discovered with/without subset selection. We used sampled event logs, just for discovery purpose and the original event logs were used for computing F-Measure values. For the cases that ranking based subset selection methods were used, we applied the sampling thresholds [0.01, 0.05, 0.1], and the average of F-Measure values is shown. For the statistical sampling method, we iterate the experiment four times, and again the average of F-Measure values are considered. According to the results, for the ILP and the Inductive miner, we always have an improvement when we use subset selection. However, for some event logs, the Split miner can discover process models with higher quality via the original event logs. Moreover, the statistical sampling method that randomly selects process instances results in process models with less quality according to the F-Measure compared to ranking based subset selection methods. Among ranking strategies, the structure and similarity-based ones result in better process models for most of the event logs. For some event logs like *BPIC-2019*, the similarity-based ranking strategy is the best choice for all of the process discovery algorithms. For *Hospital* event log, the *structure*-based ranking strategy results in process models with the highest F-Measure. As there are frequent variants in *Road*, the best subset selection method for this event log is the *frequency* one. Results show that length-based methods are not suitable to sample event logs if the purpose is to have process models with high quality.

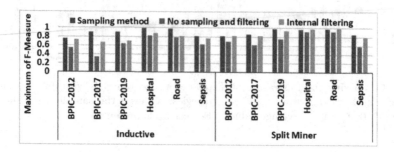

Fig. 9. The best F-Measure of discovered process models when using sampled event logs compared to applying embedded filtering mechanisms of discovery algorithms.

In this experiment, we just used the basic versions of the process discovery algorithms. However, in practice users can apply their embedded filtering and select the best discovered process model. In the next experiment, we applied the inductive miner and the Split miner with 100 different filtering thresholds and find best process models according to the F-Measure values. Figure 9 compares the best F-Measure value that discovered via ranking-based subset selection methods compared to the case that we used embedded filtering mechanisms of process discovery algorithms on original event data. As mentioned in Subsect. 4.2, by applying subset selection methods, we will lose the frequency of variants. As a result, some embedded filtering mechanisms in process discovery algorithms become unusable. However, the results of this experiment show that we may discover process models with high quality from sampled event logs, even without these filtering methods. If for any reason we need to use the frequency of variants, it is recommended to apply trace-based subset selection methods or other standards to store event logs like [17]. Also, all of the proposed subset selection methods needs to load the original event log (like the statistical sampling) that is a limitation for dealing big event logs when memory is a constraint. Moreover, much of preprocessing time in ranking-based strategies consumed to traverse the event log and find possible variants. This information is already recorded in some event log standards like MXML [17]. Using these standards leads to decrease the preprocessing time of the proposed approaches.

Note that we did not use filtering mechanisms of the process discovery algorithm for sampled event logs. Results show that subset selection methods can increase the quality of discovered process models; specifically, if we use the Inductive Miner. It shows the weakness of process discovery algorithms in dealing with infrequent behavior [11].

6   Conclusion

In this paper, we proposed some subset selection strategies to increase the performance of process discovery procedure. We recommend to apply process discovery algorithms on sampled event logs when dealing with large data sets. We implemented different ranking strategies in ProM and ported this functionality into RapidProM and applied it on some real event logs using different process discovery algorithms.

Experimental results show that event log subset selection methods decrease the required time used by state-of-the-art process discovery algorithms. We found that ranking-based strategies mostly increase the performance of process discovery by reducing the amount of behavior and the number of traces. Therefore, by applying these methods, we are able to discover an acceptable approximation of the final process model in a shorter time. Moreover, results show that for some event logs, subset selection methods can improve the quality of discovered process models according to the F-Measure metric. Results show that to have higher F-Measure value, it is better to use the structure and similarity-based strategies. However, by using random trace sampling methods, e.g., the statistical method, we can discover process models in a shorter time. As future work, we aim to find out what the best subset selection method is due to the available time and event data.

References

1. van der Aalst, W.M.P.: Process Mining - Data Science in Action, 2nd edn. Springer, Berlin (2016). https://doi.org/10.1007/978-3-662-49851-4
2. van der Aalst, W.M.P., et al.: Process mining manifesto. In: Business Process Management BPM Workshops, Clermont-Ferrand, France, pp. 169–194 (2011)
3. Verbeek, H.M.W., Buijs, J.C.A.M., van Dongen, B.F., van der Aalst, W.M.P.: XES, XESame, and ProM 6. In: Soffer, P., Proper, E. (eds.) CAiSE Forum 2010. LNBIP, vol. 72, pp. 60–75. Springer, Heidelberg (2011). https://doi.org/10.1007/978-3-642-17722-4_5
4. van der Aalst, W.M.P., Bolt, A., van Zelst, S.: RapidProM: mine your processes and not just your data. CoRR abs/1703.03740 (2017)
5. van der Aalst, W.M.P., Weijters, T., Maruster, L.: Workflow mining: discovering process models from event logs. IEEE Trans. Knowl. Data Eng. **16**(9), 1128–1142 (2004)
6. van der Werf, J., van Dongen, B., Hurkens, C., Serebrenik, A.: Process discovery using integer linear programming. Fundam. Inf. **94**(3–4), 387–412 (2009)
7. van Zelst, S., van Dongen, B., van der Aalst, W.M.P., Verbeek, H.M.W.: Discovering workflow nets using integer linear programming. Computing **100**, 529 (2018)
8. Leemans, S.J.J., Fahland, D., van der Aalst, W.M.P.: Discovering block-structured process models from event logs - a constructive approach. In: Colom, J.-M., Desel, J. (eds.) PETRI NETS 2013. LNCS, vol. 7927, pp. 311–329. Springer, Heidelberg (2013). https://doi.org/10.1007/978-3-642-38697-8_17
9. Suriadi, S., Andrews, R., ter Hofstede, A., Wynn, M.T.: Event log imperfection patterns for process mining: towards a systematic approach to cleaning event logs. Inf. Syst. **64**, 132–150 (2017)

10. Andrews, R., Suriadi, S., Ouyang, C., Poppe, E.: Towards Event Log Querying for Data Quality: Let's Start with Detecting Log Imperfections (2018)
11. Sani, M.F., van Zelst, S.J., van der Aalst, W.M.P.: Improving process discovery results by filtering outliers using conditional behavioural probabilities. In: Business Process Management BPM Workshops, Barcelona, Spain, pp. 216–229 (2017)
12. Sani, M.F., van Zelst, S.J., van der Aalst, W.M.P.: Repairing outlier behaviour in event logs. In: Abramowicz, W., Paschke, A. (eds.) BIS 2018. LNBIP, vol. 320, pp. 115–131. Springer, Cham (2018). https://doi.org/10.1007/978-3-319-93931-5_9
13. Mannhardt, F., de Leoni, M., Reijers, H.A., van der Aalst, W.M.P.: Data-driven process discovery - revealing conditional infrequent behavior from event logs. In: Dubois, E., Pohl, K. (eds.) CAiSE 2017. LNCS, vol. 10253, pp. 545–560. Springer, Cham (2017). https://doi.org/10.1007/978-3-319-59536-8_34
14. Bauer, M., Senderovich, A., Gal, A., Grunske, L., Weidlich, M.: How much event data is enough? A statistical framework for process discovery. In: Krogstie, J., Reijers, H.A. (eds.) CAiSE 2018. LNCS, vol. 10816, pp. 239–256. Springer, Cham (2018). https://doi.org/10.1007/978-3-319-91563-0_15
15. Berti, A.: Statistical sampling in process mining discovery. In: The 9th International Conference on Information, Process, and Knowledge Management, pp. 41–43 (2017)
16. Weijters, A.J.M.M., Ribeiro, J.T.S.: Flexible heuristics miner (FHM). In: CIDM (2011)
17. van Dongen, B.F., van der Aalst, W.M.P.: A meta model for process mining data (2005)
18. Leemans, S.J.J., Fahland, D., van der Aalst, W.M.P.: Discovering block-structured process models from event logs containing infrequent behaviour. In: Lohmann, N., Song, M., Wohed, P. (eds.) BPM 2013. LNBIP, vol. 171, pp. 66–78. Springer, Cham (2014). https://doi.org/10.1007/978-3-319-06257-0_6
19. Augusto, A., Conforti, R., Dumas, M., La Rosa, M., Polyvyanyy, A.: Split miner: automated discovery of accurate and simple business process models from event logs. Knowl. Inf. Syst. **50**, 1–34 (2019)
20. Conforti, R., La Rosa, M., ter Hofstede, A.: Filtering out infrequent behavior from business process event logs. IEEE Trans. Knowl. Data Eng. **29**(2), 300–314 (2017)
21. Weerdt, J.D., Backer, M.D., Vanthienen, J., Baesens, B.: A robust F-measure for evaluating discovered process models. In: Proceedings of the CIDM, pp. 148–155 (2011)
22. Fani Sani, M., van Zelst, S.J., van der Aalst, W.M.P.: Applying sequence mining for outlier detection in process mining. In: Panetto, H., Debruyne, C., Proper, H., Ardagna, C., Roman, D., Meersman, R. (eds.) OTM 2018. LNCS, vol. 11230, pp. 98–116. Springer, Cham (2018). https://doi.org/10.1007/978-3-030-02671-4_6

Exploiting Event Log Event Attributes in RNN Based Prediction

Markku Hinkka[1,3(✉)], Teemu Lehto[1,3], and Keijo Heljanko[2,4]

[1] Department of Computer Science, School of Science,
Aalto University, Espoo, Finland
[2] Department of Computer Science, University of Helsinki, Helsinki, Finland
[3] QPR Software Plc, Helsinki, Finland
[4] HIIT Helsinki Institute for Information Technology, Espoo, Finland
markku.hinkka@aalto.fi, teemu.lehto@qpr.com, keijo.heljanko@helsinki.fi

Abstract. In predictive process analytics, current and historical process data in event logs are used to predict future. E.g., to predict the next activity or how long a process will still require to complete. Recurrent neural networks (RNN) and its subclasses have been demonstrated to be well suited for creating prediction models. Thus far, event attributes have not been fully utilized in these models. The biggest challenge in exploiting them in prediction models is the potentially large amount of event attributes and attribute values. We present a novel clustering technique which allows for trade-offs between prediction accuracy and the time needed for model training and prediction. As an additional finding, we also find that this clustering method combined with having raw event attribute values in some cases provides even better prediction accuracy at the cost of additional time required for training and prediction.

Keywords: Process mining · Predictive process analytics · Prediction · Recurrent neural networks · Gated Recurrent Unit

1 Introduction

Event logs generated by systems in business processes are used in Process Mining to automatically build real-life process definitions and as-is models behind those event logs. There is a growing number of applications for predicting the properties of newly added event log cases, or process instances, based on case data imported earlier into the system [2,3,8,13]. The more the users start to understand their own processes, the more they want to optimize them. This optimization can be facilitated by performing predictions. In order to be able to predict properties of new and ongoing cases, as much information as possible should be collected that is related to the event log traces and relevant to the properties to be predicted. Based on this information, a model of the system creating the event logs can be created. In our approach, the model creation is performed using supervised machine learning techniques.

© Springer Nature Switzerland AG 2019
T. Welzer et al. (Eds.): ADBIS 2019, CCIS 1064, pp. 405–416, 2019.
https://doi.org/10.1007/978-3-030-30278-8_40

In our previous work [5] we have explored the possibility to use machine learning techniques for classification and root cause analysis for a process mining related classification task. In the paper, experiments were performed on the efficiency of several feature selection techniques and sets of structural features (a.k.a. activity patterns) based on process paths in process mining models in the context of a classification task. One of the biggest problems with the approach is that finding of the structural features having the most impact on the classification result. E.g., whether to use only activity occurrences, transitions between two activities, activity orders, or other even more complicated types of structural features such as detecting subprocesses or repeats. For this purpose, we have proposed another approach in [6], where we examined the use of recurrent neural network techniques for classification and prediction. These techniques are capable of automatically learning more complicated causal relationships between activity occurrences in *activity sequences*. We have evaluated several different approaches and parameters for the recurrent neural network techniques and have compared the results with the results we collected in our work. In both the previous publications [5,6], focusing on boolean-type classification tasks based on the *activity sequences* only.

In this work we build on our previous work to further improve the prediction accuracy of prediction models by exploiting additional event attributes often available in the event logs while also taking into account the scalability of the approach to allow users to precisely specify the event attribute detail level suitable for the prediction task ahead. Our goal is to develop a technique that would allow the creation of a tool that is, based on a relatively simple set of parameters and training data, able to efficiently produce a prediction model for any case-level prediction task, such as predicting the next activity or the final duration of a running case. Fast model rebuilding is also required in order for a tool to be able to also support, e.g., interactive event and case filtering capabilities.

To answer these requirements, we introduce a novel method of exploiting event attributes into RNN prediction models by clustering events by their event attribute values, and using the cluster labels in the RNN input vectors instead of the raw event data. This makes it easy to manage the input RNN vector size no matter how many event attributes there are in the data set. E.g., users can configure the absolute maximum length of the one-hot vector used for the event attribute data which will not be exceeded, no matter how many actual attributes the dataset has.

Our prediction engine source code is available in GitHub[1].

The rest of this paper is structured as follows: Sect. 2 is a short summary of the latest developments around the subject. In Sect. 3, we present the problem statement and the related concepts. Section 4 presents our solution for the problem. In Sect. 5 we present our test framework used to test our solution. Section 6 describes the used datasets as well as performed prediction scenarios. Section 7 presents the experiments and their results validating our solution. Finally Sect. 8 draws the final conclusions.

---

[1] https://github.com/mhinkka/articles.

2   Related Work

Lately there has been a lot of interest in the academic world on predictive process monitoring which can clearly be seen, e.g., in [4] where the authors have collected a survey of 55 accepted academic papers on the subject. In [12], the authors have compared several approaches spanning three different research fields: Machine learning, process mining and grammar inference. As result, they have found that overall, the techniques from machine learning field generate more accurate predictions than grammar inference and process mining fields.

In [13] the authors used Long Short-Term Memory (LSTM) recurrent neural networks to predict the next activity and its timestamp. They use one-hot encoded activity labels and three numerical time-based features: duration between the current activity and the previous activity, time within the day and time within the week. Event attributes were not considered at all. In [2] the authors trained LSTM networks to predict the next activity. In this case however, network inputs are created by concatenating categorical, character string valued event attributes and then encoding these attributes via an embedding space. They also note that this approach is feasible only because of the small number of unique values each attribute had in their test datasets. Similarly, in [11], the authors take a very similar approach based on LSTM networks, but this time also incorporate both discrete and continuous event attribute values. Discrete values are one-hot encoded, whereas continuous values are normalized using min-max normalization and added to the input vectors as single values.

In [9] the authors use Gated Recurrent Unit (GRU) recurrent neural networks to detect *anomalies* in event logs. One one-hot encoded vector is created for activity labels and one for each of the included string valued event attributes. These vectors are then concatenated in similar fashion to our solution into one vector representing one event, that is then given as input to the network. We use this approach for benchmarking our own clustering based approach (labeled as *Raw* feature in the text below). The system proposed in their paper is able to predict both the next activity and the next values of event attributes. Specifically, it does not take case attributes and temporal attributes into account.

In [14] the authors trained a RNN to predict the most likely future activity sequence of a running process based only on the sequence of activity labels. Similarly our earlier publication [5] used sequences of activity labels to train a LSTM network to perform a boolean classification of cases. None of the mentioned earlier works present a solution that is scalable for datasets having lots of event- or case attributes and unique attribute values.

3   Problem

Using RNN to perform case-level predictions on event logs has lately been studied a lot. However, there has not been any scalable approach on handling event attributes in RNN setting. Instead, e.g., in [9] authors used separate one-hot encoded vector for each attribute value. Having this kind of an approach when you have, e.g., 10 different attributes, each having 10 unique values would already require a vector of 100 elements to be added as input for every event. The longer the input vectors become, the more time and memory it gets for the model to create accurate models from them. This increases the time and memory required to use the model for predictions.

Table 1. Feature input vector structure

f_{11}	f_{12}	$\cdots$	f_{1m_1}	f_{21}	$\cdots$	f_{2m_2}	$\cdots$	f_{n1}	$\cdots$	f_{nm_n}

Table 2. Feature input vector example content

row	$activity_{eat}$	$activity_{drink}$	$food_{salad}$	$food_{pizza}$	$food_{water}$	$food_{soda}$	$cluster_1$	$cluster_2$
1	1	0	1	0	0	0	1	0
2	0	1	0	0	1	0	1	0
3	1	0	0	1	0	0	0	1
4	1	0	0	1	0	0	0	1
5	0	1	0	0	0	1	0	1

4   Solution

We decided to include several feature types into the input vectors of the RNN. Input vectors are formatted as shown in Table 1, where each column represents one feature vector element f_{ab}, where a is the index of the feature and b is the index of the element of that feature. In the table, n represents the number of feature types used in the feature vector and m_k represents the number of elements required in the input vector for feature type k. Thus, each feature type produces one or more numeric elements into the input vector, which are then concatenated together into one actual input vector passed to RNN both in training and in prediction phases. Table 2 shows an example input vector having three different feature types: activity label, raw event attribute values (only single event attribute named *food* having four unique values) and the event attribute cluster where clustering has been performed separately for each unique activity.

For this paper, we encoded only event activity labels and event attributes into the input vectors. However, this mechanism can easily incorporate also other types of features not described here. The only requirement for added features is that it needs to be able to be encoded into a numeric vector as shown in Table 1 whose length must be the same for each event.

4.1   Event Attributes

Our primary solution for incorporating information in event attributes into input vectors is to cluster all the event attribute values in the training set and then use a one-hot encoded cluster identifier to represent all the attribute values of the element. The used clustering algorithm must be such that it tries to automatically find the optimal number of clusters for the given data set within the range of 0 to N clusters, where N can be configured by the user. By changing N, the user can easily configure the maximum length of the one-hot-vector as well as the precision of how detailed attribute information will be tracked. For this paper, we experimented with slightly modified version of Xmeans-algorithm [10].

It is very common that different activities get processed by different resources yielding a completely different set of possible attribute values. E.g., different departments in a hospital have different people, materials and processes. Also in the example feature vector shown in Table 2, *food*-event attribute has completely different set of possible values depending on the *activity* since it is forbidden by, e.g., the external system to not allow activity of type *eat* to have *food* event attribute value of *water*. If we cluster all the event attributes using single clustering, we would easily lose this activity type specific information.

In order to retain this activity specific information, we used separate clustering for each unique activity type. All the event attribute clusters are encoded into one one-hot encoded vector representing only the resulting cluster label for that event, no matter what its activity is. This is shown in the example table as $cluster_N$, which represents the row having N as clustering label. E.g., in the example case, $cluster_1$ is 1 in both rows 1 and 2. However, row 1 is in that cluster because it is in the 1st cluster of the $activity_{eat}$ activity, whereas row 2 is in that cluster because it is in the 1st cluster of the $activity_{drink}$ activity. Thus, in order to identify the actual cluster, one would require both the activity label and the cluster label. For RNN to be able to properly learn about the actual event attribute values, it needs to be given both the activity label and the cluster label in input vector. Below, this approach is labeled as *ClustN*, where N is the maximum cluster count.

For benchmarking, we also experimented with a *raw* implementation where event attributes were used so that every event attribute is encoded into its own one-hot encoded vector and then concatenated into the actual input vectors. This method is lossless, since every unique event attribute value has its own indicator in the input vector. Below, this approach is referred to as *Raw*. Finally, we experimented also using both *Raw* and *Clustered* event attribute values. Below, this approach is referred to as *BothN*, where N is the maximum cluster count.

5   Test Framework

We have performed our test runs using an extended Python-based prediction engine that was used in our earlier work [5]. The engine is still capable of supporting most of the hyperparameters that we experimented with in our earlier work, such as used RNN unit type, number of RNN layers and the used batch size. The prediction engine we built for this work takes a single JSON configuration file as input and outputs test result rows into a CSV file.

Tests were performed using a commonly used 3-fold cross-validation technique to measure the generalization characteristics of the trained models. In

3-fold cross-validation the input data is split into three subsets of equal size. Each of the subsets is tested one by one against models trained using the other two subsets.

5.1   Training

Training begins by loading the event log data contained in the two of the three event log subsections. After this, the event log is split into actual training data and validation data that used to find the best performing model out of all the model states during all the test iterations. For this, we picked 75% of the cases for the training and the rest for the validation dataset. After the this, we initialize event attribute clusters as described in Sect. 4.1.

The actual prediction model and the data used to generate the actual input vectors is performed next. This data initialization involves splitting cases into prefixes and also taking a random sample of the actual available data if the amount of data exceeds the configured maximum amount of prefixes. In order to avoid running out of memory during any of our tests, these limits were set to 75000 for training data and 25000 for validation data. We also had to filter out all the cases having more than 100 events.

Finally after the model is initialized, we start the actual training in which we concatenate all the requested feature vectors as well as the expected outcome into the RNN model repeatedly for the whole training set until 100 test iterations have passed. The number of actual epochs trained in each iteration is configurable. In our experiments, the total number of epochs was set to be 10. After every test iteration the model is validated against the validation set. In order to improve validation performance, if the size of the validation set is larger than separately specified limit (10000), a random sample of the whole validation set is used. These test results, including additional status and timing related information, is written into resulting test result CSV file. If the prediction accuracy of the model against the validation set is found to be better than the accuracy of any of the models found thus far, then the network state is stored for that model. Finally after all the training, the model having the best validation test accuracy is picked as the actual result of the model training.

5.2   Testing

In the testing phase, the third subset of cross-validation folding is tested against the model built in the previous step. After initializing the event log following similar steps as in the training phase, the model is asked for a prediction for each input vector built from the test data. In order to prevent running out of memory and to ensure tests are not taking exceedingly long time to run, we limited the number of final test traces to 100000 traces and used random sampling when needed. The prediction result accuracy, as well as other required statistics are written to the resulting CSV file.

6 Test Setup

We performed our tests using several different data sets. Some details of the used data sets can be found in the Table 3.

Table 3. Used event logs and their relevant statistics

Event log	# Cases	# Activities	# Events	# Attributes	# Unique values
BPIC12[a]	13087	24	262200	1	3
BPIC13, incidents[b]	7554	13	65533	8	2890
BPIC14[c]	46616	39	466737	1	242
BPIC17[d]	31509	26	1202267	4	164
BPIC18[e]	43809	41	2514266	5	360

[a] https://doi.org/10.4121/uuid:3926db30-f712-4394-aebc-75976070e91f
[b] https://doi.org/10.4121/uuid:500573e6-accc-4b0c-9576-aa5468b10cee
[c] https://doi.org/10.4121/uuid:c3e5d162-0cfd-4bb0-bd82-af5268819c35
[d] https://doi.org/10.4121/uuid:5f3067df-f10b-45da-b98b-86ae4c7a310b
[e] https://doi.org/10.4121/uuid:3301445f-95e8-4ff0-98a4-901f1f204972

For each dataset, we performed next activity prediction where we wanted to predict the next activity of any ongoing case. In this case, we split every input case into possibly multiple *virtual* cases depending on the number of events the case had. If the length of the case was shorter than 4, the whole case was ignored. If the length was equal or higher, then a separate *virtual* case was created for all prefixes at least of length 4. Thus, for a case of length 6, 3 cases were created: One with length 4, one with length 5 and one with length 6. For all these prefixes, the next activity label was used as the expected outcome. For the full length case, the expected outcome was a special *finished*-token.

7 Experiments

For experiments, we have used the same system that we used already in our previous work [5]. The system had Windows 10 operating system and its hardware consisted of 3.5 GHz Intel Core i5-6600K CPU with 32 GB of main memory and NVIDIA GeForce GTX 960 GPU having 4 GB of memory. Out of those 4 GB, we reserved 3 GB for the tests. The testing framework was built on the test system using Python programming language. The actual recurrent neural networks were built using Lasagne[2] library that works on top of Theano[3]. Theano was configured to use GPU via CUDA for expression evaluation.

---

[2] https://lasagne.readthedocs.io/.
[3] http://deeplearning.net/software/theano/.

We used one layer GRU [1] as the RNN type. Adam [7] was used as gradient descent optimizer. 256 was used as hidden dimension size and 0.01 as learning rate even though it is quite probable that more accurate results could have been achieved by selecting, e.g., different hidden dimension sizes depending on the size of the input vectors. However, since that would have made the interpretation of the test results more complicated, we decided to use the constant hidden dimension size on all the tests.

We performed next activity predictions using all the four combinations of features, five data sets and three different maximum cluster counts: 20, 40, and 80 clusters. The results of these runs are shown in Table 4. In the table, *Features*-column shows the used set of features. *S.rate* shows the achieved prediction success rate. *In.v.s.* shows the size of the input vector. This column can be used to give some kind of indication on the memory usage of using that configuration. Finally, *Tra.t.* and *Pred.t.* columns tell us the time required for performing the training and the prediction for all the cases in the test dataset. In both the cases, this time includes the time for setting up the neural network, clusterings and preparing the dataset from JSON format. Sample standard deviation has been included into both *S.rate* and *Tra.t* in parentheses to indicate how spread out the measurements are within all the three test runs. Each row in the table represents three cross validation runs with unique combination of dataset and feature that was tested. Rows having a best prediction accuracy within a dataset are shown using bold font. *None*-feature represents the case in which there were no event attribute information at all in the input vector, *ClustN* represents a test with one-hot encoded cluster labels of event attributes clustered into maximum of N clusters, *Raw* represents having all one-hot encoded attribute values individually in the input vector, and finally *BothN* represents having both one-hot encoded attribute values and one-hot encoded cluster labels in the input vector.

We also aggregated some of these results over all the datasets using maximum cluster size of 80 clusters. Figure 1 shows average success rates of different event attribute encoding techniques over all the tested datasets. Figure 2 shows the average input vector lengths. Figures 3 and 4 shows the averaged training and prediction times respectively.

Based on all of these results, we can see that having event attribute values included clearly improved the prediction accuracy over not having them included at all in all datasets. The effect ranged from 0.5% in BPIC12 model to 8.5% in BPIC18. As shown in Fig. 1, very similar success rates were achieved using *ClustN* features as with *Raw*. However, model training and the actual prediction can be performed clearly faster using *ClustN* approaches than either *Raw* or *BothN*. This effect is the most prominently visible in BPIC13 results, where, due to the model having large amount of unique attribute values, the size of the input vector is almost 68 times bigger and the training time almost 14 times longer using *Raw* feature than *Clust20*. At the same time, the accuracy is still clearly better than not having event attributes at all (about 3.9% better) and only slightly worse (about 1.4%) than when using *Raw* feature. This clearly indicates that clustering can be really powerful technique for minimizing the time required for training

Table 4. Statistics of next activity prediction using different sets of input features

Dataset	Features	S.rate (σ)	In.v.s.	Tra.t. (σ)	Pred.t.
BPIC12	None	85.8% (0.3%)	25.7	489.0 s (7.0 s)	35.1 s
	Clust20	86.0% (0.4%)	30.0	500.6 s (2.5 s)	31.6 s
	Clust40	85.8% (0.3%)	30.0	499.7 s (1.3 s)	31.9 s
	Clust80	86.2% (0.1%)	30.0	502.1 s (2.3 s)	7.5 s
	Raw	85.9% (0.3%)	29	504.3 s (0.5 s)	38.9 s
	Both20	86.0% (0.2%)	33	515.3 s (2.6 s)	40.4 s
	Both40	86.0% (0.4%)	33	517.7 s (3.6 s)	40.4 s
	Both80	**86.3% (0.1%)**	**33**	**518.2 s (4.0 s)**	**40.7 s**
BPIC13	None	62.9% (0.3%)	13.7	165.6 s (21.2 s)	3.5 s
	Clust20	66.8% (0.3%)	34.7	188.0 s (22.4 s)	4.7 s
	Clust40	67.2% (0.7%)	54.7	214.8 s (3.1 s)	5.4 s
	Clust80	67.0% (0.6%)	94.7	258.4 s (4.7 s)	6.0 s
	Raw	68.2% (1.1%)	2353.7	2611.7 s (44.7 s)	74.8 s
	Both20	**69.1% (0.6%)**	**2359.3**	**2464.6 s (309.0 s)**	**94.4 s**
	Both40	68.9% (0.5%)	2395.7	2687.1 s (227.3 s)	106.6 s
	Both80	68.4% (0.7%)	2429.3	2821.8 s (33.5 s)	194.3 s
BPIC14	None	37.8% (1.5%)	40.3	488.1 s (5.3 s)	36.1 s
	Clust20	39.9% (0.5%)	61.7	523.3 s (3.5 s)	40.4 s
	Clust40	40.0% (0.3%)	80.3	553.5 s (3.8 s)	43.6 s
	Clust80	40.2% (0.1%)	84.7	556.8 s (10.5 s)	43.6 s
	Raw	39.7% (1.4%)	272.0	825.7 s (2.8 s)	68.0 s
	Both20	40.6% (0.6%)	292.3	907.1 s (7.5 s)	78.6 s
	Both40	**40.6% (0.6%)**	**309.3**	**943.3 s (10.6 s)**	**82.0 s**
	Both80	37.3% (4.2%)	305.0	935.1 s (26.9 s)	156.7 s
BPIC17	None	86.4% (0.4%)	27.7	518.7 s (2.8 s)	107.7 s
	Clust20	**90.8% (0.3%)**	**48.7**	**556.3 s (3.7 s)**	**132.4 s**
	Clust40	90.2% (1.4%)	68.3	637.5 s (58.3 s)	143.9 s
	Clust80	90.2% (0.4%)	108.7	647.3 s (3.7 s)	142.8 s
	Raw	89.9% (0.5%)	190	816.4 s (5.2 s)	164.9 s
	Both20	89.9% (0.5%)	211.0	867.8 s (3.5 s)	188.0 s
	Both40	90.2% (0.3%)	230.3	910.9 s (19.3 s)	193.7 s
	Both80	89.6% (0.6%)	271.3	986.5 (4.4 s)	197.7 s
BPIC18	None	71.3% (9.3%)	43	516.0 s (9.5 s)	197.0 s
	Clust20	79.0% (0.9%)	64.0	588.7 s (13.7 s)	268.7 s
	Clust40	**79.9% (0.2%)**	**84.0**	**628.4 s (2.8 s)**	**286.1 s**
	Clust80	79.5% (0.1%)	124.0	701.3 s (7.4 s)	306.9 s
	Raw	79.3% (0.4%)	349.7	1173.7 s (83.1 s)	381.2 s
	Both20	79.7% (0.5%)	377.7	1213.1 s (48.1 s)	463.2 s
	Both40	79.9% (0.5%)	401.0	1301.9 s (82.9 s)	540.3 s
	Both80	79.3% (0.5%)	425.7	1405.4 s (87.2 s)	619.9 s

especially when there are a lot of unique event attribute values in the used event log. Even when using the maximum cluster count of 20, prediction results will be either not affected or improved with relatively small impact to the training and prediction time. In all the datasets, the best prediction accuracy is always achieved either by using only clustering, or by using both clustering and raw attributes at the same time.

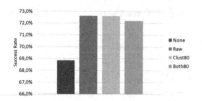

Fig. 1. Average prediction success rate over all the datasets

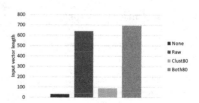

Fig. 2. Average length of the input vector over all the datasets

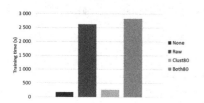

Fig. 3. Average training time over all the datasets

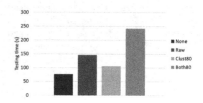

Fig. 4. Average prediction time over all the datasets

7.1 Threats to Validity

As threats to validity of the results in this paper, it is clear that there are a lot of variables involved. As initial set of parameter values, we used parameters that were found good enough in our earlier work and did some improvement attempts based on the results we got. It is most probable that the set of parameters we used were not optimal ones in each test run. We also did not test all the parameter combinations and the ones we did, we tested often only once, even though there was some randomness involved, e.g., selecting the initial cluster centers in the XMeans algorithm. However, we think that since we tested the results in several different datasets using 3-fold cross validation technique, our results can be used at least as a baseline for further studies. All the results generated by the test runs, as well as all the source data and the test framework itself, are available in support materials[4].

Also, we did not really test with datasets having really many event attribute values, the maximum amount tested being 2890. However, it can be seen that since the size of the input vectors is completely user configurable when performing event attribute clustering, the user him/herself can easily set limits to the input vector length which should take the burden off from the RNN and move the burden to the clustering algorithms, which are usually more efficient in handling lots of features and feature values. When evaluating the results of the performed tests and comparing them with other similar works, it should be taken into account that data sampling was used in several phases of the testing process.

---

[4] https://github.com/mhinkka/articles.

8   Conclusions

Clustering can be applied on attribute values to improve accuracy of predictions performed on running cases. In four of the five experimented data sets, having event attribute clusters encoded into the input vectors outperforms having the actual attribute values in the input vector. In addition, due to raw attribute values having direct effect to input vector lengths, the training and prediction time will also be directly affected by the number of unique event attribute values. Clustering does not have this problem: The number of elements reserved in the input vector for clustered event attribute values can be adjusted freely. The memory usage is directly affected by the length of the input vector. In the tested cases, the number of clusters to use to get the best prediction accuracy seemed to depend very much on the used datasets, when the tested cluster sizes were 20, 40 and 80. In some cases, having more clusters improved the performance, whereas in others, it did not have any significant impact, or even made the accuracy worse. We also found out that in some cases, having attribute cluster indicators in the input vectors improved the prediction even if the input vectors also included all the actual attribute values.

As future work, it would be interesting to test this clustering approach also with other machine learning model types such as more traditional random forest and gradient boosting machines. Similarly it could be interesting to first filter out some of the most rarely occurring attribute values before clustering the values. This could potentially reduce the amount of noise added to the clustered data and make it easier for the clustering algorithm to not be affected by noisy data. Finally, more study is required to understand whether similar clustering approach performed for event attributes in this work could be applicable also for encoding case attributes.

Acknowledgments. We want to thank QPR Software Plc for funding our research. Financial support of Academy of Finland project 313469 is acknowledged.

References

1. Cho, K., van Merrienboer, B., Bahdanau, D., Bengio, Y.: On the properties of neural machine translation: encoder-decoder approaches. In: Wu, D., Carpuat, M., Carreras, X., Vecchi, E.M. (eds.) Proceedings of SSST@EMNLP 2014, Eighth Workshop on Syntax, Semantics and Structure in Statistical Translation, Doha, Qatar, 25 October 2014, pp. 103–111. Association for Computational Linguistics (2014)
2. Evermann, J., Rehse, J., Fettke, P.: Predicting process behaviour using deep learning. Decis. Support Syst. **100**, 129–140 (2017)
3. Francescomarino, C.D., Dumas, M., Maggi, F.M., Teinemaa, I.: Clustering-based predictive process monitoring. CoRR, abs/1506.01428 (2015)
4. Francescomarino, C.D., Ghidini, C., Maggi, F.M., Milani, F.: Predictive process monitoring methods: which one suits me best? In: Weske, et al. [15], pp. 462–479

5. Hinkka, M., Lehto, T., Heljanko, K., Jung, A.: Structural feature selection for event logs. In: Teniente, E., Weidlich, M. (eds.) BPM 2017. LNBIP, vol. 308, pp. 20–35. Springer, Cham (2018). https://doi.org/10.1007/978-3-319-74030-0_2
6. Hinkka, M., Lehto, T., Heljanko, K., Jung, A.: Classifying process instances using recurrent neural networks. In: Daniel, F., Sheng, Q.Z., Motahari, H. (eds.) BPM 2018. LNBIP, vol. 342, pp. 313–324. Springer, Cham (2019). https://doi.org/10.1007/978-3-030-11641-5_25
7. Kingma, D.P., Ba, J.: Adam: a method for stochastic optimization. CoRR, abs/1412.6980 (2014)
8. Navarin, N., Vincenzi, B., Polato, M., Sperduti, A.: LSTM networks for data-aware remaining time prediction of business process instances. In: 2017 IEEE Symposium Series on Computational Intelligence, SSCI 2017, Honolulu, HI, USA, 27 November–1 December 2017, pp. 1–7. IEEE (2017)
9. Nolle, T., Seeliger, A., Mühlhäuser, M.: BINet: multivariate business process anomaly detection using deep learning. In: Weske, et al. [15], pp. 271–287
10. Pelleg, D., Moore, A.W.: X-means: extending k-means with efficient estimation of the number of clusters. In: Langley, P. (ed.) Proceedings of the Seventeenth International Conference on Machine Learning (ICML 2000), Stanford University, Stanford, CA, USA, 29 June–2 July 2000, pp. 727–734. Morgan Kaufmann (2000)
11. Schönig, S., Jasinski, R., Ackermann, L., Jablonski, S.: Deep learning process prediction with discrete and continuous data features. In: Damiani, E., Spanoudakis, G., Maciaszek, L.A. (eds.) Proceedings of the 13th International Conference on Evaluation of Novel Approaches to Software Engineering, ENASE 2018, Funchal, Madeira, Portugal, 23–24 March 2018, pp. 314–319. SciTePress (2018)
12. Tax, N., Teinemaa, I., van Zelst, S.J.: An interdisciplinary comparison of sequence modeling methods for next-element prediction. CoRR, abs/1811.00062 (2018)
13. Tax, N., Verenich, I., La Rosa, M., Dumas, M.: Predictive business process monitoring with LSTM neural networks. In: Dubois, E., Pohl, K. (eds.) CAiSE 2017. LNCS, vol. 10253, pp. 477–492. Springer, Cham (2017). https://doi.org/10.1007/978-3-319-59536-8_30
14. Verenich, I., Dumas, M., Rosa, M.L., Maggi, F.M., Chasovskyi, D., Rozumnyi, A.: Tell me what's ahead? Predicting remaining activity sequences of business process instances, June 2016
15. Weske, M., Montali, M., Weber, I., vom Brocke, J. (eds.): BPM 2018. LNCS, vol. 11080. Springer, Cham (2018). https://doi.org/10.1007/978-3-319-98648-7

Design and Implementation of a Graph-Based Solution for Tracking Manufacturing Products

Jorge Martinez-Gil[1(✉)], Reinhard Stumpner[2], Christian Lettner[1],
Mario Pichler[1], and Werner Fragner[3]

[1] Software Competence Center Hagenberg GmbH,
Softwarepark 21, 4232 Hagenberg, Austria
jorge.martinez-gil@scch.at
[2] FAW GmbH, Softwarepark 35, 4232 Hagenberg, Austria
[3] STIWA Group, Softwarepark 37, 4232 Hagenberg, Austria

Abstract. One of the major problems in the manufacturing industry consists of the fact that many parts from different lots are supplied and mixed to a certain degree during an indeterminate number of stages, what makes it very difficult to trace each of these parts from its origin to its presence in a final product. In order to overcome this limitation, we have worked towards the design of a solution aiming to improve the traceability of the products created by STIWA group. This solution is based on the exploitation of graph databases which allows us to significantly reduce response times compared to traditional relational systems.

Keywords: Data engineering · Graph databases · Manufacturing

1 Introduction

Quality related errors in manufacturing create a lot of problems for the industrial sector mainly because they lead to a great waste of resources in terms of time, money and effort spent to identify and solve them [4]. For this reason, researchers and practitioners aim to find novel solutions capable of tracking and analyzing manufacturing products in an appropriate and easy to use manner.

One of the challenges of modern manufacturing is that a final product normally consists of different components which themselves can also consist of different components, and so forth. At the lowest level, there is raw material, like steel coils, which is also very relevant to the quality of the final product. Tracking and connecting all the data of the different manufacturing stages is crucial to finding the causes to quality related errors in the final product.

In case the components and raw materials have a one-to-one or one-to-many relationship to the final products, the tracking is quite straightforward and is already implemented satisfactorily in the existing solutions. However, manufacturing products can also be made up of parts that are in lots, and lots can be

© Springer Nature Switzerland AG 2019
T. Welzer et al. (Eds.): ADBIS 2019, CCIS 1064, pp. 417–423, 2019.
https://doi.org/10.1007/978-3-030-30278-8_41

combined to make other lots. In addition, many different lots can be combined into a new one that can be part of many other ones. This means that we have to deal with a problem involving many-to-many relationships with blurred relationships among the single parts. But providing these relationships is crucial to the end-users, so they can drill down from the final product to the assembled components and ultimately to the used raw materials and can identify causes to problems which are not obvious at the first glance. Unfortunately, the existing solutions based on relational databases are not very useful for the people who are in charge of examining these lots. On the one hand, the existing solutions have bad response times and on the other hand, there are no meaningful probability values available. So the end-users must invest a lot of time to analyze all possibly related data and cannot focus on the data which is really relevant.

In order to alleviate this problem, we have looked for a solution so that it can be possible to track all items from different lots that were used in final products. Our proposed solution is based on the exploitation of graph databases. The major advantage of our approach is that it allows for informed queries, i.e. queries that can lead to an early termination if nodes with no compatible outgoing relations are found. As a result, we have got a system that presents lower execution time for a number of use cases concerning the tracking of items. Therefore, the major contribution of this work can be summarized as the design and implementation of a solution for tracking the manufacturing products created by STIWA. This solution is intended to outperform traditional systems based on relational databases in the specific context of tracking defective items in lots of manufacturing products.

The rest of this work is structured in the following way: Sect. 2 presents the state-of-the-art concerning the current solutions; Sect. 3 describes the design and implementation of our solution and a use case whereby our solution outperforms the traditional tracking systems. Finally, we remark the conclusions and lessons learned from this work.

2   Preliminaries

There is a lack of solutions in this domain despite the fact that most of quality assurance processes require controlling and supervising the whole production chain in order to timely detect human errors and defective materials. But even in the case of passing all quality assurance controls, products can be rejected by end users or other manufacturers if unknown problems appear. Therefore, the capability to track each part of a lot from its origin is very important.

Traditional relational database models are unable to tackle this problem in an effective manner. Therefore, we have focused our research on graph databases [3]. The idea behind graph databases is their capability to store data in nodes and edges versus tables, as found in relational databases. Each node represents an entity, and each edge represents a relationship between two nodes.

It is generally assumed that graph databases have some key advantages over relational databases in this context. The reason is that unlike relational databases, graph databases are designed to store interconnected data what makes it easier to work with these data by not forcing intermediate indexing at every time, and also making it easier to facilitate the evolution of the data models.

In fact, we have identified three major disadvantages of traditional relational databases in comparison with graph databases to tackle this problem:

(1) One of the advantages when dealing with a graph is that in the relational world, foreign key relationships are not relationships in the sense of edges of a graph.

(2) Another drawback of relational databases is that it is not possible to assign properties or labels to relationships. It is possible to give them a name in the database, but it is not possible to visualize them.

(3) Last, but not least, relational databases are not able to scale well when dispatching relationship-like queries [2].

The application domain of graph databases is wide [1]. In fact, many organizations are already using databases of this kind for detecting fraud in monetary transactions, providing product and service recommendations, documenting use cases and lessons learned in a wide range of domains, managing access control in restricted places, network monitoring to identify potential risks and hazards, and so on.

3   Contribution

In order to illustrate the problem with an example, let us think on a situation where a finished part is rejected by the customer because of product quality errors. In that case, the manufacturer of the finished part must make a statement within 24 hours if this error can be restricted to the single rejected product or if a greater amount of parts is affected. In case a greater amount of parts is affected, then the exactly affected lots must be reported to the customer.

The first task of the manufacturer is to find the cause of the quality error. Therefore it must analyze the captured data of the finished part but also the captured data of all assembled components and raw materials. In case the cause of the quality error lies in a specific component or raw material lot the manufacturer must find all finished parts that contain this lot. So finding the right affected lots in a short time can decide whether the manufacturer must recall millions of finished parts or just a few. This means that if we are able to identify the affected lots quickly, it is not necessary to recall all delivered lots.

In order to perform this analytic task, the manufacturer needs to search back and forth across all the data of the finished parts. If a relational database is used, this means that many queries and their corresponding responses would need to be combined. In our approach, the solution is much more intuitive since it is in general possible to easily write queries capable of running over the data in any direction. In fact, the capability to discover and see the connections between different parts of a product allows us performing this tracking in an effective manner.

3.1   Notation

Let L_m^t be a lot of parts produced on machine m at time step t. Every machine m has a buffer b_m where lots to be processed at this machine are poured into, i.e. they are getting blurred. Then, the relation $usage : (L_m^{t2}, L_n^{t1})$ defines that at time slice $t2$ the lot L_m^{t1} has been poured into the buffer of machine m for processing. So beginning at time slice $t2$ parts of lot L_n^{t1} are installed with a certain probability into parts of lot L_m^{t2}. The relation $pred : \{(L_m^{t2}, L_m^{t1})\}$ defines that lot L_m^{t1} is produced before lot L_m^{t2} on machine m and the buffer b_m of machine m was not empty when the production of L_m^{t2} started. Lots that are delivered by other suppliers, i.e. raw materials, are treated the same way. They will be assigned a virtual production machine number and time slice which uniquely identifies the batch number from the supplier.

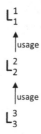

Fig. 1. One-to-one relationship between lots.

Using this notation, simple examples for one-to-one and one-to-many relationship between lots without blurring are depicted in Figs. 1 and 2. An example with a many-to-many relationship that includes blurring of lots is given in Fig. 3. By following the edges of the graph, the lots that are built in other lots can be determined easily, i.e. for lot L_3^4 parts of the lots $\{L_1^1, L_1^2, L_2^2, L_2^3\}$ may be included, or for lot L_3^5 parts of the lots $\{L_1^2, L_1^3, L_2^3, L_2^4\}$ may be included. The missing relation $pred$ between L_3^4 and L_3^5 indicates that the buffer of machine 3 was empty before the production of lot L_3^5 started so no blurring of lots could have occurred. The distance between lots can be used as a basis to determine the probability a certain part of a lot is used in another lot. A more precise determination of probabilities would also require to consider buffer levels during manufacturing but that is beyond of the scope of this work.

3.2   Implementation

A prototypical solution to the proposed manufacturing product tracking system has been implemented using the graph database OrientDB[1]. OrientDB is a multimodal NoSQL database that combines properties of document-oriented and

---

[1] https://orientdb.com/.

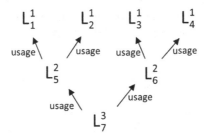

Fig. 2. One-to-many relationship between lots.

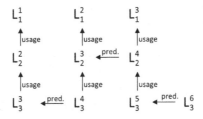

Fig. 3. Many-to-many relationship between lots, i.e. blurred lots.

graph databases. It allows to define graph structures using concepts for nodes and edges, but also allows to append complex data to nodes in the form of documents. Nodes and edges can have attributes (e.g. edge weight or similar). Moreover, inheritable classes can be defined for the nodes and edges which can be extended flexibly. The query language is an adapted form of SQL. Compared to implementations based on relational systems, using a graph database leads to an efficient and scalable solution in which the problem at hand can be modeled easily. Traversing graphs modeled in relational systems would require to write nested and recursive queries that are difficult to maintain and provide bad comparable performance.

3.3 Use Case

Based on the implementation that has been described above, such as graph-based approach is considered to have a positive impact on the operations of STIWA group. In particular, the depth calculation, i.e. the distance between lots, could be used as a basis for the construction probability of a part in a product. The search with BREADTH FIRST returns the real depth in the graph. The greater the depth, the less likely it is that the lot has been incorporated into the end product.

An example query to get the shortest path between lot 33 : 27050 and 29 : 2667 can be easily written as:

```
SELECT expand(path) FROM (
SELECT shortestPath(#33:27050, #29:2667, 'OUT') AS path UNWIND path);
```

The result of this query is depicted in Fig. 4. The number of edges between the lots gives a basic probability that parts of lot 29 : 2667 are built in parts of lot 33 : 27050.

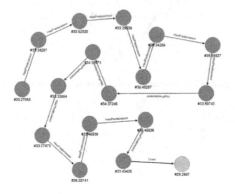

Fig. 4. Shortest path between nodes #33:27050 and #29:2667 in edge direction

4   Conclusions and Future Work

We have presented a solution for tracking each component that comprises manufacturing products from STIWA group through diverse stages of the production chain. Our solution has been modeled using a graph database. In this way, our approach provides an improved level of both transparency and traceability. These factors are able to facilitate the analysis of all manufacturing products as well as the capability to look for final products that could be affected by some specific problem. In this way, our approach presents more efficient modeling and querying mechanisms than traditional approaches. We envision that graph databases holds lot of unrealized potential in the next years as companies will be moving towards approaches being able of better data analysis and exploration.

Acknowledgments. This work has been supported by the Austrian Ministry for Transport, Innovation and Technology, the Federal Ministry of Science, Research and Economy, and the Province of Upper Austria in the frame of the COMET center SCCH.

References

1. El Abri, M.: Probabilistic relational models learning from graph databases (Apprentissage des modèles probabilistes relationnels à partir des bases de données graphe). Ph.D. thesis, University of Nantes, France (2018)
2. Angles, R., Arenas, M., Barceló, P., Hogan, A., Reutter, J.L., Vrgoc, D.: Foundations of modern query languages for graph databases. ACM Comput. Surv. **50**(5), 68:1–68:40 (2017)

3. Fernandes, D., Bernardino, J.: Graph databases comparison: AllegroGraph, ArangoDB, InfiniteGraph, Neo4J, and OrientDB. In: Proceedings of the 7th International Conference on Data Science, Technology and Applications, DATA 2018, Porto, Portugal, 26–28 July 2018, pp. 373–380 (2018)
4. Lou, K.-R., Wang, L.: Optimal lot-sizing policy for a manufacturer with defective items in a supply chain with up-stream and down-stream trade credits. Comput. Ind. Eng. **66**(4), 1125–1130 (2013)

ADBIS 2019 Workshop: International Workshop on BI and Big Data Applications – BBIGAP

An XML Interchange Format for ETL Models

Judith Awiti[✉] and Esteban Zimányi

Université Libre de Bruxelles, Brussels, Belgium
{judith.awiti,ezimanyi}@ulb.ac.be

Abstract. ETL tools are responsible for extracting, transforming and loading data from data sources into a data warehouse. Each ETL tool has its own model for specifying ETL processes. This makes it difficult to interchange ETL designs. Entire ETL workflows must be redesigned in order to migrate them from one tool to another. It has therefore become increasingly important to have a single conceptual model for ETL processes which is interchangeable between tools. Business Process Model and Notation (BPMN) has been widely accepted as a standard for specifying business processes. For this reason, it has been proposed as an efficient conceptual model of ETL processes. In this paper, we present BEXF, an Extensible Markup Language (XML) interchange format for BPMN4ETL, an extended BPMN model for ETL. It is a format powerful enough the express and interchange BPMN4ETL model information across tools that are compliant with BPMN 2.0. This XML interchange format does not only describe the control flow of ETL processes but also the data flow.

Keywords: BEXF · BPMN · BPMN4ETL · ETL · XML

1 Introduction

ETL process development is one of the complex and costly part of any data warehouse project. It involves a lot of time and resources since in practice, ETL processes are designed with a specific tool. SQL Server Integration Services (SSIS), Oracle Warehouse Builder, Talend Open Studio and Pentaho Data Integration (PDI) are examples of such tools. Each ETL tool has its own model for specifying ETL processes which requires a development team to develop ETL processes according to the capabilities of their chosen tool. This makes it difficult to interchange ETL designs. Entire ETL workflows must be redesigned in order to migrate them from one tool to the other. For this reason, several conceptual [1,6,7,9,11] and logical models [8,12] have been proposed for ETL process design.

The work in [1,6] proposes BPMN4ETL, a vendor-independent conceptual metamodel for designing ETL processes based on BPMN[1]. Using BPMN to specify ETL processes makes them simple and easy to understand. It hides technical

---

[1] https://www.omg.org/spec/BPMN/2.0/About-BPMN/.

© Springer Nature Switzerland AG 2019
T. Welzer et al. (Eds.): ADBIS 2019, CCIS 1064, pp. 427–439, 2019.
https://doi.org/10.1007/978-3-030-30278-8_42

details and enables stakeholders to focus on essential characteristics of the ETL processes. BPMN is accepted in the field of modelling business processes and thus provides well-known constructs for any process that has a starting point and an ending point.

Recently, relational algebra (RA) has been introduced as a logical model for ETL processes [8]. RA provides a set of operators that manipulates relations to ensure that there is no ambiguity. It can also be directly translated into SQL to be executed in any Relational Database Management System (RDBMS). The authors of [4] extend RA to model complex ETL scenarios like Slowly Changing Dimensions with Dependencies as well as provide a translation of BPMN4ETL to RA. They propose an ETL development approach which begins with a BPMN4ETL conceptual model translated into RA extended with update operations at the logical level. This approach is a Model Driven Architecture (MDA) approach where platform-independent models (BPMN4ETL and RA) can be implemented as SQLs on any RDBMS platform. In view of this, BPMN4ETL diagrams must have interchangeable formats that can be transformed into the above-mentioned extended RA logical model [4].

Unfortunately, there does not exist an interchange format for BPMN4ETL. In this paper, we present BEXF (BPMN4ETL interchange Format), an XML-based model interchange format for BPMN4ETL that expresses and interchange BPMN4ETL model information across tools. BEXF does not only interchange graphical design of the BPMN4ETL but also attributes and manipulations of the diagram. This way, a BPMN4ETL diagram can be reproduced in another system with all its hidden details. To the best of our knowledge, BEXF is the first step to providing an XML interchange format for BPMN4ETL.

The paper is organized as follows. In Sect. 2 we discuss related work in ETL design. We present our running example in Sect. 3 and explain briefly the already proposed BPMN4ETL model in Sect. 4. In Sect. 5, we explain the fundamentals of BEXF and show how a BPMN4ETL model can be translated into BEXF using an example in Sect. 6. Lastly, we conclude and mention ways by which this work can be pursued in Sect. 7.

2   Related Work

BPMN4ETL [1,6] conceptual model combines two perspectives, a control process view, and a data process view. A control process view consists of all the data processes in the ETL workflow, while the data process view provides a more detailed information of the input and output data of each data process in the control process view. BPMN4ETL enables easy communication and validation between an operational database designer, an ETL designer and a business intelligence analyst. It enables stakeholders to see the manipulation of data from one ETL task to the other. Also, BPMN4ETL can be translated directly into relational algebra (RA), XML[2], Structured Query Language (SQL), or even customized models of vendor tools. In this approach, well-known BPMN operators are customized for

---

[2] https://www.w3.org/TR/2008/REC-xml-20081126/.

ETL design. BPMN *gateways* are used to control the sequence of activities in the ETL workflow based on conditions. *Events* show the start and end of the workflow and are also used to handle errors. An *activity* describes an ETL task that is not further subdivided, whereas a *subprocess* represents a collection of activities.

Several platform-independent conceptual [1,6,7,9,11] and logical models [8, 12] have been proposed. The authors of [3] presents a survey of current trends in designing and optimizing ETL processes. On the other hand, each ETL tool has its own specific model for designing ETL processes. Therefore, there is the need to harmonize the ETL process development with a common and integrated development strategy. One way to do this is to apply an MDA approach to its development. MDA[3] is an approach to software design, development and implementation that separates business and application logic from underlying platform technology. Applying the MDA approach to ETL process development means developing platform-independent conceptual and logical models and then implementing them with vendor-specific technologies.

In an attempt to provide a single agreed-upon development strategy for ETL processes with BPMN, the authors of [2], introduced a framework for model-driven development (MDD) of ETL processes. In this framework, BPMN4ETL is used to specify ETL processes in a vendor-independent way and is automatically transformed into vendor-specific implementations like SSIS. Transformations between a vendor-independent model and this vendor-specific code are formally established by using model-to-text transformations[4], an Object Management Group (OMG) standard for transformations from models to text.

The authors of [5] describe an integration of BPMN and profiled UML languages. They extend UML is to accommodate BPMN notations and define transforms between BPMN and UML in cases where the profiles cannot support. One of the transforms is expressed in XSLT (eXtensible Stylesheet Language Transformations) that operates on BPMN and UML interchange files expressed in XML. The transform files translate BPMN interchange files to UML interchange files and vice versa. The paper focuses on generic BPMN specifications and does not address how the XML interchange file is obtained from the BPMN or UML diagram.

3   Running Example

This section presents an example to be used throughout this paper to illustrate how ETL designs in BPMN4ETL can be translated into BEXF. We reuse a part of the example shown in [10] which described the ETL process that loads the City dimension table of a data warehouse called NorthwindDW. Figure 1a shows the schema of a data source text file, TempCities.txt whereas Fig. 1b shows the schemas of the dimension tables mentioned above. TempCities.txt file contains three fields City, State, and Country with a few rows shown below.

---

[3] https://www.omg.org/mda/.

[4] https://www.omg.org/spec/MOFM2T/1.0.

Aachen - North Rhine-Westphalia - Germany
Albuquerque - New Mexico - USA
Sevilla - Madrid - Spain
Singapore - *NULL* - Singapore
Southfield - Michigan - United States of America

In the case of cities located in countries that do not have states, as it is the case of Singapore, a null value is found in the second field. A temporary table in the data warehouse, denoted TempCities, is used for storing the contents of this file. The schema of the table is the same as the text file given in Fig. 1a. The goal of our ETL is to load the City dimension table with a StateKey and a CountryKey, one of which in null depending on whether the country is divided into states or not. Note that, states and countries come in different forms. For example, the country United States of America can be written as its country codes USA or US, or even in other languages. Therefore in order to retrieve the CountryKey or StateKey of a particular city, we need to match the different representations of the (State,Country) pairs in TempCities to values in the State and Country tables. Finally, we store city records into the City dimension table and store records for which no state and/or country is found into a text file (BadCities.txt) for future investigation.

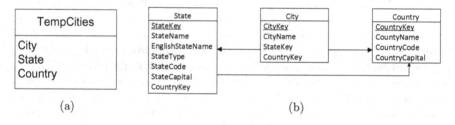

(a)   (b)

Fig. 1. (a) Schema of the TempCities.txt file; (b) Schema of the City-State-Country dimension tables in NorthwindDW.

4   BPMN4ETL

In this section, we briefly introduce BPMN4ETL by means of our running example. Additional description of the model can be found in [1,6]. The input data task to insert records from TempCities into the ETL flow. The first exclusive gateway (G1) tests whether the State attribute is null or not (recall that this is the optional attribute). In the first case, for records with a null value for the State attribute, a lookup obtains the CountryKey. In the second case, we must match (State,Country) pairs in TempCities to values in the State and Country tables. However, as we have explained, states and countries can come in many forms; thus, we need a number of lookup tasks, as shown in the annotations in Fig. 2. Due to space limitation, we only show three lookups are as follows:

- The first lookup process records where State and Country correspond, respectively, to StateName and CountryName. An example is state Loire and country France.
- The second lookup process records where State and Country correspond, respectively, to EnglishStateName and CountryName. An example is state Lower Saxony, whose German name is Niedersachsen, together with country Germany
- Finally, the third lookup process records where State and Country correspond, respectively, to StateName and CountryCode. An example is state New Mexico and country USA.

The SQL query associated with these lookups is as follows:
SELECT S.*, CountryName, CountryCode FROM State S JOIN Country C ON S.CountryKey = C.CountryKey
A union task combines the results of the four flows and the City dimension is populated with an insert data task. Recall that in the City table, if a state was not found in the initial lookup (Input1 in Fig. 2), the attribute State will be null; on the other hand, if a state was found, it means that the city will have an associated state; therefore, the Country attribute will be null (Input2, Input3, and Input4 in Fig. 2). Finally, records for which the state and/or country are not found are stored into a BadCities.txt file.

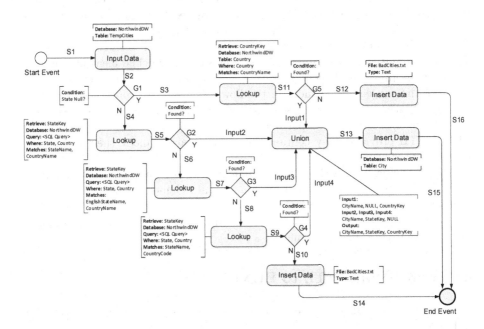

Fig. 2. Load of City dimension table

5   BEXF Fundamentals

In exchanging ETL processes modeled with BPMN 2.0 to an XML format, we consider two types of interchangeable information.

- Semantics information: This comprises the building blocks or objects of the graphical model. The activities, events, control objects, connecting objects and swimlanes together with the attributes and manipulations of them.
- Visual appearance information: This comprises information about the layout of the graphical model. The shapes and positions of the graphical elements.

In this paper, we concentrate on the interchange of semantics information from BPMN4ETL to BEXF. Note that visual appearance information can be described by XSL (eXtensible Stylesheet Language), a styling language for XML.

XML is designed to carry data with the use of arbitrary tags. Users are at liberty to define their own tags and document structure. Therefore, it has been widely applied in different fields including the Mathematical Markup Language (MathML) for describing mathematical notation, and Open Financial interchange (OFX), a data-stream format for exchanging financial information. XML simplifies data sharing, data transport, platform changes, and data availability hence providing an efficient interchange format.

Table 1 describes the basic BPMN4ETL objects and their BEXF representation. Attributes of each BPMN object are mapped to attributes of the corresponding XML element. Each BEXF element is identified by an id and a `name` attribute. The previous and next elements of each BEXF element in an ETL flow are referenced by their id attributes. The `<inRefId>` child element contains the id of the previous element whereas the `<outRefId>` child element contains the id of the subsequent element in the flow. As a naming construct, we begin all id attributes with `_id`. The sequence of all elements at the same level of a BEXF tree does not follow a particular order as elements can be reproduced in BPMN4ETL provided they contain the id information about their immediate surrounding elements.

BEXF being an XML-based language exposes all information in a BPMN4ETL object through attributes and child elements which allows us to achieve data flow of a BPMN4ETL conceptual model. For example, the expression that calculates values of an added column in a BPMN4ETL task is stored in an attribute of a BEXF element.

6   From BPMN4ETL to BEXF

We describe below BEXF elements corresponding to BPMN4ETL objects found in Fig. 2.

Process and Subprocess. An ETL process describes a sequence of flow of activities, events, gateways, and sequence flows with the objective of carrying out work. A typical example is shown in Fig. 2 where the main objective is to load

Table 1. BPMN4ETL objects and their BEXF representation

BPMN4ETL Object	Element	BEXF Representation
ETL Process	Process	`<ETLProcess>`
ETL Activities	Task	`<ETLTask>`
	Subprocess	`<ETLSubprocess>`
Control Objects	Exclusive Gateway	`<ExclusiveGateway>`
	Parallel Gateway	`<ParallelGateway>`
	Inclusive Gateway	`<InclusiveGateway>`
Events	Start Event	`<StartEvent>`
	End Event	`<EndEvent>`
	Message	`<Message>`
	Cancel	`<Cancel>`
	Compensate	`<Compensate>`
	Terminate	`<Terminate>`
	Time	`<Time>`
Connecting Objects	Sequence Flow	`<SequenceFlow>`
	Conditional Flow	`<ConditionalFlow>`
	Default Flow	`<DefaultFlow>`
	Message Flow	`<MessageFlow>`
	Association	`<Association>`
Swimlanes	Pool	`<Pool>`
	Lane	`<Lane>`

the City dimension table. In BPMN4ETL, some activities of an ETL process can be encapsulated in order to hide their details. Such parts are called subprocesses. For instance, Fig. 2 is a subprocess in the overall ETL process that loads NorthwindDW. We represent processes and subprocesses in BEXF as shown in Table 1 as `<ETLProcess>` and `<ETLSubProcess>` elements respectively. The BEXF of the entire process of Fig. 2 is as follows:

```
<ETLProcess id="_idProcess" name="Load of City dimension table">
...
</ETLProcess>
```

Control Objects. These are Gateways which control the sequence of activities in an ETL flow. The BEXF of an Exclusive Gateway is represented in Table 1 as `<ExclusiveGateway>`. It has a child element called `<condition>` which contains the condition to be checked. An Exclusive Gateway can have several output connecting objects. These links are represented by two or more `<outRefId>` child elements. Below, we show the BEXF of the exclusive gateway G1 of Fig. 2 where State = NULL is the condition, _idS2 is the id of the incoming sequence flow S2, and _idS3 and _idS4 are the outgoing sequence flows S3 and S4, respectively.

```
<ExclusiveGateway id="_idG1" name="G1">
  <condition>State = NULL</condition>
  <inRefId>_idS2</inRefId>
  <outRefId>_idS3</outRefId>
  <outRefId>_idS4</outRefId>
</ExclusiveGateway>
```

Events. They are happenings that affect the sequence of flow in an ETL processes. The (`<StartEvent>`) element contains one or more `<outRefId>` child elements. More than one reference `<outRefId>` represent scenarios where at the start, the ETL flow is divided into different paths. The `<EndEvent>` contains one or more `<inRefId>` child elements can also be used to model scenarios where several paths of an ETL flow end at the same time. The BEXF code below shows the Start Event and the End Event of Fig. 2.

```
<StartEvent id="_idStartEvent"  name="Start Event">
  <outRefId>_idS1</outRefId>
</StartEvent>
...
<EndEvent id="_idEndEvent" name="End Event">
  <inRefId>_idS14</inRefId>
  <inRefId>_idS15</inRefId>
  <inRefId>_idS16</inRefId>
</EndEvent>
```

Connecting Objects. These are mostly arrows representing the links between BPMN4ETL objects. An example is the Sequence Flow which represents the sequencing constraint between ETL flow objects. The Sequence Flow, S2 of Fig. 2 is represented in BEXF as shown below by a `<SequenceFlow>` element that has one `<inRefId>` and one `<outRefId>`.

```
<SequenceFlow id="_idS2" name="S2">
  <inRefId>_idInputData</inRefId>
  <outRefId>_idG1</outRefId>
</SequenceFlow>
```

ETL Activities and Tasks. An activity is a work performed during an ETL process. Activities are either single tasks or subprocesses. An ETL Task is a simple, atomic unit of work. ETL processes and subprocesses contain several ETL tasks. In Table 1, we represent an ETL task with the element `<ETLTask>`. The type of task is specified by its type attribute. Each ETL task has some peculiarities that distinguishes it from other tasks. We describe below, the BEXF representations of the ETL tasks in Fig. 2 as well as some other common ETL tasks.

Input Data: The Input Data task insert records into the ETL flow from a data source. `<Database>` and the `<table>` child elements of the Input Data task contain information about incoming data from a database. Both child elements can be replaced by a `<file>` child element with a type attribute showing the type of file if the data is from a file source. This is true for the Insert Data task as well. Below is the BEXF representation of the Input Data task of Fig. 2. With this representation, `<Database name="NorthwindDW"/>` and

`<Table name="TempCities"/>` specify the location of the TempCities table. All input columns (City, State, Country) are specified in the `<inputs>` child element.

```
<ETLTask id="_idInputData"  name="Input Data" type="Input Data">
  <Database name="NorthwindDW"/>
  <Table name="TempCities"/>
  <inputs>
    <inputColumn name="City"/>
    <inputColumn name="State"/>
    <inputColumn name="Country"/>
  </inputs>
  <inRefId>_idS1</inRefId>
  <outRefId>_idS2</outRefId>
</ETLTask>
```

Insert Data: Recall that this task inserts records from the ETL flow into a destination file or database table depending on the child elements available. The DbCol attribute of the `<outputColumn>` child element provides the destination column names. Note that in our running example (Fig. 2), the column names of the ETL flow and that of the City dimension table are the same. Therefore, the values of name and DbCol attributes for each `<outputColumn>` subchild element are the same.

```
<ETLTask id="_idInsertData3" name="Insert Data" type="Insert Data">
  <Database name="NorthwindDW"/>
  <Table name="City"/>
  <outputs>
    <outputColumn name="City" DbCol="City"/>
    <outputColumn name="StateKey" DbCol="StateKey"/>
    <outputColumn name="CountryKey" DbCol="CountryKey"/>
  </outputs>
  <inRefId>_idS13</inRefId>
  <outRefId>_idS15</outRefId>
</ETLTask>
```

Lookup: Several types of Lookup tasks exist depending on where the lookup data is comes from. `<Database>` and `<Query>`, `<Database>` and `<Table>`, and `<file>` child elements are used if the lookup column(s) is/are from the results of a query, a database table, or an external file, respectively. The MatchCol attribute of the `<inputColumn>` child element provide the corresponding column name of the Lookup task. This attribute does not exist in unmatched columns. The BEXF of the lookup task to obtain CountryKey of Fig. 2 is shown below. The BEXF code `<inputColumn name="Country" MatchCol="CountryName"/>` specifies that the Country column is matched with the CountryName column whereas `<outputColumn name="CountryKey"/>` specifies a new column CountryKey derived at the output.

```
<ETLTask id="_idL1" name="Lookup"    type="Lookup">
  <Database name="NorthwindDW"/>
  <table>Country<table/>
  <inputs>
    <inputColumn name="City"/>
    <inputColumn name="State"/>
    <inputColumn name="Country" MatchCol="CountryName"/>
    <inputColumn name="CountryName" MatchCol="Country"/>
  </inputs>
  <outputs>
    <outputColumn name="City"/>
    <outputColumn name="State"/>
```

```
  <outputColumn name="Country"/>
  <outputColumn name="CountryKey"/>
</outputs>
<inRefId>_idS3</inRefId>
<outRefId>_idS11</outRefId>
</ETLTask>
```

Union: This task combines data from all incoming paths. Each `<input>` subchild element contains the input columns of one input path. The default value of input columns can be set by the `default` attribute. The BEXF representation of the Union task of our running example is shown below. Note that in the first `<input>` element, we set the default value of StateKey to null. This specifies the path with records that has no StateKey value. For all other input paths, the CountryKey column is set to null to specify the reverse.

```
<ETLTask id="_idUnion" name="Union" type="Union">
  <inputs>
    <input>
      <inputColumn name="City"/>
      <inputColumn name="State"/>
      <inputColumn name="Country"/>
      <inputColumn name="StateKey" default="NULL"/>
      <inputColumn name="CountryKey"/>
    </input>
    ...
    <input>
      <inputColumn name="City"/>
      <inputColumn name="State"/>
      <inputColumn name="Country"/>
      <inputColumn name="StateKey"/>
      <inputColumn name="CountryKey" default="NULL"/>
    </input>
  </inputs>
  <outputs>
    <output>
      <outputColumn name="City"/>
      <outputColumn name="State"/>
      <outputColumn name="Country"/>
      <outputColumn name="StateKey"/>
      <outputColumn name="CountryKey"/>
    </output>
  </outputs>
  <inRefId>_idInput1</inRefId>
  <inRefId>_idInput2</inRefId>
  <inRefId>_idInput3</inRefId>
  <inRefId>_idInput4</inRefId>
  <outRefId>_idS13</outRefId>
</ETLTask>
```

Rename Column: This task adds new derived columns to an ETL flow. Assuming in Fig. 2, there exist an Rename Column task that renames the Country column of the ETL flow into Ctry. As shown below in BEXF, the value of the newname attribute of Country is Ctry.

```
<ETLTask id="_idRenameColumn" name="Rename Country" type="Rename Column">
  <Column name="Country" newname="Ctry"/>
  <inRefId></inRefId>
  <outRefId></outRefId>
</ETLTask>
```

Aggregate: This task adds new columns to an ETL flow computed by applying an aggregate function such as Count, Min, Max, Sum, or Avg on input columns.

This is done after partitioning tuples in groups that have the same values in some columns. We show below the BEXF representation of an aggregate task. Assume that there exist an aggregate ETL task that counts the cities of each state in Fig. 2. <AggColumn/> specifies the column to group by. In this example, the aggregate column is State. The order attribute specifies the order of the grouping columns. This attribute is needed for cases where the aggregate columns are more than one. <NewColumn/> specifies the new column CityCount, which is to be added to the flow as a result of the function count(City).

```
<ETLTask id="_idAggregate" name="Aggregate" type="Aggregate">
  <AggColumn name="State" order="1"/>
  <NewColumn name="CityCount" function="count(City)"/>
  <inRefId></inRefId>
  <outRefId></outRefId>
</ETLTask>
```

Add Column: This task adds new derived columns to an ETL flow. Assume in Fig. 2, there exist an Add Column task that adds a column called CityStatus with a value of NEW to show that this city was just added to the dimension table. We show the BEXF representation of this task below. The name attribute contains the name of the added column whereas the expression attribute contains the expression that computes the value of the newly added column.

```
<ETLTask id="_idAddColumn" name="Add Column" type="Add Column">
  <Column name="CityStatus" expression="CityStatus = NEW"/>
  <inRefId></inRefId>
  <outRefId></outRefId>
</ETLTask>
```

Convert Column: This task changes the data type of columns in an ETL flow. A task to convert the data type of a column called CityKey to an integer is specified as follows:

```
<ETLTask id="_idConvertColumn" name="Convert Column" type="Convert Column">
  <Column name="CityKey" DataType="INTEGER"/>
  <inRefId></inRefId>
  <outRefId></outRefId>
</ETLTask>
```

Update Column: This task replaces column values in the flow. Assume that there exist a column called NetWorth in the ETL flow of Fig. 2 which stores the amount of money each city has. To deduct 10,000 euros from this amount for cities whose NetWorth values are greater than 1,000,000 euros, we will need an Update Column task with a condition as shown in BEXF below.

```
<ETLTask id="_idUpdateColumn" name="Update Column" type="Update Column">
  <Column name="NetWorth" expression="NetWorth = NetWorth * 2"/>
  <Condition>NetWorth > 1000000<Condition/>
  <inRefId></inRefId>
  <outRefId></outRefId>
</ETLTask>
```

Another type of Update Column task replaces column values of a database table or file that corresponds to the records in the ETL flow. For such tasks, we add a <file> child element or a <database> and <table> child elements to specify the location of the column to update.

7 Conclusion and Future Work

BPMN4ETL is a conceptual model for designing ETL processes. This paper presents BEXF, an XML interchange format that can be used to interchange and reuse BPMN4ETL models. BEXF expresses both the control flow and the data flow of ETL processes through attributes and manipulations. Being a platform-independent language, it can be translated into various implementation platforms such as ETL tools (e.g., Microsoft SSIS or Pentaho PDI) or SQL dialects (e.g., PL/pgSQL for PostgreSQL).

One direction for our future work is to use the Model-Driven Architecture for translating BEXF models into such implementation platforms. We envision to use Model-to-Model and Model-to-Text transformations to generate, e.g., DTSX (an XML-based file format for SSIS) or PL/pgSQL. Another future direction is to add visual appearance information like position, size, color and shape to BEXF elements. Finally, we will tackle the issue of evolving ETL flows upon data source schema changes.

Acknowledgements. The work of Judith Awiti is supported by the European Commission through the Erasmus Mundus Joint Doctorate project *Information Technologies for Business Intelligence-Doctoral College* (IT4BI-DC).

References

1. Akkaoui, Z.E., Zimányi, E.: Defining ETL worfklows using BPMN and BPEL. In: Proceedings of the 12th ACM International Workshop on Data Warehousing and OLAP, pp. 41–48. ACM, Hong Kong (2009)
2. Akkaoui, Z.E., Zimányi, E., Mazón, J., Trujillo, J.: A model-driven framework for ETL process development. In: Proceedings of the 14th International Workshop on Data Warehousing and OLAP, DOLAP 2011, pp. 45–52. ACM, Glasgow (2011)
3. Ali, S.M.F., Wrembel, R.: From conceptual design to performance optimization of ETL workflows: current state of research and open problems. VLDB J. **26**(6), 777–801 (2017)
4. Awiti, J., Vaisman, A., Zimányi, E.: From conceptual to logical ETL design using BPMN and relational algebra. In: Proceedings of the 21st ACM International Conference on Big Data Analytics and Knowledge Discovery, DAWAK 2019. Springer, Linz (2019, forthcoming)
5. Bock, C., Barbau, R., Narayanan, A.: BPMN profile for operational requirements. J. Object Technol. **13**(2), 1–35 (2014)
6. El Akkaoui, Z., Zimányi, E., Mazón, J.N., Trujillo, J.: A BPMN-based design and maintenance framework for ETL processes. Int. J. Data Warehousing Mining **9**(3), 46–72 (2013)
7. Muñoz, L., Mazón, J.-N., Pardillo, J., Trujillo, J.: Modelling ETL processes of data warehouses with UML activity diagrams. In: Meersman, R., Tari, Z., Herrero, P. (eds.) OTM 2008. LNCS, vol. 5333, pp. 44–53. Springer, Heidelberg (2008). https://doi.org/10.1007/978-3-540-88875-8_21
8. Santos, V., Belo, O.: Using relational algebra on the specification of real world ETL processes. In: Proceedings of the 13th IEEE International Conference on Dependable, Autonomic and Secure Computing, DASC 2015, pp. 861–866. IEEE, Liverpool (2015)

9. Trujillo, J., Luján-Mora, S.: A UML based approach for modeling ETL processes in data warehouses. In: Song, I.-Y., Liddle, S.W., Ling, T.-W., Scheuermann, P. (eds.) ER 2003. LNCS, vol. 2813, pp. 307–320. Springer, Heidelberg (2003). https://doi.org/10.1007/978-3-540-39648-2_25

10. Vaisman, A.A., Zimányi, E.: Data Warehouse Systems: Design and Implementation. Springer, Heidelberg (2014). https://doi.org/10.1007/978-3-642-54655-6

11. Vassiliadis, P., Simitsis, A., Skiadopoulos, S.: Conceptual modeling for ETL processes. In: Proceedings of the 5th ACM International Workshop on Data Warehousing and OLAP, DOLAP 2002, pp. 14–21. ACM, McLean (2002)

12. Vassiliadis, P., Simitsis, A., Skiadopoulos, S.: Modeling ETL activities as graphs. In: Proceedings of the 4th International Workshop on Design and Management of Data Warehouses, DMDW 2002, pp. 52–61. CEUR-WS.org, Toronto (2002)

Metadata Systems for Data Lakes: Models and Features

Pegdwendé N. Sawadogo[1]([✉]), Étienne Scholly[1,2], Cécile Favre[1], Éric Ferey[2], Sabine Loudcher[1], and Jérôme Darmont[1]

[1] Université de Lyon, Lyon 2, ERIC EA 3083, Lyon, France
pegdwende.sawadogo@univ-lyon2.fr
[2] BIAL-X, Limonest, France
https://eric.ish-lyon.cnrs.fr/
https://www.bial-x.com/

Abstract. Over the past decade, the data lake concept has emerged as an alternative to data warehouses for storing and analyzing big data. A data lake allows storing data without any predefined schema. Therefore, data querying and analysis depend on a metadata system that must be efficient and comprehensive. However, metadata management in data lakes remains a current issue and the criteria for evaluating its effectiveness are more or less nonexistent.

In this paper, we introduce MEDAL, a generic, graph-based model for metadata management in data lakes. We also propose evaluation criteria for data lake metadata systems through a list of expected features. Eventually, we show that our approach is more comprehensive than existing metadata systems.

Keywords: Data lakes · Metadata modeling · Metadata management

1 Introduction

Since the beginning of the 21st century, the usages of organizations in decision-making processes have been disrupted by the availability of large amounts of data, i.e., big data. Mainly issued from social media and the Internet of things, big data bring about great opportunities for organizations, but also issues related to data volume, velocity and variety, which surpass the capabilities of traditional storage and data processing systems [20].

In this context, the concept of data lake [6] appears as a solution to big data heterogeneity problems. A data lake provides integrated data storage without predefined schema [11]. In the absence of a data schema, an effective metadata system becomes essential to make data queryable and thus prevent the lake from turning into a data swamp, i.e., an inexploitable data lake [1,11,26].

While the literature seems unanimous about the importance of the metadata system in a data lake, questions and uncertainties remain about its implementation methodology. Several approaches help organize metadata, but most concern

© Springer Nature Switzerland AG 2019
T. Welzer et al. (Eds.): ADBIS 2019, CCIS 1064, pp. 440–451, 2019.
https://doi.org/10.1007/978-3-030-30278-8_43

only structured and semi-structured data [8,11,17,22]. Moreover, the effectiveness of a metadata system is difficult to measure because, to the best of our knowledge, there are no widely shared and accepted evaluation criteria.

To address these issues, we first identify a set of features that should ideally be proposed by the metadata system of a data lake. By comparing several metadata systems with respect to these features, we hint that none of them offers all expected features. Thus, we propose a new metadata model that is more complete and generic. Our graph-based metadata model is named MEtadata model for DAta Lakes (MEDAL). It is also based on a typology distinguishing intra-object, inter-object and global metadata.

The remainder of this paper is organized as follows. Section 2 introduces the concept of data lake. Section 3 details the expected features of a data lake's metadata system and compares several works on the organization of metadata w.r.t. these features. Section 4 presents the metadata typology on which MEDAL is based. Section 5 formalizes MEDAL and introduces its graph representation. Finally, Sect. 6 concludes the paper and hint at research perspectives.

2   Data Lake Concept

2.1   Definitions from the Literature

James Dixon introduces the data lake concept as an alternative to data marts, which are subsets of data warehouses that store data into silos. A data lake is a large repository of heterogeneous raw data, supplied by external data sources and from which various analyses can be performed [6].

Thereafter, data lakes are associated with the Hadoop technology [7,21]. Data lake design may notably be viewed as a methodology for using free or low-cost technologies, typically Hadoop, for storing, processing and exploring raw data within a company [7]. However, this view is becoming minority in the literature, as the data lake concept is now also associated with proprietary solutions such as Azure or IBM [18,25].

A more consensual definition is to see a data lake as a central repository where data of all formats are stored without a strict schema [14,16,19]. This definition is based on two key characteristics of data lakes: data variety and the schema-on-read (or late binding) approach, which consists in defining the data schema at analysis time [20].

However, the variety/schema-on-read definition provides little detail about the characteristics of a data lake. Thus, a more complete definition by Madera and Laurent views data lake as a logical view of all data sources and datasets in their raw format, accessible by data scientists or statisticians for knowledge extraction [18]. This definition is complemented by a list of key features: (1) data quality is provided by a set of metadata; (2) the lake is controlled by data governance policy tools; (3) usage of the lake is limited to statisticians and data scientists; (4) the lake integrates data of all types and formats; (5) the data lake has a logical and physical organization.

2.2   Discussion and New Definition

Some points in Madera and Laurent's definition of data lakes [18] are debatable. The authors indeed reserve the use of the lake to data specialists and, as a consequence, exclude business experts for security reasons. Yet, in our opinion, it is entirely possible to allow controlled access to this type of users through a navigation or analysis platform.

Moreover, we do not share the vision of the data lake as a logical view over data sources, since some data sources may be external to an organization, and therefore to the data lake. Dixon also specifies that lake data come from data sources [6]. Including data sources into the lake may therefore be considered contrary to the spirit of data lakes.

Finally, although quite complete, Madera and Laurent's definition omits an essential property of data lakes: scalability [20]. Since a data lake is intended for big data storage and processing, it is indeed essential to address this issue.

Thence, we amend Madera and Laurent's definition to bring it in line with our vision and introduce scalability.

Definition 1. *A data lake is a scalable storage and analysis system for data of any type, retained in their native format and used* mainly *by data specialists (statisticians, data scientists or analysts) for knowledge extraction. Its characteristics include: (1) a metadata catalog that enforces data quality; (2) data governance policies and tools; (3) accessibility to various kinds of users; (4) integration of any type of data; (5) a logical and physical organization; (6) scalability.*

3   Basic Features of a Metadata System

3.1   Expected Features

We identify in the literature six main functionalities that should ideally be provided by the metadata system of a data lake.

Semantic enrichment (SE), also called semantic annotation [11] or semantic profiling [2], consists in generating a description of the context of data, e.g., with tags, to make them more interpretable and understandable [27]. It is done using knowledge bases such as ontologies. Semantic annotation plays a key role in data exploitation, by summarizing the datasets contained in the lake so that they are more understandable to the user. It can also be used as a basis for identifying data links. For instance, data associated with the same tags can be considered linked.

Data indexing (DI) consists in setting up a data structure to retrieve datasets based on specific characteristics (keywords or patterns). This requires the construction of forward or inverted indexes. Indexing makes it possible to optimize data querying in the lake through keyword filtering. It is particularly useful for textual data management, but can also be used in a semi-structured or structured data context [24].

Link generation and conservation (LG) is the process of detecting similarity relationships or integrating preexisting links between datasets. The integration of data links can be used to expand the range of possible analyses from the lake by recommending data related to those of interest to the user [17]. Data links can also be used to identify data clusters, i.e., groups where data are strongly linked to each other and significantly different from other data [9].

We define **data polymorphism (DP)** as storing multiple representations of the same data. Each representation corresponds to the initial data, modified or reformatted for a specific need. For example, a textual document can be represented without stopwords or as a bag of words. It is essential in the context of data lakes to at least partially structure unstructured data to allow their automated analysis [5]. Simultaneously storing several representations of the same data notably avoids repeating preprocessings and thus speeds up analyses.

Data versioning (DV) refers to the ability of the metadata system to support data changes while conserving previous states. This ability is essential in data lakes, as it ensures the reproducibility of analyses and supports the detection and correction of possible errors or inconsistencies. Versioning also allows to support a branched evolution of data, especially in their schema [13].

Usage tracking (UT) records the interactions between users and the data lake. Interactions are generally operations of data creation, update and access. The integration of this information into the metadata system makes it possible to understand and explain possible inconsistencies in the data [3]. It can also be used to manage sensitive data, by detecting intrusions [26].

Usage tracking and data versioning are closely linked, because interactions lead in some cases to the creation of new versions or representations of the data. Thus, such features are often integrated together in a provenance tracking module [12,13,27]. Yet, we still consider that they remain different features since they are not systematically proposed together [3,5,26].

3.2   Comparison of Metadata Systems

We consider in this comparison two types of metadata systems: metadata models and data lake implementations. Metadata models refer to conceptual systems for organizing metadata. They have the advantage of being more detailed and more easily reproducible than data lake implementations, which lie at a more operational level. Data lake implementations focus on operation and functionality, with little detail on the conceptual organization of metadata. Eventually, we include in this study systems (models or implementations) not explicitly associated with the concept of data lake by their authors, but that may be used in a data lake context, e.g., the Ground metadata model [13].

Table 1 provides a synthetic comparison of 15 metadata systems for data lakes (and assimilated). It shows that the most complete systems in terms of functionality are the GOODS and CoreKG data lakes, with five out of six features available. These systems notably support polymorphism and data versioning. However, they are also black boxes providing little detail on the conceptual organization of metadata. Ground may therefore be preferred, since it is much more detailed and almost as complete (4/6).

Table 1. Features provided by data lake metadata systems

System	Type	SE	DI	LG	DP	DV	UT
SPAR (Fauduet and Peyrard, 2010) [10]	♦♯	✓	✓	✓			✓
Alrehamy and Walker (2015) [1]	♦	✓		✓			
Terrizzano et al. (2015) [27]	♦	✓	✓			✓	✓
Constance (Hai et al., 2016) [11]	♦	✓	✓				
GEMMS (Quix et al., 2016) [22]	◇	✓					
CLAMS (Farid et al., 2016) [8]	♦	✓					
Suriarachchi and Plale (2016) [26]	◇				✓		✓
Singh et al. (2016) [24]	♦	✓	✓	✓	✓		
Farrugia et al. (2016) [9]	♦			✓			
GOODS (Halevy et al., 2016) [12]	♦	✓	✓	✓		✓	✓
CoreDB (Beheshti et al., 2017) [3]	♦		✓				✓
Ground (Hellerstein et al., 2017) [13]	◇♯	✓	✓			✓	✓
KAYAK (Maccioni and Torlone, 2018) [17]	♦	✓	✓	✓			
CoreKG (Beheshti et al., 2018) [4]	♦	✓	✓	✓	✓		✓
Diamantini et al. (2018) [5]	◇	✓		✓	✓		

♦ : Data lake implementation ◇ : Metadata model
♯ : Model or implementation assimilable to a data lake

In terms of functionalities, we note an almost unanimous agreement on the relevance of semantic enrichment, with 12 out of 15 systems offering this feature and, to a lesser extent, of data indexing (9/15) and data link generation (8/15). On the other hand, other features are much less shared, especially data polymorphism (4/15) and data versioning (3/15). In our opinion, this rarity does not indicate a lack of relevance, but rather implementation complexity. Such features are indeed mainly found in the most complete systems (GOODS, CoreKG and Ground) and can therefore be considered as advanced features.

4 Metadata Typology

The comparison results from Sect. 3.2 show that no metadata system offers all expected functionalities. To bridge this gap, we propose in the following a metadata model that supports all six key functionalities. Beforehand, we need a generic concept that represents any set of homogeneous data that the model can process. In the literature, we find data units [22], entities [3], datasets [17] and objects [5]. We adopt objects, which seem more appropriate to represent a dataset in an abstract way. More precisely, an object may be a relational table or a physical file (spreadsheet document, XML or JSON document, text document, tweet collection, image, video, etc.).

The definition of a metadata model for data lakes also involves identifying the metadata to be considered. To this end, we extend a medatata typology that categorizes metadata into intra-object, inter-object and global metadata [23] with new types of inter-object (relationships) and global (index, event logs) metadata.

4.1 Intra-object Metadata

This category refers to metadata associated with a given object.

Properties provide a general description of an object, in the form of key-value pairs. Such metadata are usually obtained from the filesystem: object title, size, date of last modification, access path, etc.

Summaries and previews provide an overview of the content or structure of an object. They can take the form of a data schema in a structured or semi-structured data context, or a word cloud for textual data.

Raw data in the lake are often modified through updates that result in the creation of new **versions** of the initial data, which can be considered as metadata. Similarly, raw data (especially unstructured data) can be reformatted for a specific use, inducing the creation of new **representations** of an object.

Semantic metadata are annotations that help understand the meaning of data. More concretely, they are descriptive tags, textual descriptions or business categories. Semantic metadata are often used for detecting object relationships.

4.2 Inter-object Metadata

Inter-object metadata account for relationships between at least two objects.

Objects groupings organize objects into collections, each object being able to belong simultaneously to several collections. Such groups can be automatically derived from semantic metadata such as tags and business categories. Some properties can also be used for generating groups, e.g., objects can be grouped w.r.t. format or language.

Similarity links reflect the strength of the similarity between two objects. Unlike object groupings, similarity relationships refer to the intrinsic properties of objects, such as their content or structure. For example, it may be the common word rate between two textual documents, a measure of the compatibility of the schemas of two structured or semi-structured objects [17], or other common similarity measures.

Parenthood relationships, which we add to our reference typology [23], reflect the fact that an object can be the result of joining several others. In such a case, there is a "parenthood" relationship between the combined objects and the resulting object, and a "co-parenthood" relationship between the merged objects. This type of relationship thus makes it possible to take advantage of the processing carried out in the data lake to identify objects that can be used together, in addition to maintaining traceability of the origin of the objects generated inside the lake.

4.3 Global Metadata

Global metadata are data structures designed to provide a contextual layer to the lake's data, to facilitate and optimize its analysis. Unlike intra and inter-object metadata, global metadata potentially concern the entire data lake. In addition to the semantic resources identified in our reference typology [23], we propose two new types of global metadata.

Semantic resources are essentially knowledge bases (ontologies, taxonomies, thesauri, dictionaries) used to generate other metadata and improve analyses. For example, a thesaurus can help extend a query by associating synonyms of the terms typed by the user. Similarly, a thesaurus can be used while generating object groupings, to merge collections from different but equivalent tags.

Semantic resources are generally coming from external sources. This is typically the case for ontologies that are provided by knowledge bases on the Internet. However, in some cases, semantic resources can be created and customized specifically for the management and analysis of lake data. For instance, a business ontology can thus be used to define abstract tags allowing to group together several equivalent or close tags during analysis.

Indexes are data structures that help find an object quickly. They establish (or measure) the correspondence between characteristics such as keywords, patterns or colors, with the objects contained in the data lake. Indexes can be simple (textual indexing) or more complex (e.g., on images or sound content). They are mainly used to search for data in the lake.

Logs are used to track user interactions with the data lake. This involves the sequential recording of events such as users logging in, viewing or modifying an object. Such metadata help analyze data lake usage by identifying the most consulted objects or studying user behaviour.

4.4 Formal Definition of a Data Lake

From the above typology, we can now formally define a data lake.

Definition 2. *A data lake is a pair $DL = \langle \mathcal{D}, \mathcal{M} \rangle$, where $\mathcal{D}$ is a set of raw data and $\mathcal{M}$ a set of metadata describing the objects of $\mathcal{D}$. Objects in $\mathcal{D}$ can take the form of structured (relational database tables, CSV files, etc.), semi-structured (JSON, XML, YAML documents, etc.) and unstructured data (images, textual documents, videos, etc.). Metadata are subdivided into three components: $\mathcal{M} = \langle \mathcal{M}_{intra}, \mathcal{M}_{inter}, \mathcal{M}_{glob} \rangle$, where $\mathcal{M}_{intra}$ is the set of intra-object metadata, $\mathcal{M}_{inter}$ the set of inter-object metadata and $\mathcal{M}_{glob}$ the set of global metadata.*

5 Metadata Model

MEDAL adopts a logical metadata representation based on the hypergraph, nested graph and attributed graph notions. We represent an object by an **hypernode** containing various elements (versions and representations, properties, etc.). Hypernodes can be linked together (similarity, parenthood, etc.).

5.1 Intra-object Metadata

Each hypernode contains **representations**, reflecting the fact that data associated with an object can be presented in different ways. There is at least one representation per hypernode, corresponding to raw data. Other representations all derive from this initial representation. Each representation corresponds to a node bearing attributes, simple or complex. These are the properties of the representation. A representation can be associated with an object actually stored in the lake or be a view calculated on demand.

The transition from one representation to another is done via a **transformation**. It takes the form of a directed edge connecting two representation nodes. This edge also bears attributes, which are the properties describing the transformation process from the first representation to the second (full script or description, in case of manual transformation).

A hypernode can also contain **versions**, which are used to manage the evolution of lake data over time. We also associate versions with nodes bearing attributes. The creation of a new version node is not necessarily systematic at the slightest change. Depending on the nature and frequency of data evolution, it is possible to implement various strategies, such as those used to manage slowly changing dimensions in data warehouses [15]. The creation of a new version is done via an **update** similar to a transformation, since it is also translated by a directed edge and possesses some attributes.

Finally, a hypernode also bears attributes such as the origin of the object or aggregates of the attributes of the representations and versions it contains (number of versions, representations, total size, etc.). Thus, a hypernode contains a tree whose nodes are representations or versions and directed edges are transformations or updates. One representation (resp. version) is derived from another by a transformation (resp. update). A version can lead to a representation via a transformation, but a version cannot be derived from a representation. Thus, the root of the tree is the initial raw representation of the hypernode and each version has its own subtree of representations.

Definition 3. *Let $\mathcal{N}$ be a set of nodes. The set of* intra-object metadata $\mathcal{M}_{intra}$ *is the set of hypernodes such that* $\forall h \in \mathcal{M}_{intra}, h = \langle N, E \rangle$, *where* $N \subset \mathcal{N}$ *is the set of nodes (representations and versions) carrying attributes of h and* $E = \{r_{(transformation \mid update)} \in N \times N\}$ *is the set of edges (transformations and updates) carrying attributes of h.*

Let us illustrate these notions with an example (Fig. 1). Imagine a company selling various products. Information on these products (name, unit price, description, etc.) is stored in the lake as an XML file. A hypernode describes this dataset and has a version node that corresponds to the initially ingested XML file. To assist in querying product information, a user decides to extract the XML file's schema. This generates a new representation. Now suppose that the price of some products changes and new products are added to the catalogue. This change in data generates a new version, linked to the first version by an update. Finally, if the user wants to obtain the schema of the most recent data, this creates a new representation coming from the second version.

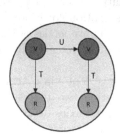

Fig. 1. Sample hypernode and its representation tree

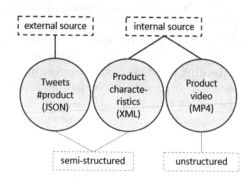

Fig. 2.   Sample   interconnected hypernodes

5.2   Inter-object Metadata

An object grouping is modeled by a set of non-oriented hyperedges, i.e., edges that can link more than two (hyper)nodes. Each hyperedge corresponds to a collection of objects. If grouping is performed on a hypernode attribute, a hypernode belongs to the hyperarc that corresponds to its value for the attribute. Thus, there are as many hyperarcs as there are distinct values of the considered attribute. Note that not all attributes are necessarily used in groupings and that groupings can be made on other elements but attributes.

A similarity link between two hypernodes is represented by a non-oriented edge with attributes: value of the similarity metric, type of metric used, date of the metric, etc. Two hypernodes connected by a similarity link must be comparable, i.e., they must each contain a representation that can be compared to the other with a similarity measure.

A hypernode can be derived from other hypernodes through a parenthood link. To translate this relationship, we use an oriented hyperedge: all the "parent" hypernodes and the "child" hypernode are connected by this oriented hyperedge toward the child hypernode. This hyperedge also bears descriptive attributes.

Definition 4. *The set of* inter-object *metadata* $\mathcal{M}_{inter}$ *is defined by three pairs* $\langle H, E_g \rangle$, $\langle H', E_s \rangle$ *and* $\langle H'', E_p \rangle$, *where* $H \subset \mathcal{M}_{intra}$, $H' \subset \mathcal{M}_{intra}$ *and* $H'' \subset$

$\mathcal{M}_{intra}$ are sets of hypernodes carrying attributes; $E_g = \{E_g^{param} \mid E_g^{param} : H \rightarrow \mathcal{P}(H)\}$ is the set of functions grouping hypernodes in collections w.r.t. a given parameter (often an attribute);

$E_s = \{s \mid s \in H' \times H'\}$ is the set of edges (similarity links) carrying attributes; and $E_p = \{(h_1, ..., h_n, h_{child}) \mid (h_1, ..., h_n, h_{child}) \in (H'')^{n+1}\}$ is the set of parenthood relationships, with $(h_1, ..., h_n)$ being the parent hypernodes $(n \geq 2)$ and h_{child} the child hypernode.

Let us pursue the example of Sect. 5.1 by adding other hypernodes: tweets related to the company and a commercial video of the products. In a grouping on the origin of data, the tweet hypernode is alone in the "external source" collection, while the other two are in the "internal source" collection. In a second grouping on the format of the initial version, the video hypernode is alone in the "unstructured" collection, while the other two hypernodes are in the "semi-structured" collection. Collections are represented by dotted rectangles in Fig. 2 (attributes of the hypernodes are not represented for simplicity).

5.3   Global Metadata

Global metadata are specific elements that are managed differently from other metadata. They "gravitate" around hypernodes and are exploited as needed, i.e., almost systematically, especially logs and indexes. We consider that semantic resources are stored in nodes, while indexes and event logs are rather physical structures and are highly dependent on the technology used to implement the data lake and the metadata system.

6   Conclusion

After an overview of the definitions of a data lake from the literature, we propose in this paper our own definition of this concept. Then, we identify the six key features that the metadata system of a data lake must provide to be as robust as possible in addressing the big data issues and the schema-on-read approach. Comparing existing metadata systems, we show that some succeed in providing most features, but none offers them all.

Hence, we propose a new metadata model, MEDAL, based on the notion of object and a typology of metadata in three categories: intra-object, inter-object and global metadata. MEDAL adopts a graph-based organization. An object is represented by a hypernode containing nodes that correspond to the versions and representations of an object. Transformation and update operations are modeled by oriented edges linking the nodes. Hypernodes can be linked in several ways: edges to model similarity links and hyperarcs to translate parenthood relationships and object groupings. Finally, global resources are also present, in the form of knowledge bases, indexes or event logs.

Thanks to all these elements, MEDAL supports all six key features we have identified, making it the most complete metadata model to the best of our knowledge. However, MEDAL is not implemented yet. It is the objective of future work in which we shall propose an application of our metadata model in a context of structured, semi-structured and unstructured data. This implementation will allow us to evaluate MEDAL in more detail, in particular by comparing it with other existing systems.

Acknowledgments. Part of the research presented in this article is funded by the Auvergne-Rhône-Alpes Region, as part of the AURA-PMI project that finances Pegdwendé Nicolas Sawadogo's PhD thesis.

References

1. Alrehamy, H., Walker, C.: Personal data lake with data gravity pull. In: BDCloud 2015, Dalian, china, vol. 88, pp. 160–167. IEEE Computer Society Washington (2015). https://doi.org/10.1109/BDCloud.2015.62
2. Ansari, J.W., Karim, N., Decker, S., Cochez, M., Beyan, O.: Extending data lake metadata management by semantic profiling. In: ESWC 2018, Heraklion, Crete, Greece, ESWC, pp. 1–15 (2018). https://2018.eswc-conferences.org/wp-content/uploads/2018/02/ESWC2018_paper_127.pdf
3. Beheshti, A., Benatallah, B., Nouri, R., Chhieng, V.M., Xiong, H., Zhao, X.: CoreDB: a data lake service. In: CIKM 2017, pp. 2451–2454. ACM, Singapore (2017). https://doi.org/10.1145/3132847.3133171
4. Beheshti, A., Benatallah, B., Nouri, R., Tabebordbar, A.: CoreKG: a knowledge lake service. Proc. VLDB Endow. **11**(12), 1942–1945 (2018). https://doi.org/10.14778/3229863.3236230
5. Diamantini, C., Giudice, P.L., Musarella, L., Potena, D., Storti, E., Ursino, D.: A new metadata model to uniformly handle heterogeneous data lake sources. In: Benczúr, A., et al. (eds.) ADBIS 2018. CCIS, vol. 909, pp. 165–177. Springer, Cham (2018). https://doi.org/10.1007/978-3-030-00063-9_17
6. Dixon, J.: Pentaho, Hadoop, and Data Lakes (2010). https://jamesdixon.wordpress.com/2010/10/14/pentaho-hadoop-anddata-lakes/
7. Fang, H.: Managing data lakes in big data era: what's a data lake and why has it became popular in data management ecosystem. In: CYBER 2015, Shenyang, China, pp. 820–824. IEEE (2015). https://doi.org/10.1109/CYBER.2015.7288049
8. Farid, M., Roatis, A., Ilyas, I.F., Hoffmann, H.F., Chu, X.: CLAMS: bringing quality to data lakes. In: SIGMOD 2016, pp. 2089–2092. ACM, San Francisco (2016). https://doi.org/10.1145/2882903.2899391
9. Farrugia, A., Claxton, R., Thompson, S.: Towards social network analytics for understanding and managing enterprise data lakes. In: ASONAM 2016, pp. 1213–1220. IEEE, San Francisco (2016). https://doi.org/10.1109/ASONAM.2016.7752393
10. Fauduet, L., Peyrard, S.: A data-first preservation strategy: data management in SPAR. In: iPRES 2010, Vienna, Austria, pp. 1–8 (2010). http://www.ifs.tuwien.ac.at/dp/ipres2010/papers/fauduet-13.pdf
11. Hai, R., Geisler, S., Quix, C.: Constance: an intelligent data lake system. In: SIGMOD, pp. 2097–2100. ACM Digital Library, San Francisco (2016). https://doi.org/10.1145/2882903.2899389
12. Halevy, A., et al.: Managing Google's data lake: an overview of the goods system. In: SIGMOD 2016, pp. 795–806. ACM, San Francisco (2016). https://doi.org/10.1145/2882903.2903730

13. Hellerstein, J.M., et al.: Ground: a data context service. In: CIDR 2017, Chaminade, CA, USA (2017). http://cidrdb.org/cidr2017/papers/p111-hellerstein-cidr17.pdf
14. Khine, P.P., Wang, Z.S.: Data lake: a new ideology in big data era. In: WCSN 2017, Wuhan, China, ITM Web of Conferences, vol. 17, pp. 1–6 (2017). https://doi.org/10.1051/itmconf/2018170302
15. Kimball, R.: Slowly changing dimensions. Inf. Manag. **18**(9), 29 (2008)
16. Laskowski, N.: Data lake governance: A big data do or die (2016). https://searchcio.techtarget.com/feature/Data-lake-governance-A-big-data-do-or-die
17. Maccioni, A., Torlone, R.: KAYAK: A framework for just-in-time data preparation in a data lake. In: Krogstie, J., Reijers, H.A. (eds.) CAiSE 2018. LNCS, vol. 10816, pp. 474–489. Springer, Cham (2018). https://doi.org/10.1007/978-3-319-91563-0_29
18. Madera, C., Laurent, A.: The next information architecture evolution: the data lake wave. In: MEDES 2016, Biarritz, France, pp. 174–180 (2016). http://dl.acm.org/citation.cfm-id=3012077
19. Mathis, C.: Data lakes. Datenbank-Spektrum **17**(3), 289–293 (2017). https://doi.org/10.1007/s13222-017-0272-7
20. Miloslavskaya, N., Tolstoy, A.: Big data, fast data and data lake concepts. In: BICA 2016, NY, USA, Procedia Computer Science, vol. 88, pp. 1–6 (2016). https://doi.org/10.1016/j.procs.2016.07.439
21. O'Leary, D.E.: Embedding AI and crowdsourcing in the big data lake. IEEE Intell. Syst. **29**(5), 70–73 (2014). https://doi.org/10.1109/MIS.2014.82
22. Quix, C., Hai, R., Vatov, I.: GEMMS: a generic and extensible metadata management system for data lakes. In: CAiSE 2016, Ljubljana, Slovenia, pp. 129–136 (2016). http://ceur-ws.org/Vol-1612/paper17.pdf
23. Sawadogo, P.N., Kibata, T., Darmont, J.: Metadata management for textual documents in data lakes. In: ICEIS 2019, Heraklion, Crete, Greece, pp. 72–83 (2019). https://doi.org/10.5220/0007706300720083
24. Singh, K., et al.: Visual Bayesian fusion to navigate a data lake. In: FUSION 2016, pp. 987–994. IEEE, Heidelberg (2016)
25. Sirosh, J.: The intelligent data lake (2016). https://azure.microsoft.com/frfr/blog/the-intelligent-data-lake/
26. Suriarachchi, I., Plale, B.: Crossing analytics systems: a case for integrated provenance in data lakes. In: e-Science 2016, Baltimore, MD, USA, pp. 349–354 (2016). https://doi.org/10.1109/eScience.2016.7870919
27. Terrizzano, I., Schwarz, P., Roth, M., Colino, J.E.: Data wrangling: the challenging journey from the wild to the lake. In: CIDR 2015, Asilomar, CA, USA, pp. 1–9 (2015). http://cidrdb.org/cidr2015/Papers/CIDR15_Paper2.pdf

A Metadata Framework for Data Lagoons

Vasileios Theodorou[1]([✉]), Rihan Hai[2], and Christoph Quix[3]

[1] Intracom Telecom, Paiania, Greece
theovas@intracom-telecom.com
[2] RWTH Aachen University, Aachen, Germany
hai@dbis.rwth-aachen.de
[3] Fraunhofer-Institute for Applied Information Technology FIT,
Sankt Augustin, Germany
christoph.quix@fit.fraunhofer.de

Abstract. In this work, we present a Metadata Framework in the direction of extending intelligence mechanisms from the Cloud to the Edge. To this end, we build on our previously introduced notion of Data Lagoons—the analogous to Data Lakes at the network edge—and we introduce a novel architecture and Metadata model for the efficient interaction between Data Lagoons and Data Lakes. We identify the service and data planes of our architecture and we illustrate the application of our framework on a use case from the TPCx-IoT benchmark. To our knowledge, our approach is the first one to examine the integration of Data Lakes with Edge components, taking under consideration data and infrastructure resources of Edge Nodes.

Keywords: Data Lagoon · Data Lake · Edge Computing · Metadata

1 Introduction

According to a projection by Gartner[1], the total number of IoT devices are expected to reach 20.4 Billion in 2020. The rise of 5G, combined with microprocessor innovations (e.g., Amazon AWS Inferentia, Google TPU, Graphcore IPU) are creating new investment opportunities, as well as research challenges in bringing complex data processing activities as close as possible to the physical world, where data is actually produced.

At the same time, the link between established Big Data technologies and the rising Edge Computing paradigm is yet to be investigated and fully exploited towards the realization of a robust and efficient data analytics continuum, extending from end devices all the way to powerful server farms. This continuum is expected to cross between an "ocean" of data and business processes spanning multiple organizational borders.

---

[1] IoT Growth Report By Gartner. http://www.gartner.com/newsroom/id/3598917. SmartMeter. 2017.

© Springer Nature Switzerland AG 2019
T. Welzer et al. (Eds.): ADBIS 2019, CCIS 1064, pp. 452–462, 2019.
https://doi.org/10.1007/978-3-030-30278-8_44

In previous work [23], we introduced the concept of *Data Lagoons*, applying principles inspired by Data Lakes that reside on resourceful data centers, to nodes at the Edge of the network. Edge Nodes can be of many types, spanning from local servers within industry plants and network Base Stations to neighborhood-level IoT gateways or even in-vehicle computers. Although infrastructure resources at the Edge are limited, data resources are context-full, taking advantage of proximity to data sources. Context can include besides time and location, also social aspects and personalization, as well as device information. It is apparent that extending intelligence to the Edge is a means of utilizing edge compute capabilities, offloading data processing tasks from centralized components, while at the same time grasping rich information and responding to observed and measured events with the lowest possible delay.

In this work, we take the next step and elaborate on the inter-operation between the Data Lagoon and Big Data technologies on the Cloud. To this end, we describe the Data Lagoon dimensions with respect to data analysis capabilities at the Edge and we introduce a novel Metadata framework for the service-based interplay between Data Lagoons and Data lakes. We describe a high level architecture of the framework and we showcase its usefulness through an illustrative use case from the IoT domain.

The main objectives of this work are the following:

1. Foster automation in optimizing distribution of analytics tasks between the Edge and the Cloud
2. Facilitate auto-discovery of data/information residing on/possible to obtain from concrete Edge nodes
3. Enable interoperability between Data Lagoons at the Edge and Data Lake(s) at the Cloud. This entails mechanisms for the exchange of metadata (upstream) and for the request/enforcement of control actions (downstream)

In Sect. 2, we present a high level view of our Metadata Framework Architecture and we describe the main components of the Service plane and the Data plane. Following in Sect. 3, we describe the Metadata Model of the Data Lagoon and its main dimensions. Subsequently, in Sect. 4 we illustrate an application of our model on a use case and in Sect. 5 we review related work mainly from the Data Lakes research area. Finally, in Sect. 6, we summarize our contributions.

2 Architecture

According to our approach, Data Lagoons expose data services that other entities can consume, by requesting access to the outcomes of running data processes or the deployment of new jobs, according to the Edge node capabilities and available datasets. Although such interactions could also take place between different Edge nodes, in this work we focus on the case where service consumers are higher in the data pipeline hierarchy, so as to facilitate the integration of the Edge with Big Data technologies commonly found on remote infrastructures.

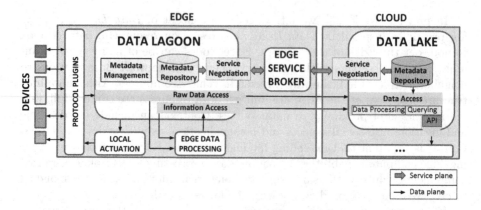

Fig. 1. Data Lagoon Metadata Framework Architecture.

In Fig. 1, we depict a high level view of the Data Lagoon Metadata Framework Architecture, illustrating the interplay between a Data Lagoon hosted at the Edge and a Data Lake, hosted either on private, or on public/remote cloud infrastructure. This interaction takes place at various planes, posing different connectivity, resource and governance requirements. The planes shown in Fig. 1 regard service level communication—i.e., the *Service plane*—and application data level communication—i.e., the *Data plane*—, whereas the infrastructure *control* and *orchestration* planes are considered to extend beyond the scope of this work and are thus omitted (more information on this topic can be found at [23]).

2.1 Service Plane

The *Service plane* involves all the necessary management processes for a Data Lake to act as a client to a Data Lagoon and to consume its provided services. The services that are provided and exposed by the Data Lagoon include *service discovery, application data and information provisioning* and *edge data processing.*

In this respect, we introduce the *Edge Service Broker* (ESB) component at the Edge, responsible for the service-level interfacing between the Edge and Data Lake(s). Hence, ESB communicates with a *Service Negotiation* (SN) interfacing component on either endpoint of the service binding—i.e., the Data Lagoon and the Data Lake. To this end, ESB maintains a registry of available services, as well as information about the status of service bindings and Role-Based Access Control (RBAC) credentials for the authentication and authorization of service interactions.

On the Data Lagoon side, the SN draws service information from the *Metadata Repository* (MR), which is managed by the *Metadata Management* (MM) component. In this regard, MR describes both the available data and information at the Data Lagoon, i.e., the currently, *as-is* produced data and information points, as well as the infrastructure resources and available data operations so as new data pipelines can be established, producing new data-points as relevant by the requesting service consumers—i.e., the Data Lakes. The dimensions of the information persisted in MR are described in more detail in Sect. 3. An analogous mechanism is assumed for the SN on the Data Lake side, interacting with the components described in [9].

2.2   Data Plane

When it comes to the *Data Plane*, the Data Lagoon is intended to complete the continuity of data and information flow between the data sources—primarily being the *Devices* connected to the Edge node—and the big-scale data repository, i.e., the Data Lake. To enable seamless integration between these endpoints, the Data Lagoon can offer multiple channels of data flow with different latency, throughput, data accuracy and data completeness specifications.

As an example, in cases where the collection of real-time information by the Data Lake is critical, the followed communication channel can support fast access to data, although possibly with some degree of missing and/or less accurate data-points. On the other hand, in cases where more importance is placed on the cautious crafting of data or the extraction of inferences and insights from available datasets, which are commonly realized with compute-intensive processes, the selected communication channel can sacrifice speed as a trade-off to more sophisticated data processing and higher degree of information quality guarantees. This differentiation could be considered as the *Lambda Architecture* [6] analogous of the Edge and is illustrated in our architecture as two different data lanes: the *Raw Data Access* lane and the *Information Access* Lane.

Information on the Data Lagoon is produced via the *Edge Data Processing* (EDP) component, applying available data operations over raw datasets, e.g., data points produced directly from temperature or humidity senors. The establishment of such data pipelines is requested by the Data Lake and agreed within a service negotiation phase between the SNs of each side. We should note at this point that the volatility of data pipelines applies only to information (e.g., "average temperature" or "pattern recognition on humidity measurements")—not to raw data collected from devices, since one main principle of the Data Lagoon architecture is the decoupling between data collection and the processing jobs, transforming data to information.

In this respect, when access to raw data is requested by the Data Lake, this translates to its connection to the relevant Raw Data Access lane already in place, and not to the establishment of a new pipeline. By default, all data from devices connected to the Edge Node are made available to the Data Lagoon, after passing through the necessary *Protocol Plugins* (PP). The *Local Actuation* component also communicates with devices via the PP, but its description does not lie within the scope of this work, which focuses mostly on the delivery of data and information from devices to the Data Lake.

To this end, we should also highlight the importance of processing resource efficiency, due to the infrastructure resource limitations of the Edge. In this

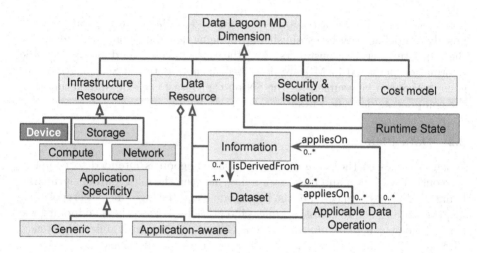

Fig. 2. Data Lagoon Metadata dimensions.

respect, data operations and algorithms should be carefully selected, for example being the lightweight implementations of their analogous Cloud versions, or optimized operations fitting to the hardware available on the Edge Node. Moreover, optimization on this level can regard the sharing of produced information among different Data Lakes, or different data consumption elements within the same Data Lake. An interesting work on this front is described in [11], where the consolidation of different data integration flows is analyzed on the basis of optimal data and code reuse.

3   Data Lagoon Metadata

It is important for the Metadata Repository to contain appropriate information, describing the capabilities of the Data Lagoon to be exposed to external entities, as well as the available data resources—both data and processes—for processing components to plan and execute data provisioning accordingly.

In Fig. 2, we show a UML class diagram with the main dimensions of the Data Lagoon Metadata model, which can be grouped as follows:

1. **Infrastructure Resource:** This dimension includes information about available and utilized Networking, Computing, Memory, Storage resources on the Edge Node. Infrastructure resources can also include cohesion and compliance information about different resources, topological and forwarding information, modeled accordingly (e.g., Heat Orchestration Template[2] or NFV Information Model[3]). In addition, what is particular about Edge nodes is that they contain information about connected devices, e.g., IoT devices/sensors,

---

[2] https://docs.openstack.org/heat/rocky/template_guide/hot_spec.html.
[3] https://www.etsi.org/technologies/nfv.

mobile devices or any type of processing units, potentially including device computing capabilities information, in case moving intelligence to the device level is applicable.

2. **Dataset:** This dimension regards information about the data models, profiling [2] and schemata of available data, available context such as reference to the devices that produced them, Data Quality attributes and Big Data Vs characterization—Volume, Variety, Velocity and Value [22].

3. **Information:** Information is derived form Datasets and contains the semantics of available datasets (for semantic description we adopt the approach described in [13]); dependency of produced information on datasets and data sources; Data Integration parameters such as aggregation levels, time windows and historical data volatility; Information Quality, and identified traffic patterns/models.

4. **Applicable data operations:** Applicable data operators refer on one hand to the existing data processing software libraries available within the Edge Node, and on the other hand, on the data processing options deriving from the semantics and the categorization of existing Datasets and Information. To this end, they include ETL logical operators [20], Analytics operations, Machine Learning/Deep Learning primitives and algorithms. Moreover, they can include quality attributes of operations, such as speed of deployment of jobs, performance and resource consumption, depending on the cost model (see below).

5. **Security & Isolation:** This dimension refers to encryption, authentication, authorization mechanisms and resilience levels of data and processes (data backup/replication). One important aspect especially for Edge Nodes acting as service providers to more than one service consumers are the available service isolation levels, e.g., physical CPU & memory versus hypervisor isolation (VMs) versus process isolation (containers).

6. **Cost model:** The Cost model is an important dimension for data service consumers, so as to assess the tradeoffs between deploying data processing jobs to the edge and the cost of doing so, leading to decisions about the extent to which task offloading or context enrichment are worth the price. Hence, it includes pricing per resource; resource & battery consumption per operation and per data unit; as well as cost per quality levels of provided services (e.g., linked to information quality for data services to security and isolation guarantees).

7. **Runtime State:** Runtime state includes the monitored information on the data service consumers. In this regard, it contains information about clients (Data Lakes), sessions status and state, transaction logs and security credentials.

Data, Information and Data Operations dimensions can be further divided to *generic*, e.g., sensor measurements for temperature and to *application-aware*, e.g., available analytics on specific multimedia content type.

3.1 Logical Data Operations

Besides typical ETL operations [20] on the raw data and information, e.g., *Join, Union, Project, Filter* etc., we identify the a set of notable data operators for the delivery of data from Data Lagoons to Data Lakes, as shown in Table 1. One can argue that these operators can derive as compositions of typical ETL operations, but the reason they are highlighted here is to illustrate the value of offloading data processing tasks from Data Lakes to Data Lagoons and the resulting synergistic benefits.

In essence, all above-mentioned data operators take as input collections of data points and produce as output collections of data points, commonly ordered by their timestamp. Thus, composite data operators are also possible, using "leaf" data operators—or other composite operators—as building blocks.

Table 1. Data Lagoon operators

Operator	Semantics
aggregate(w,t)	Aggregate data based on window w. Aggregation type t can be MIN, MAX, SUM, AVERAGE, CONCATENATION or any arbitrary user-defined function (UDF). w can be time window, number of data points or derive form arbitrary logic
cleanse(t, p)	Apply data cleansing logic of type t, e.g., filtering, data imputation, interpolation, extrapolation, with parameters p. p include information about referenced datasets, predicates and operations, as applicable
ctx_enrich(t, p)	Enrich datasets with contextual information, e.g., device id, network state information, information from sensors, or even user profile id from social media application. As evident, type t of contextual enrichment covers a broad spectrum and parameters p define the necessary data sources and operations
profile(a,m)	Profile data based on identified patterns on selected aspects a, and using defined models m, e.g., statistical models or custom models, comparing against a given knowledge base. The outcome of this profiling is used by the Metadata Management to populate segments of the Metadata Repository

Aggregation and profiling of data on the Edge node entails less bandwidth consumption—and consequently, lower cost—for the delivery of data form the Edge to the Cloud. Cleansing data as close as possible to the data sources is a documented best practice [1], to avoid the propagation of errors and to efficiently apply cleansing mechanisms taking advanctage of context-awareness. Same rationale applies to context enrichment.

4   Use Case

In Fig. 3, we illustrate the application of our framework on an exemplary use case inspired by the TPCx-IoT[4] benchmark for IoT Gateway Systems [21]. In this domain, data sources are different sensors within a Power Substation, sending data to an IoT Gateway. In our example, there are Phase Measurement Units (PMU) commonly producing data of extreme velocity (subsecond values) and sensors measuring oil temperature and relative humidity of in-service transformers.

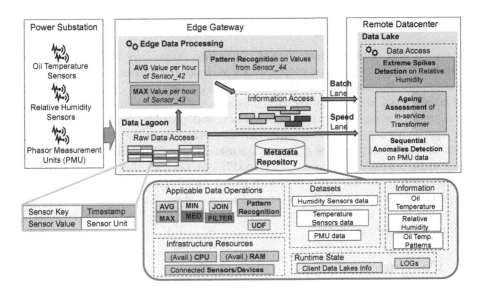

Fig. 3. Data Lagoon Use Case inspired by TPCx-IoT.

The IoT Gateway is responsible for filtering and aggregating data points, each containing a key of the concrete sensor it was produced by, a timestamp, a measured value and a unit of measurement. These datasets are directly fed into the Data Lagoon, after passing through the necessary protocol translations, and stored into the *Raw Data Access*.

The *Edge Data Processing* unit executes all the data operations, which are assumed to have already been ordered by the Data Lake. To this end, the operations—*AVG*, *MAX* and *Pattern Recognition* in our example—are implementations of the *Applicable Data Operations* of the *Metadata Repository* over the existing Datasets, also described in the Metadata Repository. The Data Lake has access to the available data over two different lanes—the *Batch Lane* and the *Speed Lane*, inspired by the Lambda Architecture layers—the first providing fast and direct access to raw data from the sensors and the second providing access to proccessed information.

---

[4] http://www.tpc.org/tpcx-iot/.

The Data Lake also contains data processing on top of the provided data from the Data Lagoon, in order to produce useful information and monitored insights over the Power Substation, such as ageing assessment (analysis inspired by [14]) and anomalies detection (extreme spikes or sequential anomalies). We note that in the example, the extreme-scale PMU data follow the Speed Lane, a decision made due to limitation of compute resources on the Edge. We also note that for the detection of Spikes, aggregated MAX values over certain time-window, produced at the Data Lagoon, are sufficient, lowering the cost of data transmission to the Data Lake.

5   Related Work

As a promising solution for big data, many organizations have initiated the study of data lake [7, 12, 15]. Some proposals are sketches from industry, e.g., IBM [7], Google [10], *Microsoft Azure* [18], *pwc* [3], *podium data*[5] and *Informatica*[6]. Such proposals often describe abstracted business data lake systems with conceptual architectures, functional requirements, and possible product components on Apache licenses. Meanwhile, there are also research prototypes focusing on a specific task of data lake, e.g., metadata management [9], data quality [8], user data analytics [4], and data preparation [17], etc. On the other hand, infrastructures applying the idea of edge-device based computing have been booming in the recent years [5, 16, 19]. However, to the best of our knowledge, a framework to facilitate the interoperability between the Edge and Data Lakes is missing from existing approaches.

6   Conclusion

In this paper, we have taken a step further in describing the Data Lagoons that we have previously introduced, as an effective counterpart of Data Provisioning at the Edge. In this regard, we have introduced a novel approach towards integrating Data Lakes at the Cloud, with Data Lagoons at the Edge, the latter acting as data service providers to the former. We have depicted the high level architecture of our framework and we have showcased its usefulness through an illustrative IoT use case example. As future work, we plan to conduct large-scale experiments with open-source Edge Gateway solutions, extended to incorporate the described Metadata Repository and Metadata Management components.

References

1. Batini, C., Cappiello, C., Francalanci, C., Maurino, A.: Methodologies for data quality assessment and improvement. ACM Comput. Surv. **41**, 1–52 (2009). https://doi.org/10.1145/1541880.1541883

---

[5] http://www.podiumdata.com/solutions/.

[6] https://www.informatica.com/de/solutions/explore-ecosystems/aws/aws-data-lakes.html.

2. Naumann, F.: Data profiling revisited. In: SIGMOD Record, vol. 42, pp. 40–49. ACM (2014). https://doi.org/10.1145/2590989.2590995
3. Stein, B., Morrison, A.: The enterprise data lake: better integration and deeper analytics. In: PwC Technology Forecast: Rethinking integration, vol. 1, p. 18 (2014)
4. Alrehamy, H., Walker, C.: Personal data lake with data gravity pull. In: Proceedings of BDCloud, pp. 160–167. IEEE (2015)
5. López, P., et al.: Edge-centric computing: vision and challenges. Comput. Commun. Rev. 45, 37–42 (2015). https://doi.org/10.1145/2831347.2831354
6. Marz, N., Warren, J.: Big data: principles and best practices of scalable realtime data systems. In: Big Data. Manning Publications Co. (2015)
7. Terrizzano, I., Schwarz, P., Roth, M., Colino, J.: The challenging yourney from the wild to the lake. In: CIDR (2015)
8. Mina H. et al.: CLAMS: bringing quality to data lakes. In: Proceedings of SIGMOD, pp. 2089–2092. (2016). https://doi.org/10.1145/2882903.2899391
9. Hai, R., Geisler, S., Quix, C.: Constance: an intelligent data lake system. In: Proceedings of SIGMOD, pp. 2097–2100 (2016). https://doi.org/10.1145/2882903.2899389
10. Alon, Y., et al.: Managing google's data lake: an overview of the goods system. In: IEEE Data Engineering Bulletin, vol. 39, pp. 5–14. IEEE (2016). https://doi.org/10.1145/1541880.1541883
11. Jovanovic, P., Romero, O., Simitsis, A., Abelló, A.: Incremental consolidation of data-intensive multi-flows. IEEE Trans. Knowl. Data Eng. 28, 1203–1216 (2016). https://doi.org/10.1109/TKDE.2016.2515609
12. LaPlante, A., Sharma, B.: Architecting Data Lakes. O'Reilly Media, Newton (2016)
13. Quix, C., Hai, R., Vatov, I.: Metadata extraction and management in data lakes With GEMMS. Complex Syst. Inform. Model. Q. 9, 67–83 (2016)
14. Tee, S.J., et al.: Seasonal influence on moisture interpretation for transformer aging assessment. IEEE Electr. Insul. Mag. 32, 29–37 (2016). https://doi.org/10.1109/MEI.2016.7527123
15. Jarke, M., Quix, C.: On warehouses lakes, and spaces: the changing role of conceptual modeling for data integration. In: Cabot, J., Gómez, C., Pastor, O., Sancho, M., Teniente, E. (eds.) Conceptual Modeling Perspectives, pp. 231–245. Springer, Cham (2017). https://doi.org/10.1007/978-3-319-67271-7_16
16. Lin, J., et al.: A survey on internet of things: architecture, enabling technologies, security and privacy, and applications. IEEE Internet Things J. 4, 1125–1142 (2017). https://doi.org/10.1109/JIOT.2017.2683200
17. Maccioni, A., Torlone, R.: Crossing the finish line faster when paddling the data lake with kayak. PVLDB 10, 1853–1856 (2017)
18. Ramakrishnan, R. et al.: Azure data lake store: a hyperscale distributed file service for big data analytics. In: Proceedings of SIGMOD, pp. 51–63. (2017). https://doi.org/10.1145/3035918.3056100
19. Satyanarayanan, M.: The emergence of edge computing. Computer 50, 30–39 (2017). https://doi.org/10.1109/MC.2017.9
20. Theodorou, V., Abelló, A., Thiele, M., Lehner, M.: Frequent patterns in ETL workflows: an empirical approach. Data Knowl. Eng. 112, 1–16 (2017)
21. Poess, M. et al.: Analysis of TPCx-IoT: the first industry standard benchmark for IoT gateway systems. In: IEEE 34th International Conference on Data Engineering (ICDE), pp. 1519–1530. IEEE (2018). https://doi.org/10.1109/ICDE.2018.00170
22. Berkani, N., Khouri, S., Bellatreche, L.: Value and variety driven approach for extended data warehouses design. In: Information Retrieval, Document and Semantic Web, vol. 2 (2019)

23. Theodorou, V., Diamantopoulos, N.: GLT: edge gateway ELT for data-driven intelligence placement. In: 2019 IEEE/ACM 1st International Workshop on Data-Driven Decisions, Experimentation and Evolution (DDrEE), Montreal, (2019, in press)

Assessing the Role of Temporal Information in Modelling Short-Term Air Pollution Effects Based on Traffic and Meteorological Conditions: A Case Study in Wrocław

Andrea Brunello[1]([✉]), Joanna Kamińska[4], Enrico Marzano[3],
Angelo Montanari[1], Guido Sciavicco[2], and Tomasz Turek[5]

[1] University of Udine, Via delle Scienze 206, 33100 Udine, Italy
{andrea.brunello,angelo.montanari}@uniud.it
[2] University of Ferrara, Via Giuseppe Saragat 1, 44122 Ferrara, Italy
guido.sciavicco@unife.it
[3] Gap Srlu, Via Tricesimo 246, 33100 Udine, Italy
e.marzano@gapitalia.it
[4] Wrocław University of Environmental and Life Sciences,
ul. Grunwaldzka 53, 50-357 Wrocław, Poland
joanna.kaminska@upwr.edu.pl
[5] Wrocław University of Science and Technology,
Wybrzeże Wyspiańskiego 27, 50-370 Wrocław, Poland
tomaszoturek@gmail.com

Abstract. The temporal aspects often play an important role in information extraction. Given the peculiarities of temporal data, their management typically requires the use of dedicated algorithms, that make the overall data mining process complex, especially in those cases in which a dataset is characterised by both temporal and atemporal information. In such a situation, typical solutions include combining different algorithms for the independent handling of the temporal and atemporal parts, or relying on an encoding of temporal data that makes it possible to apply classical machine learning algorithms (such as with the use of lagged variables). This work investigates the management of temporal information in an environmental problem, that is, assessing the relationships between concentrations of the pollutants NO_2, NO_X, and $PM_{2.5}$, and a set of independent variables that include meteorological conditions and traffic flow in the city of Wrocław (Poland). We show that taking into account temporal information by means of lagged variables leads to better results with respect to atemporal models. More importantly, an even higher performance may be achieved by making use of a recently proposed decision tree model, called J48SS, that is capable of handling heterogeneous datasets consisting of static (i.e., categorical and numerical) attributes, as well as sequential and time series data. Such an outcome highlights the importance of proper temporal data modelling.

© Springer Nature Switzerland AG 2019
T. Welzer et al. (Eds.): ADBIS 2019, CCIS 1064, pp. 463–474, 2019.
https://doi.org/10.1007/978-3-030-30278-8_45

Keywords: Urban air pollution · Machine learning · Temporal data

1   Introduction

Temporal data plays an important role in the extraction of information in many domains, and may be encoded in at least two different ways: it can be represented by a discrete sequence of finite-domain values (e.g., a sequence of purchases), as well as by a real-valued time series (e.g., a stock price history). Temporal information may also be complemented by other, static (numerical or categorical) types of data. For instance, in the medical domain, a patient may be described by: (*i*) categorical attributes, conveying information about the gender or the blood group; (*ii*) numerical features, storing information about the age and weight; (*iii*) time series data, tracking the oxygen saturation level over time, and (*iv*) discrete sequences, describing the symptoms that the patient has experienced and the medications that have been administered in response. In such an heterogeneous context, the various information sources may interact in a non trivial way, making classical problems, such as classification (e.g., determine the disease affecting the patient), quite complex, especially because different kinds of data typically require different kinds of preprocessing techniques and non trivial combinations of classification algorithms to be managed properly. In some cases, transformation techniques may be applied to temporal data to encode them by means of atemporal features, so that they can be handled by classical machine learning algorithms; nevertheless, such transformations typically entail an information loss, and they may reduce the interpretability of the results.

In this work, we focus on assessing how different ways of handling temporal information may bring to different results in a real classification scenario. To do that, we consider an environmental problem, that is, the problem of identifying the relationships between the concentrations of the pollutants NO_2 (nitrogen dioxide), NO_X (nitrogen oxides), and $PM_{2.5}$ (particulate matter), and a set of variables, describing, among others, meteorological conditions and traffic flow, in the city of Wrocław in Poland. Specifically, we build on a previous work [10], where the authors tried to model such relationships by means of atemporal regression models. Instead, we consider the task as a classification one, which allows us to apply the recently proposed J48SS decision tree inducer [6], that generates models capable of handling both temporal (sequences and time series) and atemporal (categorical and numerical) attributes. We compare its performance with those reported by J48 (WEKA's [16] implementation of Quinlan's C4.5 [12]) and Random Forest [5] runs on both an atemporal, as well as a transformed version of the original dataset, that makes use of lagged variables, showing that, at least in this case, time series and temporal sequences give rise to models with superior performance.

The paper is organized as follows: Sect. 2 briefly discusses some previous work, and presents the original problem and dataset, as considered in [10]. Section 3 describes the data preparation step, and the differences with respect to the original work. Section 4 discusses the results that have been obtained with J48 and

Random Forest neglecting temporal information, which are useful to establish a baseline. In Sect. 5, we formulate the problem with classical models, based on the use of lagged variables. In Sect. 6, we give a short account of J48SS, and then we discuss the performance of native temporal models on this problem. Finally, in Sect. 7 we provide an assessment of the work done and point out some possible future research directions.

2   Background

Over the recent years, machine learning-based environmental pollution studies have been gaining more and more traction in the scientific community. For instance, in [7], decision trees are used to predict $PM_{2.5}$ levels in the city of Quito, the capital of Ecuador, based on wind (speed and direction) and precipitation levels. In [17], the correlation between wind data and a wide range of pollutants in China's Pearl River delta region is investigated. In [14], Classification and Regression Trees (CART) and Ensemble Extreme Learning Machine (EELM) algorithms are used to model hourly $PM_{2.5}$ concentrations in the city of Yancheng, China. Finally, in [11], a machine learning approach based on two years of meteorological and air pollution data analysis is built to predict the concentrations of $PM_{2.5}$ from wind and precipitation levels in Northern Texas.

The present work builds on the findings presented in [10], where an environmental study conducted in Wrocław (Poland) is reported. The overall goal of the study was that of determining how the levels of specific pollutants, namely, NO_2, NO_X, and $PM_{2.5}$, are related to the values of other attributes, such as weather conditions and traffic intensity, with the purpose of building an *explanation* model (in opposition to a *prediction* model). In such an explanation model, the value of a pollutant at a certain time instant is linked to the value of the predictor attributes from the same time instant; discovering such a relationship is extremely important, as it may allow the city's government to identify the most critical factors that influence pollution levels, and to take them into consideration when developing proper corrective measures.

The considered dataset spans over the years 2015–2017, and it records information at one hour granularity. The dataset attributes are listed in Table 1: they are all numerical, with the exception of *holiday_or_not*, which has a binary categorical value, and *wind_direction_categ*, that may take the values N, NE, E, SE, S, SW, W, NW, and W. The attribute *traffic* refers to the number of vehicle crossings recorded at a large intersection equipped with a traffic flow measurement system, whereas the air quality information have been recorded by a nearby measurement station. We may broadly classify the predictors into three categories:

- *traffic intensity*: traffic;
- *timestamp*[1]: year, month, dom, dow, hour, date, holiday_or_workday;

---

[1] We use the term *timestamp* to refer to the kind of variables that identify a specific time instant, to distinguish them from those we will consider to be proper temporal features, i.e., the ones that encode historical values.

Table 1. Features in the original dataset.

Feature	Description
year	year which the instance refers to
month	month which the instance refers to
dom	day of the month which the instance refers to
dow	day of the week which the instance refers to
hour	hour which the instance refers to
date	full date which the instance refers to
holiday_or_workday	whether the day which the instance refers to is a workday or not
traffic	hourly sum of vehicle numbers at the considered intersection
air_temp	hourly recording of the air temperature
wind_speed	hourly recording of the wind speed
wind_direction_num	hourly recording of the wind direction (numeric, degrees)
wind_direction_categ	hourly recording of the wind direction (discretized)
rel_humidity	hourly recording of the relative air humidity
air_pressure	hourly recording of the air pressure
NO2_conc	hourly recording of the NO_2 concentration level
NOX_conc	hourly recording of the NO_X concentration level
PM25_conc	hourly recording of the $PM_{2.5}$ concentration level

- *weather conditions*: air_temp, wind_speed, wind_direction (categorical and numerical), rel_humidity, air_pressure.

Three atemporal Random Forest models have been trained on such data to perform a regression task on each of the three pollutants: NO_2, NO_X and $PM_{2.5}$. The conclusion was that in modelling nitrogen oxides concentration at a certain time, the most important variable is the traffic flow at the same time, while for $PM_{2.5}$ meteorological conditions seem to play a more predominant role. Starting from such a result, we now want to asses if more reliable explaining models can be found by taking into account the historical values of the predictor variables.

3 Data Preparation

Let us now turn our attention to the description of the dataset used in this work. Although data are essentially the same as the one described in Sect. 2, it is important to underline once more that there is a fundamental difference between our work and the work reported in [10]: we consider a classification problem, instead of a regression one. In that respect, the first step has been the choice of a sensible discretization technique for the pollutant attributes. To this

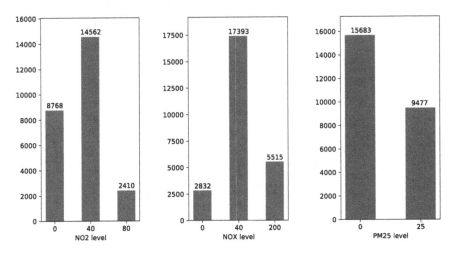

Fig. 1. Class distributions (in number of instances) for the pollutants NO_2, NO_X, and $PM_{2.5}$. Labels on the x axis refer to the lower bounds of each discretization interval.

end, we relied on official European Union directives [1,2], that led us to define the following classes:[2]

- NO_2: intervals $[0, 40)$, $[40, 80)$, $[80, \infty)$;
- NO_X: intervals $[0, 40)$, $[40, 200)$, $[200, \infty)$;
- $PM_{2.5}$ intervals $[0, 25)$, $[25, \infty)$.

Such a discretization led to the class distributions depicted in Fig. 1, which, as can be seen, are rather unbalanced. As a consequence, in order to evaluate the performances of the classifiers developed in this work, we shall use the *F1 score* [13], instead of the more common *accuracy* index. In short, F1 is defined as the harmonic mean of *precision* and *recall*:

$$F1 = \frac{2PR}{P + R}. \tag{1}$$

Thus, an F1 score reaches its best value at 1 (perfect precision and recall) and its worst at 0. Intuitively, it allows one to evaluate the balance between precision and recall, which is extremely useful for evaluating classifiers performance in situations where there is an uneven class distribution, and false positive kinds of error have the same importance as false negative ones.

Let us now focus on the predictors, which have also undergone some changes with respect to the original features. First, a *season* attribute has been added,

---

[2] Note that the European directives identify the two relevant values of 0 and 200 for NO_2 concentrations. However, we chose here to rely on different interval boundaries, since in the considered data there are just 4 instances with values over 200. Although this is a rather arbitrary choice, it does not compromise the goal of the work, namely, assessing the role played by temporal information in the overall classification task.

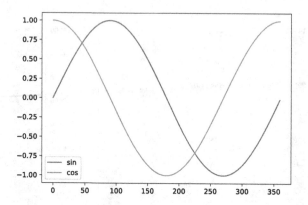

Fig. 2. Trigonometric transformations applied to the attribute *wind_direction_num*.

that (as it can be expected) tracks the season to which an instance refers (Winter, Spring, Summer, Autumn). Moreover, the timestamp attributes *month*, *dom*, *dow*, and *hour* have been replaced by two attributes each, based on the two trigonometric transformations:

$$SIN(2 * \pi * x/\delta) \ , \ COS(2 * \pi * x/\delta), \tag{2}$$

where x represents the original value, and δ is the length of the period, e.g., 24 for the attribute *hour*, and 7 for the attribute *dow*. Relying on the two trigonometric transformations allows us to take into account the periodicity of the timestamp attributes. In this way, for instance, the value 11 pm hours becomes close to that of midnight and of 1 am hours. A similar motivation is behind replacing *wind_direction_num* by its two trigonometric transformations, although the idea here is not that of dealing with periodicity, but, again, that of treating in the same way wind directions that are close to each other, e.g., 359° and 1°. Figure 2 shows the result of such a transformation on the values of the attribute *wind_direction_num*. Observe that both trigonometric transformations are necessary in order to derive the original value, e.g., to distinguish 100° from 260°.

The final step consists of identifying which attributes might carry a temporal content, that is, which attributes can be interpreted as collection of historical values. They are *traffic*, *air_temp*, *wind_speed*, *wind_direction_num* (trig. transforms), *wind_direction_categ*, *rel_humidity*, and *air_pressure*. Sections 5 and 6 present two different approaches by which it is possible to model such temporal aspects.

4   Classification with Atemporal Data

Before delving into how temporal data may be handled, let us establish a baseline by developing a set of atemporal classification models. We consider WEKA's J48 decision tree learner and Random Forest ensemble technique, that will be

Table 2. Hyperparameter search space and best parameters for Scikit-learn's RandomForestClassifier algorithm, over the three datasets.

Parameter	Search space	Atemporal	Lag_1_2_3_4	Lag_6_12_18_24
n_estimators	10, 25, 50, 75, 100, 125, 150, 175, 200, 250, 300, 400, 500, 700, 1000	NO_2: 150	NO_2: 1000	NO_2: 175
		NO_X: 300	NO_X: 700	NO_X: 300
		$PM_{2.5}$: 1000	$PM_{2.5}$: 1000	$PM_{2.5}$: 1000
criterion	gini, entropy	NO_2: gini	NO_2: gini	NO_2: entropy
		NO_X: entropy	NO_X: entropy	NO_X: entropy
		$PM_{2.5}$: entropy	$PM_{2.5}$: entropy	$PM_{2.5}$: entropy
min_samples_split	2, 4, 6, 8, 10, 15, 20	NO_2: 15	NO_2: 20	NO_2: 20
		NO_X: 8	NO_X: 10	NO_X: 15
		$PM_{2.5}$: 2	$PM_{2.5}$: 2	$PM_{2.5}$: 4
max_features	sqrt, log2	NO_2: log2	NO_2: sqrt	NO_2: sqrt
		NO_X: log2	NO_X: sqrt	NO_X: sqrt
		$PM_{2.5}$: log2	$PM_{2.5}$: sqrt	$PM_{2.5}$: sqrt

compared to single J48SS trees and ensembles of J48SS trees, respectively. The considered predictors are thus *season*, *month* (trig. tranforms), *dom* (trig. tranforms), *dow* (trig. tranforms), *hour* (trig. tranforms), *holiday_or_workday*, *traffic*, *air_temp*, *wind_speed*, *wind_direction_num* (trig. tranforms), *wind_direction_categ*, *rel_humidity*, and *air_pressure*, for a total of 18 attributes. The dataset has been partitioned into a training (66%) and a test (33%) set according to a stratified approach. Then, to account for the uneven class distribution, proper instance weights have been derived by means of Scikit-learn's compute_class_weight function [3], and have been used for model learning.

Three Random Forest models have been considered (Scikit-learn's implementation [4]), each performing classification on a different pollutant. For each of them, we proceeded in the following way. To start with, we performed a hyperparameter tuning step by means of 3-fold cross-validation on training data. Table 2 reports the hyperparameters search space and the best performing ones, that have been then used to train the final model (column *Atemporal*). Observe that the Random Forest classifier does not have a deterministic behaviour, but relies on an initial seed to guide some random choices during the learning process. Thus, in order to get some confidence intervals over the test set performance, we trained 10 different models by varying the initial seed. We then considered the average and standard deviation of the F1 score, as shown in Figs. 3, 4 and 5 (label *RF_atemporal*). As a side note, observe also that the F1 score returned by each model is actually a macro average of the F1 scores calculated for each class. Similarly, we trained three single J48 decision trees over the same data. However, unlike the previous case, we did not perform any tuning over these models, because, as we shall see in Sect. 6, this allows us to perform a fair comparison with J48SS. Performance results for J48 are also reported in Figs. 3, 4 and 5 (label *J48_atemporal*). Note that, being J48 a deterministic algorithm, no standard deviation is observed.

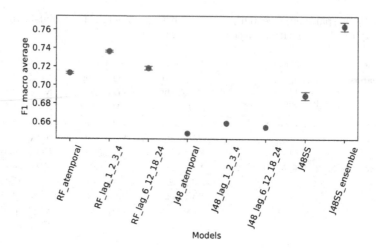

Fig. 3. Average and standard deviation of the F1 macro average scores over the dataset NO_2.

As it could be expected, Random Forest is capable of obtaining F1 scores considerably higher than J48, in all three classification tasks. Moreover, standard deviations exhibited by Random Forest are quite small, which is the result of them being composed of a relatively large number of trees.

5 Classification with Classical Models and Lagged Variables

In this section, we report the outcomes of the classification by means of classical models and lagged variables. Recall that, in Sect. 3, we identified the attributes that may be susceptible of carrying temporal information, namely, *traffic, air_temp, air_pressure, wind_speed, wind_direction_num* (trig. transforms), *wind_direction_categ*, and *rel_humidity*.

We want to assess now the impact on the classification performances of models learned by J48 and Random Forest using the historical data of these attributes. We modeled the temporal aspect by means of lagged variables, which is a commonly used technique in the literature (see, for instance, [15]). In short, a lagged variable is a delayed variable that can be used to keep track of historical values for a given attribute. For instance, given the amount of traffic at the current instant, we may as well be interested in knowing how many vehicles crossed the same intersection two hours ago.

We started from the same set of attributes as described in Sect. 4 and we added, for each of the 8 mentioned attributes, a set of lagged variables. Specifically, we considered two different datasets, characterized by the use of either: *(i)* lagged variables for 1, 2, 3, and 4 hours before the current value, or *(ii)* lagged variables for 6, 12, 18, and 24 hours before. In both cases, the resulting dataset encompasses 50 attributes (18 atemporal + 32 lagged).

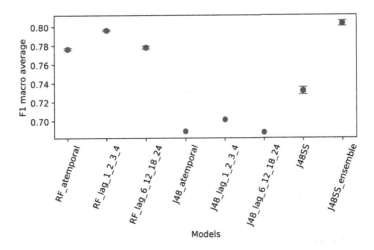

Fig. 4. Average and standard deviation of the F1 macro average scores over the dataset NO_X.

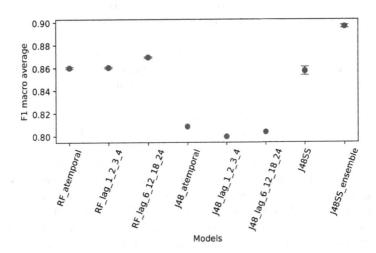

Fig. 5. Average and standard deviation of the F1 macro average scores over the dataset $PM_{2.5}$.

We trained three Random Forest and three J48 models, following the same tuning, training, and test protocol followed in Sect. 4. The results of the tuning phase are presented in Table 2 (columns $Lag\_1\_2\_3\_4$, and $Lag\_6\_12\_18\_24$), while their performances on the test set are depicted in Figs. 3, 4 and 5 (labels $RF\_lag\_1\_2\_3\_4$, $RF\_lag\_6\_12\_18\_24$, $J48\_lag\_1\_2\_3\_4$, and $J48\_lag\_6\_12\_18\_24$).

We observe that the performance of the models trained on the lagged datasets tend to be higher than or equal to those of the ones built on the atemporal features only, with the exception of J48 on the $PM_{2.5}$ classification task; such a behaviour might be explained by the high number of attributes that the decision tree has to consider, and might be improved with an additional, intermediary feature selection step.

It should finally be observed that the choice of the number and granularity of lagged variables to employ is a rather arbitrary choice which may also be susceptible of tuning. As we shall see in the next section, relying on J48SS allows one to deal with temporal information in a much more natural way.

6 Classification with J48SS

J48SS [6] is a novel decision tree learner based on WEKA's J48 (a Java implementation of Quinlan's C4.5 [12]). The key characteristic of the algorithm is that it is capable of naturally exploiting numerical and categorical attributes as well as sequential and time series data during the same execution cycle, relying on the concepts of frequent pattern extraction and time series shapelet generation via evolutionary computation. Briefly, a shapelet is a contiguous, arbitrary-length time series subsequence that is in some sense maximally representative of a class, and is aimed at capturing local features of the time series. In [6], the resulting decision tree models are shown to be intuitively interpretable, meaning that, for instance, a domain expert may easily read and validate them.

J48SS makes it much easier to integrate temporal information starting from the atemporal dataset discussed in Sect. 4: for each of the 8 temporal attributes identified in Sect. 3, we simply build a string storing the past 24 values. Thus, we end up with a total of 26 features (18 atemporal + 8 temporal). J48SS makes use of a further parameter, denoted by $W \in [0, 1]$ (weight), to evaluate the *abstraction* level of each temporal pattern extracted from either a sequence or a time series and, thus, control overfitting (as suggested in [6]). Intuitively, larger values of W should lead to more powerful patterns being extracted, in terms of class discrimination performance (at least, on the training set). Conversely, smaller values of W should result in less complex, more general features being selected. We tuned W over a stratified training data split (considering 66% as actual training data, and 33% as validation data), finding that its best values are 0.9, 1.0, and 0.75 for NO_2, NO_X, and $PM_{2.5}$ classification tasks, respectively. Observe that we did not perform any tuning on the atemporal part of J48SS, which behaves exactly in the same was as the original J48. Then, for each prediction task, we averaged the performance on the test set of 10 single J48SS models; this is necessary since, as mentioned before, J48SS extracts time series shapelets by means of an evolutionary algorithm, which in turn is non deterministic and makes use of an initial random seed to guide the search. Results are depicted in Figs. 3, 4 and 5 (label *J48SS*).

As a final step, we considered ensembles of J48SS models. Specifically, we built three models, each composed of 10 (tuned) J48SS trees, relying on the WEKA's RandomSubSpace method [9]. In short, RandomSubSpace tries to build a decision tree ensemble that maintains the highest possible accuracy on training data, and improves on generalization accuracy as it grows in complexity. The trees in the ensemble are constructed systematically by pseudorandomly selecting random samples of features instead of the entire feature set, that is, the trees are constructed in randomly chosen subspaces. We did not perform any tuning over the ensemble method hyperparameters, relying on the default choice for the subspace size, i.e., 50%. As usual, the results, computed over the execution of 10 models, are presented in Figs. 3, 4 and 5 (label *J48SS_ensemble*).

As it can be seen, single J48SS trees score better than plain J48 models, although they typically perform worse than Random Forests. The observed standard deviation is higher than that of Random Forest; however this is expected, being these models composed of just a single decision tree. More importantly, small ensembles of J48SS trees are capable of achieving results higher than large Random Forest ensembles, that may in turn include up to 1000 trees. This is a clear indication that frequent patterns and shapelets are capable of extracting more information than lagged variables. Moreover, observed standard deviations are also reduced, which is a trend that is expected to continue as the ensemble size grows. Finally, using time series and sequences in J48SS allows one to reduce the arbitrariness related to the lagged variable approaches, although, as we have witnessed, one has still to decide the appropriate length for the temporal attributes' histories.

7   Conclusions

In this paper, we considered how different ways of encoding temporal information may impact on the performance of a classification task. We considered a real case scenario, that is, that of assessing the relationships between concentrations of the pollutants NO_2, NO_X, and $PM_{2.5}$, and a set of variables describing, among others, meteorological conditions and traffic flow in the city of Wrocław, in Poland. Through a series of experiments, we showed that accounting for the historical values of the features by means of lagged variables helps in improving the overall accuracy results. Moreover, an ever higher performance is obtained by relying on J48SS, a recently introduced decision tree model that is capable of handling heterogeneous datasets, that may be composed of static, i.e., categorical and numerical, attributes, as well as sequential and time series data. Although the J48SS approach to the management of temporal data is quite natural and less complex than relying on lagged variables, some degree of arbitrariness still remains for what concerns the choice of the length of the histories for the temporal attributes.

Possible future work includes an adaptation of J48SS to deal with regression tasks, and the extension of an algorithm such as PART [8] to allow for the extraction of a set of highly interpretable rules from J48SS. At the moment, a pure lagged regression problem can be dealt with standard algorithms only, that is, by first creating (an arbitrary number of arbitrary) lagged variables and, then, running a regression algorithm. We are currently exploring the possibility of approaching this problem via dynamic preprocessing (which generalizes the concept of wrapper) instead, to obtain a more performing learning algorithm.

References

1. European Union air quality standards. http://ec.europa.eu/environment/air/quality/standards.htm. Accessed 21 May 2019
2. NOx level objectives. http://www.icopal-noxite.co.uk/nox-problem/nox-level-objectives.aspx. Accessed 21 May 2019
3. Scikit-learn's compute_class_weight function. https://scikit-learn.org/stable/modules/generated/sklearn.utils.class_weight.compute_class_weight.html. Accessed: 22 May 2019
4. Scikit-learn's RandomForestClassifier. https://scikit-learn.org/stable/modules/generated/sklearn.ensemble.RandomForestClassifier.html. Accessed 22 May 2019
5. Breiman, L.: Random forests. Mach. Learn. **45**(1), 5–32 (2001)
6. Brunello, A., Marzano, E., Montanari, A., Sciavicco, G.: J48SS: a novel decision tree approach for the handling of sequential and time series data. Computers **8**(1), 21 (2019)
7. Deters, J.K., Zalakeviciute, R., Gonzalez, M., Rybarczyk, Y.: Modeling $PM_{2.5}$ urban pollution using machine learning and selected meteorological parameters. J. Electr. Comput. Eng. **2017**, 5106045:1–5106045:14 (2017)
8. Frank, E., Witten, I.H.: Generating accurate rule sets without global optimization. In: Proceedings of the 15th International Conference on Machine Learning (ICML), pp. 144–151. Morgan Kaufmann (1998)
9. Ho, T.K.: The random subspace method for constructing decision forests. IEEE Trans. Pattern Anal. Mach. Intell. **20**(8), 832–844 (1998)
10. Kamińska, J.A.: The use of random forests in modelling short-term air pollution effects based on traffic and meteorological conditions: A case study in Wrocław. J. Environ. Manag. **217**, 164–174 (2018)
11. Mbarak, A., Yetis, Y., Jamshidi, M.: Data - based pollution forecasting via machine learning: case of Northwest Texas. In: Proceedings of the 2018 World Automation Congress (WAC), pp. 1–6 (2018)
12. Quinlan, R.: C4.5: Programs for Machine Learning. Morgan Kaufmann Publishers, San Mateo (1993)
13. Sasaki, Y.: The truth of the F-measure. Teach Tutor Mater **1**(5), 1–5 (2007)
14. Shang, Z., Deng, T., He, J., Duan, X.: A novel model for hourly $PM_{2.5}$ concentration prediction based on CART and EELM. Sci. Total Environ. **651**, 3043–3052 (2019)
15. Wilkins, A.S.: To lag or not to lag?: Re-evaluating the use of lagged dependent variables in regression analysis. Polit. Sci. Res. Methods **6**(2), 393–411 (2018)
16. Witten, I.H., Frank, E., Hall, M.A., Pal, C.J.: Data Mining: Practical Machine Learning Tools and Techniques. Morgan Kaufmann, Burlington (2016)
17. Xie, J., et al.: The characteristics of hourly wind field and its impacts on air quality in the pearl river delta region during 2013–2017. Atmos. Res. **227**, 112–124 (2019)

Extreme Climate Event Detection Through High Volume of Transactional Consumption Data

Hugo Alatrista-Salas[(⊠)] , Mauro León-Payano,
and Miguel Nunez-del-Prado

Universidad del Pacífico, Av. Salaverry, 2020 Lima, Peru
{h.alatristas,ma.leonp,m.nunezdelpradoc}@up.edu.pe

Abstract. Extreme weather events cause irreparable damage to society. At the beginning of 2017, the coast of Peru was hit by the phenomenon called "El Niño Costero", characterized by heavy rains and floods. According to the United Nations International Strategy for Disasters ISDR, natural disasters comprise a 5-step process. In the last stage - recovery - strategies are aimed at bringing the situation back to normality. However, this step is difficult to achieve if one does not know how the economic sectors have been affected by the extreme event. In this paper, we use two well-known techniques, such as Autoregressive integrated moving average (ARIMA) and Kullback-Leibler divergence to capture a phenomenon and show how the key economic sectors are affected. To do this, we use a large real dataset from banking transactions stored in a Massively Parallel Processing (MPP). Our results show the interest of applying these techniques to better understand the impact of a natural disaster into economic activities in a specific geographical area.

Keywords: Time series · Transactional banking data ·
Extreme climate event detection · Parallel processing

1 Introduction

The ENSO (El Niño - Southern Oscillation) is a climatic phenomenon that consists of the increase of the temperature in the Equatorial Pacific. ENSO has a 2 to 7 years fluctuation period, with a warm phase known as El Niño and a cold phase called La Niña. One of the crucial indicators of the presence of El Niño is the variation of the surface temperature of the SST (Sea Surface Temperature), which causes changes in the climate in the world.

By the end of 2016 and early 2017, the ENSO has an abrupt change exposing a new atypical phenomenon called "El Niño Costero" characterized by heavy

---

Authors are in alphabetical order and contributed equally to the present paper.

© Springer Nature Switzerland AG 2019
T. Welzer et al. (Eds.): ADBIS 2019, CCIS 1064, pp. 475–486, 2019.
https://doi.org/10.1007/978-3-030-30278-8_46

rains and floods in the coastal zone of Peru. This phenomenon was reflected in an unusually amount of human and material loss in large part of the northern departments and also in some provinces of the capital Lima. According to official reports of the Peruvian government, up to May 2017, $1'129,013$ affected, and 143 deaths have been reported. As well as, $25,700$ constructions have collapsed, $258,545$ are touched, and $23,280$ are uninhabitable [9].

This atypical event has affected not only the population and its housing but also the global economy of the country. The first channel where El Niño Costero hit in the economy is the increase in prices of basic food basket products due to the collapse of the key supply chains. For example, the price of lemons, essential product in the famous Peruvian gastronomy, increased considerably from US\$ 1.00 to more than US\$ 8.00 per kilo [10].

Concerning the phenomenon itself, it can be captured using meteorological data sensors, both on land and at sea. For example, an increase in seawater temperature could indicate the beginning of a possible atypical event. On the other hand, on land, we can use the data associated with rainfall (in cubic meters) or data related to river flows. However, it is challenging to perform these measurements within social structures to attempt to capture the event by studying the consumption patterns of citizens impacted by such events. Also, once the extreme event detected thought the variation of socioeconomic indicators, it could be interesting to know which economic sectors should be prioritized in the process of reconstruction after the disaster.

In this article, we use a vast amount of transactional banking data from a financial institution in Peru to capture extreme climate events through a simple but effective process. Data at our disposal was sanitized to ensure users privacy as well as anonymity and stored in a Massively Parallel Processing (MPP). Then, the entire dataset was used to study the impact of the phenomenon on the spending habits in all Peruvian departments. Later, we choose the area hard hit by the phenomena, and we used two techniques to capture the variation of banking transactions: (1) the time series analysis (ARIMA); and, (2) the Kullback-Leibler divergence. Finally, the results of both experiments were visualized and show the period in which the phenomenon appeared and how the event touched the inhabitants of the affected areas.

The present work is organized as follows. Section 2 reviews the literature. Then, Sect. 3 describes the two method used in this article. Later, Sect. 4 describes the data and the experiments respectively. Finally, Sect. 5 concludes the present work and proposes some perspectives.

2   State of the Art

In the present section, we present some related works to quantify socioeconomic indicators of populations using transactional banking data. The work of Yannick *et al.* [3] computes the average monthly purchase and debt of people to infer their social status. Based on these indicators, authors were able to analyze the social structure of around six millions of people in Mexico. The dataset was gathered during eight months from November 2014 to June 2015.

Using the same data, Yannick *et al.* [4] use, besides of transactional banking data, mobile data to infer about the kind of categories in which people spend and how similar spend their friends. Authors rely on the average monthly purchase and the purchase vector, which is the amount of money spend by category by a person. Based on these indicators, they can estimate the socioeconomic status of individuals as well as their purchase patterns. Finally, the authors detect a positive correlation between categories and groups of people that buy similar sets of categories.

In the same spirit, Di Clemente *et al.* [2] quantify the lifestyles in urban populations. Authors used the chronological sequence of purchase characterized by the MCC (Merchant Category Classification) to obtain a sequence. Then, they applied the Sequitur algorithm [5] to infer a sequence pattern. To detect significant purchases, they generate 1000 randomized code sequences for each user to apply the Sequitur algorithm. Therefore, they extract for each user the set of significant purchase with *z-score*. Thus, they were able to model the lifestyle by taking purchases bigger than two times the computed z-score. To perform the experiments, they rely on a credit card dataset composed of 150 000 users observed over ten weeks in a major city in Latin America. The dataset contains age, gender, and zip code.

In other works, the authors study the impact of economic diversity on economic development. For instance, in [8], the authors present a model to predicts how individual business types systematically change with the city size, shedding light on processes of innovation and economic differentiation with scale. To build the model, the authors used the National Establishment Time-Series dataset. Besides, the authors aggregate information about 366 cities (called metropolitan statistical areas MSAs). To summarize, the authors highlight a systematic behaviour similar to all studied cities.

To summarize, we have presented some works to characterize spending patterns of inhabitants. The idea behind our work is to rely on these kinds of indicators to detect anomalous behaviour, which captures some extreme climate event indirectly and which economic sector was the most affected.

3   Methods

In the present section, we describe the methods used in this article: (1) the time series analysis ARIMA; and, (2) the Kullback-Leibler divergence.

3.1   Auto-Regressive Integrated Moving Average

$ARIMA(p, d, q)$ is the combination of the Auto-Regressive (AR) and the Moving Average (MA) models. ARIMA is also known as Box-Jenkins models. In this model, the integrity term (I) is defined by the number of differences d, which are needed to transform a non-stationary time series to a stationary one [1]. The ARIMA model is expressed in Eq. 1.

$$y_t = c + \sum_{i=1}^{p} \varphi_i y_{t-i} + \varepsilon_t + \sum_{i=1}^{q} \theta_i \varepsilon_{t-i} \tag{1}$$

In Eq. 1, ε_t is the white noise; $\varphi_1, ..., \varphi_p, \theta_1, ..., \theta_q$ are the parameters of the $AR(p)$ and $MA(q)$ respectively; and c is the constant depending on d parameter.

To study the time series with ARIMA, four stages of the Box-Jenkins methodology are applied; namely: (a) the *identification* of the stationarity process using the Dickey-Fuller test. If the time series is not stationary in average or standard deviation a differences or logarithmic transformation is applied, respectively; (b) the *estimation* of the q and p parameters using the auto-correlation function (ACF) and partial autocorrelation function ($PACF$); (c) the *diagnosis*, to detect the model with the lowest *Akaike* information criterion (AIC) among the ARIMA(p,d,q) model candidates passing the Ljung-Box test; and, (d) the *prediction* using the best ARIMA model.

3.2   Kullback-Leibler Divergence (KL Divergence)

This metric measures the difference between two probability distributions. The KL divergence is a non-symmetric metric, *i.e.*, the divergence between $KL(p(x), q(x))$ is not equal to $KL(q(x), p(x))$. Indeed, the divergence measures the information lost when using $p(x)$ to approximate $q(x)$ distribution. The KL divergence is defined by Eq. 2.

$$KL(p(x)|q(x)) = \sum_{x \in X} p(x) ln \frac{p(x)}{q(x)} \qquad (2)$$

KL divergence is a non-negative measure. Thus, $KL(p(x)|q(x)) \geq 0$ and $KL(p(x)|q(x)) = 0$ if and only if $p(x) = q(x)$. It is worth noting that $\lim_{p \to 0} p log(p)$. Nevertheless, when $p \neq 0$ and $q = 0$, $KL(p(x)||q(x))$ is defined as ∞. Hence, the two distributions are absolutely different [7]. To solve this problem, it is possible to add a constant ($\epsilon = 10^{-3}$) to smooth the the distribution.

4   Experiments

This section describes the steps developed in this article and show our findings.

4.1   Data Description and Broad Analysis

In the present effort, we use banking transactions dataset belonging credit and debit card payments. This data is composed of the Merchant Category Code (MCC)[1]; the timestamp of the transaction (Timestamp); spending in USD; and the district of the transaction.

Data at our disposal is stored in Greenplum Database[2], which is a Massively Parallel Processing (MPP) database server with an architecture specially designed to manage large-scale analytic databases. The dataset was gathered

---

[1] VISA   Merchant   Category   Classification   (MCC),   https://www.dm.usda.gov/procurement/card/card_x/mcc.pdf.
[2] https://greenplum.org.

from June 2016 to May 2017 containing more than 16 millions of transactions from both credit and debit cards in Peru. Table 1 shows some charaterisctics of the provided dataset.

Table 1. Dataset example

Characteristics	Values
Total number of transactions	16 857 449
Total number of clients	1 809 925
Total number of merchants	182 830
Size of the dataset (measure in the DBMS)	30 GB

The overall dataset was used to capture the impact of the "El Niño Costero" phenomenon throughout the Peruvian territory. In this regard, Fig. 1 shows the diary spending in USD - in logarithmic shape - by the department during June 01, 2016 and October 31, 2017 (blue line). In Fig. 1, the red ribbon shows the period in which the phenomenon took place. This figure allows us to have a general overview of the spending behaviour and the impact of the phenomenon in all departments belonging the Peruvian territory. On one hand, concerning the spending, the citizens of Lima spend much more than the inhabitants in Pasco. On the other hand, the impact of the phenomenon in Lima was more devastating than in Pasco. On an overall basis, we can observe that the phenomenon impacted more than 85% of the departments of Peru in different intensities as depicted in Fig. 1.

4.2   Targeted Analysis

Let analyze a specific area touched by the studied phenomenon. For this purpose, we use customer consumption registered in commercial premises located in one of the most affected districts of Lima (*San Juan de Lurigancho* district) during the floods in February 2017. Concerning the bank transactions, we focus only on the top ten businesses categories with the highest frequency of consumption (number of transactions), to avoid noise from unusual types of shops. The top ten businesses categories are shown in Table 2. Finally, the resulting database contains 1.1 millions of records.

The time series methods require that each category has a complete series. The absence of periods in the series can be interpreted as the absence of consumption by the users. To avoid divided-by-zero problems, we assign a consumption of 0.0 USD when data is missing. Also, we replace outlier values with the median consumption on the same day of the week. For instance, the consumption on December 24, 2016 (Saturday) is replaced by the median consumption consumed every Saturday.

For analysis facilities, we group the daily consumption into weekly consumption for each MCC. In this way, we obtain 53 weekly observations instead of 365

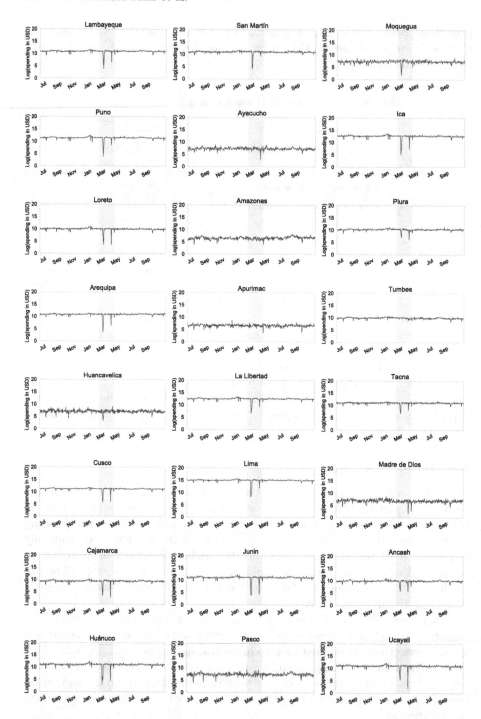

Fig. 1. Total spending in USD for each department in Peru during June 01, 2016 and October 31, 2017 (Color figure online)

Table 2. The top ten businesses categories

Code	Name	Count
5812	Eating Places, Restaurants	137,912
5814	Fast Food Restaurants	83,469
5541	Service Stations	70,342
5912	Drug Stores and Pharmacies	56,933
5411	Grocery Stores, Supermarkets	46,799
7995	Betting/Casino Gambling	46,644
5311	Department Stores	41,372
5813	Drinking Places	36,885
5661	Shoe Stores	9,425
8220	Colleges, Universities	8,965

daily observations per MMC. Later, to analyze the effects of "El Niño", each weekly series is divided into two parts:

– Before the "El Niño Costero" phenomenon (from June 2016 to January 2017). Data associated with this period is used as a training series.
– During the "El Niño Costero" phenomenon (from February 2017 to May 2017). Data related to this period is used as a series to compare the prediction results of the model.

4.3   Studying of the MCC Seasonality

From the training series, we used the methodology proposed by Box-Jenkins, in which the seasonality of each MCC series is validating by applying the Dickey-Fuller test. Indeed, we apply successive differences until achieving effective stationarity. As a result, the T-statistics, the p-values and the number of differences performed are shown in Table 3.

Regarding the results of the Dickey-Fuller test, all series ensure the seasonality of the data. In two cases, no differences were applied (MCC 5912 and MCC 7995), while in the case of MCC 5311, two differences were applied. To summarize, all series have a p-value less than 0.05; thus, we reject the null hypothesis with a significance level value less than 1%.

Later, we calculate the autocorrelation and partial autocorrelation functions of each MCC series. Finally, values of the parameters p, d and q of the ARIMA model were computed.

4.4   Estimation and Check-In

The parameters of each proposed MCC are estimated. Besides, it is necessary to verify the compliance of the white residue assumption, *i.e.*, if the residues of

Table 3. Test Dickey-Fuller

MCC	T-Statistic	p-value	N difference
5812	$-1.236862e+01$	$5.343719e-23$	1
5814	$-8.490075e+00$	$1.320743e-13$	1
5541	$-9.071323e+00$	$4.294390e-15$	1
5912	$-5.671192e+00$	$8.914266e-07$	0
5411	$-5.766309e+00$	$5.522354e-07$	1
7995	-3.733986	0.003660	0
5311	-3.449616	0.009379	2
5813	$-1.222750e+01$	$1.072175e-22$	1
5661	$-6.697515e+00$	$3.965052e-09$	1
8220	$-6.922218e+00$	$1.138785e-09$	1

each proposed model have a normal distribution. For this, we used the Shapiro-Wilk Test for the small size of weekly data [6]. For all models that pass the test, the Akaike Selection Criterion (AIC) and the Mean Absolute Percentage Error (MAPE) were used to select the best model for each MCC. The models whose residues meet the Shapiro-Wilk test and have the lowest AIC and MAPE are shown in Table 4.

Table 4. Test Shapiro-Wilk, Valores MAPE y AIC

MCC	Model	p-value	MAPE	AIC
5812	(1, 1, 0)	0.38194	1.889	-14.36
5814	(1, 1, 0)	0.72236	1.646	-32.125
5541	(0, 1, 1)	0.49689	5.481	-33.037
5411	(3, 1, 1)	0.7177	5.411	-25.314
7995	(3, 0, 0)	0.92763	1.539	-6.823
5813	(1, 1, 0)	0.97314	2.519	6.464
5661	(0, 1, 0)	0.09234	3.105	23.928
8220	(3, 1, 2)	0.89752	7.386	46.238

It is important to notice that eight MCCs have an ARIMA model whose residues have a normal distribution. On the opposite, MCCs 5912 (Drug Stores and Pharmacies) and 5311 (Department Stores) do not meet the criteria (see Table 3).

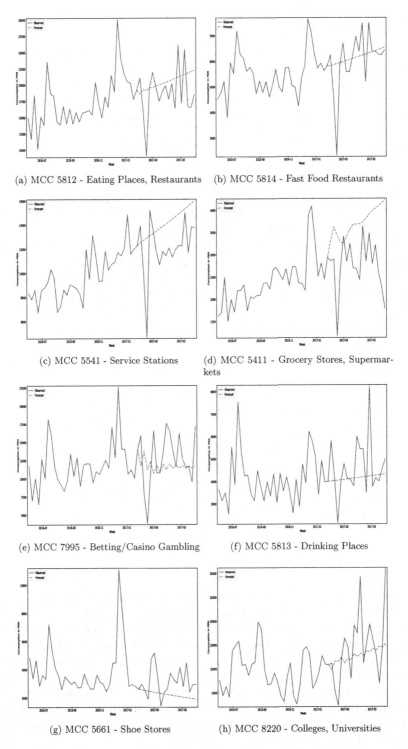

(a) MCC 5812 - Eating Places, Restaurants

(b) MCC 5814 - Fast Food Restaurants

(c) MCC 5541 - Service Stations

(d) MCC 5411 - Grocery Stores, Supermarkets

(e) MCC 7995 - Betting/Casino Gambling

(f) MCC 5813 - Drinking Places

(g) MCC 5661 - Shoe Stores

(h) MCC 8220 - Colleges, Universities

Fig. 2. Consumption of 18 weekly during the phenomenon of "El Nino", (from 2017/02/01 to 2017/05/31) (Color figure online)

4.5   Prediction Phase

Figure 2 shows the histograms representing the forecasts for each MCC (*c.f.*, Table 4). In Fig. 2 every single image represents the customer consumption within 18 weeks in the district of *San Juan de Lurigancho*, in which the solid blue line represents the fluctuations of the consumers' weekly consumption while the dotted green line represents the expected consumption obtained by the models.

4.6   Performing the Kullback-Leibler Method

In this step, we compute the information gain of changing from an a priori distribution of a dataset s_t to an a posteriori distribution of a data set s_{t+1} by performing the Eq. 2. Given the time series T, such that $S = \{S_1, ..., S_k\}$ is the set of possible daily consumption sequences and $c = \{c_i | c_i \in S, i = 1, ..., n\}$ is the daily consumption sequence of length n. It is important to notice that the sum of all the values of the daily consumption should be 1.

From the time series T, 175 sets of consumption sequences (represented by k) of size seven days (denoted by d) were formed. The 175 Kullback-Leibler divergence coefficients are shown in Fig. 3.

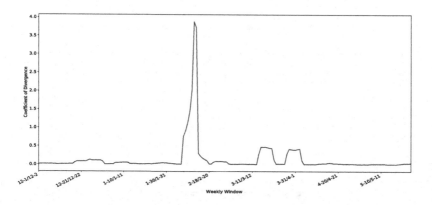

Fig. 3. Kullback-Leibler coefficient with smoothing

Regarding Fig. 3, the coefficients of the Kullback-Leibler divergence show values close to zero until the first week of February 2017. In the second week of February, specifically from February 9 to 14, noticed an increase in the coefficient due to the overflow of rivers *Rimac* and *Huaycoloro* that affected the district of *San Juan de Lurigancho*. The rise registered in the week of February 15 relieves the effects of the phenomenon of "El Niño Costero", generating a significant contrast between the sets of sequences. Finally, during March, there was a substantial increase in consumption due to the over-supply and the over-valuation of necessities. Indeed, the price of products such as mineral water, rice, soft meats, was triggered.

It is important to stress that, the phenomenon touch the *San Juan de Lurigancho* district on 14 and 15 February 2017, and the consumption in these two dates was 0.0 USD. This fact, probably generated by a lack of electricity, resulting in an inconsistency in the calculation of divergence in the datasets where there is at least one absence (consumption equal to 0.0 USD). Following the recommendation provided in [7], we replace the lack of consumption with a small constant ϵ. This value was distributed on the other values of the dataset. In this work, the ϵ value is fixed in 10^{-4}.

5   Conclusions and Future Works

The economic sector is always affected when a natural disaster strikes a country. Besides, some commercial activities are more disadvantaged than others; for example, the pharmaceutical industry may be more impacted than the entertainment sector in the face of an outbreak. In this paper, we analyze this phenomenon using a massive dataset from banking transactions recovered between June 2016 and May 2017, which was conveniently stored in a Massively Parallel Processing (MPP). This dataset contains more than 16 millions of purchases made in different establishments which were grouped by categories. Firstly, we selected the categories with the highest flow of transactions. Later, used ARIMA to show how the categories were imported by the phenomenon predicting the values of the transactions in a regular period (without phenomenon). Finally, we use the Kullback-Leibler divergence measure to be able to measure the information gain from an a-priori state to an a-posteriori state. Our results show the interest of applying these techniques to better understand the impact of an extreme phenomenon on economic activities and how they impact on the consumption behaviour of society.

As future work, we want to study the impact of the phenomenon in other neighbouring areas, *i.e.*, how a localized phenomenon impacts the economy of the other regions. For example, shortages of some necessities in the impacted area can generate the movement of people to other neighbouring areas to buy the missing products. Moreover, we are interested in integrating different types of data into the study. For example, we could use weather data to look for correlations between climate changes and banking transactions. Also, we can use product price data in the markets to measure the impact of the phenomenon on the increase of the prices of essential products and to correlate them with the banking transactions.

References

1. Brockwell, P.J., Davis, R.A.: Forecasting techniques. Introduction to Time Series and Forecasting. STS, pp. 309–321. Springer, Cham (2016). https://doi.org/10.1007/978-3-319-29854-2_10

2. Di Clemente, R., Luengo-Oroz, M., Travizano, M., Xu, S., Vaitla, B., González, M.C.: Sequences of purchases in credit card data reveal lifestyles in urban populations. Nat. Commun. **9**(1), 3330 (2018). https://doi.org/10.1038/s41467-018-05690-8

3. Leo, Y., Fleury, E., Alvarez-Hamelin, J.I., Sarraute, C., Karsai, M.: Socioeconomic correlations and stratification in social-communication networks. J. R. Soc. Interface **13**(125), 20160598 (2016). https://doi.org/10.1098/rsif.2016.0598

4. Leo, Y., Karsai, M., Sarraute, C., Fleury, E.: Correlations of consumption patterns in social-economic networks. In: Proceedings of the 2016 IEEE/ACM International Conference on Advances in Social Networks Analysis and Mining, ASONAM 2016. pp. 493–500. IEEE Press, Piscataway (2016). http://dl.acm.org/citation.cfm?id=3192424.3192516

5. Nevill-Manning, C.G., Witten, I.H.: Identifying hierarchical structure in sequences: a linear-time algorithm. J. Artif. Int. Res. **7**(1), 67–82 (1997). http://dl.acm.org/citation.cfm?id=1622776.1622780

6. Pedrosa, I., Juarros-Basterretxea, J., Robles-Fernández, A., Basteiro, J., García-Cueto, E.: Goodness of fit tests for symmetric distributions, which statistical should i use? Universitas Psychologica **14**, 245–254 (2015). http://www.scielo.org.co/scielo.php?script=sciarttext&pid=S1657-92672015000100021&nrm=iso

7. Yan, X., Cheng, H., Han, J., Xin, D.: Summarizing itemset patterns: a profile-based approach. In: Proceedings of the Eleventh ACM SIGKDD International Conference on Knowledge Discovery in Data Mining, KDD 2005, pp. 314–323. ACM, New York (2005). https://doi.org/10.1145/1081870.1081907

8. Youn, H., Bettencourt, L.M.A., Lobo, J., Strumsky, D., Samaniego, H., West, G.B.: Scaling and universality in urban economic diversification. J. R. Soc. Interface **13**(114), 20150937 (2016). https://doi.org/10.1098/rsif.2015.0937

9. Ministry for Primary Industries, ERCC PORTAL Emergency Response Coordination Centre (ERCC) European Civil Protection and Humanitarian Aid Operations (2017). https://erccportal.jrc.ec.europa.eu/Preparedness/Country-profiles/country/Peru/iso3/PER. Accessed 4 Apr 2019

10. Sabes cuánto está el precio del limón y de otros productos en los mercados? (2017). https://peru21.pe/lima/precio-limon-otros-productos-mercados-video-69493. Accessed 8 Apr 2019

ADBIS 2019 Workshop: International Workshop on Qualitative Aspects of User-Centered Analytics – QAUCA

Data Quality Alerting Model for Big Data Analytics

Eliza Gyulgyulyan[1,2], Julien Aligon[2(✉)], Franck Ravat[2],
and Hrachya Astsatryan[2]

[1] IIAP, National Academy of Science of Republic of Armenia,
1 Paruyr Sevak str., Yerevan, Armenia
[2] IRIT-CNRS (UMR 5505), Université Toulouse 1 Capitole,
2 Rue du Doyen Gabriel Marty, 31042 Toulouse Cedex 9, France
{eliza.gyulgyulyan, julien.aligon, franck.ravat,
hrachya.astsatryan}@irit.fr

Abstract. During Big Data analytics, correcting all the problems of large, heterogeneous and swift data, in a reasonable time, is a challenge and a costly process. Therefore, organizations are confronted with performing analysis on massive data, potentially of poor quality. This context is the starting point of our current research: how to identify data quality issues and how to notify users without solving these quality issues in advance? To this end, we propose a quality model, as the main component of an alert system, which allow to inform users about data quality issues, during their analysis. This paper discusses about the conceptual and implementation frameworks of the quality model, as well as examples of usage.

Keywords: Quality model · Data quality · Big Data analytics

1 Introduction

Big Data analytics has undeniable vast importance as it has been absorbed into almost all aspects of scientific or industrial activities. In today's digital world an enormous amount of data is collected, categorized, stored for further analysis, with increasing speed. Decision-makers or knowledge-workers analyze data coming from different sources in order to make a fast and fair analysis.

The tremendous volume of data changing with high speed, the consideration of Data Quality (DQ) requires a higher amount of time and higher processing resources [1, 2].

Moreover, the context of Big Data induces always more diversity of data sources, data types or data structures. This diversity increase the difficulty of correcting every type of quality issue [1]. If handling traditional DQ problems is a challenge, then handling them, facing with the Big Data analytics context, is even more challenging. On the other hand, doing analysis on faulty, poor and untrustworthy data can have considerable and negative consequences for companies. According to [3], quality problems cost to US businesses around 600 billion dollars annually. Thus, performing

© Springer Nature Switzerland AG 2019
T. Welzer et al. (Eds.): ADBIS 2019, CCIS 1064, pp. 489–500, 2019.
https://doi.org/10.1007/978-3-030-30278-8_47

analysis using rough data and obtaining a reliable result is a real issue in the context of Big Data.

This paper considers analysis in a Big Data context, without using any correction processes, beforehand. Indeed, producing *Value* from analysis results, using a minimum of resources for solving quality problems, is our main motivation. This work can be viewed as a continuation of the impact study of the classical 5V's over the Big Data *Value* [2]. To this end, our main research questions refer to the three following challenges:

- How is it possible to identify DQ problems during the analysis?
- How the user can know the quality of the data he/she is analyzing?
- Can the user obtain a good analysis result without solving all the quality problems on the data beforehand?

The first solution to face our problematic is to alert about poor quality when an analysis is being done. An alert system is suggested to notify users about the quality problems along their analysis. The system gives the possibility to refer to DQ problems (or part of them) described in [4] during the analysis process only if the problems are relevant for the ongoing analysis. The notion of alerting about the quality problems can make the process of Big Data analysis more accurate as the analysis is being done without data correction and the result can be biased. Moreover, it will prevent companies from spending resources on solving all the quality problems before the analysis.

The aim of this paper is to describe the main components of the alert system, in particular the quality model. We provide the characteristics of this model thanks to a conceptual schema and an implementation of it, illustrated with examples. Referring to the Big Data context, the model is designed considering quality characteristics and data source diversity. Thanks to the quality model, the alert system can offer to the user the opportunity to ask quality questions over the data sources he/she wants to analyze.

The rest of the paper is organized as follows: Sect. 2 describes the related work of the DQ field, both in a general approach and in the context of Big Data, problems and our suggestions. Section 3 presents our main contribution i.e. the quality model, to be used in our future alert system. This model is defined by a conceptual model and a physical representation of it. The rules of transformation between the conceptual components and the implementation are also given. The implementation is realized in a graph database, in particular using Neo4j[1] database. The last section concludes our paper and gives several perspectives of work.

2   Related Work

DQ itself is considered as a measurable notion describing the level of a set of qualitative and quantitative dimensions and metrics describing it. DQ dimensions are widely and differently discussed and described in the literature [1, 5–13]. Most of the articles identifies various DQ dimensions like consistency, accuracy, timeliness, completeness,

---

[1] https://neo4j.com/developer/get-started/.

etc. There are more than 170 dimensions specified for the last 20 years. Some concepts are very close to each other. In the Big Data analytics context, the majority of identified quality dimensions coincide with the traditional DQ dimensions in database field.

In the context if DQ, metrics are classically required for each dimension. A quality metric is a standard of measurement to compute the dimension. A wide used method for choosing appropriate metrics has been suggested in [14], called the Goal-Question-Metric approach. The idea of this approach allows a user to ask a Question, over his data and related to his own quality goal, the answer of which provides the corresponding metric. This approach is used in the quality meta-model of the ESPRIT project [15] and quality assessment meta-model of the QUADRIS project [7].

In the Big Data context, the majority of possible data quality dimensions coincide with the traditional DQ dimensions, such as *consistency* [1, 8, 9, 16–18], *uniqueness* [6, 8, 10, 18], *accuracy* [1, 8, 16–19], *completeness* [1, 16–19], *timeliness* [3, 5, 6, 11–13]. Still, literature provide discussions of Big Data quality dimensions and metrics [1, 3, 16–21]. When dealing with different information sources, complementary values related to a same entity (for instance, the name of a person is present in one source, and the surname in another one) should also be considered. Thus, considering *uniqueness*, not only purging duplicate values but also merging complementary values is essential [8]. Also, *synchronization* is important to obtain a consistent data [1]. *Interpretability* [6, 9, 12, 18] delivers the notion of extracting a good *Value* from Big Data. *Data trustworthiness* is one of the major attributes of data Veracity, which is one of the 7V's of Big Data related to quality [2]. It is defined by a number of factors including data origin, collection and processing methods, including trusted infrastructure and facility [22]. Reputation and credibility of data source is considered a highly regarded level of trustable data [1, 9, 16, 18, 19].

The general notion of quality in Big Data has also been widely used in the literature. In the context of Big Data, quality is not limited to data: quality dimensions are discussed also for system quality [23, 24] or quality for analysis platform [25]. Moreover quality models for Big Data initiatives are presented in the literature to handle quality issues and deal with the quality evaluation, assessment, and management. A discovery model of quality constraints for lake data has been suggested in [26] and a quality-in-use models (3As and 3Cs) have been suggested in [19, 20]. The latest papers describe quality assessment in big data projects based on ISO/IEC standards. However, it considers the quality-in-use approach which assess the quality of data for Big Data projects providing the appropriate data quality for Big Data analysis. Nevertheless, [16] gives a good literature review about the notion of quality in Big Data context. However, these works consider data preprocessing as quality provider where the analysis are done on already processed data. In other words, in the literature, analyzing data quality only implies on pre-processing of Big Data analytics [16]. Also, the quality models reflected in the literature mainly consider data quality on a single source. Another flaw is that literature does not take into account user preferences sufficiently.

We propose a more complete model (compared to the literature), which is applicable on multiple data sources of even different types. In addition, we believe there are organizations that cannot afford (in terms of time and resources) [3] data quality corrections before the analysis process and consider all of the existing quality problems in their data. These problems are described by the level of quality dimensions e.g. how consistent data is. Thus, we ambition to alert about data quality problems during the analysis stage without any preprocessing. Moreover, the alert system takes into account user preferences. It should directly give an opportunity to a user interact with it and the quality model via an interface.

3   The Quality Model

In the following figure (Fig. 1), we present a roadmap of the alert system describing the interactions of the alert system with the users and the Big Data sources. The user can either directly query the data through the alert system or interact with the quality model through a user interface to choose the quality questions. This section is dedicated to the description of the quality model.

3.1   Conceptual Model

The quality model of our alert system represents the quality assessment of a data source in the context of Big Data Analytics. It is based on quality dimensions and metrics described in Sect. 2. The user is able to use this model through quality questions related to the analysis he/she performs. The main concepts of the quality model and the interactions between them are illustrated in Fig. 2. The assessment of a metric on a data source and/or attribute is done thanks to a predefined measurement method. A central concept of this model is the consideration of "Quality Question". This concept is seen as a "negotiator" between the user and the quality assessment of the model.

Fig. 1. The roadmap of the alert system.   **Fig. 2.** Main entities of quality model.

The conceptual model is represented in Fig. 3. In order to be independent of any implementation considerations, we model our solution with a class diagram. The description of each class is detailed below:

Quality Question: This class expresses the user requirements about the quality assessment. The user chooses the dimensions and/or the metrics he wants to be alerted about from a predefined list (this list can be extended later due to the Neo4J functionalities such as adding nodes and relationships easily), according to his/her analysis. This class refers to data sources and attributes. If the user is unable to express particular dimensions or metrics, the model considers all the dimensions and metrics by default. The latest option is not preferable if the user is sophisticated enough in the quality domain.

Data Source and Attribute: These two classes describe the Big Data substance on which the user performs analysis. The user can perform an analysis on one or several attributes in order to have a specific quality assessment or consider a global analysis by analyzing the whole data source. Here, different types of data sources and attribute formats are supported by the model. Our quality model should be able to operate on the data sources regardless of their type, meaning all types of data sources need to be supported by the model.

Source Set: When querying multiple data sources and/or attributes, the user queries a set of them. Here, data structure should be considered. In a single source set the data structure need to be uniform.

User: This class indicates who performs the analysis and who, if able, chooses quality questions on data source, attribute, and/or source set. This class is essential as it defines user rights in terms of "role". In other words, this class can personalize the data quality information, of a same data sources, to different users. The "role" property considers two types of users: (1) a user not sophisticated enough in the domain of quality (for instance, a decision-maker or analyst): this user does not choose quality questions but the system chooses instead by default, (2) a user sophisticated enough to choose quality questions, and set quality limit as a threshold.

Quality Dimension: This class considers the dimensions listed in Sect. 3. The dimension is considered as a view of quality assessment for the user. When checking the quality, the user is able to select one or several quality dimensions in his Quality question. The relationships between dimensions (such as improving completeness can have negative impact on uniqueness.) should also be reflected by the model, for suggesting quality improvements by the system. It has been already noted, that discovering and analyzing relationships of quality dimensions plays a significant role when rationing an analysis process [27], and even a dependency discovery model has been suggested in [28]. Particularly, negative (inverse) relationships are important to be considered in our model. These relationships will help user to make a decision about the improvement of a particular dimension as he/she will also be informed about possible deviations on other dimensions in case of modification. The direct relationships should not be considered as there is nothing to alert about. It is even better that one more dimension is going to be improved in case of the improvement of the detected dimension.

Quality Metric: This class refines the Quality question and is a way of quality computation addressing the Quality Dimension. For instance, "NullValues" is a metric of the dimension "completeness" [6]. In case where this metric is specified in the Quality question, the model should check the level of "null values", over the data, to alert about completeness. If no metric is specified in the Quality question (only dimensions are expressed), the model considers all the metrics of that dimensions.

Measurement Method: This class defines a quality formula which measures the problem of quality for a specific metric. The measurement method represents the implementation of the metric computation (a formula, an algorithm) on data source, attribute, or source set. In the literature there are several algorithms already developed for quality dimension evaluation such as [17], which can be a base for this class to compute the level of a quality metric. For instance, to compute the dimension of "completeness", the model needs to compute the metric "Number of NullValues" using the "CheckNULL" function. Thus, the formula "[(1-Number of not null values)/total number of values]" is calculated to alert about the "completeness" dimension. The complete set of measurement methods (such as CheckNULL, CheckRule, CheckReferential, Aggregation, LookUp, Count, Ratio, Max, Min, etc. [7] need to be implemented.

Quality Limit: This class considers a limit value as a threshold for which the system alerts the user. This value is specific for each measurement method and can be entered by the user. Of course, this value must depend of the measurement domain. For instance, if the domain of values for the measurement "null values" is between [0,1] (0-worst case, 1-best case), the user could specify a value of 0.9, and the system alerts only if the result is out of this limit. If the user does not select a limit, the model considers the limit as the lowest value of the domain for the concerned quality problem.

3.2   Quality Model Example

We illustrate the quality model of the previous section, considering the classical example of Sales, when a user wants to analyze product sales. Let us consider the object diagram of Fig. 4. Without going deep into the analysis process, we consider the scenario when the user is sophisticated enough to handle the quality model by choosing particular quality questions. In the suggested example, a knowledge-worker Arsen (U1: User) analyzing sales of a product. The data are stored in a local server via a HDFS (Hadoop Distributed File System). He performs the analysis on the attribute "productSale" (A1:Attribute). Arsen prefers to check over the completeness and consistency dimensions to be sure that the attribute he is analyzing has no quality problem. Besides, Arsen understands that if there will be duplicates on the data, this number may be exaggerated. That is why he also decides to check over the uniqueness of the attribute "productID" (A2:Attribute). Thus, during his analysis, Arsen chooses the quality questions (Q1) and (Q2). Please note, in the object diagram, everything concerning the completeness part is colored pink, consistency – yellow, and uniqueness - green:

– (Q1) considers the completeness and consistency dimensions. Because Arsen didn't specified metrics for these dimensions, the model considers all the metrics for each of these dimensions (of course the full list of dimensions and metrics is stored in the model, beforehand). Thus, the corresponding metrics for completeness are (Com1), (Com2), and for consistency (Con1). For each quality metric two measurement methods are predefined. The method (MCom1) only counts the number of null values and returns it as a result, whereas the other method (MCom2) checks the number of "null values" and computes a ratio by giving the proportion of "not null values". Thus, when alerting the alert system presents both results to Arsen. Then, he knows that there are 500 null values in the attribute "productSale" and the proportion of "not null values" is 0.5. Now, let us discuss (Com2). In this case the metric considers the whole data source and not a specific attribute. Arsen selects quality limit [0.75;1]. This means that Arsen does not mind to analyze the sales even if the amount of data is 75% of the needed data amount (e.g. usually X amount of data is needed but this time 0.75*X is enough). That is why, the system will not alert, though the measurement method (MCom3) returns a result of 0.82.

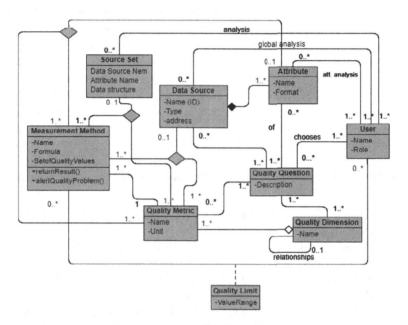

Fig. 3. Conceptual model of the quality model.

- (Q2) Arsen mentions that only the duplicates need to be checked. This prevents the model to check over other quality metrics of the dimension "uniqueness" that are not specified. And thus the system will only alert about the number of duplicate values. Also, when suggesting an improvement of completeness, the system will consider the negative relationship (see Sect. 3.1) between the completeness and uniqueness and will notify the user about it. This alert message should also contain an information about the level by which the uniqueness will suffer if the completeness is improved.

3.3 Physical Representation of the Quality Model

We intend to implement the quality model through a NoSQL database. Due to the fact that different concepts of the model are interrelated, we choose a graph database. For example, a same metric may be shared by several measurement methods and may be

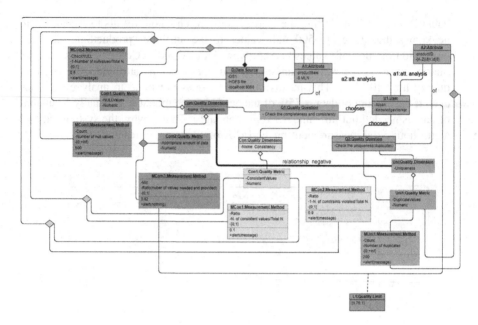

Fig. 4. The object diagram of the example when user analyzes sales. (Color figure online)

applied on several data sources. Thus, these concepts can easily be queried using a graph query language.

Various definitions of a graph exist in the literature. By combining the common definition of a graph presented in [29] and the basics of Neo4J the graph can be defined as follows: the graph G is an ordered pair of (N(G), E(G)) consisting of a nonempty set N(G) of nodes and a set E(G) of edges, which represents the relationship between an unordered pair of (not necessary distinct) nodes of G. Thus a graph database considers as equally important the relationships as the data itself.

The delightful part of the graph database is also the possibility to extend the existing model, by adding new nodes (e.g. set of quality questions) and/or relationships continuously, without changing the complexity of the queries. While classical relational databases compute relationships at query time through expensive JOIN operations, a graph database stores connections alongside the data in the model. This is a good way for our quality model to be enhanced and supplemented over the time. There will be a possibility to add new dimensions, metrics, relationships and data source types thanks to our model. Also, retrieving nodes and relationships in a graph database is an efficient, constant-time operation, which traverse numerous of connections per second per core.

We use Neo4J community version (see Fig. 5) as the graph database for our model, using the query language - Cypher. The main concepts of Neo4J are labels, nodes, properties and relationships. In order to implement our conceptual model, translation rules are required. Methods for translating conceptual schema of data model exist in the literature [30, 31]. The translation of our class diagram in Neo4J is done by considering the mapping from [31]. From the global point of view, the classes of class diagram are "label" node in Neo4J, the attributes are "properties", and the associations are "relationships".

Considering the translation the data is imported into Neo4J database. An extract of the database may be seen in Fig. 5. This schema presents the labels of the quality model (not the values). All the labels from Fig. 5 can be linked with the classes of Fig. 3. We present a similar case to the example from Sect. 4.1 as a graph (see Fig. 6). In this case a user Arsen decides to check only over completeness and consistency dimensions of a data source without mentioning any metric in the quality question. This time Arsen wants, to be sure that the data source has no problem with these two quality dimensions as they are the most commonly discussed quality dimensions in the literature.

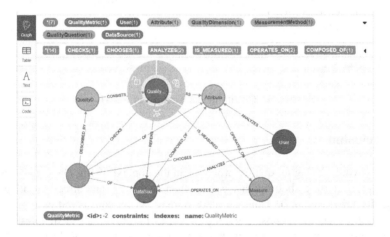

Fig. 5. The quality model implemented into a database schema in Neo4J

The Fig. 6 shows the result of the following Cypher query in Neo4J:

```
MATCH (n:QualityQuestion)<-[:CHOOSES]-(u:User)-[:ANALYZES]-
>(d:DataSource)WHERE n.description=~'Check over Dim. Complete-
ness.*' OR n.description=~'Check over Dim. Cosistency.*'OPTIONAL
MATCH (n)-[:DESCRIBED_BY]->(q:QualityDimension)-[:CONSISTS_OF]-
>(m:QualityMetric)OPTIONAL MATCH (m)-[:IS_MEASURED]-
>(mm:MeasurementMethod)RETURN n, u, d,q,m,mm
```

Because there is no information about the metrics in quality question, all the metrics of the chosen dimensions are considered. Thus, as completeness has three metrics and consistency has one metric, the model considers four quality questions for each metric (despite Arsen chooses only one quality question, the model needs to consider all of the possible quality questions containing the metrics). For each metric the appropriate measurement methods are identified. At the top of the Fig. 6, all the nodes "label" can be seen (at the bottom, the properties of the selected node are visualized).

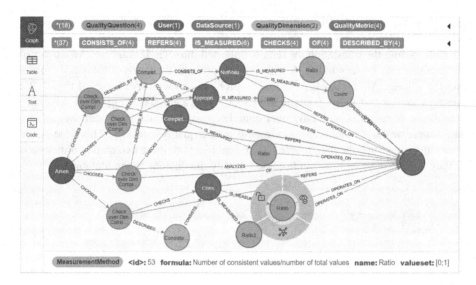

Fig. 6. Quality model example in Neo4J

4   Conclusion

In order to avoid solving all quality problems in the context of Big Data analytics, this paper suggests alerting about quality issues when analyzing data directly (without any data quality preprocessing). Thus, an alert system can be considered as a good candidate. To this end, we propose a quality model as the main part of this system. The model allows users to control the quality of their data during the analysis process and be informed about the problems on it. Thanks to the quality metrics and their measured values, the system can be able to alert about poor data quality, according to user requirements. This solution relieves the load of data correction before the analysis and consider them during the analysis. Then, it is up to a user to decide whether the problems need to be improved or not. A Neo4J implementation of quality model is presented with a query example.

The future work includes the definition of new quality dimensions dedicated to the data analysis step [32]. In particular, these dimensions should be able to alert a user about a trustful/untruthful analysis. The integration of relationships between the quality dimensions should enhance the capabilities of it. The system should also alert about the quality of analysis based on the alerted data quality by considering the relationships between the dimensions (data quality dimensions and analysis quality dimensions). Then, the next step will be to design and implement the complete alert system. A long term perspective is also to support and solve automatically the quality problems that are detected (under the supervision of the user, of course).

References

1. Cai, L., Zhu, Y.: The challenges of data quality and data quality assessment in the big data era. Data Sci. J. **14**, 2 (2015)
2. Khan, M.A., Uddin, M.F., Gupta, N.: Seven V's of big data understanding big data to extract value. In: Proceedings of the 2014 Zone 1 Conference of the American Society for Engineering Education, pp. 1–5. IEEE, Bridgeport (2014)
3. Saha, B., Srivastava, D.: Data quality: the other face of big data. In: 2014 IEEE 30th International Conference on Data Engineering, pp. 1294–1297 (2014)
4. Erl, T., Khattak, W., Buhler, P.: Big data analytics lifecycle. In: Big Data Fundamentals, pp. 65–87. Arcitura Education Inc. (2016)
5. Lee, Y.W., Pipino, L.L., Funk, J.D., Wang, R.Y.: Journey to Data Quality. The MIT Press, Cambridge (2006)
6. Batini, C., Cappiello, C., Francalanci, C., Maurino, A.: Methodologies for Data Quality Assessment and Improvement (2009)
7. Berti-Équille, L., et al.: Assessment and analysis of information quality: a multidimensional model and case studies. Int. J. Inf. Qual. **2**, 300–323 (2011)
8. Rahm, E., Do, H.H.: Data cleaning: problems and current approaches. IEEE Data Eng Bull. **23**, 3–13 (2000)
9. Wang, R.Y., Strong, D.M.: Beyond accuracy: what data quality means to data consumers. J Manag. Inf. Syst. **12**, 5–33 (1996)
10. Akoka, J., Berti-Equille, L., Boucelma, O., Bouzeghoub, M., Wattiau, I., Cosquer, M.: A framework for quality evaluation in data integration systems (2007)
11. Pipino, L.L., Lee, Y.W., Wang, R.Y.: Data quality assessment. Commun. ACM **45**, 211–218 (2002)
12. Hazen, B.T., Boone, C.A., Ezell, J.D., Jones-Farmer, L.A.: Data quality for data science, predictive analytics, and big data in supply chain management: an introduction to the problem and suggestions for research and applications. Int. J. Prod. Econ. **154**, 72–80 (2014)
13. Ballou, D.P., Tayi, G.K.: Enhancing data quality in data warehouse environments. Commun. ACM **42**, 73–78 (1999). https://doi.org/10.1145/291469.291471
14. Oivo, M., Basili, V.R.: Representing software engineering models: the TAME goal oriented approach. IEEE Trans. Softw. Eng. **18**, 886–898 (1992)
15. Jeusfeld, M.A., Quix, C., Jarke, M.: Design and analysis of quality information for data warehouses. In: Ling, T.-W., Ram, S., Li Lee, M. (eds.) ER 1998. LNCS, vol. 1507, pp. 349–362. Springer, Heidelberg (1998). https://doi.org/10.1007/978-3-540-49524-6_28
16. Taleb, I., Serhani, M.A., Dssouli, R.: Big data quality: a survey. In: 2018 IEEE International Congress on Big Data (BigData Congress), pp. 166–173 (2018)
17. Taleb, I., Kassabi, H.T.E., Serhani, M.A., Dssouli, R., Bouhaddioui, C.: Big data quality: a quality dimensions evaluation. In: 2016 International IEEE Conferences on Ubiquitous Intelligence Computing, Advanced and Trusted Computing, Scalable Computing and Communications, Cloud and Big Data Computing, Internet of People, and Smart World Congress (UIC/ATC/ScalCom/CBDCom/IoP/SmartWorld), pp. 759–765 (2016)
18. Arolfo, F., Vaisman, A.: Data quality in a big data context. In: Benczúr, A., Thalheim, B., Horváth, T. (eds.) ADBIS 2018. LNCS, vol. 11019, pp. 159–172. Springer, Cham (2018). https://doi.org/10.1007/978-3-319-98398-1_11
19. Caballero, I., Serrano, M., Piattini, M.: A data quality in use model for big data. In: Indulska, M., Purao, S. (eds.) ER 2014. LNCS, vol. 8823, pp. 65–74. Springer, Cham (2014). https://doi.org/10.1007/978-3-319-12256-4_7

20. Merino, J., Caballero, I., Rivas, B., Serrano, M., Piattini, M.: A data quality in use model for big data. Future Gener. Comput. Syst. **63**, 123–130 (2016)
21. Sidi, F., Panahy, P.H.S., Affendey, L.S., Jabar, M.A., Ibrahim, H., Mustapha, A.: Data quality: a survey of data quality dimensions. In: 2012 International Conference on Information Retrieval Knowledge Management, pp. 300–304 (2012)
22. Demchenko, Y., Grosso, P., de Laat, C., Membrey, P.: Addressing big data issues in scientific data infrastructure. In: 2013 International Conference on Collaboration Technologies and Systems (CTS), pp. 48–55 (2013)
23. Choi, S.-J., Park, J.-W., Kim, J.-B., Choi, J.-H.: A quality evaluation model for distributed processing systems of big data. J. Digit. Contents Soc. **15**, 533–545 (2014)
24. Canalejo, O., Isabel, M.: A Quality Model for Big Data Database Management Systems (2018)
25. Lee, J.Y.: ISO/IEC 9126 quality model-based assessment criteria for measuring the quality of big data analysis platform. J. KIISE **42**, 459–467 (2015)
26. Farid, M., Roatis, A., Ilyas, I.F., Hoffmann, H.-F., Chu, X.: CLAMS: bringing quality to data lakes. In: Proceedings of the 2016 International Conference on Management of Data - SIGMOD 2016, pp. 2089–2092. ACM Press, San Francisco (2016)
27. Berner, E.S., Kasiraman, R.K., Yu, F., Ray, M.N., Houston, T.K.: Data quality in the outpatient setting: impact on clinical decision support systems. In: Proceedings of AMIA Annual Symposium, pp. 41–45 (2005)
28. Barone, D., Stella, F., Batini, C.: Dependency discovery in data quality. In: Pernici, B. (ed.) CAiSE 2010. LNCS, vol. 6051, pp. 53–67. Springer, Heidelberg (2010). https://doi.org/10.1007/978-3-642-13094-6_6
29. Bondy, J.A., Murty, U.S.R.: Graph theory with applications (1976)
30. Daniel, G., Sunyé, G., Cabot, J.: UMLtoGraphDB: mapping conceptual schemas to graph databases. In: Comyn-Wattiau, I., Tanaka, K., Song, I.-Y., Yamamoto, S., Saeki, M. (eds.) ER 2016. LNCS, vol. 9974, pp. 430–444. Springer, Cham (2016). https://doi.org/10.1007/978-3-319-46397-1_33
31. Delfosse, V., Billen, R., Leclercq, P.: UML as a schema candidate for graph databases. In: NoSQL Matters 2012, pp. 1–8 (2012)
32. Djedaini, M., Furtado, P., Labroche, N., Marcel, P., Peralta, V.: Benchmarking exploratory OLAP. In: Nambiar, R., Poess, M. (eds.) TPCTC 2016. LNCS, vol. 10080, pp. 61–77. Springer, Cham (2017). https://doi.org/10.1007/978-3-319-54334-5_5

Framework for Assessing the Smartness Maturity Level of Villages

Jorge Martinez-Gil[1]([✉]), Mario Pichler[1], Tina Beranič[2], Lucija Brezočnik[2],
Muhamed Turkanović[2], Gianluca Lentini[3], Francesca Polettini[3],
Alessandro Lué[3], Alberto Colorni Vitale[3], Guillaume Doukhan[4],
and Claire Belet[4]

[1] Software Competence Center Hagenberg GmbH,
Softwarepark 21, 4232 Hagenberg, Austria
jorge.martinez-gil@scch.at
[2] Faculty of Electrical Engineering and Computer Science,
University of Maribor, Koroška cesta 46, 2000 Maribor, Slovenia
[3] Poliedra-Politecnico di Milano, via G. Colombo 40, 20133 Milano, Italy
[4] ADRETS, 69 rue Carnot, 05000 Gap, France

Abstract. In this work, we have developed the first version of a smartness assessment framework that allows the representatives from a village to make a self-evaluation of its current status based on smartness criteria identified by an international group of experts. The framework allows a detailed evaluation of six different aspects including Mobility, Governance, Economy, Environment, Living, People, with weightings of the criteria using the multi-criteria analysis Electre Tri. In addition, the results enable further data analysis and offers an input for different functionality like identification of best practices and collaboration and matchmaking among potential stakeholders. In addition, we show the effectiveness of the proposed framework by means of a case study on a test area around the European Alpine space.

Keywords: Rural development · Smart villages ·
User-centered data analytics

1 Introduction

The digital revolution and the possibilities offered by the new technologies have radically transformed the way we live in the last decades. As a result, for the first time in many years there is a hope to overcoming very negative trends such as the rural depopulation. In fact, there is an increasing amount of young people that prefer to leave the comfort of large conurbations and exploit their knowledge in the villages by creating innovative development models. In this way, the new generations are increasingly turning places that seemed doomed to be abandoned into nodes of attraction of great dynamism and employment. This could put an end to the high rates of unemployment, the aging of the population, talent drain, and the loss of public services.

© Springer Nature Switzerland AG 2019
T. Welzer et al. (Eds.): ADBIS 2019, CCIS 1064, pp. 501–512, 2019.
https://doi.org/10.1007/978-3-030-30278-8_48

Contrary to popular opinion, digital advances are not only reserved for large megalopolis and conurbations; there are also small villages that bet on disruptive technologies to improve the lives of their inhabitants, boost the local economy and promote themselves as a tourist destination [12]. One of the ways to implement these technologies is through the concept of Smart Villages that, although with specific nuances of each place, is widely spread in the five continents. In fact, areas which do not have many inhabitants have now the opportunity to embrace intelligent modernization and information technology based on Artificial Intelligence, Big Data, Blockchain, Internet-of-Things, Energy Informatics, Digital Health, Collaborative online tools, Open Source, Civic Tech, and so on, with the goal of turning the old village into a "smart" village [4].

In addition, the smart villages' paradigm aims to transform traditionally rural sectors such as agriculture, livestock, fishing, mining, etc. by applying novel methods for intelligent data management and robotics. In this way, new concepts linked to sustainability and competitiveness gain a lot of importance and give rise to new forms of development such as agriculture 4.0, agribusiness, or other kinds of business models.

In this work, we present our framework for the automatic assessment of the smartness maturity level in villages. This framework mainly consists of a software system that assists to analyze the smart functionality that a given village implements at a given time. The framework aims at providing a smartness score, a.k.a. maturity level score, based on the users' perceptions of how this smartness fulfills their expectations. At the same time, new opportunities for improvement or development of new concepts can be found by looking for gaps with the help of the smartness reports provided within the system. As a side effect, the tool can be used for the identification and sharing of best practices as well as for stakeholder collaboration and matchmaking. The implemented framework is offered to the general public as a part of the Digital Platform developed within the SmartVillages project [11].

The rest of this work is structured in the following way. Section 2 describes the State-of-the-Art in relation to smartness assessment frameworks. Section 3 describes the design, implementation, and exploitation of our framework. In Sect. 4, we show preliminary results obtained from the exploitation of the framework in the context of an Italian municipality belonging to the European Alpine space. Finally, we present the major conclusions and future lines of research.

2   State-of-the-Art

New trends on technologies and smart infrastructure such as connectivity can transform the old notion of village whereby the unequal property distribution and lack of opportunities were assumed. In this context, the development of novel initiatives that address territorial development and innovation is of vital importance. For example, in recent years, much has been done towards the goal of creating smart spaces for living [9]. However, there are still regions where proper infrastructure and services are lacking, not so much the basic Internet, but other necessary aspects such as broadband and 4G and 5G connectivity are still very deficient [8]; so it is clear that there are still blank areas that need further development.

Moreover, this a common problem in all the continents of the world, although perhaps it is aggravated in low-income countries where the lack of monetary resources for infrastructures does not facilitate the reduction of the gap between cities and villages. However, not everything is dependent on the available budget since the situation is not very different in places with larger public investments such as the European Union. In fact, according to figures provided by the European Commission, there are large differences in smart facilities between urban and rural areas across all the continent. Therefore, developing the concept of smart villages by providing new business models and supporting infrastructures to the rural world are some of the challenges that public and private organizations have on the table to make this notion a reality.

It is necessary to remark that there are many worldwide initiatives in this context. For example, the IEEE Smart Village program has been calling for new technologies based on smart village thinking to bridge the urban-rural breach [2]. With a view to reducing the digital divide between rural and urban areas and promoting the rural economy, the aforementioned European Commission has also given priority to the development of Smart Villages within its community agricultural policies, as well as in other plans related to research programs including: the European Rural Parliament, the SIMRA project and the ERUDITE project or some specific calls from the Horizon 2020 program. Moreover, Policy on Digital India has envisaged national level focus in rendering services to citizens in India. This focus plans for convergence of all possible services through a digital backbone [5]. Also, the South African government is constantly developing new ICT projects which are initiated by individuals, government and private organizations within the context of the ongoing SEIDET Digital Village [7].

In the literature, there are some works on smart cities, for example [1] and [10]. However, there is a lack of field-oriented systematic tools to guide and monitor the evolutionary process of the villages to higher smartness maturity levels. In fact, at present, this process is so unstructured that most local authorities do not have a starting point and guidelines that support them in making adequate progress in terms of smartness maturity. Our work aims to overcome that limitation.

3 Framework for Smartness Maturity Level

A maturity model is defined as a set of practices which is considered as a development path or an improvement tool for public or private organizations. The maturity level indicates in which exact part of that path we are at a given moment. However, it is important to note that the question of how smart a village is, does not only have to do with the degree of advanced facilities that have been deployed but also on how people perceive it. In order to develop this notion, we have created a framework to audit the current village status and guide local authorities to higher levels of smartness.

3.1 Introduction to the Framework

To shed light on this context, we have developed a framework for the assessment of the smartness maturity level. What we present here is an intermediate version, while in the future, new functionality will be added. This framework works around an advanced questionnaire that allows obtaining information directly from the people involved in the village, together with other complementary information that can be compiled from open sources available on the web. The general architecture of the smartness assessment framework is presented in Fig. 1.

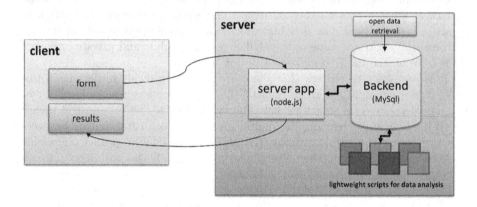

Fig. 1. General architecture of the smartness assessment framework

The framework is integrated into the Digital Platform developed within the SmartVillages project from where it can be accessed through a regular web browser. An online questionnaire allows users to fill in all the information concerning the smartness indicators. The questionnaire consists of 24 questions with multiple choice answers grouped into six sets representing six smartness domains: Smart People, Smart Governance, Smart Living, Smart Environment, Smart Economy, and Smart Mobility. In addition, text fields are provided to allow users providing comments inputted in natural language that it deems appropriate. When the online questionnaire is complete, the data, e.g. the village name, the answers to the questions, and the comments provided are sent to the server and analyzed according to the developed methodology. The results are then calculated and plotted on the screen so that the user can view and analyze them. All the data generated during the process are stored in the server in order to proceed with further analysis. In addition, every entry is appropriately timestamped in order to enable functionality that allow monitoring the evolution of the test areas along the time.

Figure 2 shows us the questionnaire which is used to collect data from stakeholders. Please note that, as requested by the stakeholders involved in the design phase, each question admits a double answer. On the one hand, a short and concise answer that our system transforms into a numerical value for later automatic

Fig. 2. Screenshot from the online questionnaire which is used to collect data from stakeholders

analysis, and an answer based on natural language, which will be analyzed by expert personnel. It is also worth mentioning that no personal data is stored and that all information complies with current EU regulations for the processing of data provided by users.

3.2 Smartness Dimensions

The smartness dimensions identified by an international group of experts are those related to Mobility, Governance, Economy, Environment, Living, and People. Below, we can see them in more detail.

Smart Mobility. Smart People is related to the quantity and quality of sustainable transport and mobility systems in the village. Examples of indicators include the number of non-conventional-fuel cars being owned or used, the presence of limited-traffic zones, the level and sustainability of public transport, etc.

Smart Governance. Smart Governance is related to the level of smartness of the governance systems, the penetration of green public procurement, e-governance, facilities to networking. Some examples of indicators include the number of electric cars used, the convenience of recycling policies, energy policies, etc.

Smart Economy. Smart Economy is measured in terms of the presence of creative and innovative enterprises and business models in the area, level of employment and unemployment, level of economic attractiveness, penetration of ICT in the local economic system. Examples of indicators include the number and density of certified enterprises, number of young and women-led enterprises, the rate of business creation, the number of patents, etc.

Smart Environment. Smart Environment involves measuring the quality of the environment in terms of air, water, and soil. Examples of indicators include the air quality, level of recycling, percentage of natural spaces in the overall area, etc.

Smart Living. Smart People is related to the quantity and quality of services to the population in the area, and the degree of satisfaction in them. Examples of indicators include the level of criminality, the level of general services such as banks, post offices, and so on, the quality health care and social care services, as well as the quality and quantity of services to the elderly, etc.

Smart People. Smart People measures the participation of local citizens to the job market, the decision-making and the involvement in associations, and the education level of people. Examples of indicators include the number of associations, policies for promoting equal opportunities, level of schooling, overall employment, degree of political engagement, etc.

3.3   Score Calculation

The calculation of scores represents the first step in the multi-criteria analysis Electre Tri (Elimination and Choice Expressing Reality) methodology that has been implemented to assess and rate the level of smartness. 24 core indicators of smartness, 4 for each of the 6 smart dimensions (Mobility, Governance, Economy, Environment, Living, and People) have been selected and presented in the form of questions in a dedicated survey. For each of the questions, 4 answers are possible, from the most negative one (scored 1) to the most positive one (scored 4). A further round of scoring allows to determine the degree of certainty with which each of the answers are given, on a scale from 1 (not very certain) to 3 (very certain): this second round of scoring allows for integrating the subjective assessment of smartness for a given compiler which is fundamental in the Electre Tri self-assessment process.

A further step to be integrated enables the creation of a system of weights capturing the perceived importance of each smart dimensions with respect to all others. This is done by creating a comparison matrix between the 6 dimensions (it can also be done at indicator level, although this would be more time and machine-consuming) in which the compiler indicates how important one dimension is with respect to all others. This creates a 6×6 matrix, the eigenvalues of which result in the assignment of multiplying factors (weights) for the previously-calculated scores.

The weighting factors are useful for the compilers to ascertain which dimensions are more critical for their own assessment and for their present and future smart transformation. The final step of the Electre Tri assessment and rating methodology entails the creation of profiles and categories: it was decided that a categorization along four categories (maturity levels) and three profiles were reasonable, in order to capture the variability between low scoring areas and very high-scoring areas; the profiles have been created with the Electre-Tri outclassing rationale, with the concept of λ-cutting levels, outclassing, winning combinations and vetoes. The four maturity levels are described in the following.

3.4   Maturity Levels

From the score calculation, we can establish four maturity levels:

Level A: High Level of Smartness or New Goals Level. It is the level whereby the majority of the dimensions registers the highest level of smartness considered reachable in this model. In other words, most indicators score between 10 and 12. It is important to highlight that these are the targets used in this survey and so, placing in this category is only a milestone in a possibly more complex process.

Level B: Good Level of Smartness or Satisfactory Level. It is the level that indicates that there are numerous activities and initiatives that focus on innovation, the services are adequate and innovative approaches are used in a lot of sectors. However, not the majority of facilities score the highest level of smartness in their services.

Level C: Medium Level of Smartness or Work in Progress Level. This level means that there are some smart services planned and people are aware of the importance of smart transformation.

Level D: Low Level of Smartness or Traditional Concept Level. This level indicates that there are very few initiatives that focus on innovation and very little is planned to improve this situation.

4   Data Analysis

One of the key modules of our framework is devoted to data analysis. The goal is to discover useful hidden information, that can derive conclusions for supporting the decision-making of the stakeholders who make use of our framework, e.g. for identification of similarities among villages, clustering of villages, allowing analysis even when data was not entered completely or by providing forecasts of smartness maturity level development.

4.1   Similarity Between Villages

Since we have several sources of information, we can have a very rich set of features that unequivocally identify a village. The appropriate processing of these features allows us to establish similarities between villages according to different criteria, based on the application of various statistical measures of similarity, distance, and correlation.

4.2 Clustering of Villages

The purpose of village clustering is to group set of villages in such a way that villages in the same cluster are more similar (according to some predefined criteria) to each other than to those in other clusters. This functionality is really useful in order to provide informed facts specifically targeted on clusters that meet some requirements.

4.3 Working with Missing Data

Many questions are difficult to answer, either because they are not easy to understand, or because the user filling the questionnaire does not have that information, or maybe they are not applicable in that context. For cases such as these, our data analysis module is able to dive into the historical record, combine these data with data retrieved by online sources such as DBpedia[1] or Wikidata[2], e.g. number of inhabitants, geographical coordinates, etc., and identify similar situations in order to predict the most likely response to a given question.

4.4 Development Forecast

In addition, the prediction functionality is also able to guess how the smartness maturity level of a given village will evolve along the time. To do that, we will use the historical record of village evolution that will help in the task of automatic learning.

4.5 Visualization

The results of statistical processes are often difficult to understand. For this reason, our framework implements a module for the adequate visualization of most outputs and reports. The idea is to offer the capability to the visual inspection of results by means of charts, plots, maps or any other means that may facilitate its understanding and/or dissemination.

4.6 Querying

In addition, the platform offers the capability to formulate complex queries. Traditional systems are based on the manual compilation of information coming from different sources. In our case, the semi-automatic integration of open data sources such as DBpedia or Wikidata along with the information entered by users through self-evaluation allows us to answer very complex questions that could not be handled otherwise.

---

[1] http://dbpedia.org.
[2] http://www.wikidata.org.

4.7 Best Practices

Best practices [6] are a set of actions that have performed very well in a given village and that are expected to perform similarly in similar villages. Users are encouraged to document their best practices through using a predefined template to gather the information, that are going to be available for other users that score high on the similarity index. Our framework is able also to support the identification and sharing of best practices with other platforms with similar interest such as CESBA[3].

4.8 Matchmaking

The matchmaking process aims to identify stakeholders who, due to a similar degree of smartness, might be willing to collaborate or exchange experiences [3]. In addition, this process is not limited only to villages, but may also be able to bring together companies, or even regional organizations.

The first version of the matchmaking functionality is using different algorithms to match queries with companies descriptions. The idea is to use open datasets of European Companies such as the ones provided by the European Business Register[4], etc.

4.9 Multi-language Support

Since the framework is developed in a multinational context around the European Alpine region (Austria, France, Germany, Italy, Slovenia, Switzerland), it is necessary to have a multilingual version of the framework. Therefore, we have worked to offer the framework in several languages and will continue to add languages as more stakeholders join the community.

5 Results

As an example of a completed self-assessment procedure, we include an use case on the Test Area of Tolmezzo (in the Friuli-Venezia Giulia region, North-Eastern Italy). This municipality has self-assessed itself as Level B- good level of smartness or satisfactory level. The overall rating of Tolmezzo is included in Fig. 3.

The Level B rating has been mostly due to the winning combination of having high scores in the dimensions of smart governance, smart living and smart people (see Fig. 3, and categories in each dimension in Table 1).

---

[3] http://wiki.cesba.eu/wiki/Greta_Best_practices.

[4] https://www.ebr.org/.

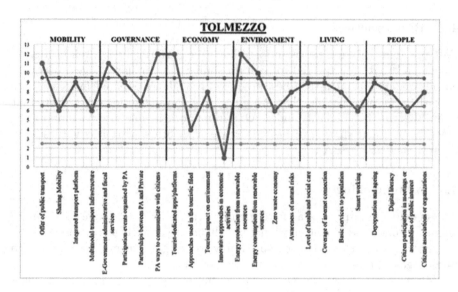

Fig. 3. Tolmezzo scores following the first two steps of the Electre tri procedure. The three horizontal profiles divide Level A (at the top, above 9.5), Level B (between 6.5 and 9.5), Level C (2.5 to 6.5), and Level D (below 2.5)

Table 1. Tolmezzo comparison matrix, step three of the Electre Tri procedure. Dimensions scored on a 1 to 6 scale, with 1 = equally important to 6 (or 1/6), 6 times more (or 6 times less) important.

Tolmezzo	Mobility	Governance	Economy	Environment	Living	People
Mobility	1	1/4	1/5	1/4	1/5	1/2
Governance	4	1	1	2	2	5
Economy	5	1	1	2	3	4
Environment	4	1/2	1/2	1	1	2
Living	5	1/2	1	1/3	1	3
People	2	1/5	1/4	1/2	1/3	1

It has however to be noted that, according to the Tolmezzo comparison matrix, the most crucially-assessed dimension has been that of smart economy (see Fig. 4): the area can therefore be inspired to further smart transition in the smart dimension locally defined as more crucial.

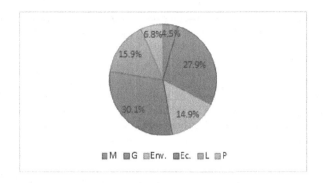

Fig. 4. Tolmezzo weights, or percentage values of the priority vector. M = Mobility, G = Governance, Env. = Environment, Ec. = Economy, L = Living, P = People

6   Conclusions and Future Work

We have presented our research towards a framework for assessing the smartness maturity level of a particular village at a given time. With this regard, many villages aim to raise their level of smartness by considering a number of aspects belonging to a wide range of thematic areas: Mobility, Governance, Economy, Environment, Living, and People. However, there is a lack of field-oriented systematic tools to help them to pilot the transition into a smart village.

In the context of this work, we have shown the design, implementation and exploitation phases of our smartness maturity assessment framework, which is intended to serve as a useful tool and decision support system for planners, administrative staff, political decision-makers, builders, and other users. Although this framework has arisen around a community belonging to the European Alpine space, the lessons learned can be easily transferred to other scenarios in which determining the degree of smartness maturity of the rural communities is a key challenge which can facilitate undertaking public or private investments.

In this context, we have been able to assess the smartness maturity level of an Italian municipality from the European Alpine space. The incremental development of our tool will even allow adding new functionality in the future, according to the feedback we receive from the stakeholders involved. But for the moment it seems clear that we need to design mechanisms to facilitate the aggregation of data from data entered by a number of people belonging to the same village, so that the data entered can reflect much better the collegiate opinion of the stakeholders from the same village.

Disclaimer

The Tolmezzo data and self-assessment has been published in a Master Dissertation Thesis in Civil Engineering entitled "Smartness Assessment of rural areas: multicriteria rating with Alpine stakeholders", defended in April 2019

512 J. Martinez-Gil et al.

by Francesca Polettini at the Politecnico di Milano. Data are to be considered preliminary and functional only to the testing of the procedure for research purposes.

Acknowledgements. This work has been developed within the SmartVillages project, Smart Digital Transformation of Villages in the Alpine Space, co-funded by Interreg Alpine Space (20182021). The authors would like to express the appreciation to SmartVillages project members for their contribution.

References

1. Bruni, E., Panza, A., Sarto, L., Khayatian, F.: Evaluation of cities' smartness by means of indicators for small and medium cities and communities: a methodology for Northern Italy. Sustain. Cities Soc. **34**, 193–202 (2017)
2. Coughlin, T.: IEEE consumer electronics society sponsors the smart village program [society news]. IEEE Consum. Electron. Mag. **4**(3), 15–16 (2015)
3. Hamdy, M., König-Ries, B., Küster, U.: Non-functional parameters as first class citizens in service description and matchmaking - an integrated approach. In: Di Nitto, E., Ripeanu, M. (eds.) ICSOC 2007. LNCS, vol. 4907, pp. 93–104. Springer, Heidelberg (2009). https://doi.org/10.1007/978-3-540-93851-4_10
4. Huang, G., Fang, Y., Wang, X., Pei, Y., Horn, B.K.P.: A survey on the status of smart healthcare from the universal village perspective. In: UV, pp. 1–6 (2018)
5. Das, R.K., Misra, H.: Digital India, e-governance and common people: how connected are these in access layer of smart village? In: ICEGOV 2017, pp. 556–557 (2017)
6. Mersand, S., Gascó-Hernández, M., Gil-García, J.R., Burke, G.B., Figueroa, M., Sutherland, M.: The role of public libraries in smart, inclusive, and connected communities: current and best practices. In: DG.O 2018, pp. 107:1–107:2 (2018)
7. Phahlamohlaka, J., Dlamini, Z., Mnisi, T., Mashiane, T., Malinga, L.: Towards a smart community centre: SEIDET digital village. In: Kimppa, K., Whitehouse, D., Kuusela, T., Phahlamohlaka, J. (eds.) HCC 2014. IAICT, vol. 431, pp. 107–121. Springer, Heidelberg (2014). https://doi.org/10.1007/978-3-662-44208-1_10
8. Pham, C., Rahim, A., Cousin, P.: Low-cost, long-range open IoT for smarter rural African villages. In: ISC2 2016, pp. 1–6 (2016)
9. Prinsloo, G., Mammoli, A., Dobson, R.: Participatory smartgrid control and transactive energy management in community shared solar cogeneration systems for isolated rural villages. In: GHTC, pp. 352–359 (2016)
10. Postránecký, M., Svítek, M.: Assessment method to measure smartness of cities. In: SCSP 2017 (2017)
11. SmartVillages. Smart digital transformation of villages in the Alpine Space. https://www.alpine-space.eu/projects/smartvillages. Accessed 21 June 2019
12. Visvizi, A., Lytras, M.D.: Rescaling and refocusing smart cities research: from mega cities to smart villages. J. Sci. Technol. Policy Manag. **9**(2), 134–145 (2018)

GameRecs: Video Games Group Recommendations

Rama Hannula, Aapo Nikkilä, and Kostas Stefanidis[(⊠)]

Tampere University, Tampere, Finland
{rama.hannula,aapo.nikkila,konstantinos.stefanidis}@tuni.fi

Abstract. Video games are a relatively new form of entertainment that has been rapidly gaining popularity in recent years. The number of video games available to users is huge and constantly growing, and thus it can be a daunting task to search for new ones to play. Given that some games are designed to be played together as a group, finding games suitable for the whole group can be even more challenging. To counter this problem, we propose a content-based video game recommender system, GameRecs, which works on open data gathered from Steam, a popular digital distribution platform. GameRecs is capable of producing both user profiles based on Steam's user data, as well as video game recommendations for those profiles. It generates group recommendations by exploiting lists aggregation methods, and focus on providing suggestions that exhibit some diversity by using a k-means clustering-based approach. We have evaluated the usability of GameRecs in terms of the user profile generation and the produced video game recommendations, both for single users and for groups. For group recommendations we compared two recommendation aggregation methods, Borda Count and Least Misery method. For diversity evaluation we compared results with and without the proposed k-means clustering method.

Keywords: Recommendations · Group recommendations ·
Game recommendations

1 Introduction

Nowadays, video games are a very popular form of entertainment and new games are getting released all the time. Since older games are also still playable, the number of games available to customers is constantly increasing. For this reason, it is not feasible for a user to manually go through every existing game when searching for new games to play. Fortunately, recommender systems can be used to help solve this problem. In general, recommender systems aim at providing suggestions to users or groups of users by estimating their item preferences and recommending those items featuring the maximal predicted preference [2,10,11].

Clearly, recommender systems can be also useful in the domain of video games. To our knowledge, the only application for recommender systems to digital games was proposed in [12], in which two different recommender systems

© Springer Nature Switzerland AG 2019
T. Welzer et al. (Eds.): ADBIS 2019, CCIS 1064, pp. 513–524, 2019.
https://doi.org/10.1007/978-3-030-30278-8_49

were proposed based on archetypal analysis. However, this work only generates recommendations for a single user at a time. For many multi-player games, it is essential to have a group of friends to play with. Thus, there is need for systems that can recommend games for groups.

In this work, we target at developing a system that could generate diverse and fair game recommendations for groups of users. In other words, the resulting recommendations should contain games that every member of the group would like, but also games that are dissimilar to each other, so as to increase user satisfaction. The system should also recommend both popular games and more obscure games, while prioritizing neither of them.

Specifically, we present a content-based method for recommending games for groups of people to play together. Our method exploits tags that the community has given to games, and using these tags it generates user profiles, and then game recommendations for these profiles. For demonstrating the effectiveness of our approach, we exploit user and game data available at the popular digital distribution platform Steam (http://store.steampowered.com/). We have evaluated the usability of our approach in terms of the user profile generation and the produced video game recommendations, both for single users and groups.

The rest of this paper is structured as follows. Section 2 presents related work, while Sect. 3 describes the users and games data in GameRecs. Section 4 introduces our approach for group games recommendations. Section 5 presents our usability evaluation results, and finally, Sect. 6 concludes the paper with a summary of our contributions.

2   Related Work

Recommender systems aim at providing suggestions to users or groups of users by estimating their item preferences and recommending those items featuring the maximal predicted preference. Typically, recommendation approaches can be classified as content-based [10], collaborative filtering [11], and hybrid ones [2]. In content-based approaches, information about the features/content of the items is processed, and the system recommends items with features similar to items a user likes. For example, if a Yelp user is always eating at sushi restaurants, he/she most likely likes this kind of food, so we can recommend him/her restaurants with the same cuisine. In collaborative filtering approaches, we produce interesting suggestions for a user by exploiting the taste of other similar users. For instance, if many users frequently go to Irish pubs after visiting an Italian restaurant, then we can recommend an Irish pub to a user that also shows preference for Italian restaurants. In knowledge-based approaches, users express their requirements, e.g., in terms of recommendation rules, and the system tries to retrieve items that are similar to the specified requirements. Finally, the hybrid recommender systems combine multiple of the aforementioned techniques to identify valuable suggestions.

Nowadays, recommendations have more broad applications, beyond products, like links (friends) recommendations [17], social-based recommendations

[14], query recommendations [5], health-related recommendations [15,16], open source software recommendations [7], diverse venue recommendations [6], or even recommendations for evolution measures [13]. There is also a lot of work on specific aspects of recommendations due to challenges beyond accuracy [1], like the cold start problem, the long tail problem and the evaluation of the recommended items in terms of a variety of parameters, like surprise, persistence [3] and serendipity [4]. More recently, many approaches that combine numerical ratings with textual reviews, have been proposed (e.g., [8]). For achieving efficiency, there are approaches that build user models for computing recommendations. For example, [9] applies subspace clustering to organize users into clusters and employs these clusters, instead of a linear scan of the database, for making predictions.

Clearly, recommender systems can be also useful in the domain of video games. Due to the large number of game releases every year, gamers can have hard time finding games fitting their interests. To our knowledge, the only application for recommender systems to digital games was proposed in [12]. Two different recommender systems were proposed based on archetypal analysis. Moving forward, in our work, we focus on group recommendations, and on how to identify a diverse set of games to propose to the group.

3   Games and Users Data

In our recommender, we pay attention on data regarding games and users. Specifically, we aim to find out the type of a game according to its tags, and the game type preferences of a user according to the tags of the games that they have played. For doing so, we employ data available at the popular digital distribution platform Steam.

3.1   Games Data

Steam's database contains several types of applications in addition to games, like media editing software and forms of extension packs for games. For example, they have media editing software, which we are not interested in. They also have DLCs (downloadable content), forms of extension packs for games. We are strictly interested in recommending games that can be acquired and played on their own, so we only consider the applications that indicate that their application type is a standalone game.

Each game in Steam has a number of tags. Tags are keywords voted for by the community, and they aim to describe the game with keywords resembling genres, categories, and others. The number of tags can vary greatly. Tags only appear on games if players have voted on them, which leads to popular games often having a greater number of tags than unpopular games. It's possible that a game has no tags at all. A common number of tags for a game would be from 10 to 20. For each game, we know the number of votes of each tag of the game. Thus, a game most likely has more votes on certain tags than others.

For example, some game might have considerably more votes for the "Third-Person Shooter" tag than for the "Crime" tag, in which case we are more interested in the former.

In our recommender, we use the tags and their votes on a game to describe the type of the game. In essence, we consider two games similar to each other if they share many tags with many votes on those tags. From all existing 339 unique tags, we hand-picked 19 tags we did not see fit for describing the type of the game. Some examples of these tags are "Co-op", "Singleplayer" and "Multiplayer", which describe how the game is played instead of what the game is like. We ignore these tags in our recommendation methods.

3.2 Users Data

For users, we are interested in what types of games they prefer. As described above, we determine the type of a game according to their tags and votes. In order to define a user's game preferences, we want to know ratings for tags measuring how much the player enjoys games with those tags. Essentially, we build user profiles containing tag ratings resembling the tag votes on games.

4 Recommendations

In this section, we introduce our method for producing group recommendations for video games. The process is divided into four parts. First, we focus on generating the user profiles, after which we compute single-user recommendations for those profiles. These recommendations are then aggregated into a single list of recommendations and finally, diversified by exploring a clustering method.

4.1 Generating User Profiles

First, our approach generates a user profile for each user. For creating the profile, we exploit the knowledge of which games the user has played and for how long. Each game the user has played for longer than a certain threshold is considered, and for each of these games the user profile's tag rating for that tag is increased. The final user profile contains scaled ratings for tags ranging from 0 to 1, higher value being better. Since some tags are more common than others, we want to prevent very frequent tags from dominating other, less common ones. For this purpose, we use inverse document frequency (IDF), which is calculated as: $IDF(t) = log\frac{N}{n(t)}$, where N is the total number of games in the dataset and $n(t)$ is the number of games that contain the tag t. This means that more frequent tags get lower IDF values than less common tags. IDF is calculated for every tag that appears in any game in the dataset.

To generate the actual user profile, we look at every game in the user's library that the user has played for more than 2 h. Some kind of playing time threshold is necessary because many users own games that they are not really interested in. It is more safe to assume that users like games they have actually played for some time. For each of these games, we iterate over every tag t the game g has, and calculate a strength for the tag with the following formula:

$$strength(t, g) = \frac{tagVotes(t, g)}{maxTagVotes(g)} \tag{1}$$

where *strength(t, g)* is the strength value for tag t for game g, *tagVotes(t, g)* is the number of votes the current game g has for tag t, and *maxTagsVotes(g)* is the highest number of votes any tag has for the game g. This results in higher strength for tags that have more votes, with the maximum strength being one and minimum strength being close to zero. It is not possible for a tag to have zero votes, since then the game would not have that tag at all.

For each of these tags that exist in any of the games the user has played, we calculate the strength and add it to the total rating of the tag. The final rating for a tag, called tag rating, is the sum of the tag strengths, multiplied by their respective IDF value squared:

$$tag\_rating(t, p) = IDF(t)^2 \times \sum_{g \in G_p} strength(t, g) \tag{2}$$

where t is the tag in question, G_p is the set of all games the user with profile p has played for more than 2 h. Finally, when the tag ratings have been calculated, for every tag that exists in any of the games user has played, the ratings are scaled between 0 and 1 so that the highest tag rating is always 1. After this we take the top 30 tags per profile and the rest are pruned.

With our method, the tags that have higher strengths and appear more frequently in the user's games get higher tag ratings. As some tags are more frequent and less informative than others, IDF balances the tag ratings between common and uncommon tags.

Our method is limited in such a way, that we only know which tags the user likes, and have no way of telling which tags they dislike. Therefore, a low tag rating does not mean that the user dislikes a tag, but instead the tag is just not as preferable as other tags that have higher tag ratings.

4.2 Generating Single-User Recommendations

After constructing the user profiles, recommendations are generated for each of these profiles separately. For each user profile, every game in the dataset is given a rating, called the game rating. This game rating is determined by the tag preferences of the profile, described by the tag ratings, and the tags the game has.

The rating algorithm works in such a way, that it increases game ratings for games that have multiple preferred tags and penalizes games for tags that the user has no preference for. Overall, the system aims to give equally good ratings for games that have few tags, all of which the profile has high tag ratings for, and games that have a large number of tags but with only moderate tag ratings.

The game rating for a single game is calculated by comparing the tags the game has, to the user profile. The general idea is that the more tags the game contains that the user likes, the better the game rating. If there are many tags but only a few of them are preferred by the user, give a small penalty. This is to avoid only recommending games that have a lot of tags. The game rating for a game g is calculated as:

$$game\_rating(g, p) = \frac{\sum\limits_{t \in T_g} strength(t, g) * tag\_rating(t, p)}{\sqrt[3]{|T_g|}} \tag{3}$$

where g is the game in question and T_g is the set of tags the game g has.

With this formula, games that have multiple tags the user likes, get higher game ratings, but the number of tags the game has also reduces the game rating. This means that games that have a large number of tags might not get better game ratings than games with only few tags. Calculating the game ratings without taking the number of tags into account would have resulted in games with large number of tags dominating the recommendations. On the other hand, also dividing by the number of tags without taking the cube root, in other words taking the average, would result in games with only few tags dominating.

4.3   Generating Group Recommendations

Individual recommendation lists are aggregated into a single group recommendation list using two alternative methods: the *Borda count* and the *least misery* method. The main idea of both methods is to take the single user recommendations of every group member and order the recommendations in such a way that it takes every group member into consideration. For this purpose, each item is given a *group score*, which is then used to rank the items to form the aggregated recommendations. In our case, users are the generated user profiles and items are the games.

Borda count is commonly used in political elections, but is also applicable to recommendations aggregation. It gives somewhat balanced results overall, and in our case games that have high ratings for every user profile get also high group scores, while games that only some of the users like get moderate group scores. Games that none of the group members like get very low group scores. With Borda count, each item is given a score depending on its position in the ranking. The last item in the ranking gets a score of 1 and the first item gets a score of n, where n is the total number of items in the ranking. The group score of an item is calculated by summing up the individual user scores for that particular item:

$$score\_borda\_count(g, i) = \sum_{u \in g} score(u, i) \tag{4}$$

where u is a user in the group g and $score(u, i)$ is the score for the item i, calculated from the ranking.

Least misery is an aggregation method that tries to sort the recommendations in such a way that everyone in the group is satisfied, in other words, causes the least misery among the group. To achieve this, the group score for an item i

is the minimum rating for that item from the individual recommendations, as described in: $score\_least\_misery(g, i) = \min_{u \in g}\{score(u, i)\}$.

The items are sorted by their group score to form the final aggregated recommendations.

4.4 Diversifying Group Recommendations

Our goal is to provide diverse results for the users, because this way the chance of finding something interesting is higher. To achieve diverse results, we utilize a k-means clustering-based method for creating clusters of the recommended games. This way, we group similar games and separate dissimilar ones, and by showing results from every one of these clusters evenly, the final result set will be diverse. Because clustering is quite an expensive operation, it is only performed for some of the top recommendations that are generated with the previously described recommendation aggregation. This way, only the games that have good ratings are considered, and there is no need to check later if the clusters contain good recommendations. In our experiments, we used top-500 game recommendations for clustering.

The clustering algorithm works as follows:

1. Get the top-500 recommendations generated with Borda, and prune everything else. These games are hence referred simply as all games.
2. Pick the top-1 recommendation and set it as the first cluster center. This way, we always get a cluster that initially centers on the best game.
3. Pick the most dissimilar game to all the current cluster centers from the top-100 recommendations and make it a new cluster center. This step helps to ensure that the clusters really are dissimilar. This step is repeated until we have 5 cluster centers.
4. Assign every game to their nearest cluster. Distance to every cluster center is measured with the Manhattan distance between their tag strengths.
5. Calculate new cluster centers for each cluster by calculating the averages of tag strengths of each game in that cluster.
6. Steps 4 and 5 are repeated until no more changes happen in the clusters or the maximum number of iterations is reached. 50 iterations is the limit used in our experiments.

5 Usability Evaluation

To evaluate the effectiveness of GameRecs, we implemented it as a web application and performed usability experiments with a number of real users[1]. Specifically, we evaluated the quality of profile generation (Sect. 5.1), the quality of single user recommendations (Sect. 5.2), and the quality of group recommendations (Sect. 5.3).

---

[1] Our application implementation and the data used in the experiments are publicly available at https://github.com/Nikkilae/group-game-recommender-test.

Table 1. Evaluation of generated user profiles

	u1	u2	u3	u4	u5	u6	u7	u8	u9	u10	u11	u12
PP	93.3%	66.7%	100.0%	93.3%	100.0%	100.0%	100.0%	100.0%	100.0%	100.0%	80.0%	93.3%
PHPP	40.0%	13.3%	66.7%	80.0%	46.7%	86.7%	53.3%	86.7%	46.7%	80.0%	40.0%	46.7%
PQ	7	4	8	8	7	8	9	8	9	9	7	7

	u13	u14	u15	u16	u17	u18	u19	u20	u21	u22	u23	u24
PP	100.0%	93.3%	100.0%	100.0%	100.0%	100.0%	100.0%	100.0%	80.0%	93.3%	100.0%	86.7%
PHPP	40.0%	60.0%	73.3%	73.3%	40.0%	26.7%	73.3%	66.7%	46.7%	26.7%	53.3%	46.7%
PQ	7	9	9	7	7	9	8	7	7	8	9	7

	PP	PHPP	PQ
average	95.0%	54.7%	7.7

5.1 Quality of Profile Generation

We asked users to evaluate the precision and quality of the profiles generated by GameRecs. We use three measures to evaluate the quality of a generated profile: Profile Precision, Profile Highly Preferred Precision and Profile Quality.

Users were asked to evaluate the quality of the 30 tags appearing in their profile. For characterizing the quality of the tags, users were asked to rate each of the tags with an interest score in the range [1, 5]. According to these ratings, we calculate **Profile Precision** $PP(u, k)$ for user u as follows:

$$PP(u, k) = \frac{relevant\_tags(u, k)}{k} 100\% \tag{5}$$

where $relevant\_tags(u, k)$ is the number of tags rated 2 or higher by the user u in the top k tags of their generated profile. In our experiments, we used $k = 15$. Furthermore, we reported the number of tags that were rated highly (interest score $>= 4$), and calculated the **Profile Highly Preferred Precision** $PHPP(u, k)$:

$$PHPP(u, k) = \frac{highly\_preferred\_tags(u, k)}{k} 100\% \tag{6}$$

where $highly\_preferred\_tags(u, k)$ is the number of tags the user u rated highly (interest score $>= 4$) in the top k tags of their generated profile. Finally, users were asked to provide an overall **Profile Quality** $PQ(u)$ in the range [1, 10] to indicate their degree of satisfaction of the overall result set including all 15 tags. A high number indicates an accurate representation of the user's taste.

The general impression is that having user profiles generated automatically makes it easier for someone to understand the main idea behind the system, since tags in the profiles act as examples of user preferences. As seen in Table 1, the PP is generally high, 95.0% on average, and in most cases even 100%. The PHPP values are varied, but on average more than half of the tags are highly preferred. Although a high PP seems to lead to a high PQ, their respective values differ from user to user. In conclusion, the profile generation seems quite effective.

5.2   Quality of Single User Recommendations

In addition, to evaluate group recommendations, we study the effectiveness of the recommender system for single users. First, we asked users to count the number of games of the top 20 recommendations that they deemed relevant or interesting. Second, we asked for a general recommendations quality rating between 1 and 10. Third, we asked for a rating of diversity among the recommended games between 1 and 10. We asked for these three ratings twice for different sets of recommendations: with clustering and without clustering. We use three different measures for evaluating the quality of single user recommendations: Recommendations Precision, Recommendations Quality and Recommendations Diversity.

Recommendations Precision $RP(u,k)$ for a user u is calculated as:

$$RP(u,k) = \frac{relevant\_games(u,k)}{k} 100\% \tag{7}$$

where $relevant\_games(u,k)$ is the number of games deemed relevant by the user u in the top k (20) recommendations. **Recommendations Quality** $RQ(u)$ represents the general satisfaction on the generated recommendations given by the user u as a number between 1 and 10. Finally, **Recommendations Diversity** $RD(u)$ represents the general recommendation diversity evaluated by the user u as a number between 1 and 10. As seen in Table 2, on average, just over half of the top recommendations were relevant to the user. As expected, applying clustering reduces RP and increases RD. However, the difference that clustering made in RD is not very impressive.

Table 2. Evaluation of single user recommendations

		u1	u2	u3	u4	u5	u6	u7	u8	u9	u10	u11	u12
With clustering	RP	60.0%	45.0%	75.0%	35.0%	50.0%	50.0%	25.0%	65.0%	25.0%	40.0%	90.0%	75.0%
	RQ	8	7	9	4	4	6	4	8	5	8	5	7
	RD	4	2	7	9	8	9	2	4	3	2	3	7
Without clustering	RP	40.0%	65.0%	65.0%	60.0%	90.0%	40.0%	45.0%	85.0%	45.0%	70.0%	90.0%	50.0%
	RQ	6	8	7	8	9	5	7	8	8	9	4	6
	RD	7	1	8	5	8	3	1	2	6	4	5	6
		u13	u14	u15	u16	u17	u18	u19	u20	u21	u22	u23	u24
With clustering	RP	55.0%	40.0%	60.0%	35.0%	55.0%	40.0%	75.0%	65.0%	25.0%	30.0%	65.0%	85.0%
	RQ	7	4	10	6	5	6	8	8	7	7	8	8
	RD	4	4	7	4	5	8	5	3	8	8	7	8
Without clustering	RP	70.0%	40.0%	80.0%	75.0%	50.0%	60.0%	80.0%	90.0%	30.0%	35.0%	55.0%	95.0%
	RQ	6	9	10	8	5	9	6	9	7	8	7	9
	RD	4	5	4	4	3	6	8	2	8	8	8	5

	With clustering			Without clustering		
	RP	RQ	RD	RP	RQ	RD
average	52.7%	6.6	5.5	62.7%	7.4	5.0

5.3 Quality of Group Recommendations

To evaluate our main focus, the group recommendations, we performed multiple experiments with groups of two and four members. The evaluation was done in four setups, with and without clustering, and using Borda count and least misery aggregation. For each of our four setups, we asked each member of a group to mark each produced recommendation as either relevant or irrelevant. These markings were then used to calculate different quality measures, which are described in more detail below.

Group Recommendations Precision $GRP(g, k)$ for a group g is calculated with the following formula:

$$GRP(g, k) = \frac{relevant\_games\_to\_all(g, k)}{k}100\% \tag{8}$$

where $relevant\_games\_to\_all(g, k)$ is the number of games in the top k recommendations that were relevant to everyone in the group individually. This means that a recommendation has to be relevant to every group member to be considered as relevant recommendation for the whole group. We used a k of 20 in our experiment.

Partial Group Recommendations Precision is similar to Group Recommendation Precision, but this time the recommendations have to be relevant to only part of the group members to be considered as relevant to the whole group. Partial Group Recommendation Precision $PGRP(g, k)$ for a group g is calculated with the following formula:

$$PGRP(g, k) = \frac{relevant\_games\_to\_half(g, k)}{k}100\% \tag{9}$$

where $relevant\_games\_to\_half(g, k)$ is the number of games relevant to at least half of the group members from group g from the top k recommendations.

To explore a potential correlation between group similarity and recommendations precision, we measure a group's similarity with Jaccard similarity of tags appearing in group members' profiles. More precisely, it's the relation of the number of common tags shared by the group members' profiles to the number of unique tags that appear in any of the group members' profiles.

The results of group recommendations evaluation can be seen in Table 3. Borda count seems to give slightly better results in both cases. Overall, similarly to single user recommendations, the precision is higher without clustering. Smaller group size seems to lead to higher similarity and better precision.

Table 3. Evaluation of group recommendations

	With clustering				Without clustering					
	Borda Count		Least Misery		Borda Count		Least Misery			
	GRP	PGRP	GRP	PGRP	GRP	PGRP	GRP	PGRP	Size	Similarity
[u1,u2,u3,u4]	5.0%	40.0%	5.0%	40.0%	0.0%	40.0%	5.0%	35.0%	4	5.8%
[u5,u6,u7,u8]	15.0%	55.0%	10.0%	45.0%	45.0%	85.0%	30.0%	85.0%	4	17.2%
[u9,u10,u11,u20]	15.0%	75.0%	10.0%	60.0%	10.0%	75.0%	15.0%	70.0%	4	1.2%
[u12,u13,u14,u15]	0.0%	15.0%	0.0%	10.0%	0.0%	5.0%	0.0%	10.0%	4	1.2%
[u16,u17]	35.0%		20.0%		30.0%		25.0%		2	57.9%
[u18,u19]	55.0%		55.0%		75.0%		80.0%		2	36.4%
[u21,u22]	20.0%		15.0%		30.0%		35.0%		2	53.8%
[u23,u24]	50.0%		45.0%		55.0%		45.0%		2	30.4%
average	24.4%	46.3%	20.0%	38.7%	30.6%	51.3%	29.4%	50.0%		

6 Summary

In this paper, we focus on group recommendations for video games. We propose
generating diverse game recommendations for groups of people to play together.
We work on open data gathered from Steam, a popular digital distribution platform, and we are capable of producing user profiles based on Steam's data, as
well as video game recommendations for those profiles. To generate group recommendations, we exploit lists aggregation methods, and we target at providing
recommendations that exhibit some diversity by using a k-means clustering-
based approach. For demonstrating the effectiveness of GameRecs, we performed
experiments with real users, and evaluated the usability of the method in terms
of the user profile generation and the produced video game recommendations,
both for single users and for groups.

References

1. Adomavicius, G., Kwon, Y.: Multi-criteria recommender systems. In: Recommender Systems Handbook, pp. 847–880 (2015)
2. Balabanovic, M., Shoham, Y.: Content-based, collaborative recommendation.
   Commun. ACM **40**(3), 66–72 (1997)
3. Beel, J., Langer, S., Genzmehr, M., Nürnberger, A.: Persistence in recommender
   systems: giving the same recommendations to the same users multiple times. In:
   Aalberg, T., Papatheodorou, C., Dobreva, M., Tsakonas, G., Farrugia, C.J. (eds.)
   TPDL 2013. LNCS, vol. 8092, pp. 386–390. Springer, Heidelberg (2013). https://
   doi.org/10.1007/978-3-642-40501-3_43
4. Desrosiers, C., Karypis, G.: A comprehensive survey of neighborhood-based recommendation methods. In: Ricci, F., Rokach, L., Shapira, B., Kantor, P.B. (eds.) Recommender Systems Handbook, pp. 107–144. Springer, Boston, MA (2011). https://
   doi.org/10.1007/978-0-387-85820-3_4
5. Eirinaki, M., Abraham, S., Polyzotis, N., Shaikh, N.: Querie: collaborative database
   exploration. IEEE Trans. Knowl. Data Eng. **26**(7), 1778–1790 (2014)

6. Ge, X., Chrysanthis, P.K., Pelechrinis, K.: MPG: not so random exploration of a city. In: MDM (2016)
7. Koskela, M., Simola, I., Stefanidis, K.: Open source software recommendations using Github. In: Méndez, E., Crestani, F., Ribeiro, C., David, G., Lopes, J.C. (eds.) TPDL 2018. LNCS, vol. 11057, pp. 279–285. Springer, Cham (2018). https://doi.org/10.1007/978-3-030-00066-0_24
8. McAuley, J.J., Leskovec, J.: Hidden factors and hidden topics: understanding rating dimensions with review text. In: RecSys (2013)
9. Ntoutsi, E., Stefanidis, K., Rausch, K., Kriegel, H.: Strength lies in differences: diversifying friends for recommendations through subspace clustering. In: CIKM (2014)
10. Pazzani, M.J., Billsus, D.: Content-based recommendation systems. In: The Adaptive Web, Methods and Strategies of Web Personalization (2007)
11. Sandvig, J.J., Mobasher, B., Burke, R.D.: A survey of collaborative recommendation and the robustness of model-based algorithms. IEEE Data Eng. Bull. **31**(2), 3–13 (2008)
12. Sifa, R., Bauckhage, C., Drachen, A.: Archetypal game recommender systems. In: Proceedings of the 16th LWA Workshops (2014)
13. Stefanidis, K., Kondylakis, H., Troullinou, G.: On recommending evolution measures: a human-aware approach. In: 33rd IEEE International Conference on Data Engineering, ICDE 2017, San Diego, CA, USA, 19–22 April 2017, pp. 1579–1581 (2017)
14. Stefanidis, K., Ntoutsi, E., Kondylakis, H., Velegrakis, Y.: Social-based collaborative filtering. In: Alhajj, R., Rokne, J. (eds.) Encyclopedia of Social Network Analysis and Mining, pp. 1–9. Springer, New York (2017). https://doi.org/10.1007/978-1-4614-7163-9
15. Stratigi, M., Kondylakis, H., Stefanidis, K.: Fairness in group recommendations in the health domain. In: ICDE (2017)
16. Stratigi, M., Kondylakis, H., Stefanidis, K.: FairGRecs: fair group recommendations by exploiting personal health information. In: Hartmann, S., Ma, H., Hameurlain, A., Pernul, G., Wagner, R.R. (eds.) DEXA 2018. LNCS, vol. 11030, pp. 147–155. Springer, Cham (2018). https://doi.org/10.1007/978-3-319-98812-2_11
17. Yin, Z., Gupta, M., Weninger, T., Han, J.: LINKREC: a unified framework for link recommendation with user attributes and graph structure. In: WWW (2010)

FIFARecs: A Recommender System
for FIFA18

Jaka Klancar[1], Karsten Paulussen[2], and Kostas Stefanidis[3]([⊠])

[1] University of Ljubljana, Ljubljana, Slovenia
`jk7808@student.uni-lj.si`
[2] Eindhoven University of Technology, Eindhoven, The Netherlands
`k.paulussen@student.tue.nl`
[3] Tampere University, Tampere, Finland
`konstantinos.stefanidis@tuni.fi`

Abstract. One of the most popular features in the FIFA18 game is the career mode, where the target of the users is to improve their teams and win as much competitions as possible. Usually, it is hard for the users to decide which players to select to buy to maximally improve their team by taking into account all different players' attributes. In this paper, we introduce an approach towards helping the users to both determine the worst player in a specific team and recommend a list of players seen as possible replacement to improve the team.

Keywords: Recommender systems · Recommendations ·
Group recommendations · FIFA18 game

1 Introduction

FIFA18 is a football simulation video game developed by Electronic Arts Sports (shortly, EA Sports[1]). It is the newest game in a series of football video games going back to 1994, with more than six million sales around the globe. EA Sports was the first label that had an official license from FIFA, namely the International Federation of Association Football. The license gave them rights to use names and presentations of most players and competitions around the world.

In FIFA18, users can play with more than 700 teams from 30 competitions. The game contains different gameplays. One of the most popular is the career mode, on which we focused in this paper. In career mode, the users start with a self-chosen team. They can decide and change which formation they want their team to play with, and then choose players for the chosen formation. The users play matches with their team in different competitions. To improve the team, they can buy new players during the transfer periods, which are two times during the season. Improving a team by buying and selling players is an important part of the career. Users in the game can build their team through the career-mode

---

[1] https://www.ea.com.

© Springer Nature Switzerland AG 2019
T. Welzer et al. (Eds.): ADBIS 2019, CCIS 1064, pp. 525–536, 2019.
https://doi.org/10.1007/978-3-030-30278-8_50

gameplay, by exploiting a big number of attributes/characteristics regarding the players, which can increase or decrease over the years. Due to the large number of players in the game and their different attributes that have be taken into account, it is typically difficult for the user to find the right player for the right position.

Users first have to decide which player is the worst in their team. One option is just to look at the overall score describing the player's performance, but this overall score can change based on the age and the potential of the players. Furthermore, it can also be that a player whose overall score is higher, but his transfer price and wage are also higher, can be rated as worse. After deciding which player is the worst the user has to find the right player to take his place. Again the age, overall and potential, together with some other attributes have to be taken into account.

It is hard for the users to decide which players to buy to maximally improve their team by taking into account all different attributes. Therefore, in this work, we provide an application that helps the users by both determining which is the worst player in their team and by recommending a list of players that could replace that player and improve the team. So far, only the Xbox game console has its own recommender system that proposes games based on the games played earlier [5]. The first application for recommender systems to digital games was proposed in 2014 [10]. Two different Top-L recommender systems were proposed based on archetypal analysis. However, in all existing approaches the concept of suggesting the video games option to improve user's experience and performance is missing. To the best of our knowledge, our approach is the first in the field that explorers how to provide in-game recommendations.

We have evaluated our approach along two perspectives: usability and performance. Our usability experiments consider both the easiness of the approach and the satisfaction of the users from the quality of the results. We used a real dataset, containing 17980 football players with more than 70 attributes, describing both personal and performance characteristics. Our performance experiments focus on evaluating the efficiency of the similarity measures used, namely Pearson correlation, cosine similarity and Minkowski similarity, versus the quality of the results thus achieved.

In a nutshell, in this paper:

- We introduce a recommenders-like approach for assisting users form their teams in FIFA18: We initially suggest the first eleven players, as well as the worst player in the team, and then recommend a list of players that can replace the latter.
- We propose an effective presentation solution for the recommended players which builds upon real needs. Our solution provides a ranked overview of the players enriched with summarized information, and can facilitate user browsing.
- We implement a proof of concept prototype for players recommendations and evaluate our approach in terms of usability and performance. Our experiments show that using FIFARecs can achieve fast and useful suggestions for the users.

The rest of this paper is structured as follows. In Sect. 2, we present related work, and in Sect. 3, we introduce our approach for producing recommendations in FIFARecs. In Sect. 4, we describe the dataset that is associated with FIFA18, while in Sect. 5, we focus on the effective presentation of FIFARecs. Section 6 presents our usability and performance evaluation results, and finally, Sect. 7 concludes the paper with a summary of our contributions.

2   Related Work

Recommender systems aim at providing suggestions to users or groups of users by estimating their item preferences and recommending those items featuring the maximal predicted preference. Typically, depending on the type of the input data, i.e., user behavior, contextual information, item/user similarity, recommendation approaches are classified as content-based [8], collaborative filtering [9], knowledge-based [2], hybrid [1], or even social ones [12]. Nowadays, recommendations have more broad applications, beyond products, like links (friends) recommendations [15], social-based recommendations [12], query recommendations [3], health-related recommendations [13,14], open source software recommendations [6], diverse venue recommendations [4], recommendations for groups [7], or even recommendations for evolution measures [11].

Clearly, recommender systems can be also useful in the domain of video games. Due to the large number of game releases every year, gamers can have hard time finding games fitting their interests. Therefore, the Xbox game console has its own recommender system that proposes games based on the games played earlier [5]. The first application for recommender systems to digital games was proposed in 2014 [10]. Two different Top-L recommender systems were proposed based on archetypal analysis. However, in all existing academic literature of recommender systems the concept of suggesting the video games option to improve user's experience and performance is missing. To the best of our knowledge, our approach is the first in the field that explorers how to provide in-game recommendations.

However, in all existing recommender systems the concept of suggesting video games is completely missing. To the best of our knowledge, our approach is the only one in this field that explores how to provide games recommendations.

3   Recommendations in FIFA18

In this section, we introduce the main steps of our approach. First, we explain how the first eleven football players are chosen and how we determine which of these players is the worst. Then, a joint profile for the team players is created, and finally, we describe how the recommendations for the players to replace the worst one are made. Each player has more than 70 attributes divided into personal attributes (e.g., age, nationality and value), performance attributes (e.g., overall, potential and aggression), and overall score for each position. More details on the FIFA18 dataset appear in Sect. 4.

```
def sort_first_eleven(players):
    true_overall = overall if age > 23 else
        (overall+potential)/2
    players_diff = true_overall - avg_overall
    players_diff = players_diff * penalty
    players_diff = normalize(players_diff)
    wage_diff = normalize(wage)
    players_diff = 0.7*players_diff + 0.3*(-1 * players_wage)
    return sort(players_diff)
```

Fig. 1. The pseudo-code for sorting the players in a team.

3.1   Selection of the First Eleven and the Worst Player

Every team in FIFA 18 contains more than 25 players in its selection. This makes the process of selecting the formation and the players fitting the formation hard. FIFARecs introduces a method that helps users make valuable selection in an easy way. The user can choose between different formations, and the best eleven players for that formation will be provided automatically. For each position in the formation, the best player who has this position as a preferred position is taken. The best player for a position is based on the positional score, i.e., the attribute players have for that position. If none of the players in the team have the specific position as preferred, then the most similar position is taken. Similar positions are positions, which should a player also be able to play. Usually football players can play a variety of positions, but only some of them are their preferred positions. For example, a central defensive midfielder can also play as a center back. After the first eleven players were selected, the application will determine the worst player in the team. Figure 1 shows the algorithm used to rank and sort the eleven players. Based on a group of experts experience with the game, we decide that age, overall and potential score, as well as player's salary are important attributes to evaluate players.

At first, a new overall score (true overall score) is computed based on the age, overall, and potential score of the players. The first score is given to each player (player diff) based on the difference between his true overall score and the average score of the team. This score is multiplied by a penalty, between 0 and 1, and is based on the age of the player. Older players get higher penalty, because most of the users prefer younger players. The score is normalized and multiplied with the normalized wage. Here, more weight is given to the combination of overall score and age than to the salary of the players. The weights are decided purely by the experience of a group of experts with the game: overall, experts support that the player's ability is more important than his wage.

3.2   Joint Profile of the Team

After the selection of the worst player of the team, a joint profile for the team is formed. This joint profile will be used to evaluate if the quality of the players that are recommended fit with the general quality of the team. For doing so, the average and standard deviations of the following attributes are computed: Age, Aggression, Ball control, Composure, Positioning, Reactions, Short passing, Sprint speed, Stamina, and Strength. These attributes are important for all field players, regardless of their positions. Average values were given weights – the more dense an attribute is, the higher weight it gets. The weights were given, because the more dense attributes appear to be more important for the selected team.

3.3   Collaborative Filtering Recommendations

The list of potential recommendations is first filtered based on self-determined constraints. First of all, a player that is recommended needs to have the same position as the player we want to replace. The score for this position has to be higher than the score of the worst player of this position, except when the player is younger than 23 years old. Then, the application take into account the potential overall score instead of the score for the specific positions. The potential score is used, because the player can still develop and become a better player in future. Furthermore, the recommended players cannot have a wage that is more than twice as high than that of the worst player – we assume that the users would not be able to afford that player for their team. In addition to the self-determined constraints, the user of the application can also choose some constraints, including age, wage, transfer price, overall.

After the list of players is filtered, a collaborative filtering algorithm is implemented in order to identify the recommendations. For doing so, we actually compute the similarity of each player in the filtered list with the joint profile we created for the first eleven team players. Such joint profile represents an artificial player. Thus, the top-k most similar players to the artificial one, represent the candidates for replacing the worst player.

Similar players are located via a *similarity function* $simP(p, p')$ that evaluates the proximity between two players p and p', by considering their shared dimensions. In our approach, we use three different similarity measures, namely: Pearson correlation, cosine similarity, and Minkowski distance. Pearson is widely used for computing similarities between users, players in our case. It actually measures the linear dependence between two players p and p': it returns a value between $+1$ and -1, where $+1$ is total positive linear correlation, 0 is no linear correlation and -1 is total negative linear correlation.

$$Pearson(p, p') = \frac{\sum_{i \in I}(r(p, i) - \mu_p)(r(p', i) - \mu_{p'})}{\sqrt{\sum_{i \in I}(r(p, i) - \mu_p)^2}\sqrt{\sum_{i \in I}(r(p', i) - \mu_{p'})^2}} \qquad (1)$$

where $r(p, i)$ (resp., $r(p', i)$) is the score that the player p (resp., p') has for attribute i, μ_p (resp., $\mu_{p'}$) is the mean of p's (p''s) scores, and I is the set of attributes for which both p and p' have scores.

In a similar way, the cosine similarity of two players can be derived by using the Euclidean dot product formula, as follows:

$$cosine(p, p') = \frac{\sum_{i \in I} r(p, i) r(p', i)}{\sqrt{\sum_{i \in I} r(p, i)^2} \sqrt{\sum_{i \in I} r(p', i)^2}} \tag{2}$$

The resulting similarity ranges from -1 meaning exactly opposite, to $+1$ meaning exactly the same, with 0 indicating orthogonality, and in-between values indicating intermediate similarity or dissimilarity.

Alternatively, via Minkowski, we can measure the numerical difference for the corresponding attributes of p and p':

$$Minkowski(p, p') = \sqrt[n]{\sum_{i \in I} |r(p, i) - r(p', i)|^n} \tag{3}$$

Intuitively, Minkowski measures the numerical difference for each corresponding attributes of player p and player p'. Then it combines the square of differences in each attribute into an overall distance.

4   Data in FIFA18

The dataset for the FIFA18 game is available on Kaggle[2]. It contains 17980 cases, where each case relates to one football player. Each football player has more than 70 attributes. These attributes can be divided into personal attributes (e.g., age, nationality and value), performance attributes (e.g., overall, potential and aggression), and overall score for each position. Those three groups, with attributes and their data types, can be seen in Table 1. Note that Table 1 shows only the attributes that are used in our framework. The rest (attributes like club logo, flag and photo) are dropped.

Most of the players have multiple preferred positions. Those positions are presented as a string and positions are separated by space. We transform this

Table 1. Player attributes in FIFA18 dataset.

Personal attribute	Performance attributes	Position scores
Name (object)	Acceleration (int64)	CAM (oat64)
Age (int64)	Aggression (int64)	CB (oat64)
Nationality (object)	Agility (int64)	CDM (oat64)
Overall (int64)	Vision (int64)	ST (oat64)
Potential (int64)	Volleys (int64)	GK (oat64)
Club (object)		
Value (oat64)		
Wage (oat64)		

---

[2] https://www.kaggle.com/thec03u5/fifa-18-demo-player-dataset.

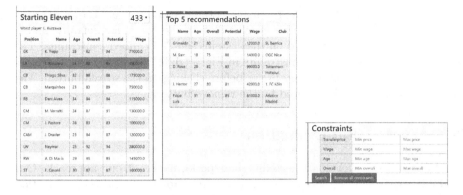

Fig. 2. The first eleven and the worst player (left), the top 5 recommendations for replacing the worst player (middle), and constraints specification (right). (Color figure online)

string to a list of positions for every player. That makes it easier to use during development. The Wage and Value attributes are transformed to numeric values. Both attributes are structured the same (e.g., €123M and €280K). To transform them to numeric values, we needed to get rid of the € symbol and transform the last letter to corresponding number of zeros. Every player has score for every position, expect position of goalkeeper (GK). We created a new attribute GK, which has a score for goalkeeper position. To get that score, we simply took the overall of the player, and check if the player has GK as a preferred position, otherwise his GK score is 0. Some players were in the dataset twice, with different IDs. Those duplicated cases where dropped. Finally, we transformed all performance attributes to numeric values.

5   Presentation of Recommendations

To facilitate the players selection towards the formation of a team, we propose to organize the user interactions in a compact yet intuitive and representative way. To this end, to demonstrate the feasibility of our approach, we have developed a research prototype for FIFARecs that employs particular constraints on specific attributes related to the football players. Let's assume that the user in question is a fan of Paris Saint-Germain. After choosing his favorite team, using a dropdown element, the application will automatically create the first eleven players of the team by following the default formation, namely 4-3-3. The user can easily change the team's formation by using the drop-down select element in the top right field of the section "Starting Eleven" (Fig. 2, left). The first eleven players are depicted also in Fig. 2 (left). The worst player of the top eleven players is marked in red. In this example, this player, L. Kurzawa, has the lowest overall score, but note that the worst player is not only determined by the overall or potential score. The top five recommendations for replacing the worst player are given in Fig. 2 (middle).

The recommendation list shows the name of the player, his age, his overall score, potential score, his wage, and his current club in the game. By default, the list of players is only filtered based on the systems self-determined constraints. The users can specify their own constraints in Fig. 2 (right). In particular, the users first have to fill constraints they prefer before clicking on the search button. The constraints can be easily cleared by the remove button.

6   Experimental Evaluation

In this section, we evaluate usability and performance of our approach. Our usability experiments consider the easiness of the approach and the satisfaction of the users from the quality of recommendations. Our performance experiments focus on evaluating the efficiency of the similarity measures used, namely Pearson correlation, cosine similarity and Minkowski similarity.

For both cases, in our experiments, we used a real dataset (Sect. 3), containing 17980 football players, each one related to more than 70 attributes, describing personal and performance characteristics[3].

6.1   Usability Evaluation

The goal of our usability study is to justify the use of FIFARecs. The objective here is to show that users get interesting and useful results in an automatic way, without spending time for manually selecting the football players. We conducted an empirical evaluation of our approach with 20 users. For all users, it was the first time that they used the system. We evaluated our approach along two lines: overhead of understanding and using FIFARecs, and quality of results.

Easiness of Approach. For counting the easiness of our approach, we consider cases in which FIFARecs is either used or not. For both cases, we counted how long (in mins) it took users to specify their teams. Specifically, when FIFARecs was not involved in the process, we counted the total time needed by the users to manually select their teams. We repeated the same with FIFARecs.

For the FIFARecs case, since for all users this was their first experience with the system, the reported time includes the time it took the user to get accustomed with the system. These results are reported in Table 2; the last line of the table summarizes the resulting time for all users. Our general impression is that FIFARecs save much time in selecting football players. Using our system makes it easier for someone to understand several ideas behind the game, since the whole process acts as example. With regards to time, there was deviation among the time users spent on selecting teams: some users were more meticulous than others, spending more time in adjusting their players.

---

[3] Data and codes can be found here: https://github.com/klancar16/FIFA18_recommender.

Table 2. Easiness of approach: Time (in minutes) for selecting teams manually and with FIFARecs.

	Selection time - manually	Selection time - FIFARecs
User 1	7	2
User 2	7	3
User 3	8	2
User 4	8	3
User 5	12	4
User 6	13	4
User 7	5	3
User 8	5	2
User 9	5	2
User 10	7	2
User 11	9	5
User 12	8	4
User 13	8	3
User 14	9	3
User 15	11	3
User 16	14	3
User 17	9	2
User 18	13	4
User 19	11	2
User 20	11	3
Average time	9	2 min & 57 s

Quality of Results. In this set of experiments, we asked the users to evaluate the quality of the FIFARecs results. First, users were asked to evaluate the quality of the 11 players appearing in the result. For characterizing the quality of the players, users marked each of them with 1 or 0, indicating whether they considered that the player should belong to the best 11 ones or not. The number of 1s corresponds to the precision of the top-11 players, or $p(11)$. Users were also asked to give a numerical interest score between 1 and 10 to each of the 11 players. This score was in the range $[1, 5]$, if the previous relevance indicator was 0 and in the range $[6, 10]$, otherwise. We reported the number of players that were rated highly (interest score ≥ 7), namely the Highly Preferred Players (HPP). Finally, users were asked to provide an Overall Score (OS) in the range $[1, 10]$ to indicate their overall satisfaction.

Next, we asked users to evaluate the result presenting the worst player. Users marked the worst player (WP) with 1 or 0, indicating whether they considered that the reported player should be the worst or not. Also, users were asked to

Table 3. Quality of results.

	p(11)	HPP	OS	WP	P	C	M
User 1	82%	9	9	1			✓
User 2	91%	10	10	1	✓		
User 3	91%	10	10	1	✓		
User 4	91%	9	9	0		✓	
User 5	64%	7	8	1	✓		
User 6	82%	9	8	1			✓
User 7	82%	8	8	0		✓	
User 8	91%	10	10	1	✓		
User 9	82%	9	8	1			✓
User 10	91%	10	9	1	✓		
User 11	82%	9	9	0	✓		
User 12	100%	10	10	1	✓		
User 13	91%	10	8	0		✓	
User 14	91%	9	9	1	✓		
User 15	100%	10	9	1			✓
User 16	91%	10	9	1	✓		
User 17	82%	9	8	1	✓		
User 18	64%	7	7	1	✓		
User 19	91%	10	8	1			✓
User 20	91%	10	9	1			✓
Avg	86,5%	9,25	8,75	80%	50%	20%	30%

evaluate the suggestions for replacing the worst player. Users were presented with three alternatives, each containing 5 player names, computed using three different measures, namely: Pearson (P), cosine (C), and Minkowski (M), and selected the one that best matches their needs.

Table 3 depicts the detailed per user scores. Our results show that using FIFARecs, we can achieve recommendations of high quality. In almost all cases, FIFARecs is able to detect the worst players, while the overall satisfaction when considering the reported suggestions is high. With regards to the employed similarity functions, users seem to prefer the results computed via the Pearson correlation. Minkowski was selected by the 30% of the users, while only the 20% of the users selected cosine. Overall, given that users many times were not aware of the whole set of available players, since they were just presented with the top 11 players, they left room in their choices for better results that could be lying in the dataset that was not presented to them. For this reason, precision can be higher than the computed one.

6.2   Performance Evaluation

In this set of experiments, we evaluate the performance of our approach in terms of the execution time. This depends on whether we use Pearson (P), cosine (C) or Minkowski (M) for locating the candidate players for replacing the worst player. Thus, we study these three cases separately. We looped through all the teams in our database, and produced the top-5 suggestions, i.e., players' names, for replacing the worst player, for all three measures. The results appear in Table 4. Given that the reported time refers to not a single prediction but to predictions for all teams, it is easy for someone to claim that for single users, and thus recommendations for single teams, the system's response time is very good for the gaming conditions. Cosine reports the best time, while Pearson needs the highest amount of time. However, according to our usability evaluation, the detailed computations of Pearson offer the best recommendations.

Table 4. Efficiency of the FIFARecs approach.

	P	C	M
Time (in secs)	236.40	6.15	64.44

7   Conclusions

FIFARecs introduces a method useful for the FIFA18 users. The goal of our recommender is to automatically pick the best first eleven players, so as to form a team. Furthermore, the recommender targets at helping users to decide which player of their team they have to replace and by whom, based on the formation and the constraints selected by the user.

Our future work focuses on how to enhance FIFARecs with online discussions between the users of FIFA18. Specifically, we focus on how users express their opinions, how opinions related to each other and how discussions around a game are formed. This process involves techniques such as text mining, topic modeling, and sentiment analysis. Moreover, we target at supporting services for retrieving opinions and providing overviews of opinions sets in a way that avoids information overload, so as to help users understand strategies of other users, as well as discover useful information about players and teams. User information needs will be expressed through keywords, and results will incorporate ranking and diversification, according to user preferences and contextual relevance. Finally, to be able to explore connections between opinions, we study navigation means and appropriate representation models.

References

1. Balabanovic, M., Shoham, Y.: Content-based, collaborative recommendation. Commun. ACM **40**(3), 66–72 (1997)
2. Bridge, D.G., Göker, M.H., McGinty, L., Smyth, B.: Case-based recommender systems. Knowl. Eng. Rev. **20**(3), 315–320 (2005)
3. Eirinaki, M., Abraham, S., Polyzotis, N., Shaikh, N.: QueRIE: collaborative database exploration. IEEE Trans. Knowl. Data Eng. **26**(7), 1778–1790 (2014)
4. Ge, X., Chrysanthis, P.K., Pelechrinis, K.: MPG: not so random exploration of a city. In: MDM (2016)
5. Koenigstein, N., Nice, N., Paquet, U., Schleyen, N.: The Xbox recommender system. In: Proceedings of the Sixth ACM Conference on Recommender Systems, RecSys 2012, pp. 281–284 (2012)
6. Koskela, M., Simola, I., Stefanidis, K.: Open source software recommendations using Github. In: Méndez, E., Crestani, F., Ribeiro, C., David, G., Lopes, J.C. (eds.) TPDL 2018. LNCS, vol. 11057, pp. 279–285. Springer, Cham (2018). https://doi.org/10.1007/978-3-030-00066-0_24
7. Ntoutsi, E., Stefanidis, K., Nørvåg, K., Kriegel, H.-P.: Fast group recommendations by applying user clustering. In: Atzeni, P., Cheung, D., Ram, S. (eds.) ER 2012. LNCS, vol. 7532, pp. 126–140. Springer, Heidelberg (2012). https://doi.org/10.1007/978-3-642-34002-4_10
8. Pazzani, M.J., Billsus, D.: Content-based recommendation systems. In: Brusilovsky, P., Kobsa, A., Nejdl, W. (eds.) The Adaptive Web. LNCS, vol. 4321, pp. 325–341. Springer, Heidelberg (2007). https://doi.org/10.1007/978-3-540-72079-9_10
9. Sandvig, J.J., Mobasher, B., Burke, R.D.: A survey of collaborative recommendation and the robustness of model-based algorithms. IEEE Data Eng. Bull. **31**(2), 3–13 (2008)
10. Sifa, R., Bauckhage, C., Drachen, A.: Archetypal game recommender systems, vol. 1226, pp. 45–56, January 2014
11. Stefanidis, K., Kondylakis, H., Troullinou, G.: On recommending evolution measures: a human-aware approach. In: 33rd IEEE International Conference on Data Engineering, ICDE 2017, San Diego, CA, USA, 19–22 April 2017, pp. 1579–1581 (2017)
12. Stefanidis, K., Ntoutsi, E., Kondylakis, H., Velegrakis, Y.: Social-based collaborative filtering. In: Alhajj, R., Rokne, J. (eds.) Encyclopedia of Social Network Analysis and Mining, pp. 1–19. Springer, New York (2017). https://doi.org/10.1007/978-1-4614-7163-9
13. Stratigi, M., Kondylakis, H., Stefanidis, K.: Fairness in group recommendations in the health domain. In: ICDE (2017)
14. Stratigi, M., Kondylakis, H., Stefanidis, K.: FairGRecs: fair group recommendations by exploiting personal health information. In: Hartmann, S., Ma, H., Hameurlain, A., Pernul, G., Wagner, R.R. (eds.) DEXA 2018. LNCS, vol. 11030, pp. 147–155. Springer, Cham (2018). https://doi.org/10.1007/978-3-319-98812-2_11
15. Yin, Z., Gupta, M., Weninger, T., Han, J.: LINKREC: a unified framework for link recommendation with user attributes and graph structure. In: WWW (2010)

ADBIS 2019 Doctoral Consortium

Algorithms and Architecture
for Managing Evolving ETL Workflows

Judith Awiti[(✉)]

Université Libre de Bruxelles, Brussels, Belgium
judith.awiti@ulb.ac.be

Abstract. ETL processes are responsible for extracting, transforming and loading data from data sources into a data warehouse. Currently, managing ETL workflows has some challenges. First, each ETL tool has its own model for specifying ETL processes. This makes it is difficult to specify ETL processes that are beyond the capabilities of a chosen tool or switch between ETL tools without having to redesign the entire ETL workflow again. Second, a change in structure of a data source leads to ETL workflows that can no longer be executed and yields errors.

Therefore, we propose a logical model for ETL processes that makes it feasible to (semi-)automatically repair ETL workflows. Our first approach is to specify ETL processes using Relational Algebra extended with update operations. This way, ETL processes can be automatically translated into SQL queries to be executed into any relational database management system. Later, we will consider expressing ETL tasks by means of an Extensible Markup Language (XML) and other programming languages. We also propose the Extended Evolving-ETL (E3TL) framework in which we will develop algorithms for (semi-) automatic repair of ETL workflows upon data source schema changes.

Keywords: ETL · E3TL · Database · Data warehouse ·
Relational algebra

1 Introduction

There is no agreed-upon conceptual or logical model to specify ETL processes. Consequently, existing ETL tools use their own specific language to define ETL workflows. For this reason, several conceptual and logical models [3,6–9] have been proposed for ETL process design. Unfortunately, each model has some limitations discussed in the related work and hence does not provide suitable solutions to the problem. Moreover, it is our aim to have a model that makes ETL workflow reparation easy and feasible.

Not only that, most of the existing ETL tools tacitly assume that the structure of every DS is static, but such an assumption is incorrect. Data source (DS) schema changes are bound to happen, and they disrupt the smooth execution of

© Springer Nature Switzerland AG 2019
T. Welzer et al. (Eds.): ADBIS 2019, CCIS 1064, pp. 539–545, 2019.
https://doi.org/10.1007/978-3-030-30278-8_51

ETL workflows resulting in data loss. For example, Wikipedia had 171 schema versions from April 2003 to November 2007 [1]. A change in structure of a DS leads to ETL workflows that can no longer be executed and yields errors. Large organizations might have thousands of deployed ETL workflows which makes it nearly impossible for ETL designers to repair ETL workflows after every DS change.

Considering this, we study BPMN4ETL [2], a conceptual model for designing ETL processes based on the Business Process Modeling Notation (BPMN) a de-facto standard for specifying business processes. The model provides a set of primitives that cover usual ETL requirements. Since BPMN is already used for specifying business processes, adopting this methodology for ETL would be smooth for users already familiar with that language. Further, BPMN provides a conceptual and implementation-independent specification of such processes, which hides technical details and allows users and designers to focus on essential characteristics of such processes. ETL processes expressed in BPMN can be translated into executable specifications for ETL tools. Out of this model we develop a logical model based on relational algebra (RA). Being a well-studied formal language, RA provides a solid basis to specify ETL processes for relational databases, and its expressiveness allows providing a detailed view of the data flow of any ETL process. We prove the efficiency of using an extended RA to specify ETL processes at a logical level as compared to commercial tools with the TPC-DI benchmark [5]. Later, we may consider expressing ETL tasks by means of an Extensible Markup Language (XML) and other programming languages.

After that, we will develop algorithms for our proposed E3TL framework to (semi-) automatic reparation of ETL workflows. In these algorithms, we will apply rule-based reasoning (RBR) and case-based reasoning (CBR) and, a combination of both. Then, we will develop a prototype system for ETL workflow reparation based on the above-mentioned framework and verify the applicability of our proposed solution with the TPC-DI benchmark. The project has two main objectives:

- To propose a methodology for designing ETL processes that will facilitate a smooth transition from gathering user requirements to the actual implementation. This requires expressing ETL processes at conceptual, logical, and physical levels.
- To develop a framework to (semi-) automatically repair ETL workflows upon DS changes.

Currently, we have been working on the first objective of the project.

2 Related Work

The conceptual model proposed in [8] analyzes the structure and data of the DSs and their mapping to a target data warehouse (DW). Attributes are treated as *first-class citizens* in the inter-attribute mappings. In [9], a logical model was developed independent of this conceptual model. The authors modelled an ETL workflow as a graph which they named the *architecture graph*. All entities in an ETL scenario including data stores, activities and their constituent parts are modelled as nodes of the graph. These models are complex and not easy to understand or design because they do not utilize a standard modelling language. Apart from that, the logical model cannot be implemented directly in a DW environment unless via a custom-built tool. Another conceptual model of ETL processes developed by [7], uses the Unified Modeling Language (UML) to design ETL processes. Each ETL process is represented by a stereotyped class which is in turn represented by an icon. Because UML is a widely accepted standard for object-oriented analysis and design, this approach minimizes the efforts of developers in learning new diagrams for It was refined in [3] with the use of UML activity diagrams instead of class diagrams and icons. Recently, RA has been introduced as a logical model for ETL processes [6]. RA provides a set of operators that manipulates relations to ensure that there is no ambiguity. Even though loops and certain complex functions cannot be specified directly with RA, it has enough operations to model common ETL processes when extended with update operations. It can also be directly translated into SQL to be executed in any relational database management system (RDBMS).

Although little research has been done in the field of evolving ETL workflows, a recent one in [10] suggest a solution to the problem by using case-based reasoning. The authors developed a framework for solving the problem of evolving ETLs by applying user-defined rules and case-based reasoning. The user-defined rules approach is independent of the case-based reasoning approach. A previous research [4] also provided a framework in which the ETL workflow is manually annotated with rules that defines its behavior in response to data source changes. Since both approaches do not provide comprehensive solutions, the problem of evolving ETL workflows still needs substantial research.

TPC-DI [5], the first industry benchmark for Data Integration (also known as ETL) provides a suitable case study for this research. The source data model used was based on the relational data of an Online Transaction Processing (OLTP) database of a fictitious retail brokerage firm and some other external sources. The target data model has a snowstorm schema. The source system is made up of an Online Transaction Protocol Database (OLTP DB), a Customer Relation System (CRM) and a Human Resource (HR) System. Some data in the source system are static and only required to be loaded during the historical load. Examples of such static data of the dataset are date/time, industry segments, trade types and tax rates. Other non-static data such as the customer data are loaded in two additional incremental loads. The source file comprises a variety of data formats including XML, CSV and data dumps.

3 Our Approach

3.1 Modelling ETL Processes

We begin our design of ETL processes with the BPMN4ETL [2] conceptual model. This model customized from BPMN improves communication and validation between various stakeholders in a DW project. It serves as a standard means of documentation. It being technology-independent provides several paths to implementing ETLs including RA (which can be directly translated into SQLs), XML (which can be extended as in interchange format), and commercial tool specifications. In our approach, we develop a logical model with RA extended with update operations. RA provides a set of operators that manipulates relations to ensure that there is no ambiguity. This unambiguity will make it easier for us to repair the logical model upon DS schema changes. RA can be directly translated into SQL queries which is faster than commercial tool implementations. Commercial tools mostly add another layer for manipulations of ETL processes before finally SQL queries are generated to be executed on RDBMSs. We intend to implement ETL processes directly in the RDBMS environment from our logical model.

We explain our modelling approach with Fig. 1, the ETL process that loads the DimBroker dimension table of the TPC-DI. Two input data tasks inserts records from an HR.csv file (a DS file) and DimDate (a dimension table in the TPCDI DW) into the ETL flow. An aggregate task filters only records with EmployeeJobCode value of 314. Records from DimDate are aggregated to obtain the record with the minimum value in the DateValue column. Records of both paths are joined by a right outer join. Note that the condition of the join is just to ensure that each record at S3 is joined to the date value. A surrogate key column with the value of the row number is added the the ETL flow and finally, an insert data task inserts the records into DimBroker.

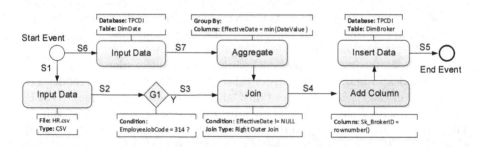

Fig. 1. Load of DimBroker dimension table

Now, we translate the BPMN model in Fig. 1 into our extended RA logical model in Fig. 2. Equations 1 and 2 stores records from HR.csv and Dim-Date into TempHR and TempDate temporary tables. Equation 3 selects records with EmployeeJobCode = '314' into Temp1. An aggregate operation retrieves the record with the minimum value in the DateValue column into Temp2 (Eq. 4). Temp1 and Temp2 are joined by a right outer join and the results are stored in Temp3 (Eq. 5). A surrogate key column (SK_BrokerID) is added in Eq. 6 and the records are inserted into DimBroker (Eq. 7).

Currently, we are working on BXF, an Extensible Markup Language (XML) interchange format for BPMN4ETL which is to express and interchange

BPMN4ETL model information across tools that are compliant with BPMN 2.0. This XML interchange format does not only describe the control flow of ETL processes but also the data flow.

3.2 ETL Workflow Reparation

RBR is a type of reasoning that uses (if-then-else) rule statements whereas CBR means understanding and solving a new problem by remembering how a similar old problem was solved. A problem and its solution could be described as a case. Case-based management systems learn by acquiring knowledge from their knowledge base and they also augment the knowledge base with newly developed knowledge. The overall system architecture may look as depicted in Fig. 3.

An ETL parser takes an ETL workflow in the form of RA or SQLs and parses the parts of each command of the workflow. The ETL manager assesses the impact of the DS change on each command of the ETL workflow. For DS changes that can be handled by applying rules stored in the rule base, the ETL Rewriter rewrites the commands in the ETL workflow by applying recommendations from the ETL manager. It iterates trough the commands and makes changes to each part. It also stores the history of versions of the ETL workflow. Therefore, it can rewrite the workflow into an old version if the new version becomes problematic. For situations that the ETL manager does not find any suitable rules, the ETL Rewriter waits for the user's input to rewrite ETL commands. The Rule Base contains distinct rules based on conditions. In the beginning of this framework implementation, there may be few rules in the rule base but with time, the rules are increased by the translation of cases into rules. Translation of cases into rules are done by the translator component of the framework. There are different ways in which rules can be inferred from cases. One of them is to make the translator check the case base anytime there is a new case, if it is possible to infer rules from the cases in the case base. Another is to wait for a specific amount of time or a specific number of new cases to be added to the case base before rules are inferred from cases. The choice of method of rule inference will depend on which one provides a more efficient and accurate framework.

For situations where no rules were found, or several solutions are applicable with the rules the user decides how the problem should be handled. The ETL rewriter rewrites the ETL commands in the ETL workflow by applying the user's

$$TempHR \leftarrow \mathcal{I}_{EmployeeID,ManagerID,FirstName,...}(HR.csv) \tag{1}$$

$$TempDate \leftarrow \mathcal{I}_{Sk\_dateid,Datevalue,Datedesc,...}(DimDate) \tag{2}$$

$$Temp1 \leftarrow \sigma_{EmployeeJobCode\ =\ '314'}(TempHR) \tag{3}$$

$$Temp2 \leftarrow \mathcal{A}_{EffectiveDate=min(DateValue)}(TempDate) \tag{4}$$

$$Temp3 \leftarrow Temp1 \bowtie_{EffectiveDate \neq NULL}(Temp2) \tag{5}$$

$$Temp4 \leftarrow \mathcal{E}_{SK\_BrokerID\ =\ rownumber()}(Temp3) \tag{6}$$

$$DimBroker \leftarrow DimBroker \cup (\pi_{SK\_BrokerID,\ EmployeeID,ManagerID,FirstName,...}(Temp4)) \tag{7}$$

Fig. 2. RA expressions to model the ETL for DimBroker

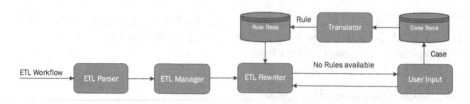

Fig. 3. Framework for ETL workflow reparation (E3TL)

decision. The DS change and its solution are stored in a knowledge base called the case base. The translator component applies algorithms to develop distinct rules from cases. These rules are stored in the rule base and are used to handle similar DS changes in the future.

3.3   Results

In this section we discuss the results of the experimental evaluation of our approach in modelling ETL processes. We implemented TPC-DI first, with our approach from BPMN4ETL into extended RA (using Postgres PLSQL) and second, with BPMN4ETL translated directly into the Pentaho Data Integration (PDI) tool. Both tests were run over an Intel i7 computer, with a RAM of 16 GB, running the Windows 10 Enterprise OS, using PostgresSQL as the data warehouse storage. The experiments showed that the SQL implementation runs orders of magnitude faster than that of PDI for all scale factors (SF). For example, the total execution time for the historical load of SF 3 in PLSQL was 12:50 min whereas that of PDI was 11 h and 23:52 min. Part of the reason for this poor performance of PDI is because it handles loops with the Copy Rows to Result step, that stores the rows in memory first and retrieves them one row at a time.

4   Conclusion

ETL process development is one of the complex and costly part of any data warehouse (DW) project. Currently, managing ETL workflows has some challenges. First, each ETL tool has its own model for specifying ETL processes making it is difficult to specify ETL processes that are beyond the capabilities of a chosen tool or switch between ETL tools without having to redesign the entire ETL workflow again. Second, a change in structure of a data source (DS) leads to ETL workflows that can no longer be executed and yields errors. This project provides a means of managing ETL processes in two ways. First, their modelling and second, their reparation upon DS schema changes. In the former, we model ETL processes with BPMN4ETL, an extended BPMN model for ETL at the conceptual level and with relational algebra (RA) extended with update operations at a logical level. With the later, we propose the E3TL framework in which we will develop algorithms for (semi-) automatic repair of ETL workflows upon DS schema changes. In order to assess the performance of our RA approach, we experimentally evaluated it on the TPC-DI benchmark. The results showed that our RA implementation is orders of magnitude faster than that of the PDI tool, for all scale factors. In future, we will evaluate the performance of our approach with other ETL tools such as Microsoft SSIS and Talend, and handle DW changes. We will also extend the E3TL framework to handle NoSQL environments.

Acknowledgements. The work of Judith Awiti is supported by the European Commission through the Erasmus Mundus Joint Doctorate project *Information Technologies for Business Intelligence-Doctoral College* (IT4BI-DC).

References

1. Curino, C., Moon, H.J., Tanca, L., Zaniolo, C.: Schema evolution in Wikipedia: toward a web information system benchmark. In: Proceedings of the 10th International Conference on Enterprise Information Systems, ICEIS 2008. pp. 323–332. Citeseer, Barcelona, Spain (2008)
2. El Akkaoui, Z., Zimányi, E.: Defining ETL worfklows using BPMN and BPEL. In: Proceedings of the ACM 12th International Workshop on Data Warehousing and OLAP, DOLAP 2009. pp. 41–48. ACM, Hong Kong, China (2009)
3. Muñoz, L., Mazón, J.-N., Pardillo, J., Trujillo, J.: Modelling ETL processes of data warehouses with uml activity diagrams. In: Meersman, R., Tari, Z., Herrero, P. (eds.) OTM 2008. LNCS, vol. 5333, pp. 44–53. Springer, Heidelberg (2008). https://doi.org/10.1007/978-3-540-88875-8_21
4. Papastefanatos, G., Vassiliadis, P., Simitsis, A., Sellis, T., Vassiliou, Y.: Rule-based management of schema changes at ETL sources. In: Grundspenkis, J., Kirikova, M., Manolopoulos, Y., Novickis, L. (eds.) ADBIS 2009. LNCS, vol. 5968, pp. 55–62. Springer, Heidelberg (2010). https://doi.org/10.1007/978-3-642-12082-4_8
5. Poess, M., Rabl, T., Jacobsen, H., Caufield, B.: TPC-DI: The first industry benchmark for data integration. Proc. of the VLDB Endowment **7**(13), 1367–1378 (2014)
6. Santos, V., Belo, O.: Using relational algebra on the specification of real world ETL processes. In: Proceedings of the 13th IEEE International Conference on Dependable, Autonomic and Secure Computing, DASC 2015. pp. 861–866. IEEE, Liverpool, UK (2015)
7. Trujillo, J., Luján-Mora, S.: A UML based approach for modeling ETL processes in data warehouses. In: Song, I.-Y., Liddle, S.W., Ling, T.-W., Scheuermann, P. (eds.) ER 2003. LNCS, vol. 2813, pp. 307–320. Springer, Heidelberg (2003). https://doi.org/10.1007/978-3-540-39648-2_25
8. Vassiliadis, P., Simitsis, A., Skiadopoulos, S.: Conceptual modeling for ETL processes. In: Proceedings of the 5th ACM International workshop on Data Warehousing and OLAP, DOLAP 2002. pp. 14–21. ACM, McLean, Virginia, USA (2002)
9. Vassiliadis, P., Simitsis, A., Skiadopoulos, S.: Modeling ETL activities as graphs. In: Proceedings of the 4th International Workshop on Design and Management of Data Warehouses, DMDW 2002. pp. 52–61. CEUR-WS.org, Toronto, Canada (2002)
10. Wojciechowski, A.: ETL workflow reparation by means of case-based reasoning. Inf. Syst. Front. **20**(1), 21–43 (2018)

Data Integration of Legacy ERP System Based on Ontology Learning from SQL Scripts

Chuangtao Ma[✉]

Department of Information Systems, Faculty of Informatics,
Eötvös Loránd University, Budapest 1117, Hungary
machuangtao@casear.elte.hu

Abstract. To tackle the problem of low-efficiency integration of heterogeneous data from various legacy ERP systems, a data integration approach based on ontology learning are presented. Considering the unavailability of database interface and diversity of DBMS and naming conventions of legacy information systems, a data integration framework for legacy ERP systems based on ontology learning from structured query language (SQL) scripts (RDB) are proposed. The key steps and technicality of the proposed framework and the process of ontology-based semantic integration are depicted.

Keywords: Data integration · Ontology learning · Knowledge extraction · SQL scripts

1 Introduction

Legacy Enterprise Resource Planning (ERP) systems is a kind of systems that were developed under various platforms in a few years ago. In spite of the technology of these systems are obsolete, they also have been playing a vital role in the process of daily management, operation, and decision [1]. In recent years, an increasing number of corporations plan to develop Business Intelligence (BI) systems for achieving central-management of data and intelligent decision-making driven by the requirement of business [2].

There are two alternative solutions: (1) Update and replace all of the existing legacy ERP systems, and then implement a brand-new BI system. (2) Integrate existing legacy ERP systems and create a central database for providing data of the BI system. However, it's not an easy task to replace all parts of legacy ERP systems, because legacy systems are critical for operation. In general, the technologies of legacy ERP systems lack documentation that fact increases the risk of upgrade project [3]. More importantly, for the majority of enterprise, especially for some Small Medium Enterprise (SMEs), they don't have enough budget for upgrading all of the legacy ERP systems [4]. Hence, it is necessary for them to build a BI system based on the integration of existing legacy ERP systems in the current dynamic and increasingly competitive business environment [5].

© Springer Nature Switzerland AG 2019
T. Welzer et al. (Eds.): ADBIS 2019, CCIS 1064, pp. 546–551, 2019.
https://doi.org/10.1007/978-3-030-30278-8_52

2 Problem Statement

The typical features of the legacy information system are unavailability of database interface and diversity of DBMS and naming conventions. In general, the legacy ERP systems were developed by various software cooperation under different (Integrated development environment) IDE, hence there is no uniform standard of APIs and DBMS. In addition, the data of legacy ERP systems regarding the same theme from different sub-systems are usually stored in different DBMS, e.g., SQL Servers, Oracle, Mysql and so forth. Meanwhile, there are various formats and naming conventions of the database table name, fields name, and so forth, hence there is a possibility of the same entity is named by various names in different legacy ERP systems [6].

Therefore, the problem of how to integrate the various data of legacy ERP systems efficiently and effectively are becoming an urgent task that should be studied. To integrate the heterogenous data from various legacy ERP systems. To tackle the above problems, a data integration approach based on ontology learning from SQL scripts is proposed. The main objective is to improve the degree of automation and efficiency of the data integration from various legacy ERP systems.

3 Related Work

3.1 Data Access

Aim to access data from legacy information systems, knowledge discovery metamodel (KDM) standard is proposed to represent the artifacts of legacy systems as entities, relationships and attributes [7]. In real industrial application scenery, a data reengineering based on common business-oriented language (COBOL) is designed for supporting the access the data from the relational database of legacy systems [8]. Considering the interoperability of ontology, ontology-based data access (OBDA) paradigm are presented to extract the log data from legacy information system [9]. In addition, a data access visualization method for legacy systems is proposed based on software clustering [10], to identify and represent the features of the entity automatically. However, ontology-based data access from distributed data-source still requires a data access interface to be available.

3.2 Knowledge Extraction

To achieve the semi-automated integration of data from legacy systems, the common RDF (Resource Description Framework) model extraction algorithm and data extraction algorithm were designed for extracting the knowledge from legacy systems [11]. Considering the limitations of static algorithms for extracting business rules from legacy systems, a dynamic knowledge extraction approach based on process mining [12, 13] and sequential pattern mining [14] are proposed respectively. In order to improve the efficiency of ontology construction, an ontology extraction approach from the relational database is proposed to construct the ontology from RDF [15].

3.3   Semantic Data Integration

Ontology-based semantic integration (OBA-SI) is proposed to tackle the semantic heterogeneity in existing rule-based data integration approaches [16]. To address the semantic consistency of enterprise application integration, an ontology-based semantic integration approach is presented in [17]. In order to improve the precision of semantic integration results, a DISFOQuE system for data integration based on fuzzy ontology is designed [18]. In addition, a semantic integration approach exploiting linked data are presented to achieve RDF data integration based on query rewriting [19]. Nevertheless, the efficiency of semantic integration is also limited by the quality of knowledge graph and ontology.

4   Data Integration Based on Ontology Learning

To improve the efficiency and accuracy of data accessing, knowledge extraction and ontology construction, ontology learning from SQL scripts is introduced to (semi-) automatically construct ontologies, which help to integrate heterogeneous data from various legacy ERP systems at semantic level.

4.1   Ontology Learning from SQL Scripts

Ontology learning is a core element in the proposed data integration framework, the detailed process of ontology learning from relational database SQL scripts could be depicted in Fig. 1.

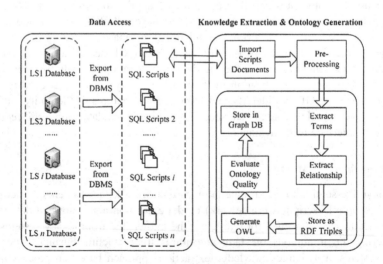

Fig. 1. Ontology learning from SQL scripts.

The specified steps of data integration based on ontology learning are depicted from the perspectives of data access, knowledge extraction, and ontology generation.

Data access from various legacy systems. To integrate the heterogeneous data from various legacy systems, it is necessary to access data. Considering the features of legacy ERP systems, we intend to directly access data through DBMS so that the SQL scripts are exported from DBMS of the various legacy ERP systems and stored in text documents.

Knowledge extraction from the relational database. Initially, the NLP techniques, e.g., sentence breaking, word segmentation, named entity recognition (NER), and so forth, are adopted to pre-process the text documents of SQL scripts. After that, the ontology learning model is employed to extract the knowledge from query RDBMS scripts. The SQL scripts as documents will be the input of ontology learning model, and the corresponding output is the elements and relationship between knowledge that stored with the form of RDF triples [20].

Generate OWL object from RDF schema. On the basis of knowledge extraction, the OWL (Web Ontology Language) object are generated for representing the elements and its relationships between various knowledge items accurately. In addition, an ontology-based knowledge graph utilizing relational database SQL query scripts is built and stored in Neo4j graph database.

4.2 Ontology-Based Semantic Data Integration

There is a critical task is to integrate the heterogeneous data from various legacy ERP systems after generating ontologies. In order to illustrate the process of ontology-based semantic data integration, the integration demo of two entities from different ERP systems are selected, and the corresponding processes are depicted in Fig. 2.

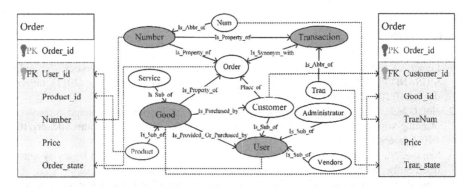

Fig. 2. Ontology-based semantic data integration of legacy ERP systems.

As shown in Fig. 2, the ontology-based knowledge graph is employed to interpret and reconcile the conflicts between different database entities. The entity is represented by ellipse filled with gray, while the relationships between various entities are represented by the solid arrow in the ontology-based knowledge graph. It is obvious that the ontology-based knowledge graph provides a bridge for integrating the heterogeneous entities from various legacy ERP systems, where the dotted arrow represents the semantic connection between heterogeneous data entity and ontology-based knowledge graph.

5 Conclusion and Further Work

This paper presents a data integration framework based on ontology learning to integrate heterogenous date from various legacy ERP systems. The main contribution is to propose a novel solution for generating ontologies by ontology learning from relational database SQL scripts, which is beneficial to integrate the heterogeneous data from legacy ERP systems efficiently and effectively.

Nevertheless, the proposed solution is currently in the initial exploration phase. The following works should be investigated and studied in the next research: (1) Knowledge extraction algorithm from SQL scripts. The knowledge extraction algorithm based on NLP will be designed to extract the knowledge from SQL scripts; (2) Ontology generation approaches from RDF schema. The ontology generation approaches from RDF schema will be studied to generate the ontology for supporting the integration of heterogeneous data; (3) Design of the tools to support data integration. It is necessary for to design and develop the data integration tool after generating the ontology from RDF schema and extracting knowledge from SQL scripts.

Acknowledgment. This work was supported by grants of the European Union co-financed by the European Social Fund (EFOP-3.6.3-VEKOP-16-2017-00002) and the China Scholarship Council (201808610145).

References

1. Lenart, A.: ERP in the cloud – benefits and challenges. In: Wrycza, S. (ed.) SIGSAND/PLAIS 2011. LNBIP, vol. 93, pp. 39–50. Springer, Heidelberg (2011). https://doi.org/10.1007/978-3-642-25676-9_4
2. Nofal, M.I., Zawiyah, M.Y.: Integration of business intelligence and enterprise resource planning within organizations. Procedia Technol. **11**, 658–665 (2013). https://doi.org/10.1016/j.protcy.2013.12.242
3. Serrano, N., Hernantes, J., Gallardo, G.: Service-oriented architecture and legacy systems. IEEE Softw. **31**(5), 15–19 (2014). https://doi.org/10.1109/MS.2014.125
4. Ahmad, M.M., Ruben, P.C.: Critical success factors for ERP implementation in SMEs. Robot. Comput. Integr. Manuf. **29**(3), 104–111 (2013). https://doi.org/10.1016/j.ijinfomgt.2009.03.001
5. Malhotra, R., Cecilia, T.: Critical Decisions for ERP Integration: small business issues. Int. J. Inf. Manage. **30**(1), 28–37 (2010)
6. Singh, R., Singh, K.: A descriptive classification of causes of data quality problems in data warehousing. Int. J. Comput. Sci. Issues **7**(3), 41–50 (2010)
7. Pérez-Castillo, R., De Guzman, I.G.R., Piattini, M.: Knowledge discovery metamodel-ISO/IEC 19506: a standard to modernize legacy systems. Comput. Stand. Interfaces **33**(6), 519–532 (2011). https://doi.org/10.1016/j.csi.2011.02.007

8. Millham, R., Yang, H.: Industrial report: data reengineering of COBOL sequential legacy systems. In: Proceedings of 33rd Annual IEEE International Computer Software and Applications Conference, vol. 1, pp. 646–647. IEEE, Seattle (2009)
9. Calvanese, D., Kalayci, T.E., Montali, M., Tinella, S.: Ontology-based data access for extracting event logs from legacy data: the onprom tool and methodology. In: Abramowicz, W. (ed.) BIS 2017. LNBIP, vol. 288, pp. 220–236. Springer, Cham (2017). https://doi.org/10.1007/978-3-319-59336-4_16
10. Yano, K., Matsuo, A.: Data access visualization for legacy application maintenance. In: Proceedings of 24th IEEE International Conference on Software Analysis, pp. 546–550. IEEE, Klagenfurt (2017). https://doi.org/10.1109/SANER.2017.7884671
11. Ilya, S., Dmitry, M.: Semi-automated integration of legacy systems using linked data. In: Proceedings of 4th International Conference on Analysis of Images, Social Networks and Texts, pp. 166–171. Ural Federal University, Yekaterinburg (2015)
12. Kalsing, A.C., do Nascimento, G.S., Iochpe, C., et al.: An incremental process mining approach to extract knowledge from legacy systems. In: Proceedings of 14th IEEE International Enterprise Distributed Object Computing Conference, pp. 79–88. IEEE, Vitoria (2010). https://doi.org/10.1109/EDOC.2010.13
13. Pérez-Castillo, R., Weber, B., de Guzman, et al.: Process mining through dynamic analysis for modernising legacy systems. IET Softw. 5(3), 304–319 (2011). https://doi.org/10.1049/iet-sen.2010.0103
14. Sartipi, K., Safyallah, H.: Dynamic knowledge extraction from software systems using sequential pattern mining. Int. J. Softw. Eng. Knowl. Eng. 20(6), 761–782 (2010). https://doi.org/10.1142/S021819401000492X
15. Santoso, H.A., Haw, S.C., Abdul-Mehdi, Z.T.: Ontology extraction from relational database: concept hierarchy as background knowledge. Knowl. Based Syst. 24(3), 457–464 (2011). https://doi.org/10.1016/j.knosys.2010.11.003
16. Gardner, S.P.: Ontologies and semantic data integration. Drug Discov. Today 10(14), 1001–1007 (2005)
17. Calhau, R.F., De Almeida Falbo, R.: An ontology-based approach for semantic integration. In: Proceedings of 14th IEEE International Enterprise Distributed Object Computing Workshop, pp. 111–120. IEEE, Vitoria (2010). https://doi.org/10.1109/EDOC.2010.32
18. Yaguinuma, C.A., Afonso, G.F., Ferraz, V., Borges, S., et al.: A fuzzy ontology-based semantic data integration system. J. Inf. Knowl. Manag. 10(3), 285–299 (2011). https://doi.org/10.1109/IRI.2010.5558938
19. Correndo, G., Salvadores, M., Millard, I., Glaser, H.: SPARQL query rewriting for implementing data integration over linked data. In: Proceedings of 2010 EDBT/Workshops, pp. 1–11. ACM, Lausanne. https://doi.org/10.1145/1754239.1754244
20. Li, Y.F., Kennedy, G., Ngoran, F., et al.: An ontology-centric architecture for extensible scientific data management systems. Future Gener. Comput. Syst. 29(2), 641–653 (2013). https://doi.org/10.1016/j.future.2011.06.007

Business Intelligence & Analytics Applied to Public Housing

Étienne Scholly[1,2(✉)]

[1] University of Lyon, Lyon 2, ERIC EA 3083, Lyon, France
[2] BIAL-X, Limonest, France
etienne.scholly@bial-x.com
https://eric.ish-lyon.cnrs.fr, https://www.bial-x.com

Abstract. Business Intelligence, with data warehouses, reporting and
OnLine Analytical Processing (OLAP) are about twenty years old technologies, they are mastered and widely used in companies. Their goal is
to collect, organize, store and analyse data to support decision-making.
In parallel, there are many algorithms from Data Science for conducting
advanced data analyses, including the ability to conduct predictive analyses. However, the reflection on the integration of Data Science methods
into reporting or OLAP analysis is relatively incomplete, although there
is a real demand from companies to integrate prediction into decision-
making processes. In the meantime, with the rise of the Internet, the
proliferation of multimedia data (sound, image, video, etc.), and the fast
development of social networks, data has become massive, heterogeneous,
of diverse and rapid varieties. The Big Data phenomenon challenges the
process of data storage and analysis and creates new research problems.

The PhD thesis is at the junction of these three main topics: Business
Intelligence, Data Science and Big Data. The objective is to propose an
approach, a framework and finally an architecture allowing prediction to
be made in a decision-making process, but with a Big Data perspective.

Keywords: Business intelligence · Data Science · Big data

1 Introduction

Business intelligence (BI) refers to all the methods and tools used to collect,
store, organize and analyze data. The objective is to provide decision-makers,
such as business leaders, with an overview of the data as well as decision support, with the ultimate aim of improving the activity managed by these decision-
makers [16]. BI is mainly based on the feeding and interrogation of a data warehouse [11]. The first phase is usually managed through ETLs (Extract - Transform - Load), which extract data from operational sources, transform them to
meet upstream expectations, and store them in the data warehouse, which is
very often a relational database. The second part consists in querying the data
warehouse: it is mainly done through OnLine Analytical Processing (OLAP) and
dashboards (reporting) [16].

© Springer Nature Switzerland AG 2019
T. Welzer et al. (Eds.): ADBIS 2019, CCIS 1064, pp. 552–557, 2019.
https://doi.org/10.1007/978-3-030-30278-8_53

Data Science allows to go further than the analyses proposed by BI. Data Science refers, in its broadest definition, to the extraction of knowledge from data. In this whole process, it is therefore necessary to clean, prepare, and analyze the data [15]. Data Science is at the intersection of several other disciplines, such as statistics, Artificial Intelligence (AI) and data visualization. One of the fields of study of AI is Machine Learning. It uses essentially statistical algorithms to give a computer the ability to "learn". Among these algorithms, we can distinguish two categories: unsupervised learning gathers observations into homogeneous groups; and supervised learning, where one of the objectives is to create a model from a sample of data with a known membership class, and then test it with new data by trying to predict their class [14].

For a few years now, we hear about the "Big Data" phenomenon. Initially, this term referred to the amount of data available that is growing exponentially, but reducing this phenomenon to volume is too narrow. The first consensual definition of Big data is the "3 V's": Volume, Variety and Velocity [7]. More recently, two V terms have been added to the definition of Big Data, thus creating the "5 V's": Value and Veracity [4]. Traditional database storage and management systems have reached their limits and do not allow Big Data to be processed, forcing researchers and industrials to rethink all data processing methods, whether in BI or Data Science.

The goal of the PhD project is to unify BI with Data Science, in the general context of Big Data. It means that we want to be able to handle Big Data and enable users to run any kind of analyses, whether BI or Data Science ones. Indeed, we think that both types of analyses are needed and will co-exist. To this end, we will need a central storage area that can handle Big Data, and that can be queried by multiple applications downstream. BI analyses will mostly be conducted on structured data, while Data Science methods can process any kind of data. Data Science analyses will either be used as ad hoc analyses, or to feed the data warehouse with advanced indicators calculated through predictive methods (for instance).

The rest of the document is organized as follows. Section 2 presents the state of the art on Business Intelligence and Analytics. Section 3 explicits the two main problems of the PhD project. Section 4 announces results obtained so far and the work that is still to be done.

2   State of the Art

The use of Data Science methods on a company's data is associated with the term *Business Analytics* (BA) [3]. Thus, by relying on the strengths of Data Science, BA makes it possible to carry out more advanced analyses than BI-specific analyses, and especially to "look forward" with supervised learning [12]. The ability to conduct predictive and prescriptive analyses opens up a wide range of new decision support applications [14].

Although the Data Science methods used by BA have been around for a long time, BA is still relatively new, particularly compared to BI. As a result, BA's definition is still rather vague, especially since the range of possibilities is very wide thanks to the diversity of existing Data Science methods. Some authors use the term *Business Intelligence & Analytics* (BI&A) to signify that the two domains, BI and BA, complement each other and form a whole [3]. Others suggest that BA is an extension of BI and tends to replace it by proposing new and more advanced analyses [14]. We believe the term is not essential because it is mainly marketing.

Authors have proposed BI&A architectures to meet the challenges of Big Data. Baars and Ereth [1] propose a BI&A platform that can manage Big Data. It relies on what they call *analytical atoms*, which can be seen as small, autonomous data warehouses ready for analysis. The atom's data has been historized, versioned, enriched, cleaned, etc. The idea is that there is an analytical atom for each incoming data source, and these atoms can be combined to create what the authors call virtual data warehouses. Gröger [8] proposes a very general data analysis platform, which is based on the *lambda architecture*. It is a standard for developing high-performance systems that can manage both batch and stream processing. Data is stored in a data lake, and numerous applications retrieve the data stored in the lake.

In addition to these BI&A architectures, a new approach to organizing and storing data is emerging: Data Lakes. This term was first introduced by Dixon [6], offering this alternative to datamarts, which are subsets of data warehouses. A data lake is a vast repository of raw data, of heterogeneous structures, and fed by several external sources. Data lakes rely on the *schema-on-read* property: data is stored in its raw format, and the schema is only specified when queried [13]. It is therefore mandatory to have an efficient metadata system in the data lake, which makes it possible to query data stored in the lake. A data lake without an efficient metadata system is called a data swamp, and is completely unusable [13]. From a technological point of view, most data lake implementations are based on Apache Hadoop, although new solutions are gradually emerging [13].

3   The PhD Project

Our objective is to combine BI and BA in the general context of Big Data, by enabling users to conduct both types of analyses, separately or together, on Big Data. We consider that BA analyses can be performed as ad hoc studies (through clustering for example), and can also be used to consolidate BI analyses, particularly through advanced indicators, for example generated by predictive algorithms. Since most companies will always have "classic" BI analysis needs, we choose to keep a BI framework in our thinking, based on OLAP - if necessary - and dashboards.

Given the current state of the art in BI, BA and Big Data, we decide to draw inspiration from the work of Baars and Ereth [1], in particular their notion of an analytical atom, which we wish to expand. We are also inspired by the idea of a general data analysis platform such as the one introduced by Gröger [8]. Finally, we use the concept of data lakes, which will be at the heart of our approach.

More concretely, two types of data will feed the data lake: internal data, which is the company's data, and external data, collected from the Internet. Internal data will, in most cases, be structured data that does not present Big Data type problems. This structured internal data represents the core of the company's BI activity and is therefore traditionally managed by an ETL, which feeds a data warehouse, and reports are created from the data warehouse. External data, coming from very heterogeneous sources, can present Big Data issues. External data is the main source of BA analyses, although internal data can also be analyzed by these techniques. We believe that the main strength of external data lies in their capacity to be associated with the company's internal data, either by crossing them to have more variables or observations available, or by generating "advanced" indicators that can then be used in BI analyses.

We believe that the analytic atom concept, combined with the object notion [5], will help us propose a metadata system for a data lake efficient enough to manage any type of data, Big Data included. Although several studies have already been carried out on metadata systems for data lakes, and some of them have proven their efficiency [2, 9, 10], we believe we can offer a more complete metadata system that offers all the features we consider essential, and thus completely meets our expectations.

This constitutes our first problem. It is vital for our work to have a storage area in which to store all data, both internal and external. Various applications are powered by this storage area (in this case, the data lake): data warehouse, predictive models, visualization, statistics, etc. The use of the lake is made possible thanks to a reliable metadata system, and the latter also helps users to better understand data and track its usage. In addition to the metadata system, the development of the entire data lake is necessary.

Once the development of the data lake and its metadata system is complete, we will have to feed it with data. This is the basis of our second problem. The PhD is conducted in partnership between the ERIC laboratory and BIAL-X through a CIFRE convention. Our work is anchored in the business issue of public housing, and more particularly in the study of the attractiveness of a social dwelling. Indeed, this is a key issue for BIAL-X because a significant part of the company's activity is dedicated to social landlords. The latter have data about their dwellings (number of rooms, surface area, energy category, construction date, etc.). However, finding information on the environment in which the dwelling is located is a more complex problem: this can be found in external data, collected from the Internet.

Combining external data with internal data can lead to the discovery of new and valuable information. For example, users can create "advanced indicators" with the help of analyses ran on external data to fine-tune indicators already existing in BI analyses. Even unstructured data can be processed: tweets to find the general opinion on a district, or pictures to determine if dwellings are similar, for instance. Thus, we would like to provide social landlords with the opportunity to simultaneously manage their activity through "classic" BI analyses, discover hidden insights in their data with BA analyses, and finally enrich BI analyses

with advanced indicators such as the attractiveness of their dwellings. All these
use cases would be derived from the data lake.

4   First Results and Future Outcomes

We have started to work on the first issue presented in Sect. 3. This work was
carried out with another doctoral student from the ERIC laboratory who works
on textual data lakes. After an overview of the definitions of a data lake from
the literature, we proposed our own definition of this concept. We then identified
six key features that the metadata system of a data lake must provide, in our
opinion, to be as robust as possible in addressing the Big Data issues and the
schema-on-read approach. Comparing existing metadata systems, we showed
that some works offer five out of six features, but none offers all six.

We proposed a metadata typology for data lakes. It is based on the object
notion, which represents any set of homogeneous data [5], and the typology
declines metadata into three categories. Intra-object metadata describe objects
through versions, representations and various properties to name a few; inter-
object metadata explain how objects are linked together; and global metadata
facilitate and improve data analyses and the use of the data lake in general.

We also introduced a graph-based model of our metadata typology named
MEDAL. An object is represented by a hypernode, and it contains nodes (representations and versions) connected via edges (updates and transformations).
Hypernodes are linked through edges or hyperedges. Theoretically, MEDAL proposes all six key features identified.

However, we have not yet implemented MEDAL. This constitutes a technological challenge: we will have to identify which technologies we think are relevant
to implement our data lake. There are many existing tools to implement a data
lake, but after a quick review, we think that we might have to develop our own
graph-based metadata management interface. At first sight, Apache Kafka seems
to us to be a suitable way to implement our data lake, although other ways are
still possible. Among our upcoming studies, a complete survey on existing tools
will be carried out.

Once the data lake is implemented and functional, it will then have to be
supplied with data. This constitutes the answer to the second issue presented in
Sect. 3. We will insert into the data lake the social landlord's data, which mainly
contains the properties of social dwellings, as well as external data retrieved
from the Internet, mainly from open data. The latter are the ones that will have
the most Big Data problems, especially in terms of data variety. Our ultimate
objective will be to exploit the capabilities of our metadata system to have both
raw and reworked data in the lake, to feed various applications (including a
data warehouse and predictive models), to try to fine-tune the attractiveness
of a home, and to enable social landlords to better manage their activity by
comparing their data to external information.

References

1. Baars, H., Ereth, J.: From data warehouses to analytical atoms-the Internet of Things as a centrifugal force in business intelligence and analytics. In: 24th European Conference on Information Systems (ECIS), Istanbul, Turkey. Research Paper 3 (2016)
2. Beheshti, A., Benatallah, B., Nouri, R., Chhieng, V.M., Xiong, H., Zhao, X.: CoreDB: a data lake service. In: 2017 ACM on Conference on Information and Knowledge Management (CIKM 2017), Singapore, Singapore, pp. 2451–2454. ACM, November 2017. https://doi.org/10.1145/3132847.3133171
3. Chen, H., Chiang, R.H., Storey, V.C.: Business intelligence and analytics: from big data to big impact. MIS Q. **36**(4), 1165–1188 (2012)
4. Chen, M., Mao, S., Liu, Y.: Big data: a survey. Mob. Netw. Appl. **19**(2), 171–209 (2014)
5. Diamantini, C., Giudice, P.L., Musarella, L., Potena, D., Storti, E., Ursino, D.: A new metadata model to uniformly handle heterogeneous data lake sources. In: Benczúr, A., et al. (eds.) ADBIS 2018. CCIS, vol. 909, pp. 165–177. Springer, Cham (2018). https://doi.org/10.1007/978-3-030-00063-9_17
6. Dixon, J.: Pentaho, Hadoop, and Data Lakes, October 2010. https://jamesdixon.wordpress.com/2010/10/14/pentaho-hadoop-and-data-lakes/
7. Gandomi, A., Haider, M.: Beyond the hype: big data concepts, methods, and analytics. Int. J. Inf. Manag. **35**(2), 137–144 (2015)
8. Gröger, C.: Building an industry 4.0 analytics platform. Datenbank-Spektrum **18**(1), 5–14 (2018)
9. Halevy, A.Y., et al.: Goods: organizing Google's datasets. In: Proceedings of the 2016 International Conference on Management of Data (SIGMOD 2016), San Francisco, CA, USA, pp. 795–806, June 2016. https://doi.org/10.1145/2882903.2903730
10. Hellerstein, J.M., et al.: Ground: a data context service. In: 8th Biennial Conference on Innovative Data Systems Research (CIDR 2017), Chaminade, CA, USA, January 2017. http://cidrdb.org/cidr2017/papers/p111-hellerstein-cidr17.pdf
11. Inmon, W.H.: Building the Data Warehouse. Wiley, New York (1996)
12. Larson, D., Chang, V.: A review and future direction of agile, business intelligence, analytics and data science. Int. J. Inf. Manag. **36**(5), 700–710 (2016)
13. Miloslavskaya, N., Tolstoy, A.: Big data, fast data and data lake concepts. Procedia Comput. Sci. **88**, 1–6 (2016). https://doi.org/10.1016/j.procs.2016.07.439. 7th Annual International Conference on Biologically Inspired Cognitive Architectures (BICA 2016), NY, USA
14. Mortenson, M.J., Doherty, N.F., Robinson, S.: Operational research from taylorism to terabytes: a research agenda for the analytics age. Eur. J. Oper. Res. **241**(3), 583–595 (2015)
15. Shmueli, G., Koppius, O.R.: Predictive analytics in information systems research. MIS Q., 553–572 (2011)
16. Watson, H.J., Wixom, B.H.: The current state of business intelligence. Computer **40**(9), 96–99 (2007)

Textual Data Analysis from Data Lakes

Pegdwendé N. Sawadogo[✉]

Université de Lyon, Lyon 2, ERIC EA 3083, 5 avenue Pierre Mendès France,
69676 Bron, France
pegdwende.sawadogo@univ-lyon2.fr
http://eric.univ-lyon2.fr/sawadogop/

Abstract. Over the last decade, the data lake concept has emerged as
an alternative to data warehouses for data storage and analysis. Data
lakes adopt a schema-on-read approach to provide a flexible and extend-
able decision support system. In absence of a fixed schema, data querying
and exploration depend on a metadata system. However, existing works
on metadata management in data lakes mainly focus on structured and
semi-structured data, with little research on unstructured data. Thence,
we propose in this thesis a methodological approach to enable textual
data analyses from data lakes through an efficient metadata system.

Keywords: Data lake · Metadata management · Olap

1 Introduction and Context

Since the beginning of the 21st century, we observe a tremendously growth of data
production in the world. So-called big data give great opportunities to organizations. Thus, in the marketing domain, big data can be used to improve customer
retention through customer profiles identification. Similarly, big data may serve
to improve the efficiency in industries and health-care through anomaly prediction. However, such applications require overcoming the challenges posed by the
volume, variety and velocity of big data.

To address these challenges, several adaptations of the data warehouse concept were proposed. That is, some approaches make it possible to ensure scalability in data warehouses and others support fast data. Nevertheless, data warehouses remain challenged by unstructured data. Yet, the majority of big data
is unstructured [13]. In addition, data warehouses hardly support heterogeneous
and changing data because of their fixed schema.

More recently, the data lake concept was proposed to address these issues. A
data lake is a large repository of raw and heterogeneous data from which various
analyses can be performed [4]. Data lakes adopt a schema-on-read approach to
ensure flexibility and to avoid data loss. That is, ingested data do not follow
any fixed schema. Therefore, data access and querying depends on a metadata
system.

© Springer Nature Switzerland AG 2019
T. Welzer et al. (Eds.): ADBIS 2019, CCIS 1064, pp. 558–563, 2019.
https://doi.org/10.1007/978-3-030-30278-8_54

However, existing works on metadata management in data lakes mostly focus on structured and semi-structured data [2,7,8,11], with little research on unstructured data. Furthermore, a substantial part of the literature restricts data lake usage to data scientists [5,10,12]. Thus, according to this conception, only on-demand analyses can be performed from a data lake. Therefore, opening a data lake to industrialized analyses is an open issue.

2   Main Purposes

This thesis aims to address data lakes issues related to textual data. This may be done through three main steps. First of all, we need to introduce a more complete and generic metadata model for data lakes than existing ones. Such a metadata model may then serve to build an efficient metadata system that allow to simultaneously deal with structured, semi-structured and especially with unstructured data in data lakes.

Afterwards, we will propose, on the basis of our metadata model, a way to activate industrialized analyses in a data lake. We shall particularly focus on textual data analyses using Text-OLAP. For this purpose, we need to propose an approach to dynamically define dimensions and hierarchies. It may also be necessary to introduce new Text-OLAP measures or re-adapt existing measures, in order to comply with the context of data lakes.

Last but not least, we should identify an efficient storage mode to ensure scalability in a data lake, still in the context of textual data. To do this, we should identify and compare different storage strategies and tools. In essence, that would give a clear guideline to build an optimized textual data lake in terms of storage and query cost.

3   Related Works

Most research on metadata management in data lakes adopts a graph approach to organize the metadata system. We particularly identify two main usages of the graph representation.

Data Provenance Tracking. Data provenance is the information about the interactions of activities, data and people in a data lake [16]. In this approach, the metadata system is viewed as a provenance graph, i.e., a directed acyclic graph (DAG) where nodes represent entities and edges express activities. Entities can be users, user profiles or datasets [2]. Activities are generally described through a timestamp and a set of additional characteristics such as the type of activity (reading, creation, modification) [2], the status of the system (CPU, RAM, bandwidth) [16] or even the script used [9].

Data provenance tracking helps ensure the reproducibility of the processes in data lakes through version management. Therefore, provenance metadata can be used to understand, explain and repair inconsistencies in the data [2]. They may also serve to deal with sensitive data, by detecting intrusions [16].

Generation of Similarities and Connections. This second approach focuses on detecting and expressing similarities between datasets. In its simplest variant, datasets are represented by nodes and the link between a couple of datasets is expressed through a weighted or unweighted edge [6,11]. In the case of unweighted edges, the edge is simply interpreted as a connection [6]. Some similarity measures are used to calculate the strength of the similarity, or just to detect whether the similarity is significant.

An enhanced variant of this approach splits each dataset into a set of nodes. Thus, a connection between a couple of datasets is expressed between two of their inherent nodes [1,3]. A simple similarity measure such as a string similarity may be sufficient for this purpose, since the nodes may express atomic data [3].

Data links are particularly useful, because they allow to give a certain structure to unstructured data, and thus to enable automatic processes [3]. More concretely, similarity metadata can be used to detect some clusters of data, i.e., groups of datasets that are strongly linked to each other [6]. Another use of data similarity may be to automatically recommend to lake users the data related to the data they use [11].

Discussion. These two approaches for metadata representation in data lakes can be considered as complementary rather than contradictory. That is, similarities and connections expand the range of available analyses in the lake, while data provenance gives a better control of data lake usage. Moreover, they all adopt a graph approach to represent the metadata system. Therefore, we envisage to combine these features to build a more comprehensive graph-based metadata management model for data lakes. However, some adaptations are required to integrate as far as possible additional features to the metadata management model.

4   First Results: A Graph-Based Model to Organize Metadata in Data Lakes [14, 15]

General Description. We propose a metadata model that distinguishes intra-object metadata, inter-object metadata and global metadata. The concept of object expresses any set of homogeneous data that can be analyzed individually, e.g., a textual document, an image, a video, a database table, etc.

Intra-object metadata express information about a specific object and its derivatives. As illustrated in Fig. 1, such metadata are organized in a DAG for each object. The origin node represents raw data and other nodes correspond to other declinations (versions or representations) of the object. A *version* (dark colored in Fig. 1) results from an *update* operation on raw data; while a *representation* (light colored in Fig. 1) is a possibly more structured data obtained after a *transformation* operation. Each transformation or update operation is translated with all its characteristics (timestamp, user, actions, etc.) through an edge going from the source node to the new node.

To illustrate these concepts, let us consider an organization where applicant resumes are stored in a data lake. The resume of an applicant is initially represented by a simple node. Then, when an updated version of a resume is submitted, a second node is created, and linked to the first one through an incoming update edge. Let us now imagine that we need to automatically match resumes with a job offer. A solution is to transform each resume as well as the job offer into a term-frequency vector, and then compute the cosine similarity. The vector representation of each resume would be expressed in our metadata model by a new representation node, linked to the raw resume with an incoming transformation (vectorization) edge.

Inter-object metadata express links and similarities between objects. We distinguish three types of inter-object metadata. Data *groupings* are a way to split the set of objects into collections w.r.t. a specific characteristic. For example, a grouping based on language may generate a collection of resumes in English and another resumes in French (Fig. 2). Each generated collection is expressed in our metadata model with an hyperedge, i.e., an edge that can connect more than two nodes.

Another way to express links between data consists in calculating *similarity measures* between objects. This may be done by applying usual similarity measures (cosine, chi-square, spearson's correlation coefficient, etc.) on equivalent representations of two objects. For example, cosine similarities may be generated between vector representations of resumes. Such metadata are expressed in the metadata model through weighted edges.

A third type of links between data is the *consanguinity* relationship that is generated when a new object is obtained from the join of others. In this case, there is a *joinability* link between the joined objects and a *consanguinity* relationship between the older objects and the newer one. This link is expressed by an oriented hyperarc.

Global metadata are data structures that are not directly associated to any specific object. They are made of semantic databases (ontologies, taxonomies, thesauri, etc.), indexes (or inverted indexes) and logs. These metadata are not expressed in our metadata model because they mainly relate to operational implementation, and therefore highly depend on the adopted technologies.

Expected Features. Our metadata model allows to manage data versioning and multiple data representations. This helps improve the data consistency and save repeated processes in data lakes.

This metadata model can be considered as a generalization of the two main approaches from the literature, because it may serve to both express links between data and track data provenance. This makes it the most complete in the literature as shown in [15].

Finally, one of the most valuable innovations expected from our metadata model is the activation of OLAP-like analysis. That is, graph-OLAP analyses may be envisaged thanks to this metadata model. Indeed, data collections can serve as dimensions during analyses. Hence, an intersection of a couple data

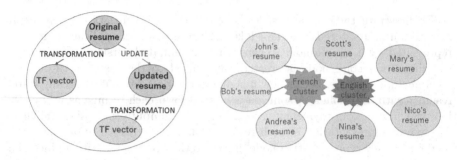

Fig. 1. Sample hypernode and its representation tree (Color figure online)

Fig. 2. Sample data grouping w.r.t. language

collections could be used to filter data, like OLAP slice and dice operations. Then, different visualizations can serve to express some aggregated indicators of the resulting data, like OLAP measures.

5 Conclusion and Future Prospects

Our first results are promising. However, they remain theoretical for now. Thence, the next step focuses on implementing our metadata system and test it on a real-life case. Thus, we could experimentally evaluate our proposition on the basis of enabled analyses.

Afterwards, we envisage to propose, from our metadata model, a methodological approach to setup industrialized text analyses from a data lake. This may be done through the conception of an efficient OLAP-like analysis platform dedicated to business users.

However, this task implies three key issues. The first is about adapting Text-OLAP measures, as most text-OLAP measures were proposed in the context of data warehouses. Some adaptations seem necessary to match with our metadata model. The context of data lakes also induces a need of dynamic dimensions and hierarchies, instead of (mostly) static ones in data warehouses, which is another key challenge. Eventually, our analysis platform would need to be scalable. Thus, we should identify the most efficient storage modes and engines for textual data.

Acknowledgments. The research presented in this article is funded by the Auvergne-Rhône-Alpes Region, as part of the AURA-PMI project that finances Pegdwendé Nicolas Sawadogo's PhD thesis.

References

1. Alrehamy, H., Walker, C.: Personal data lake with data gravity pull. In: BDCloud 2015, Dalian, China. IEEE Computer Society Washington, vol. 88, pp. 160–167, August 2015. https://doi.org/10.1109/BDCloud.2015.62

2. Beheshti, A., Benatallah, B., Nouri, R., Chhieng, V.M., Xiong, H., Zhao, X.: CoreDB: a Data Lake Service. In: CIKM 2017, Singapore, Singapore. pp. 2451–2454. ACM (November 2017). https://doi.org/10.1145/3132847.3133171

3. Diamantini, C., Giudice, P.L., Musarella, L., Potena, D., Storti, E., Ursino, D.: A new metadata model to uniformly handle heterogeneous data lake sources. In: Benczúr, A., et al. (eds.) ADBIS 2018. CCIS, vol. 909, pp. 165–177. Springer, Cham (2018). https://doi.org/10.1007/978-3-030-00063-9_17

4. Dixon, J.: Pentaho, Hadoop, and Data Lakes, October 2010. https://jamesdixon.wordpress.com/2010/10/14/pentaho-hadoop-and-data-lakes/

5. Fang, H.: Managing data lakes in big data era: what's a data lake and why has it became popular in data management ecosystem. In: CYBER 2015, Shenyang, China, pp. 820–824. IEEE, June 2015. https://doi.org/10.1109/CYBER.2015.7288049

6. Farrugia, A., Claxton, R., Thompson, S.: Towards social network analytics for understanding and managing enterprise data lakes. In: ASONAM 2016, San Francisco, CA, USA. pp. 1213–1220. IEEE, August 2016. https://doi.org/10.1109/ASONAM.2016.7752393

7. Hai, R., Geisler, S., Quix, C.: Constance: an intelligent data lake system. In: SIGMOD 2016, San Francisco, CA, USA, pp. 2097–2100. ACM Digital Library, July 2017. https://doi.org/10.1145/2882903.2899389

8. Halevy, A., et al.: Managing Google's data lake: an overview of the GOODS system. In: SIGMOD 2016, San Francisco, CA, USA, pp. 795–806. ACM, July 2016. https://doi.org/10.1145/2882903.2903730

9. Hellerstein, J.M., et al.: Ground: a data context service. In: CIDR 2017, Chaminade, CA, USA, January 2017. http://cidrdb.org/cidr2017/papers/p111-hellerstein-cidr17.pdf

10. Khine, P.P., Wang, Z.S.: Data lake: a new ideology in big data era. In: WCSN 2017, Wuhan, China. ITM Web of Conferences, vol. 17, pp. 1–6, December 2017. https://doi.org/10.1051/itmconf/2018170302

11. Maccioni, A., Torlone, R.: Crossing the finish line faster when paddling the data lake with KAYAK. VLDB Endowment 10(12), 1853–1856 (2017). https://doi.org/10.14778/3137765.3137792

12. Madera, C., Laurent, A.: The next information architecture evolution: the data lake wave. In: MEDES 2016, Biarritz, France, pp. 174–180, November 2016. http://dl.acm.org/citation.cfm?id=3012077

13. Miloslavskaya, N., Tolstoy, A.: Big data, fast data and data lake concepts. In: BICA 2016, NY, USA. Procedia Computer Science, vol. 88, pp. 1–6, December 2016. https://doi.org/10.1016/j.procs.2016.07.439

14. Sawadogo, P.N., Kibata, T., Darmont, J.: Metadata management for textual documents in data lakes. In: ICEIS 2019, Heraklion, Crete, Greece, pp. 72–83, May 2019. https://doi.org/10.5220/0007706300720083

15. Sawadogo, P.N., Scholly, E., Favre, C., Ferey, É., Loudcher, S., Darmont, J.: Metadata systems for data lakes: models and features. In: ADBIS 2019 Short Papers and Workshops, Bled, Slovenia, September 2019

16. Suriarachchi, I., Plale, B.: Crossing analytics systems: a case for integrated provenance in data lakes. In: e-Science 2016, Baltimore, MD, USA, pp. 349–354, October 2016. https://doi.org/10.1109/eScience.2016.7870919

A Dockerized String Analysis Workflow for Big Data

Maria Th. Kotouza[1(✉)] [iD], Fotis E. Psomopoulos[2] [iD],
and Pericles A. Mitkas[1]

[1] Electrical and Computer Engineering, Aristotle University of Thessaloniki,
54124 Thessaloniki, Greece
maria.kotouza@issel.ee.auth.gr, mitkas@auth.gr
[2] Institute of Applied Biosciences, Centre for Research and Technology Hellas,
57001 Thessaloniki, Greece
fpsom@certh.gr

Abstract. Nowadays, a wide range of sciences are moving towards the Big Data era, producing large volumes of data that require processing for new knowledge extraction. Scientific workflows are often the key tools for solving problems characterized by computational complexity and data diversity, whereas cloud computing can effectively facilitate their efficient execution. In this paper, we present a generative big data analysis workflow that can provide analytics, clustering, prediction and visualization services to datasets coming from various scientific fields, by transforming input data into strings. The workflow consists of novel algorithms for data processing and relationship discovery, that are scalable and suitable for cloud infrastructures. Domain experts can interact with the workflow components, set their parameters, run personalized pipelines and have support for decision-making processes. As case studies in this paper, two datasets consisting of (i) Documents and (ii) Gene sequence data are used, showing promising results in terms of efficiency and performance.

Keywords: Workflow · Docker · Big data analytics · String analysis

1 Introduction

Data Science refers to the manipulation of data using mathematical and algorithmic methods to solve complex problems in an analytical way. It includes models, tools and techniques for data of various types, including biological data, documents, energy consumption, environmental measurements, financial data and more. The large amount of data that are available for analysis in combination with the lack of systematic and generalized methods in the aforementioned areas constitute the ideal environment for applying machine learning techniques on large-scale computational infrastructures.

Big data analysts have to deal with many challenges including the high dimensionality, complexity and diversity of the data, the limited resources and the varying structures of the available analytic tools [1]. Scientific workflows are often the key tools for combining heterogeneous components to solve problems characterized by data

© Springer Nature Switzerland AG 2019
T. Welzer et al. (Eds.): ADBIS 2019, CCIS 1064, pp. 564–569, 2019.
https://doi.org/10.1007/978-3-030-30278-8_55

diversity and high computational demands. Workflows organize computational steps into a logical series to prove a scientific hypothesis [2] or support data exploration. Generative approaches explore the domain space semi-autonomously, suggest parameters and create adaptive processes according to users' specifications, dealing with data diversity and overcoming the lack of abstraction in general-purpose systems.

Cloud computing is a popular way of acquiring computing and storage resources on demand through virtualization technologies. An increasing number of researchers tend to shift workflows from traditional high-performance infrastructures to Cloud Computing platforms [3]. Many works in recent literature have addressed the opportunity of deploying scientific workflows to cloud infrastructures, experimenting with well-known workflows and proving that execution costs can be decreased [4].

In this paper, we develop a generative Dockerized String Analysis workflow (DSA) consisting of scalable algorithmic modules, that transforms datasets coming from various scientific fields into character vectors and provides big data analytics services tailored to the domain. The main objectives of DSA are summarized below.

1. Transform input data into internal format, considering domain specific features;
2. Create custom pipelines based on the user preferences;
3. Provide analytics services integrating new scalable tools;
4. Provide visualization services that can support decision-making;
5. Be available in both script-based format and in a graphical interface;
6. Be suitable for cloud infrastructures.

2   DSA Workflow

The proposed big data analysis workflow, presented on Fig. 1, consists of two phases. The first phase includes all the modules needed for data importing and transformation, whereas the second one integrates machine learning-based modules for data analysis. All the modules take into account domain-specific characteristics and the users' preferences to build a custom pipeline.

2.1   Preparation Phase

The preparation phase includes data importing and transformation, in order for the input data to be reformatted as a set of *Character vectors + meta-data*.

Data Importer. The first step of the data analysis workflow is to acquire the data to be analyzed. Data can be acquired in specific supported formats based on their domain.

Preprocessing Module. The second step is to clean the input data and transform them into a general format which is required by the analysis phase. Data are transformed into vectors of values accompanied with the appropriate meta-data depending on the domain. The processes followed for three different types of data supported by the proposed system are described below.

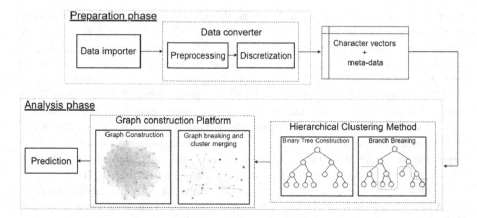

Fig. 1. The proposed DSA workflow architecture.

Documents. Document-structured data are characterized by sets of words. After applying a word-based preprocessing (removing stop-words, special characters and acronyms), each document is transformed into a numeric vector using topic modeling and especially using LDA [5]. LDA builds a set of L thematic topics, each expressed with a set of N_W terms. Parameters L and N_W are user selected.

Gene Sequence Data. This kind of data are preprocessed by our proposed Antigen receptor gene profiler (ARGP[1]), a software framework that provides analytics services on antigen receptor. The tool takes as input the output files of the IMGT[2] tool, the global reference for immunogenetics, and provides analytic services including data cleaning, clustering, combination and visualization.

Time Series Data. A time series is usually a series of numeric data points (measurements over time), so the preprocessing module can apply techniques for data cleaning, normalization, missing value handling etc.

Discretization Module. The numeric vectors created by the preprocessing module are discretized into partitions of length B by assigning each value into a bin based on the closed interval where it belongs to. By making use of alphabetic letters to represent the bins, the numeric vectors are converted into strings. Gene sequence data are already in the appropriate sequence format, so this module is skipped in this instance.

2.2   Analysis Phase

The second phase of the workflow involves modules that make use of new machine learning-based algorithms to provide analysis, clustering and prediction services.

---

[1] https://github.com/mariakotouza/ARGP-Tool/wiki/Antigen-receptor-gene-profiler-(ARGP).

[2] http://www.imgt.org

Clustering Module. In this module, a new scalable multi-metric algorithm for hierarchical clustering is applied. This is a frequency-based clustering algorithm (FBC) [6] that overcomes limitations of baseline hierarchical clustering algorithms regarding memory usage and computational time, making it suitable for big data analytics. The innovation of our proposed algorithm lies in the fact that, instead of performing pairwise comparisons between all the items of the dataset, a low dimensional frequency matrix for the root cluster is constructed, which is split recursively as one goes further away from the root. This procedure forms the *Binary Tree Construction* algorithm.

The frequency matrix is a two-dimensional matrix (B x L). Each element (i,j) of the matrix corresponds to the number of times bin i is present in the j-th position of the string for all vectors. The three metrics that are used to form the clusters are: (i) Identity (I), which indicates how compact the cluster is and it is expressed as the percentage of strings contained in the cluster with an exact alignment, (ii) Entropy (H), which represents the diversity of each column of the frequency matrix, and (iii) Bin Similarity (BS), which is a weighted version of Identity.

After the application of the *Binary Tree Construction* algorithm, the user may select to apply the *Branch Breaking* algorithm: for each branch of the tree, the appropriate level to be cut is examined by recursively comparing the parent cluster with its two children using the values of the evaluation metrics and user selected thresholds.

Graph Mining Module. The hierarchical structure that is created by the clustering module is ideally suited for applying graph-based analysis methods. Using clustering results in combination with graph construction techniques, we provide information about the data relationships in a graphical interactive environment. Graph mining metrics and graph clustering algorithms for sub-graph creation are also utilized [7].

Prediction Module. This module integrates the results from the previous modules to train a model that can make predictions for missing connections of data and classify new items. The model can be trained using state-of-the-art machine learning techniques, after emending [8] the network created at the previous module.

2.3   Software Implementation

The modules that participate in the workflow system described in this paper are implemented in R. All of them are available in both script-based format, where parameters can be selected by the user through the command line and the execution process can be quicker, and through a graphical user interface, where the parameters can be selected through an interactive environment suitable for domain experts with limited technical experience. The graphical modules are implemented in R Shiny, an R package that allows to build web applications in R.

Since the proposed workflow is oriented on analyzing big data that cannot fit in a single machine, the workflow components are dockerized [9], in order to be able to run in cloud infrastructures, including machines with different software and hardware specifications.

3   Results

Most of the modules of the proposed workflow have been applied on various datasets coming from different domains to ensure their efficiency and performance. In this paper results from two different case studies are presented.

3.1   Case Study 1: Documents

In the first case study [6] we aimed to form clusters on benchmark data consisting of 27,000 movies. We created 20-length string vectors after applying LDA at the preparation phase. Through the application of the FBC algorithm a hierarchy structure with 19 levels and 53 leaf clusters were constructed.

The results of the FBC algorithm were compared with those obtained by a Baseline Divisive Hierarchical Clustering[3] (BHC) algorithm, using the evaluation metrics [6]: I, H, BS. Table 1 depicts the average values of the evaluation metrics that were computed for each cluster, using three different number of clusters: 23, 53, 125. High values of I and BS, and low values for H indicate high quality clustering results, so this table makes clear that the FBC algorithm outperforms the baseline algorithm. Regarding the performance results, our method achieved a 98% reduction in memory usage and a 99.4% reduction in computational time.

Table 1. Evaluation of the clusters created using the FBC and the BHC algorithms [6].

#C	Algorithm	I	H	BS
23	BHC	13.696	0.167	85.769
	FBC	74.783	0.081	93.264
53	BHC	35.189	0.139	89.847
	FBC	80.849	0.066	94.237
125	BHC	53.080	0.120	92.886
	FBC	90.600	0.038	96.981

3.2   Case Study 2: Gene Sequence Data

In the second case study [10] we aimed to extract groups of patients based on a biologically important gene region of immunoglobulin. We used a real-world dataset comprising of 123 amino acid sequences of length 20, from patients with chronic lymphocytic leukemia.

The dataset was preprocessed using the ARGP tool at the preparation phase and the FBC algorithm was applied to the amino acid sequences, constructing a binary tree with 19 levels. The clustering results were assessed using the biological groups each sequence comes from. It is important to note that most of the sequences that belonged to a well know biologically important group (93/101, 92%), the largest subset in the present data, formed cluster 13, at level 4 of the clustering process.

---

[3] https://www.rdocumentation.org/packages/cluster/versions/2.0.7-1/topics/diana.

4   Conclusion and Discussion

In this study, we present a generative workflow of scalable algorithmic modules that provides big data analytic services in datasets coming from various scientific fields, by transforming the source data into character vectors, considering domain specific features. Most of the modules of the workflow were applied on two practical case studies, showing promising results in terms of efficiency and performance. Future plans involve adding further functionality on the graph mining module and the development of the prediction module. Moreover, further expansion of the work in more application fields is needed, emphasizing in the source data transformation and the accurate representation of them. Two use cases that we are planning to test include time-series data and data characterized by both numerical and verbal features. Finally, all the modules should be combined and described together using a workflow language in a way that it will be portable and scalable across different environments.

References

1. Lu, S., et al.: A framework for cloud-based large-scale data analytics and visualization: case study on multiscale climate data. In: 2011 IEEE Third International Conference on Cloud Computing Technology and Science, pp. 618–622. IEEE, November 2011
2. Caíno-Lores, S., Lapin, A., Carretero, J., Kropf, P.: Applying big data paradigms to a large scale scientific workflow: lessons learned and future directions. Future Gen. Comput. Syst. (2018)
3. Zhao, Y., Fei, X., Raicu, I., Lu, S.: Opportunities and challenges in running scientific workflows on the cloud. In: 2011 International Conference on Cyber-Enabled Distributed Computing and Knowledge Discovery, pp. 455–462. IEEE, October 2011
4. Berriman, G.B., Deelman, E., Juve, G., Rynge, M., Vöckler, J.S.: The application of cloud computing to scientific workflows: a study of cost and performance. Philos. Trans. Roy. Soc. A: Math. Phys. Eng. Sci. **371**(1983), 20120066 (2013)
5. Blei, D.M., Ng, A.Y., Jordan, M.I.: Latent dirichlet allocation. J. Mach. Learn. Res. **3**, 993–1022 (2003)
6. Kotouza, M., Vavliakis, K., Psomopoulos, F., Mitkas, P.: A hierarchical multi-metric framework for item clustering. In: 2018 IEEE/ACM 5th International Conference on Big Data Computing Applications and Technologies (BDCAT), pp. 191–197. IEEE, December 2018
7. Getoor, L., Diehl, C.P.: Link mining: a survey. ACM Sigkdd Explor. Newslett. **7**(2), 3–12 (2005)
8. Cui, P., Wang, X., Pei, J., Zhu, W.: A survey on network embedding. IEEE Trans. Knowl. Data Eng. **31**, 833–852 (2018)
9. Merkel, D.: Docker: lightweight Linux containers for consistent development and deployment. Linux J. **2014**(239), 2 (2014)
10. Tsarouchis, S.F., Kotouza, M.T., Psomopoulos, F.E., Mitkas, P.A.: A multi-metric algorithm for hierarchical clustering of same-length protein sequences. In: Iliadis, L., Maglogiannis, I., Plagianakos, V. (eds.) AIAI 2018. IFIPAICT, vol. 520, pp. 189–199. Springer, Cham (2018). https://doi.org/10.1007/978-3-319-92016-0_18

Author Index